GEOENVIRONMENTAL ENGINEERING

Ground contamination:
pollutant management and remediation

edited by R. N. Yong and H. R. Thomas

Proceedings of the second conference organized by the British Geotechnical Society and the Cardiff School of Engineering, University of Wales, Cardiff, and held in London on 13–15 September 1999

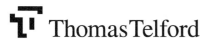

Published by Thomas Telford Publishing, Thomas Telford Ltd, 1 Heron Quay, London E14 4JD.
URL: http://www.t-telford.co.uk

Distributors for Thomas Telford books are
USA: ASCE Press, 1801 Alexander Bell Drive, Reston, VA 20191-4400, USA
Japan: Maruzen Co. Ltd, Book Department, 3–10 Nihonbashi 2-chome, Chuo-ku, Tokyo 103
Australia: DA Books and Journals, 648 Whitehorse Road, Mitcham 3132, Victoria

First published 1999

Organizing Committee
Professor R. N. Yong, Cardiff University (Co-Chairs)
Professor H. R. Thomas, Cardiff University (Co-Chairs)
Dr J. Hellings, Dames and Moore
Dr S. Jefferis, Golder Associates
Dr M. J. Carter, Marion Carter Associates
Dr D. G. Griffiths, Welsh Development Agency
Dr J. M. Reid, Transport Research Laboratory
Ms C. Summers (Secretariat)

A catalogue record for this book is available from the British Library

ISBN: 0 7227 2840 7

Printed and bound in Great Britain by Redwood Books, Trowbridge, Wiltshire

Preface

This book contains peer reviewed papers presented at the 2nd British Geotechnical Society Geoenvironmental Engineering Conference held at the Institution of Civil Engineers, Westminster, London, UK, in September, 1999. The first conference in this series was held in Cardiff exactly two years ago. Issues related to Geoenvironmental Engineering are now well established in importance throughout the world. For example, in the UK, the Government is actively encouraging greater use of brownfield sites for urban developments to ease the pressure on the countryside. Pollution of the natural environment continues to occur as a result of both industrial activity and natural disasters.

The conference is intended to facilitate the exchange of information on the latest developments in research and practice in relation to the main theme of the conference, Ground Contamination: Pollutant Management and Remediation. The papers in this book have been grouped together into the eight major areas considered at the conference: *Landfilling and Engineered Barriers; Detection and Monitoring; Transport, Persistence and Fate of Pollutants; Containment and Remediation; Contaminated Ground and Constructed Facilities; Risk Assessment and Management, Recycling and Reuse of Waste Materials; and Contaminated Water.* The papers in this volume have been submitted by authors from around the world, reflecting the worldwide interest in geoenvironmental engineering.

The conference was organised by the Geoenvironmental Engineering Research Centre at Cardiff University and the British Geotechnical Society. We would like to thank the members of the organising committee, Dr Marion Carter, Dr Gwyn Griffiths, Professor Stephan Jefferis, Dr John Powell, and Dr Murray Reid, for their assistance in organising the conference and reviewing papers. We would also like to thank the Conference Secretariat, Ms Cherrie Summers and Mrs Aderyn Reid, for their contribution to the organisation of the conference.

R N Yong
H R Thomas

Contents

CONTAMINATED GROUND AND CONSTRUCTED FACILITIES

DETECTION AND MONITORING

TRANSPORT, PERSISTENCE AND FATE OF POLLUTANTS

CONTAINMENT AND REMEDIATION

RISK ASSESSMENT AND MANAGEMENT

RECYCLING AND REUSE OF WASTE MATERIALS

CONTAMINATED WATER

Landfilling and Engineered Barriers

Landfill Leachate Management: Flaws in the Containment Strategy

DR. ALISTAIR ALLEN,
Dept of Geology, University College Cork, Cork, Ireland

INTRODUCTION

Current EU landfill regulations, now enacted into law by all member states, have made the installation of artificial lining systems and impermeable cappings mandatory for all landfills, except for sites possessing a suitable *in situ* low permeability ($<10^{-9}\,ms^{-1}$) natural liner, which can also ensure complete containment of landfill emissions. Thus containment is now the only permissible landfill leachate management strategy within the EU. Other leachate management strategies, such as hydraulic traps and dilute and disperse, which take advantage of the natural characteristics and properties of the subsurface, and which, in appropriate circumstances, could be developed and operated at a fraction of the cost of a containment landfill, will not in future even be considered by planners, since, under current legislation, they will no longer be granted a licence. Thus the EU has favoured, to the exclusion of all other strategies, an expensive purely technological approach to landfill leachate management, at the expense of cheaper natural solutions.

THE CONTAINMENT STRATEGY

The new EU regulations are based on the premise that artificial lining systems can wholly contain all leachate produced during degradation of landfill waste, and so provide complete protection to all groundwater, i.e. the concentrate and contain method of leachate control (Gray et al. 1974). However, due to unremitting leakage problems, the requirement to contain all leachate within the landfill, has necessitated the design of more and more elaborate liner systems, so that now it is standard to install composite two, three, or four layer multibarrier clay-membrane systems (Tchobanoglous et al. 1993; Cossu 1995). These typically consist of sheets of synthetic membrane, most commonly high density polyethylene (HDPE), interlayered with clay-rich mineral material, usually bentonite or a bentonite-enriched soil (BES). Two layer systems consist of a sheet of HDPE overlying a 1m thick mineral layer, three layer systems are composed of a 1m thick mineral layer sandwiched between two sheets of the synthetic membrane, whilst four layer systems are represented by two sheets of membrane alternating with two 1m thick mineral layers. Leak detection and leachate collection systems are also generally built into the lining designs. Daily covering of the waste with a clay-rich soil, in order to reduce windblown litter, odours, birds, vermin, flies and visual intrusion, is a further requirement. On closure of the landfill, an impermeable capping is installed to prevent infiltration of rainwater. The cap commonly consists of a sheet of flexible membrane such as HDPE, or a sufficiently thick layer of clay-rich soil with a permeability of at least $10^{-7}cm\ s^{-1}$ (Cossu 1995). A landfill gas collection system is usually installed immediately beneath the capping material, and completed cells are generally landscaped, to ensure that virtually all rainwater runs off the surface.

Geoenvironmental engineering, Thomas Telford, London, 1999, 3–12

The containment strategy thus employs a purely engineering solution to leachate management, and represents a high cost technological approach, which is not only expensive to construct, but also involves costly levels of maintenance (Mather 1995).

FLAWS IN THE CONTAINMENT STRATEGY
The high level of confidence that has been accorded the containment principle of landfill leachate management may be severely misplaced. There are a number of fundamental flaws in the containment approach, some of which have serious long-term environmental implications, but which have tended to be either ignored or played down. The main problems are discussed below.

• Artificial landfill liners are of uncertain long-term durability. Apart from the leakage problems, which have plagued artificial lining systems from the outset, and have led to the development of more and more complex lining systems, of major concern is the long-term durability of artificial lining systems, which as yet is unproven.

Numerous recent studies have drawn attention to some of the deficiencies associated with artificial lining systems, particularly the synthetic membranes. The behaviour of synthetic materials (e.g. CPE, PVC, EPDM, PP and HDPE) subjected over long timescales to the corrosive effects of leachate, and to the elevated temperatures generated by the exothermic processes operating within landfills, is extremely uncertain. The polymer membranes (e.g. HDPE) are generally regarded as being more chemically and biologically resistant than other synthetics (Cossu 1995). However, HDPE membranes have been shown to be prone to stress cracking (Rollin et al. 1991, Thomas & Woods-DeSchepper 1993) and are also known to crack under cold conditions (Thomas & Kolbasuk 1995, Thomas et al. 1995). Nonwoven textiles such as PET and PP appear to be highly prone to ageing during exposure to the natural elements, leading to severe embrittlement (Cazzuffi, et al. 1995), and PVC is known to degrade when exposed to gasoline products (Surmann, et al. 1995). Bituminous membranes (e.g. SPS) may also be sensitive to stress cracking, and have been shown to be subject to ageing, particularly at elevated temperatures (Duquennoi et al. 1995).

Synthetic membranes are also prone to damage (Artieres & Delmas 1995, Colucci & Lavagnolo 1995), particularly due to poor dumping procedures, or failure of the membrane near welded seams (Surmann et al. 1995). Furthermore, extreme care and suitable weather conditions are essential during installation of these lining systems, because they are susceptible to failure if strict quality controls are not adhered to during installation (Averesch 1995). Thus, apart from their high costs of production and installation, and the need for long-term maintenance, the durability of synthetic membranes remains highly suspect.

Mineral layers within the liner system are usually situated below the synthetic membranes, and are thus isolated from the landfill leachate. These necessitate emplacement and compaction at optimum moisture contents (Majeski & Shackleford 1997), and even if this is adhered to, they will readily desiccate under the elevated temperatures generated within landfills. Bentonitic mineral layers have been shown to be susceptible to severe dessication cracking due to inaccessibility of moisture (Meggyes et al. 1995, Holzlöhner & Ziegler 1995), and elevated temperatures (Holzlöhner 1989). Joseph & Mather (1995) have also shown that the method of emplacement of the mineral liner in "lifts" can create horizontal migration pathways which can connect with vertical migration pathways of the desiccation type. Thus, in the event of failure of the synthetic membrane, the mineral layer may have a significantly reduced potential to inhibit leachate migration.

Landfill waste degradation is a long-term process, and even under wet conditions, stabilisation of waste to an inert state ('final storage quality') has not occurred in most landfills 20 years after completion and capping (Belevi & Baccini 1989). However, landfill liner systems have only been in use for about 30 years, so their long-term performance is uncertain.

Ultimately, the key to the containment method of leachate control, i.e. the effectiveness of composite artificial liner systems in preventing leachate migration from the landfill, will be almost solely dependant on the performance of the synthetic membrane member(s). It is unlikely that any synthetic membrane is completely free of defects, regardless of quality control and, whilst leakage may be minimal initially, it is the long-term durability of the membrane(s) over periods of tens or possibly even hundreds of years, under conditions that are ultimately unpredictable, that leaves grounds for concern. In the light of the 'precautionary principle', the wisdom of placing such long-term reliance on an as yet unproven technology is short-sighted, and may ultimately be to our detriment.

- Little account is taken of local hydrogeological conditions in siting landfills, so that not only are the most suitable sites from a geological/hydrogeological perspective ignored, but sites are selected regardless of negative geological factors. Confidence in the containment concept is such, that little account is generally taken of the geological/ hydrogeological characteristics of a proposed site, which are generally seen as no more than the basis for an elaborate engineering plan, based on the attitude that any site can be engineered for landfilling (O'Sullivan 1995).

For example, landfills are frequently sited in pre-excavated holes such as quarries and gravel pits, chosen because a hole already exists, thus reducing the cost of site development. Rocks forming the floors and sides of quarries are typically highly fractured due to blasting operations, so present no barrier to leachate migration, and similarly gravel provides little natural attenuation to migrating leachate. In such quarry sites, the lining system is placed against the bedrock, usually with only a very minimal thickness layer of gravel or soil beneath it as protection for the liner. Quarry sites selected may even be filled with water, and thus must be temporarily drained to lower the local water table in order to install the lining system and emplace the refuse.

At one such site in Ireland, the bedrock is highly karstified limestone, which is first quarried for road metal down to the level of the water table, and then lined and used for landfill. The site is coastal, occupying a small peninsula, which juts into a tidal lagoon with mudflats used for oyster farming, and overlying a major limestone syncline, representing an important regional aquifer. Not only is there the potential to pollute an ecologically sensitive area, in the event of leakage or liner failure, but widespread contamination of groundwater in this important regional aquifer would be at serious risk. However, the site is regarded as a good site on the basis that because of its seclusion, it has attracted little local opposition apart from the oyster farm, although from a geological/hydrogeological perspective it represents a worst case scenario.

The fact that at such sites there is no underlying geological barrier to leachate migration to protect the groundwater in the event of liner failure seems to have been of little consequence in selecting the site. Indeed at some sites, overburden with a high attenuation potential has been stripped off during site development. Clearly the need for a secondary natural barrier is regarded as unnecessary, thus placing total reliance on the artificial lining system, which in view of leakage problems and uncertainties as to its long-term durability, is somewhat ill-advised.

• Natural hydrogeological solutions and the attenuation properties of subsurface materials are totally ignored in current landfill management strategies. As indicated above, there are no provisions in EU regulations for landfill leachate management strategies other than containment. Two types of natural leachate management solution are available. The first, 'dilute and disperse', employs the natural confinement potential of primarily low permeability clay-rich overburden, and, to a lesser extent, bedrock, to impede the migration of leachate from the landfill, whilst at the same time attenuating and purifying it by processes of filtration, sorption and ion exchange. Such natural processes are in continuous and effective operation in the purification of groundwater, which in normal circumstances requires no treatment for use as household water supply.

The attenuation potential of overburden is dependant principally on the proportion of clay minerals and iron and manganese oxides present in the deposit, and also the types of clay minerals present, because the various clay mineral groups possess different sorption and cation exchange capacities (CEC). Of the major clay mineral groups, kandites possess the smallest activity (sorptive capacity) and also the smallest CEC, followed by illites, whereas smectites have the highest CEC and sorptive capacities due to their ability both to adsorb ions on to their external surfaces and also to absorb ions between their lattice sheets. Bentonite-based mineral layers are used in artificial lining systems on the basis of these properties, because bentonite is largely composed of the smectite-group mineral montmorillonite, which has the highest sorption and cation exchange capacities of the common clay minerals.

The dilute and disperse principle has been militated against by current legislation that requires all leachate emanating from the landfill to be collected and treated. These regulations have been introduced despite the fact that field and laboratory studies to assess the effectiveness of natural processes in attenuating leachate concentrations (e.g. DOE 1978), have led to the conclusion that, in appropriate situations, the dilute and disperse method would be effective enough to prevent the pollution of water resources, and could be used as a leachate management strategy. The dilute and disperse principle of leachate control has been unfairly maligned, much of the criticism being that it represents no control whatsoever and relies on, largely unknown, subsurface characteristics at any individual site. However, failure of this approach has stemmed largely from the fact that numerous landfill sites in which the strategy was employed, were chosen without an adequate geological/hydrogeological investigation and were totally inappropriate for this method of leachate management, due to the absence of a suitable geological barrier to attenuate the leachate.

Hydraulic traps, the other type of natural solution are hydrogeological situations where, instead of the leachate migrating outwards from the landfill into the surrounding ground were it allowed to do so, the groundwater surrounding the landfill migrates into the landfill. This reversal of the migration path not only suppresses outwards migration of leachate from the landfill, but the addition of groundwater to the leachate produced in the landfill dilutes it, rendering it less harmful. It is necessary to collect the leachate and dispose of it, otherwise the build-up of leachate plus ingressing groundwater could ultimately overtop the landfill.

Natural hydraulic traps are quite common and usually associated with hollows, often with accompanying swampy ground. It is also possible to artificially create a hydraulic trap, by siting the landfill within a pit excavated in the subsurface to a depth below that of the local water table or potentiometric surface, and controlling the leachate head within the landfill, so as to maintain it at a lower level than that of the water table in the surrounding ground. This creates a negative hydraulic head towards the landfill, which thus acts as a sink towards which groundwater will migrate, so preventing leachate migration from the landfill. Water-filled quarries represent examples of holes below groundwater level, but the permeability of the surrounding rocks make this type of hydraulic trap unsatisfactory, since they must be drained to enable waste to be emplaced, temporarily lowering the local water table; during this period leachate can migrate outwards into the surrounding groundwater.

Most landfills employing hydraulic traps, be they natural or artificially created (Rowe, 1988), are lined with a synthetic lining system, so no cost benefit is derived from the hydraulic trap. Whilst lining may be advantageous to reduce ingress of groundwater from the surroundings during emplacement of the waste, the whole advantage of the hydraulic trap is that leachate cannot migrate outwards regardless of whether the landfill is lined or not, so lining the landfill renders the hydraulic trap superfluous. Natural clay-rich overburden would behave as a natural barrier to groundwater movement, impeding ingress of groundwater during waste emplacement activities, but still allow operation of the hydraulic trap, both during the operational phase of the landfill, and after waste emplacement has ceased.

Although the negative hydraulic head induced by a hydraulic trap suppresses advective flow of leachate from the landfill, diffusional flow may result in migration of contaminants outwards from the landfill against the hydraulic gradient in response to a concentration gradient. Furthermore, (Rowe, 1994) has pointed out that diffusion of contaminant can even take place through synthetic landfill liner membranes. However, the clay layers in the lining system have the potential to attenuate contaminants, provided they are of sufficient thickness, and similarly natural clay–rich overburden will also perform an attenuating role with respect to diffusing contaminants. Thus an artificial liner system gives no added protection in a hydraulic trap situation, and is less suitable than a natural clay-rich geological barrier.

- The containment strategy employs a costly high technology engineering solution, which puts severe constraints on the economics of the landfill operation. Because of the high cost of preparation of the site and installation of the lining system, it has become uneconomic to develop small landfills and the trend is now towards huge superdumps serving large catchment areas. Because of their remoteness from the source of much of the waste arising, these superdumps generate further costs, as waste has to be transported often over great distances, with the inevitable pressure on road networks and the potential for en route traffic accidents and waste spillages. In order to reduce the volumes of waste being transported, construction of a series of transfer stations where waste is compressed and baled, are an essential additional component of the superdump landfill management strategy. All of this adds to the overall costs of the landfilling operation.

Furthermore, local communities typically feel threatened by such superdumps, which generates intense resistance to the siting of such dumps (the NIMBY syndrome), compounded by the fact that little of the waste is of local origin. This invariably gives rise to an adversarial and often acrimonious relationship between advocates and opponents of any given superdump, leading to costly review and licensing procedures, commonly involving court proceedings. The loss of social harmony within communities confronted by the prospect of a superdump in their backyard is a cost that cannot be quantified.

The hugely increased costs associated with the use of artificial lining systems as opposed to natural liners is highlighted by the case of a small landfill in the south-west of Ireland. Ballygyroe in north County Cork is situated on 21-30m of very low permeability red lateritic clay (1 x 10^{-9} m/sec), representing a tropical weathering profile, which overlies Old Red Sandstone bedrock. Initially opened in 1990 as a stop gap measure following a court order to close the existing landfill, whilst the most suitable site in this administrative district was sought, this landfill was initially operated using the natural geological liner, and employing a cellular system at a cost of IR£100,000 per annum. Subsequently it was concluded that this was the best site available, and a licence was sought from the EPA, which insisted on the installation of an artificial lining system, despite the geological evidence of the suitability of the natural overburden as a liner. A cellular system is still in operation, but the cells have had to be increased in size, and slopes considerably reduced in order to accommodate the lining system, with significant loss of landfilling space. The annual cost of operation of the landfill is now IR£1,000,000, a tenfold increase. Sadly, this landfill, an example of the optimum natural landfill site, and probably the best site in Ireland, is to close due to an injunction obtained by opponents, but it would inevitably have been forced to close anyway, as the operational costs make it uneconomic to run, given that the large rural area it serves, supports a rather sparse population of only 70,000.

A further negative impact of the increased costs of operation of the Ballygyroe landfill, is that high charges of IR£300 per truck load of refuse now levied to private refuse collectors or private individuals delivering waste to the landfill is likely to have the effect of encouraging illegal dumping.

• The high technology containment approach is beyond the financial and technological resources of all but the richest nations. Given the difference in operational costs in installing an artificial liner as opposed to using a natural geological barrier, as illustrated above, the promotion by the richer western nations, of a containment strategy based on artificial lining systems, must have a profoundly detrimental impact on the economies of poorer developing countries, placing unnecessary demands on their very limited financial resources.

Landfill is critical to most waste management strategies, because it is the simplest, cheapest and most cost-effective method of disposing of waste. Proportions of waste going to landfill in 1989, ranged from about 60% in OECD countries (Stanners & Bourdeau 1995) to 100% in developing countries. Although in the future, waste minimisation and recycling programmes may reduce waste volumes, and other waste treatment solutions may be developed, at the end of the day landfills will still be required to accommodate residual wastes (Allen et al. 1997). In the developing world, a general lack of education and social and technological infrastructure mitigates against the initiation of waste reduction programmes, or the development of alternative waste treatment options, thus ensuring that in these regions, for the foreseeable future, landfills will continue to be the major method of waste disposal (Allen, 1998).

In at least one developing nation, South Africa, opposition to unlined landfills has arisen, on the mistaken belief that since the richer western nations are pursuing this approach, then it must be the best available technology, i.e. the BATNEEC principle. This is unfortunate, since, whilst for the more overcrowded nations of western Europe, the shortage of sites with adequate natural geological barriers may make the use of landfill liners an unavoidable necessity in most instances, in developing countries with less infrastructure, it is probable that there exist numerous sites with suitable natural liners, both for containment or dilute and disperse strategies, or sites with natural hydraulic traps.

In framing landfill site selection and management policies, it is incumbent upon richer nations to take cognisance of the resources of poorer nations in order that universal pollution control standards can be applied that are within the technological and financial capabilities of the poorer nations. Pollution transcends national boundaries, and given the recent trend of economic groupings such as the EU to develop standard pollution control legislation, it would seem logical to attempt, where possible, to develop standards that, whilst being acceptable to the more developed nations, are relatively simple and inexpensive to implement, so making them achievable by third world nations.

- Encapsulation of waste inhibits waste degradation and considerably prolongs the activity of the waste. The most critical flaw in the current containment landfill ethos, is the misconception that encapsulation of landfill waste within artificial liner systems will, by minimising leachate and gas production, protect the environment (Joseph & Mather 1993). In fact the opposite is more likely to be the case. By isolating the waste from the natural agents of degradation, particularly water (i.e. keeping the waste dry), rates of degradation within the waste will be minimised, thereby prolonging the activity of the waste and inhibiting its stabilisation to an inert state. Stabilisation of waste results from degradation processes which, whether they occur over a period of decades or centuries, involve the production of the same amount of leachate and gas (Joseph & Mather 1995). Permanently isolating the waste, with the resultant long-term threat to the environment, will necessitate an infinite period of monitoring (Carter 1993, Stegmann 1995). Furthermore, prevention of rainwater infiltration, designed to minimise the production of leachate, leads to the generation of a highly concentrated, toxic leachate, which in contact with the artificial membrane over a long time-span, may have an extremely corrosive effect on the membrane leading to its degradation.

Thus, on the one hand, encapsulation of waste in landfills within artificial or natural containment systems supposedly reduces the potential for environmental pollution by leachate in the short and medium term. On the other hand, however, minimisation of the production of leachate resulting from 'dry entombment' of the waste, inhibits its degradation, delaying its stabilisation to an inert state certainly by decades, and perhaps even centuries.

- Long term post-closure maintenance and monitoring of landfills may be financially unacceptable. The new EU regulations and national legistration of member states holds landfill operators responsible for aftercare and monitoring of landfills after completion and capping, and require the license holders to post bonds to cover financial aspects of the discharge of their responsibilities under the terms of the license. Furthermore, the licensee will not be able to surrender the license until the regulatory agency is satisfied that the facility concerned is not causing, and is unlikely to cause future environmental pollution. This aspect of landfill regulations has major implications for landfill operators in that the landfill operators, be they local authorities or private contractors, will be responsible for the landfill for as long as the waste is active and has a potential to cause pollution. Thus a scenario of long-term, largely unpredictable, maintenance and monitoring costs following completion and capping of the landfill (after revenue earnings have ceased) looms for landfill operators (Mather 1995). It therefore becomes incumbent on landfill operators to ensure that the rate of degradation of waste in landfills is optimised in order to reduce the time-scale of their liability.

The threat of long-term liability also has serious implications both for landfill operators and regulatory agencies. Should the liner system fail before the waste is stabilised to an inert state, leading to leachate or gas migration and environmental pollution, then under the 'polluter pays' principle, the landfill operator will be liable. The economic ramifications of this for landfill operators, given the unpredictability of the costs of mitigation of environmental damage possibly decades into the future, would seem an unacceptable risk, regardless of long-term liability insurance cover. Furthermore, legal difficulties in enforcing such a principle several decades or even centuries into the future, may be daunting, and pose a major problem for regulatory agencies.

- The present generations waste problems will be left for the next generation to deal with. Perhaps the most fundamental impact of the containment strategy is that in encapsulating the waste and reducing the degradation rate, this generations waste will still be active and posing problems certainly for the next generation, and even perhaps for several future generations. Given that future waste production is unlikely to decrease, and waste management problems are also unlikely to diminish, it seems morally indefensible that, in addition to having to deal with their own waste problems, future generations may have to deal with waste problems created by this generation.

Conclusions
Flaws in the current containment strategy, outlined above, are :-

- leakage problems and major uncertainties as to the long-term durability of landfill lining systems;
- the total reliance placed on the lining system, with little account taken of geological/ hydrogeological characteristics of sites being selected, and commonly no secondary geological barrier to protect groundwater in the event of liner failure;
- the failure to take advantage of natural hydrogeological solutions to leachate migration, or the natural filtration, sorption and ion exchange properties of clay-rich overburden to attenuate leachate;
- excessive costs in development and operation of containment landfills, making the whole strategy uneconomic and financially unsustainable;
- the unsuitability of such a high-technology, high-cost waste management strategy to the financial and technological resources of all but the richest nations of the developed world;
- encapsulation of waste in a synthetic lining/capping system, so inhibiting waste degradation and thus prolonging the activity of the waste, possibly for many decades;
- the financial burden of long-term, post-closure maintenance and monitoring of landfills;
- the present generations waste problems being left for future generations to deal with.

The alternative landfill strategies can be represented as, on the one hand, high technology solutions offering favourable short-term protection to the environment, but less certainty of long-term protection, possibly resulting in serious environmental pollution in the long-term, as opposed to natural solutions, which offer possibly less guarantee of environmental protection in the short term, but less likelihood of serious long-term environmental pollution. Earth scientists (e.g. Mather 1995, Allen, 1998) favour the latter approach, whereas the engineering community, in the belief that an engineering solution is superior to a natural approach, have promoted the current policy - which is being followed without due regard to long-term cost or environmental impact.

The containment strategy employs a purely technological approach to the management of leachate, ignoring the potential of natural solutions based on the confinement and attenuation properties of the subsurface. High technology engineering solutions to pollution control are usually expensive and rarely completely successful, and frequently have negative impacts, the tendency being that the more sophisticated the solution, the greater the cost and maintenance that they entail (Mather 1995). A much more sensible and cost effective approach typically involves some form of enhancement of natural processes by the integration of a cheap, simple technology.

Landfill management options are curtailed by the inflexibility of the current EU landfill regulations and national legislation of member states, which not only makes a containment approach mandatory to the exclusion of all other strategies, but militates against the use of natural geological liners in the form of clay-rich overburden. The current legislation reflects the triumph of the engineering solution over the natural solution in landfill management strategies and represents an extreme approach to the protection of groundwater.

Finally, it should be pointed out that the current regulations render protection of all groundwater as a mandatory requirement, regardless of whether the groundwater being protected constitutes a material resource or not. Not all groundwater can be regarded as a substantive resource, since a real resource only exists where it is readily available and extractable in sufficient quantity at an acceptable cost. Groundwater only constitutes a resource provided the porosity and hydraulic conductivity of the subsurface are sufficient to provide an adequate supply at a sufficient yield for the purpose for which the groundwater is being sought, which at the lowest common denominator could represent household supply to a single dwelling. Commonly, subsurface characteristics do not fulfil these requirements, so in many areas groundwater cannot be regarded as a resource. If the groundwater does not constitute a resource, then protection of such groundwater becomes a very costly and futile exercise.

References

Allen, A. R. (1998) Sustainability in landfilling: Containment versus dilute and disperse. In *Engineering Geology: A Global View from the Pacific Rim*. D. P. Moore & O. Hungr (Eds.) 8th Congress of the International Association of Engineering Geologists, Vancouver, Canada, Vol. IV, 2423-2431

Allen , A. R., Dillon, A. M. & O'Brien, M. (1997) Approaches to landfill site selection in Ireland. In Marinos, P. G., Koukis, G. C., Tsiambaos, G. C. & Stournaras, G. C. (eds), Engineering Geology and the Environment: 1569-1574. Rotterdam, Balkema.

Artieres, O. & Delmas, P. (1995) Puncture resistance of geotextile-geomembrane lining systems. In T. H. Christensen, R. Cossu & R. Stegmann (eds), *Proceedings Sardinia 95, Fifth International Landfill Symposium*: 2: 213-224. Cagliari: CISA Publisher.

Averesch, U. (1995) Specific problems in the construction of composite landfill liner systems. In T. H. Christensen, R. Cossu & R. Stegmann (eds), *Proceedings Sardinia 95, Fifth International Landfill Symposium*: 2: 115-130. Cagliari: CISA Publisher.

Belevi, H. & Baccini, P. (1989) Long-term assessment of leachates from municipal solid waste landfills. In T. H. Christensen, R. Cossu & R. Stegmann (eds), *Proceedings Sardinia 89, Second International Landfill Symposium*: 8pp. Porto Conte, Sardinia.

Carter, M. J. (1993) The impact of landfill on groundwater. In *Second Annual Groundwater Pollution Conference*: 15pp. London: IBC Technical Services,

Cazzuffi, D., Corbett, S. & Rimoldi, P. (1995) Compressive creep test and inclined plane test for geosynthetics in landfills. In T. H. Christensen, R. Cossu & R. Stegmann (eds), *Proceedings Sardinia 95, Fifth International Landfill Symposium*: 2: 477-491. Cagliari: CISA Publisher.

Colucci, P. & Lavagnolo, M. C. (1995) Three years field experience in electrical control of synthetic landfill liners. In T. H. Christensen, R. Cossu & R. Stegmann (eds), *Proceedings Sardinia 95, Fifth International Landfill Symposium*: 2: 437-452. Cagliari: CISA Publisher.

Cossu, R. (1995) The multi-barrier landfill and related engineering problems. In T. H. Christensen, R. Cossu & R. Stegmann (eds), *Proceedings Sardinia 95, Fifth International Landfill Symposium*: 2: 3-26. Cagliari: CISA Publisher.

Department of the Environment (DOE) (1978) *Cooperative Programme of Research on the Behaviour of Waste in Landfill Sites*: 169pp. London: HMSO

Duquennoi, C., Bernhard, C. & Gaumet, S. (1995) Laboratory ageing of geomembranes in landfill leachates. In T. H. Christensen, R. Cossu & R. Stegmann (eds), *Proceedings Sardinia 95, Fifth International Landfill Symposium*: 2: 397-404. Cagliari: CISA Publisher.

Gray, D. A., Mather, J. D. & Harrison, I. B. (1974) Review of groundwater pollution from waste disposal sites in England and Wales, with provisional guidelines for future site selection. *Quart. J. Eng. Geol.* 7: 181-196.

Holzlöhner, U. (1994) Moisture behaviour of soil liners and subsoil beneath landfills. In T. H. Christensen, R. Cossu & R. Stegmann (eds), *Landfilling of Waste: Barriers*: 247-258. E & Fn Spon.

Holzlöhner, U. & Ziegler, F. (1995) The effect of overburden pressure on dessication cracking of earthen liners. In T. H. Christensen, R. Cossu & R. Stegmann (eds), *Proceedings Sardinia 95, Fifth International Landfill Symposium*: 2: 203-212. Cagliari: CISA Publisher.

Joseph, J. B. & Mather, J. D. (1993) Landfill - does current containment practice represent the best practice? In T. H. Christensen, R. Cossu & R. Stegmann (eds), *Proceedings Sardinia 93, Fourth International Landfill Symposium*: 2: 99-107. Cagliari: CISA Publisher.

Joseph, J. B,. Mather, J. D. (1995) Landfill regulation: The need for a scientific framework. In T. H. Christensen, R. Cossu & R. Stegmann (eds), *Proceedings Sardinia 95, Fifth International Landfill Symposium*: 2: 141-148. Cagliari: CISA Publisher.

Majeski, M. J. & Shackleford, C. D. (1997) Evaluating alternative water content – Dry unit weight criteria for compacted clay liners. In Marinos, P. G., Koukis, G. C., Tsiambaos, G. C. & Stournaras, G. C. (eds), Engineering Geology and the Environment: 1989-1995. Rotterdam, Balkema.

Mather, J. D. (1995) Preventing groundwater pollution from landfilled waste – is engineered containment an acceptable solution? In H. Nash & G. J. H. McCall (eds), *Groundwater Quality*: 191-195. London: Chapman & Hall

Meggyes, T., Holzlöhner, U. & August, H. (1995) Improving the technical barrier for landfills. In T. H. Christensen, R. Cossu & R. Stegmann (eds), *Proceedings Sardinia 95, Fifth International Landfill Symposium*: 2: 89-102. Cagliari: CISA Publisher.

O'Sullivan, S. (1995) Landfill studies: the role of the hydrogeologist. In *The Role of Groundwater in Sustainable Development*: Proceedings of 15th Annual Groundwater Seminar: 6pp. Portlaoise: International Association of Hydrogeologists (Irish Group)

Rollin, A., Mlynarek, J., Lafleur, J. & Zanescu, A. (1991) An investigation of a seven years old HDPE geomembrane used in a landfill. In T. H. Christensen, R. Cossu & R. Stegmann (eds), *Proceedings Sardinia 91, Third International Landfill Symposium*: 667-678. Cagliari: CISA Publisher.

Rowe, R. K. (1988) Contaminant migration through groundwater: The role of modelling in the design of barriers. *Canadian Geotechnical Journal*, 25, 778-798

Rowe, R. K. (1994) Design options for hydraulic control of leachate diffusion. In T. H. Christensen, R. Cossu & R. Stegmann (eds), Landfilling of Waste: Barriers: 101-113. E & FN Spon, London.

Stanners, D. & Bourdeau, P. (eds.) Europe's environment: The Dobris assessment. *European Environment Agency:* Copenhagen.

Stegmann, R. (1995) Concepts of waste landfilling. In T. H. Christensen, R. Cossu & R. Stegmann (eds), *Proceedings Sardinia 95, Fifth International Landfill Symposium*: 1, 3-12. Cagliari: CISA Publisher.

Surmann, R., Pierson, P. & Cottour, F. (1995) Geomembrane liner performance and long term durability. In T. H. Christensen, R. Cossu & R. Stegmann (eds), *Proceedings Sardinia 95, Fifth International Landfill Symposium*: 2: 405-533-547414. Cagliari: CISA Publisher.

Tchobanoglous, G., Thiesen, H., Vigil, S. A. (1993) Integrated Solid Waste Management: engineering principles and management issues. McGraw Hill, New York, 978pp

Thomas, R. W., & Kolbasuk, G.M. (1995) Lessons learned from a cold crack in an HDPE geomembrane. In J.P. Giroud (ed), *Geosynthetics: Lessons Learned from Failures*: 251-254. Industrial Fabrics Association International.

Thomas, R.W., & Woods-DeSchepper, B. (1993) The environmental stress crack behavior of coextruded geomembranes and seams. In *Proceedings of the Fifth International Conference on Geotextiles, Geomembranes and Related Products*, Singapore: 945-948.

Thomas, R.W., Kolbasuk, G. & Mlynarek, J. (1995) Assessing the quality of HDPE double track fusion seams. In T. H. Christensen, R. Cossu & R. Stegmann (eds), *Proceedings Sardinia 95, Fifth International Landfill Symposium*: 2: 415-427. Cagliari: CISA Publisher.

The influence of chemical solutes on the behaviour of engineered clay barriers.

P. J. CLEALL, H. R. THOMAS and R. N. YONG
Geoenvironmental Research Centre, Cardiff School of Engineering, Cardiff University, Newport Road, Cardiff, CF2 1XH, Wales.

ABSTRACT
The use of clays as engineered barriers, in particular their use in landfill sites and high level nuclear waste repositories, has been the focus of much research. The development of suitable models to represent the highly complex coupled behaviour exhibited by such materials is investigated in this paper. In particular the extension of thermal-hydro-mechanical (THM) models to a thermal-hydraulic-chemical-mechanical (THCM) model is discussed. The new model is applied to simulate the behaviour of a clay when subjected to changes in osmotic potential and suction. The results presented show the new model's ability to qualitatively represent the observed experimental behaviour.

INTRODUCTION
This paper describes recent developments carried out to investigate the thermal/hydro/mechanical behaviour of swelling clays. The work has been performed within the context of the development of a model of coupled thermo/hydraulic/mechanical behaviour of unsaturated clay, COMPASS. In particular the inclusion of the osmotic potential both in terms of flow and deformation is addressed.

The flow and deformation behaviour of unsaturated soils has received much attention in recent years. The effect of osmotic potential has been largely neglected in such analyses of soils, as it is often considered negligible in comparison to the effects of matrix potential or gravitational potential. However, a number of soils show considerable influence of the osmotic potential on both the flow and deformation behaviour. In particular expansive / active clays exhibit this feature of behaviour, (Yong and Warkentin, 1975).

These expansive soils, such as bentonite, are often used as part of an engineered barrier. These barriers form an integral component of many containment schemes, such as landfill projects and nuclear waste repositories. In both of these applications it is necessary to include the effects of temperature on the material, as elevated temperature levels are generated by the waste.

The stress-strain behaviour of unsaturated soil has been the subject of numerous experimental and theoretical investigations. The measurement and description of volumetric change of unsaturated soil under a variety of stress and suction paths have received considerable attention. For unsaturated slightly swelling soils, these have led to the proposition that any pair of stress fields among the three stress states, the suction, the excess of total stress over pore air pressure and the excess of total stress over water pore pressure, may represent a suitable stress framework to describe stress-strain-strength behaviour. The second author has developed a numerical solution of the state surface constitutive relationship, Thomas and He (1995), and has fully coupled this with a flow model for heat, mass and air transfer. This model was shown capable of producing qualitatively physically correct results.

The formulations described above do not take account of the effect of the osmotic potential on the stress-strain behaviour of unsaturated soils. In particular they are not able to realistically reproduce the large swelling strains or swelling pressures observed experimentally in bentonitic soils. Experimental evidence suggests that in addition to suction and net stress, osmotic potential is another important driving force for highly active clays (Bolt, 1956; Yong and Warkentin, 1975; Mitchell, 1993). The evidence leads to the proposition of a development of the existing non-linear elastic model of heat, mass and air flow in a deformable soil (HMAE) presented by Thomas and He (1995) to include the effect of osmotic potential on both the flow and stress-strain fields.

In this paper, a description of the behaviour is based on a state surface approach for the stress-strain behaviour of the soil. Pore water pressure, pore air pressure, temperature, chemical solute concentration and displacement are chosen as the basic variables in the model. The evaluation of volumetric strain has been achieved via a combination of a state surface approach and a suitable relationship between osmotic potential changes and strains. A framework of numerical techniques is then proposed that will enable a numerical solution of the complete model to be achieved.

THEORY

Unsaturated soil can be considered a three-phase porous medium consisting of solid, liquid and gas. In this paper, the liquid phase is considered to be pore water with a chemical solute concentration and the gas phase as pore air. A set of coupled governing differential equations can now be developed to describe the flow and deformation behaviour of the soil. The development of these equations is dealt with below. The primary variables of the model are pore water pressure, u_l, pore air pressure, u_a, temperature, T, chemical solute concentration c_s and displacement, $\mathbf{u}$.

According to Yong and Warkentin (1975), the osmotic potential is defined as the work required to transfer water reversibly and isothermally from a reference pool of pure water to a pool of the soil water at a specific elevation at atmospheric pressure. The osmotic potential, π, can be defined by,

$$\pi = n_m k T c_s \tag{1}$$

where n_m is number of molecules per mole of chemical and k is Boltzmann's constant.

Heat transfer
Conservation of heat energy can be defined via a classical conservation equation.

$$\frac{\partial \Omega_H}{\partial t} = -\nabla . \mathbf{Q} \tag{2}$$

where the heat content of unsaturated soil per unit volume, Ω_H, is defined as

$$\Omega_H = H_c \left(T - T_r\right) + L n S_a \rho_v \tag{3}$$

L is the latent heat of vaporisation, n is the porosity, S_a is the degree of pore air saturation and ρ_v is the density of water vapour.

Following the approach presented by Ewen and Thomas (1989) the heat capacity of unsaturated soil H_c at reference temperature, T_r, can be expressed as

$$H_c = (1-n)C_{ps}\rho_s + n(C_{pl}S_l\rho_l + C_{pv}S_a\rho_v + C_{pda}S_a\rho_{da})$$ (4)

where C_{ps}, C_{pl}, C_{pv} and C_{pda} are the specific heat capacities of solid particles, liquid, vapour and dry air respectively, ρ_s is the density of the solid particles, ρ_{da} is the density of the dry air and S_l is the degree of saturation .

The heat flux per unit area, $\mathbf{Q}$, can be defined as (Thomas and King, 1991)

$$\mathbf{Q} = -\lambda_T\nabla T + (\mathbf{v}_v\rho_v + \mathbf{v}_a\rho_v)L + (C_{pl}\mathbf{v}_l\rho_l + C_{pv}\mathbf{v}_v\rho_l + C_{pv}\mathbf{v}_a\rho_v + C_{pda}\mathbf{v}_a\rho_{da})(T-T_r)$$ (5)

where λ_T is the coefficient of thermal conductivity of unsaturated soil and $\mathbf{v}_l$, $\mathbf{v}_v$ and $\mathbf{v}_a$ are the velocities of liquid, vapour and air respectively.

The governing equation for heat transfer can be expressed, in primary variable form as

$$C_{Tl}\frac{\partial u_l}{\partial t} + C_{TT}\frac{\partial T}{\partial t} + C_{Ta}\frac{\partial u_a}{\partial t} + C_{Tu}\frac{\partial \mathbf{u}}{\partial t} = \nabla[K_{Tl}\nabla u_l] + \nabla[K_{TT}\nabla T] + \nabla[K_{Ta}\nabla u_a] +$$
$$V_{Tl}\nabla u_l + V_{TT}\nabla T + V_{Ta}\nabla u_a + V_{Tc_s}\nabla c_s + J_T$$ (6)

where C_{Tj}, V_{Tj}, K_{Tj} and J_T are coefficients of the equation ($j = l, T, a, c_s, \mathbf{u}$).

Moisture transfer

The governing equation for moisture transfer in an unsaturated soil can be expressed as

$$\frac{\partial(\rho_l nS_l)}{\partial t} + \frac{\partial(\rho_v n(S_l-1))}{\partial t} = -\rho_l\nabla.\mathbf{v}_l - \rho_l\nabla.\mathbf{v}_v - \nabla.\rho_v\mathbf{v}_a$$ (7)

The velocities of pore liquid and pore air are based on a generalised Darcy's law and are defined as follows

$$\mathbf{v}_l = -K_l\left(\nabla\left(\frac{u_l}{\gamma_l}\right) + \nabla z\right) + K_\pi(\nabla\pi)$$ (8)

$$\mathbf{v}_a = -K_a\nabla u_a$$ (9)

where K_l is the hydraulic conductivity, K_π is coefficient of osmotic permeability, K_a is the conductivity of the air phase and γ_l is the specific weight of water. The inclusion of an osmotic flow term in the liquid velocity allows the representation of liquid flow behaviour found in some highly compacted clays, (Yong and Warkentin, 1975).

The definition of vapour velocity follows the approach presented by Thomas and King (1991) which was based on the flow law proposed by Philip and deVries (1957).

The governing equation for moisture transfer can be expressed, in primary variable form as

$$C_{ll}\frac{\partial u_l}{\partial t} + C_{lT}\frac{\partial T}{\partial t} + C_{la}\frac{\partial u_a}{\partial t} + C_{lu}\frac{\partial \mathbf{u}}{\partial t} = \nabla[K_{ll}\nabla u_l] + \nabla[K_{lT}\nabla T] +$$
$$\nabla[K_{la}\nabla u_a] + \nabla[K_{lc_s}\nabla c_s] + J_l$$ (10)

where C_{lj}, K_{lj} and J_l are coefficients of the equation ($j = l, T, a, c_s, \mathbf{u}$).

Dry air transfer
Air in unsaturated soils is considered to exist in two forms: bulk air and dissolved air. In this approach the proportion of dry air contained in the pore liquid is defined using Henry's law.

$$\frac{\partial[S_a + H_s S_l]n\rho_{da}}{\partial t} = -\nabla.[\rho_{da}(\mathbf{v_a} + H_s \mathbf{v_1})]$$ (11)

where H_s represents Henry's coefficient of solubility and the dry air density can be defined via the use of Daltons law of partial pressures.

The governing equation for dry air transfer can be expressed, in primary variable form as

$$C_{al}\frac{\partial u_l}{\partial t} + C_{aT}\frac{\partial T}{\partial t} + C_{aa}\frac{\partial u_a}{\partial t} + C_{au}\frac{\partial \mathbf{u}}{\partial t} = \nabla[K_{al}\nabla u_l] + \nabla[K_{aa}\nabla u_a] + \nabla[K_{ac_s}\nabla c_s] + J_a$$ (12)

where C_{aj}, K_{aj} and J_a are coefficients of the equation ($j = l, T, a, c_s, \mathbf{u}$).

Chemical solute transfer
Where a chemical solute is considered non-reactive and sorption onto the soil surface is ignored the governing equation for conservation of chemical solute can be defined as (Thomas and Cleall, 1997; Cleall, 1998),

$$\frac{\partial(nS_l c_s)}{\partial t} = -\nabla.[c_s \mathbf{v_1}] + \nabla.[D_h \nabla(nS_l c_s)]$$ (13)

where c_s is the chemical solute concentration. D_h is the hydrodynamic dispersion coefficient defined as (Bear and Verruijt, 1987)

$$D_h = D + D_d$$ (14)

which includes both molecular diffusion, D_d, and mechanical dispersion, D.

The governing equation for chemical solute transfer can be expressed, in primary variable form as

$$C_{c_s l}\frac{\partial u_l}{\partial t} + C_{c_s a}\frac{\partial u_a}{\partial t} + C_{c_s c_s}\frac{\partial c_s}{\partial t} + C_{c_s u}\frac{\partial \mathbf{u}}{\partial t} = \nabla[K_{c_s l}\nabla u_l] + \nabla[K_{c_s T}\nabla T] +$$
$$\nabla[K_{c_s a}\nabla u_a] + \nabla[K_{c_s c_s}\nabla c_s] + J_{c_s}$$ (15)

where $C_{c_s j}$, $K_{c_s j}$ and J_{c_s} are coefficients of the equation ($j = l, T, a, c_s, \mathbf{u}$).

Stress-strain relationship
A non-linear elastic model for expansive soils, including the effects of suction, chemical solute and net mean stress is proposed here. For problems in unsaturated expansive clays, the total strain can be given in an incremental form, without loss of generality, as

$$d\varepsilon = d\varepsilon_\sigma + d\varepsilon_T + d\varepsilon_s + d\varepsilon_\pi$$ (16)

where the subscripts σ, T, s and π refer to net stress, temperature, suction and osmotic potential contributions.

Extending the state surface concept and based on the definition of osmotic potential, chemical solute concentration is included here as a new primary variable. With the inclusion of the traditional net mean stress and suction effects, the volumetric deformation of unsaturated soil is defined. The new state surface of void ratio required in the model, is therefore of the form

$$e = f(e_0, p, s, c_s)$$ (17)

where s is the suction, c_s is the chemical solute concentration and e_0 is the initial void ratio for a reference temperature T_r.

The stress strain relationship can be expressed as

$$d\sigma'' = \mathbf{D}(d\underline{\varepsilon} - d\underline{\varepsilon}_s - d\underline{\varepsilon}_T - d\underline{\varepsilon}_\pi) = \mathbf{D}(d\underline{\varepsilon} - \mathbf{A}_s ds - \mathbf{A}_T dT - \mathbf{A}_{c_s} dc_s)$$ (18)

where $\mathbf{A}_s$ and $\mathbf{A}_{c_s}$ are derived from the state surface for void ratio. $\mathbf{A}_T$ is defined to include the thermal expansion coefficient and the change in surface energy and osmotic potential due to a temperature variation. σ'' is the net stress and $\mathbf{D}$ is the elastic matrix.

The governing equation for stress-strain behaviour can be expressed, in primary variable form as

$$C_{ul}du_l + + C_{uT}dT + C_{ua}du_a + C_{uc_s}dc_s + C_{uu}d\mathbf{u} + d\mathbf{b} = 0$$ (19)

where C_{uj} are coefficients of the equation $(j = l, T, a, c_s, u)$.

Numerical solution
Through the spatial and temporal discretisation of the variables, the differential equations derived above can be transferred into a system of algebraic equations in residual form. In particular, with finite element discretisation in space, the semi-discretised system can be obtained. For temporal discretisation a fully implicit scheme is employed. The coupled processes are solved simultaneously for a given time step size and iterated on within each time step until convergence is achieved.

EXAMPLE
This example describes the modelling of a swelling test of a Pleistocene clay by means of an empirical relationship to describe the volume change behaviour due to osmotic potential and suction changes.

An unsaturated Pleistocene clay was investigated by Richards et al, (1984), who presented a series of results describing the development of strain due to changes in osmotic potential and suction. Although the reported results are not comprehensive enough to allow the complete development of a state surface for void ratio, there are enough data to allow the description of $\mathbf{A}_{c_s}$ and $\mathbf{A}_s$ at an average vertical stress of 72kPa.

As a test example, a load path has been specified, shown in Figure 1 and COMPASS has been used to predict the deformation response of the soil. The sample was assumed to have an initial osmotic potential of 3.2 MPa (chemical solute concentration of 1.3 kMol/m^3) and suction of 1.0 MPa. The finite element mesh used and the displacement boundary conditions applied are shown in Figure 2. Both the osmotic potential and suction were reduced linearly to values of 12.8 kPa (chemical solute concentration of 5 Mol/m^3) and 100 kPa respectively.

The predicted deformation response of the soil is shown in Figure 3. It can be observed that the total volumetric strain of the sample at the conclusion of the load path was 23.8%. The total strain can be separated into a strain of 13.9% due to osmotic potential changes and a strain of 9.9% due to the suction change. Figure 1 also shows the increase in the volume of water in the sample in terms of specific water volume (defined by Wheeler, 1991). The flow of water can be observed to increase as the swelling of the sample increase as the test progresses.

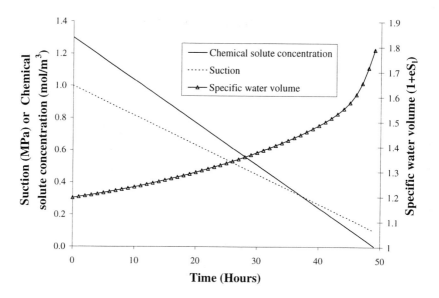

Figure 1 Load path and specific water volume during test.

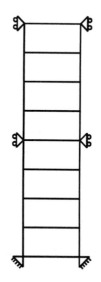

Figure 2 Finite element mesh and displacement boundary conditions for example

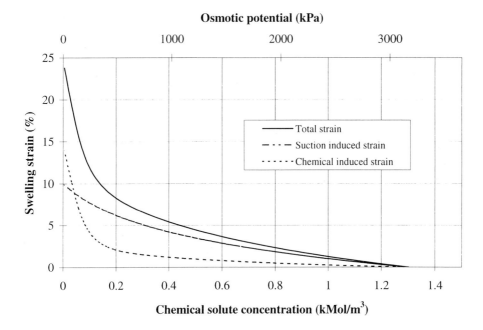

Figure 3 Development of strain against osmotic potential and chemical solute concentration in example.

This example illustrates that experimental results can be used to predict the stiffness of the soil due to changes in chemical solute concentration. It is noted that further experimental results would be required to form a complete state surface for this clay.

CONCLUSIONS

The work presented in this paper describes the development of a flow and deformation model for the analysis of the coupled thermo /mechanical / hydraulic and osmotic behaviour of unsaturated clays. In particular an existing thermal hydraulic mechanical model has been extended to include the effects of osmotic potential on both flow and deformation behaviour. This has been achieved via the inclusion of chemical solute concentration as an additional primary variable. Theoretical and numerical formulations that can accommodate a non-linear elastic constitutive relationship based on the state surface approach have been presented.

An example has been presented which shows the capability of the model to simulate the typical deformation behaviour of unsaturated soil while experiencing changes in suction and osmotic potential. A relationship to describe the volume change behaviour due to changes in osmotic potential was required and for the example an empirical relationships was used. The results from the effects of osmotic potential change have been illustrated.

The work presented in this paper is an initial attempt at the modelling the influence of osmotic potential on unsaturated soil. While it is recognised that further experimental work is required to develop more sophisticated models and produce qualitatively accurate numerical results, this approach provides an important development by introducing a osmotic potential to describe an important aspect of the physical behaviour of soils.

REFERENCES

Bear, J. and A.Verruijt (1987) *Modelling groundwater flow and pollution.* Dordrecht, D. Reidel Publishing Company.

Bolt, G.H. (1956) "Physico-chemical analysis of the compressibility of pure clay". *Geotechnique.* **6**(2): 86-93.

Cleall, P.J., (1998) *An investigation of the thermo/hydraulic/mechanical behaviour of unsaturated soils, including expansive soils.* Ph.D. Thesis, University of Wales, Cardiff, U.K.

Ewen, J. and H.R. Thomas (1989) "Heating unsaturated medium sand". *Geotechnique* **39**(3): 455-470.

Mitchell, J.K. (1993) *Fundamentals of soil behaviour.* John Wiley & sons, New York:

Philip, J.R. and D.A. deVries (1957) "Moisture movement in porous materials under temperature gradients". *Trans. American Geophysical Union.* **38**(2): 222-232.

Richards, B.G., P. Peter and R. Martin (1984) "The determination of volume change properties in expansive soils". *Proc. 5th. Int. Conf. Exp. Soils.* Adelaide: 179-186.

Thomas, H.R. and P.J. Cleall (1997) "Chemico-osmotic effects on the behaviour of unsaturated expansive clays". *Proc. Int. Conf on Geoenviromental Engineering.* Cardiff. Thomas Telford, London: 272-277.

Thomas, H.R and Y. He (1995) "Analysis of coupled heat, moisture and air transfer in a deformable unsaturated soil". *Geotechnique.* **45**(4): 677-689.

Thomas, H.R. and S.D. King (1991) "Coupled temperature/capillary potential variations in unsaturated soil". *ASCE, Journal of Eng. Mech.* **117**(11): 2475-2491.

Wheeler, S.J. (1991) "An alternative framework for unsaturated soil behaviour". *Geotechnique.* 41(3): 257-261.

Yong, R.N. and B.P Warkentin (1975) Soil properties and behaviour. Elsevier, Amsterdam.

ACKNOWLEDGEMENTS
The work presented has been carried out as a part of a research programme supported by the Higher Education Funding Council for Wales and the Engineering and Physical Science Research Council. This support is gratefully acknowledged.

Performance-Based Method for Analyzing Landfill Liners

Takeshi Katsumi[1], Craig H. Benson[2], Gary J. Foose[3], and Masashi Kamon[1]
[1] Kyoto University, Uji, Kyoto, Japan
[2] University of Wisconsin-Madison, Madison, Wisconsin, USA
[3] Florida State University-Florida A&M University, Tallahassee, Florida, USA

ABSTRACT: Factors affecting chemical transport in geomembrane, clay and composite liners are reviewed, and a simplified performance-based method for evaluating landfill bottom liners is presented. For single geomembrane liners, mass transport of inorganic chemicals is calculated from the leakage rate from holes under an assumed frequency of hole occurrence. Transport of organic chemicals is obtained by accounting for molecular diffusion through the intact geomembrane. Release of inorganic and organic chemicals from compacted clay liners is calculated using a solution of the steady-state 1-D advection-diffusion-reaction equation. For composite liners consisting of a geomembrane and a clay liner, 3-D flow and transport of inorganic chemicals is computed using an equivalent 1-D model of transport through an effective area of transport that is based on results of 3-D analyses that have been conducted for a variety of cases. Release of organic chemicals through composite liners is calculated using a 1-D diffusion model. Applicability of the method is described by using it to evaluate the relative performance of several different liner systems.

INTRODUCTION

Bottom lining systems consisting of a geomembrane, a clay layer, or both are commonly used to contain wastes in landfills in many countries. To compare the effectiveness of different liner systems or alternatives to a regulatory prescribed liner, a performance-based analysis is necessary. Performance-based analysis is difficult, however, because chemical transport in landfill liners is complex and difficult to model. Nevertheless, simplified methods can be used that have sufficient accuracy for making engineering judgments, such as choosing an appropriate type of liner from the several possible options. In this paper, factors affecting chemical transport in geomembrane, clay, and composite liners are reviewed, and a simplified performance-based method for evaluating landfill bottom lining systems is presented. The proposed method is readily applied by engineering practitioners; Calculations are made using typical spreadsheet applications, such as Microsoft Excel®.

PRINCIPLE OF PERFORMANCE-BASED DESIGN METHOD

Design methods can be generally classified as construction-based design, product-based design, and performance-based design. Regulations are generally based on product-based design where properties of the materials are prescribed (e.g., hydraulic conductivity and thickness of a clay liner). Performance-based design considers the surrounding environment (i.e., use of the land and water and its impact on the effluent standard), and is more logical way to design landfill liners. Figure 1 is a schematic showing how performance-based design is used to select an appropriate landfill liner. Possible performance criteria include: (1) a specified value for the rate of leakage, Q, (2) a specified maximum value for the solute flux of an individual solute species, J, (3) a specified maximum value for the concentration of the chemical, C, (4) time to reach a specified leakage rate, T_Q, (5) time to reach a specified solute flux, T_J, and (6) time to reach a specified solute concentration, T_C. Regardless of the criterion selected, the basis for specifying the criterion is to maintain groundwater quality. A liner is then selected that ensures that the criterion selected is met.

Geoenvironmental engineering, Thomas Telford, London, 1999, 21–28

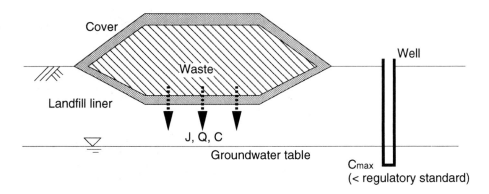

Figure 1. Basic concept of the performance-based design of landfill liner

SIMPLIFIED PERFORMANCE-BASED METHOD

Accurate design of landfill liners using performance-based methods is difficult because chemical transport in landfill liners is difficult to model. However, one of the purposes of performance-based methods is to make a rational engineering judgment such as choosing an appropriate type of liner from several possible options. In such cases, a simplified method for performance-based design can be used effectively. A simplified method is proposed below with a review of the performance of landfill liners. The calculations can be done using a spreadsheet application such as Microsoft Excel®.

Geomembrane liners

Two primary mechanisms for contaminant transport through geomembranes are "leakage" through holes and molecular "diffusion" through the intact geomembrane.

Holes and defects in geomembranes are caused by defects in geomembrane seams, punctures caused by sharp materials beneath the geomembrane liner and installation tools, tension forces induced by placing waste on the liner, and material failure induced by creep or cyclic loading. Giroud and Bonaparte (1989) investigated the occurrence of defects in geomembrane liners, and concluded that 8-10 holes/ha are usually present with good quality assurance and 17 holes/ha are typically present when quality assurance is poor. Even if quality assurance is excellent, 1-2 holes/ha are unavoidable. Equations for calculating the rate of leakage from geomembrane defects are proposed by Giroud and Bonaparte (1989) and Giroud et al. (1998).

The leakage rate, Q, can be obtained two different ways. If the hole is small enough such that the diameter of the hole is less than the thickness of geomembrane, the leakage rate is a function of the viscosity and Q can be expressed using Poiseuille's equation (Giroud and Bonaparte 1989):

$$Q = \frac{\pi \rho_w h_w d^4}{128 \mu t_g} \qquad (1)$$

where ρ_w = the density of the leachate, μ = viscosity of the leachate, h_w = water head above the geomembrane, t_g = thickness of geomembrane, and d = diameter of the hole. From a practical perspective, ρ_w and μ for water can be used in Equation (1). If the hole diameter is larger than the thickness of geomembrane, Bernoulli's equation for free flow through an orifice can be used (Giroud and Bonaparte 1989);

$$Q = C_B a \sqrt{2gh_w} \qquad (2)$$

where C_B is a dimensionless coefficient which can be obtained empirically, a = area of the hole,

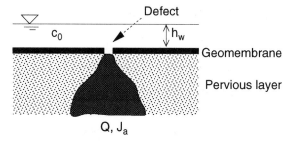

Figure 2. Leakage through defects of geomembrane

and g = acceleration due to gravity. Experiments conducted by Benson et al. (1995) indicate that $C_B = 0.6$ is appropriate for most geomembrane holes. The solute flux, J_a, due to leakage can be calculated by (Figure 2);

$$J_a = NQc_0 \tag{3}$$

where N = number of the geomembrane holes per area, and c_0 = concentration of the solute chemical in the leachate in waste layer above geomembrane. The number of the holes, N, depends on the level of quality assurance as mentioned previously.

Organic chemicals diffuse at the molecular level through geomembranes. The basic transport mechanisms are illustrated in Figure 3. An organic chemical having concentration, c_0, first partitions into the geomembrane ($K_g c_0$), then diffuses downward, and then partitions back into the pore water at the base of the liner (c_e). Park et al. (1996) illustrate that molecular diffusion of organic chemicals is more significant than leakage through geomembrane defects.

Since geomembranes are thin enough such that steady-state condition are quickly reached, the concentration gradient ($\partial c/\partial x$) is constant throughout the geomembrane, and the mass flux of organic chemical can be expressed as (Park et al. 1996);

$$J_d = D_g K_g \frac{c_0 - c_e}{t_g} \tag{4}$$

where D_g = diffusion coefficient for a geomembrane, K_g = partition coefficient, and c_e = concentration of the organic beneath the geomembrane. Values of diffusion and partition coefficients for several organic chemicals in geomembranes are summarized by Rowe (1998).

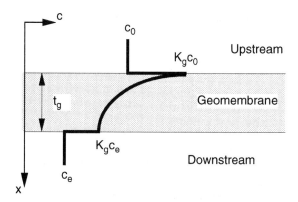

Figure 3. Schematic diagram of organic mass flux through geomembrane

Clay liners

Since clay liners generally have low hydraulic conductivity, diffusive and advective transport must be taken into account. The one-dimensional advection-diffusion equation which accounts for adsorption can be expressed as:

$$\left(1 + \frac{\rho_d K_p}{n}\right)\frac{\partial c}{\partial t} = D\frac{\partial^2 c}{\partial x^2} - v_s\frac{\partial c}{\partial x} \tag{5}$$

where c = the concentration of the chemical, ρ_d = the dry density, n = the porosity of the soil, K_p = the partition coefficient, D = the diffusion coefficient, and v_s = the seepage velocity. The term $(1 + \rho_d K_p / n)$ in Equation (5) is called retardation factor, R.

Clay liners placed above the groundwater table are generally unsaturated. However, if seepage is assumed to be steady-state and any suction existing at the bottom of the liner is ignored (Figure 4, i.e., the level of the groundwater table is at the bottom of the liner), the transport calculations can be performed relatively easily. These assumptions generally result in conservative predictions. Also, if the soil properties (i.e., ρ_d, n, K_p, D) are assumed to be homogenous and time invariant, and no chemical reactions occur, then the mass flux at the bottom of the liner at time, t, can be obtained by (Ogata and Banks 1961, Shackelford 1990):

$$\frac{c(x = L, t)}{c_0} = 0.5\left\{ erfc\left[\frac{1 - T_R}{2\sqrt{T_R / P_L}}\right] + \exp(P_L)erfc\left[\frac{1 + T_R}{2\sqrt{T_R / P_L}}\right]\right\} \tag{6}$$

$$\frac{J(t)}{v_s n c_0} = 0.5 erfc\left[\frac{1 - T_R}{2\sqrt{T_R / P_L}}\right] + \frac{1}{\sqrt{\pi P_L T_R}}\exp\left[-\frac{(1 - T_R)^2}{4T_R / P_L}\right] \tag{7}$$

$$T_R = \frac{v_s t}{RL} \tag{8}$$

$$P_L = \frac{v_s L}{D} \tag{9}$$

In Equations (6)-(9), L = the thickness of the clay liner. The initial and boundary conditions used to obtain Equations (6) and (7) are $c(0, t) = c_0$, $c(x, 0) = 0$ (for x>0), and $\partial c(\infty, t)/\partial x = 0$. The parameter T_R is the dimensionless time factor and P_L is the Peclet number, which represents the ratio of advective transport to diffusive transport. Smaller values of P_L correspond to diffusive transport being more dominant. Calculations using Equations (6) and (7) can be made with applications where the error function is available, such as Microsoft Excel®, or using an electronic calculators if a table of the error function is available.

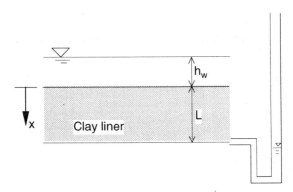

Figure 4. Concept of chemical transport through clay liner

Composite liners

A composite liner exploits the advantages of geomembrane and clay liners; the geomembrane restricts the area through which leakage occurs and the clay liner beneath the geomembrane minimizes leakage from the geomembrane defects. As a result, leakage from composite liners is often orders of magnitude less than leakage from single geomembrane liners and clay liners.

Several equations for calculating the rate of leakage through geomembrane defects in composite liners have been proposed (Giroud and Bonaparte 1989, Giroud et al. 1998, Foose 1997). To calculate the leakage rate, contact between the geomembrane and clay liner, the size and the shape of geomembrane hole, and the thickness of clay liner must be considered. Good contact minimizes leakage since a smaller area of the clay liner is exposed to the flow, whereas poor contact permits greater leakage because liquid can freely penetrate in the space between the geomembrane and clay. Poor contact can be caused by geomembrane wrinkles or an uneven surface of the clay liner beneath the geomembrane.

When analyzing transport of inorganic chemicals through composite liners, leakage through geomembrane defects is the primary transport mechanism. Because 3-D flow and transport in composite liners (Figure 5 (a)) is difficult to simulate, the 3-D system is equated to an equivalent 1-D uniform steady-state flow (Q_e) through an area A_e where transport can be simulated using the 1-D advection-diffusion equation (Figure 5 (b)). The relationship between Q_e and A_e is:

$$Q_e = kiA_e \qquad (10)$$

The magnitude of Q_e depends on the shape of the defects, contact between the soil and geomembrane, and the depth of leachate. Equations by Giroud et al. (1998), Rowe (1998), or Foose (1997) can be used to calculate Q_e. For circular defect, Q_e is calculated as (Foose 1997):

$$Q_e = F_r kh_t r \qquad (11)$$

and for linear defects, Q_e is calculated as (Foose 1997):

$$Q_e = F_w kh_t \qquad (12)$$

where h_t = the total head drop across the composite liner, r = the radius of the defect, and F_r and F_w are flow factors obtained from 3-D FDM analysis as shown in Table 1 (Foose et al. 1999). The mass flux and the concentration can be calculated by Equations (6) and (7). Since the mass flux obtained, J_e, is per A_e instead of per total area, the mass flux for the liner is obtained as:

$$J_{total} = J_e NA_e \qquad (13)$$

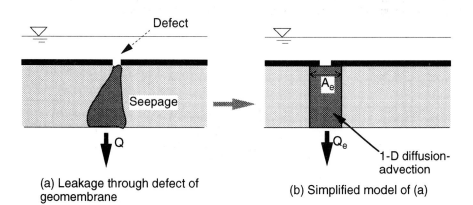

(a) Leakage through defect of geomembrane

(b) Simplified model of (a)

Figure 5. Concept for analizing inorganic chemical transport through composite liners

Table 1   Equations to calculate Flow Factors (after Foose 1997)

Contact condition	Circular defect F_r	Long defect F_w
Perfect contact	$F_r = 4 + 3.35(r/L)$	$F_w = 1/(0.52 - 0.76\log(w/L))$
Good contact	$F_{r.g} = 172\ r^{-0.8}\ F_r$	$F_{w.g} = 6.45\ F_w$
Poor contact	$F_{r.p} = 5.48\ F_{r.g}$	$F_{w.p} = 2.35\ F_{w.g}$

For organic chemicals, the following assumptions are made: (1) the contribution of leakage through the geomembrane defects is negligible because molecular diffusion through geomembrane is far more significant, (2) diffusion through the geomembrane is ignored because the geomembrane is significantly thinner than the clay liner, and (3) advection is zero because the geomembrane limits leakage to very small quantities. As a result, only diffusion in clay layer is taken into account. Equations (6) and (7) are used for the calculations. Because $v_s = 0$, these equations can be simplified as follows:

$$\frac{c(x = L, t)}{c_0} = erfc\left[\frac{1}{2\sqrt{Dt/L^2R}}\right] \qquad (14)$$

$$\frac{J(t)}{nc_0} = \frac{1}{\sqrt{\pi t/DR}}\exp\left[-\frac{1}{4Dt/L^2R}\right] \qquad (15)$$

Predictions made using this above method are in excellent agreement with the results from a more exact 3-D FDM analysis (Foose et al. 1999).

PARAMETRIC STUDY
Results for several liner systems obtained from the aforementioned analysis are summarized in Table 2. Although a single geomembrane liner may have leakage (1.58×10^6 L/ha/y) similar to that from a single 60-cm-thick clay liner having hydraulic conductivity of 10^{-7} cm/s (1.18×10^6 L/ha/y), the geomembrane liner releases a much greater mass of organic chemicals because molecular diffusion occurs across the entire area of the thin geomembrane liner. Increasing the thickness of clay liner from 60 cm to 120 cm decreases the release of chemicals (4.73×10^{-1} kg/ha/y to 3.94×10^{-1} kg/ha/y for both inorganic and organic chemicals), and increases the retardation effect. However, placing a geomembrane above the 60-cm-thick clay liner has a

Table 2   Calculated results of leakage and chemical release of liners

Type of Liner	Geo-membrane Thickness (cm)	Clay layer Hydraulic conductivity (cm/s)	Clay layer Thick-ness (cm)	Leakage (L/ha/y)	Release of Inorganics $T_{0.1}$ (y)	Release of Inorganics $T_{0.9}$ (y)	Release of Inorganics J_{max} (kg/ha/y)	Release of organics $T_{0.1}$ (y)	Release of organics $T_{0.9}$ (y)	Release of organics J_{max} (kg/ha/y)
Geomenbrane liner	0.1	-		1.58×10^6	-	-	1.58×10^0	-	-	8.22×10^1
Clay liner	-	1×10^{-7}	60	1.18×10^6	5.6	16	4.73×10^{-1}	2.8	7.9	4.73×10^{-1}
Clay liner	-	1×10^{-7}	120	9.87×10^5	16	35	3.94×10^{-1}	7.7	18	3.94×10^{-1}
Composite liner	0.1	1×10^{-7}	60	1.36×10^3	5.6	16	1.36×10^{-3}	11	1808	2.03×10^{-2}
Composite liner	0.1	1×10^{-6}	60	1.36×10^4	0.85	1.2	1.36×10^{-2}	11	1808	2.03×10^{-2}

It was assumed that $c_0 = 1$ mg/l and $h_w = 30$ cm for leachate above liner, circular defect, N = 10 holes/ha, d = 2 mm, $D_g = 2 \times 10^{-8}$ cm^2/s, $K_g = 130$ l/kg and $t_g = 0.1$ cm for geomembrane, and $D = 2 \times 10^{-6}$ cm^2/s and R = 2 (for inorganics), R = 1 (for organics), and n = 0.4 for clay liner.
$T_{0.1}$ and $T_{0.9}$ stand for the time to reach $0.1\ c_0$ and $0.9\ c_0$ concentrations of leaked water respectively.

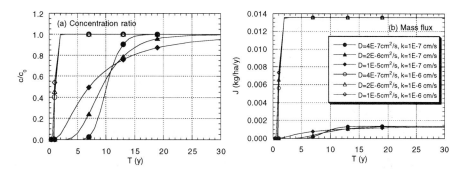

Figure 6. Contaminant release at the bottom of a composite liner for different diffusion coefficients and hydraulic conductivities

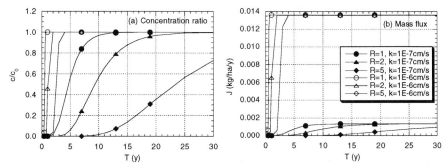

Figure 7. Contaminant release at the bottom of a composite liner for different retardation factors and hydraulic conductivities

more significant effect on decreasing the release of chemicals (4.73×10^{-1} kg/ha/y to 1.36×10^{-3} kg/ha/y for inorganics; 4.73×10^{-1} kg/ha/y to 2.03×10^{-2} kg/ha/y for organics). In addition, adding the geomembrane increases the retardation effect. Even the composite liner consisting of geomembrane and clay liner having higher hydraulic conductivity of 10^{-6} cm/s releases a smaller mass of inorganic chemicals than a single clay liner having lower hydraulic conductivity of 10^{-7} cm/s.

Results for inorganic transport through a composite liner are shown in Figure 6 for three different values of D, two different hydraulic conductivities, k, and R = 2 for the clay liner. Thickness of the clay liner is 60 cm. The hydraulic conductivity of 10^{-6} cm/s represents the composite liner prescribed by the Japanese government in 1998 (although the prescribed thickness is 50 cm instead of 60 cm), while 10^{-7} cm/s corresponds to the liner prescribed by the US and some European governments. When the hydraulic conductivity is 10^{-7} cm/s, the effect of diffusion coefficient is significant. The largest value of D (1×10^{-5} cm^2/s) results in an increase in concentration and mass flux in only 2 years, whereas release does not occur until 7 years for the smallest value of D (4×10^{-7} cm^2/s). In contrast, if the hydraulic conductivity of the clay liner is 10^{-6} cm/s, c/c_0 reaches 1 in only 2 years regardless of the diffusion coefficient. The mass flux also directly reflects the hydraulic conductivity; the mass flux is an one-order of magnitude higher when the hydraulic conductivity is 10^{-6} cm/s.

Results for three different values of R, D = 2×10^{-6} cm^2/s, and k = 10^{-6} or 10^{-7} cm/s are shown in Figure 7. When the hydraulic conductivity is 10^{-7} cm/s, R significantly affects the release of inorganic chemicals. Chemicals are released in 2 years and c/c_0 reaches almost 1 after 12 years when R = 1 (no adsorption). When R = 5, chemicals are not released for 10 years because of the adsorptive capacity of the soil. However, if the hydraulic conductivity of clay liner is 10^{-6} cm/s, c/c_0 reaches 1 in only 3 years even if R = 5, and in one year if R = 1. The magnitude of the mass flux is similarly reflected by the hydraulic conductivity. Much lower mass flux is obtained when the hydraulic conductivity is 10^{-7} cm/s.

CONCLUSIONS
Transport mechanisms and the performance of landfill liners are summarized in this paper and a simplified performance-based method to design landfill liners was introduced. Discussion was presented based on results obtained using the simplified design method. The proposed method can be readily used to compare the effectiveness of different liner systems (geomembrane liner, clay liner, and composite liner) based on a performance measure such as mass flux of a chemical of concern. The significant advantage of the composite liner relative to geomembrane liners and clay liners was presented by comparing the mass flux of chemicals. The diffusion coefficient and retardation factor affect the performance of liners when the hydraulic conductivity of the clay liner is low (10^{-7} cm/s), whereas these parameters have much smaller effect when the hydraulic conductivity is higher (10^{-6} cm/s)

ACKNOWLEDGMENT
Financial support for Dr. Katsumi was provided by The Kajima Foundation, which has supported his leave at the University of Wisconsin-Madison from January to December 1998. This support is gratefully acknowledged. Support for Dr. Foose and Dr. Benson was provided by the Wisconsin Groundwater Research Advisory Council.

REFERENCES
Benson, C.H., Tinjun, J.M. and Hussin, C.J. (1995): Leakage rates from geomembrane liners containing holes, *Geosynthetics'95,* IFAI, Vol.2, pp.745-758.

Foose, G.J. (1997): Leakage rates and chemical transport through composite landfill liners, Ph.D. Thesis, University of Wisconsin-Madison.

Foose, G.J., Benson, C.H. and Edil, T.B. (1999): Equivalency of composite geosynthetic clay liners as a barrier to volatile organic compounds, *Geosynthetics'99,* IFAI (in press).

Giroud, J.P. and Bonaparte, R. (1989): Leakage through liners constructed with geomembranes, *Geotextiles and Geomembranes,* Vol.8, pp.27-67, 71-111.

Giroud, J.P., Soderman, K.L., Khire, M.V. and Badu-Tweneboah, T. (1998): New development in landfill liner leakage evaluation, *Proceedings of the Sixth International Conference on Geosynthetics,* IFAI, Vol.1, pp.261-268.

Manassero, M., van Impe, W.F. and Bouazza, A. (1997): Waste disposal and containment, *Environmental Geotechnics,* M. Kamon, Ed., Balkema, Vol.3, pp.1425-1474.

Ogata, A. and Banks, R.B. (1961): A solution of the differential equation of longitudinal dispersion in porous media, *U.S. Geologic Survey Professional Paper,* 441-A.

Park, J.K., Sakti, J.P. and Hoopes, J.A. (1996): Transport of aqueous organic compounds in thermoplastic geomembranes. II: Mass flux estimates and practical implications, *Journal of Environmental Engineering,* ASCE, Vol.122, No.9, pp.807-813.

Rowe, R.K. (1998): Geosynthetics and the minimization of contaminant migration through barrier systems beneath solid waste, *Proceedings of the Sixth International Conference on Geosynthetics,* IFAI, Vol.1, pp.27-102.

Shackelford, C.D. (1990): Transit-time design of earthen barriers, *Engineering Geology,* Vol.29, pp.79-94.

Remediation of the old embankment sanitary landfills

DR E. KODA, Department of Geotechnics, Warsaw Agricultural University, POLAND

INTRODUCTION
In Poland there are a few thousands of old municipal landfills, where remediation works protecting against progressive environmental degradation should be carried out immediately. Having considered the complexity and the weight of problems concerning remediation works of old sanitary landfills, interdisciplinary engineering and ecological solutions are recommended [ISSMFE, 1993; Daniel, 1993]. The paper presents threats to the environment posed by the old landfill and polluted areas. These threats among others, relate to environmental, location and hydrogeological conditions. The scope of investigations, necessary for the hazards identification and for the design of remediation works is discussed.

The examples of engineering solutions used for remediation of the large embankment sanitary landfill in Radiowo and Łubna are presented in the paper. Additional problem to be solved on these sites, was the necessity of extending the duration of the exploitation of these landfills, i.e. till the introduction of alternative utilisation methods concerning the municipal waste management for Warsaw. On Radiowo and Łubna landfills basic remediation problems were connected with protection against leachate migration and groundwater pollution as well as waste body formation (safe stability of high slopes of large inclination).

Field investigations consisting of morphological analysis, settlement measurements, WST and CPT soundings, back analysis (as well as slope failure tests) were carried out [Koda, 1997; 1998]. The aim of the investigations was the determination of mechanical characteristics of municipal solid wastes for remediation works design.

OLD LANDFILLS THREAT
The threat resulting from old landfills depends mainly on: hydrogeographical and hydrogeological conditions, kind of wastes and the technics of waste disposal. Effective remediation works should be based on precise definition of the landfill condition as well as its closer and further vicinity.

Basic investigations of old sanitary landfills should mainly include:
- analysis of exploitation files (if any) defining the kind and age of wastes as well as unpredictable events (fire, landslides, hazardous wastes, etc.);
- waste disposal survey;
- hydrogeological conditions in the landfill subsoil and its vicinity;
- determination of waste mechanical characteristics;
- description of protected areas and objects (if any);
- assessment of gas volume.

The above mentioned information is necessary for designing the remediation works.

Advantageous landfill location conditions are such, when in the subsoil there is a natural continuous aquitard layer (thickness min. 3 m, permeability coefficient $k<10^{-9}$ m/s). In the case of old landfills, this kind of isolation is hardly possible. The protection of groundwater against leachate can be realised by the cut-off wall and pheripherical drainage (Fig.1). When an old landfill is located in the opencast and there is no natural aquitard, limitation of the leachate influence can be achieved by pumping well system (Fig. 2 - scheme for Stalowa Wola old landfill).

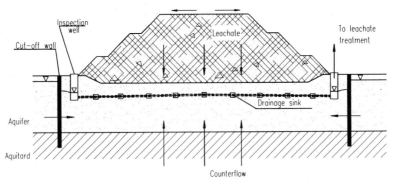

Figure 1. Scheme of the groundwater protection system against the transport of pollutants from the Łubna old landfill (cut-off wall, drainage).

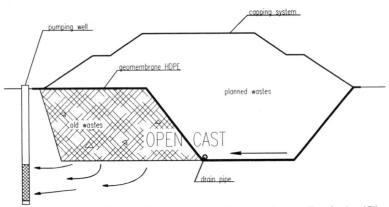

Figure 2. Scheme of the landfill protection and intermediate sealing for landfill extension.

The influence of landfills on groundwater pollution was described by many authors for a large number of sites. The leachate from sanitary landfills is ranked as strong sewage with considerable contents, diversity and concentration of pollution. Pollution characteristics in the leachate depend on many factors, such as: landfill construction, sealing system, waste oxygenation, water balance in the waste body. Pollution concentration in the leachate is generally much higher than in the sanitary sewage. The very high level of oxidizability and salinity is to be underlined (Table 1).

Wastes, even well compacted, are characterised by considerably high permeability. It is assured that on landfills of poor compaction 25 to 60% of precipitation is infiltrated into the waste body, while on the landfills of good compaction only 10 to 25%.

Table 1. Comparision of chemical parameters of sanitary sewage and leachate from landfills [Altlasten..., 1987; Koda et all, 1998].

Parameter		Sanitary sewage	Leachate from landfills				
		Poznań	Szczecin - Sierakowo	Wrocław	Łubna	Radiowo	Germany
Oxidizability	mg O₂/l	132-335	320	3304-4000	10-1500	66-800	100-45000
Chlorides	mg/l	96-132	5320	-	200-3500	600-4000	10-420
Sulphates	mg/l	154-184	98	-	10-200	230-750	-
Ammonium nitrogen	mg/l	28.5-44.2	1100	224	25-900	70-1200	30-3000
Detergents	mg/l	-	40	-	20-70	15-30	-
Phenols	mg/l	0.32-4.35	0.15	-	0.002-0.05	0.003-0.08	-
Lead	mg/l	0.05-0.14	0.083	0.2-0.3	0.15-3.0	0.02-5.0	0.02-1.0
Chromium	mg/l	0.12-0.34	n.w.	1.25-1.50	0.12-0.60	0.01-0.60	0.02-15.0
Cadmium	mg/l	0.01-0.07	0.019	-	0.06-0.15	0.002-0.02	0.001-0.1
Nickel	mg/l	0.02-3.47	0.15	to 0.4	0.1-20	0.02-50	0.02-2.0

The transport of pollutants into the subsoil depends on hydrogeological conditions and migration from the old landfills is often very difficult to define precisely. The situation is more complicated when the landfill is located in the river valley. The spread of pollution in the valley also depends on landfill location in relation to the main river bed. Transport of pollutants can also be changed as a result of the underground water intake localization (Fig. 3).

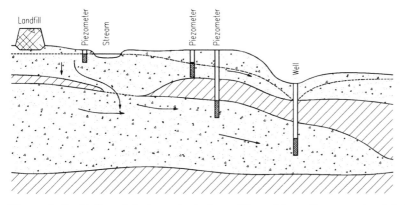

Figure 3 Sophisticated hydrogeological conditions (discontinuous natural isolation layer, stream, underground intake).

SITE DESCRIPTION

Radiowo landfill (embankment type) was established in 1962. It covers the area of approximately 15 ha and it is more than 55 m high. No protection system against environmental pollution was introduced on the surrounding area. Municipal solid wastes were disposed there till the beginning of the 90-ies. Improvement and reinforcement measures were recommended due to local landslides on the landfill. Since 1993 only non-composted wastes from a compostory plant have been disposed there, i.e. approximately 300 tons per a day. It is expected that the landfill will be utilised for least for the next 5 years. Since 1994 remediation works have been carried out on the landfill. They include among others: forming and planting of the slopes, stability reinforcement solution (lateral reinforcements and berms), mineral capping, bentonite cut-off wall limitation of groundwater pollution (under construction) and pheripherical drainage (Fig.4).

Łubna landfill (embankment type) has been in existence since 1978 and also has no environmental protection system. It covers the area of approximately 20 ha, and it is almost 50 m high. Łubna is the only sanitary landfill where all kinds of municipal wastes from Warsaw are stored, i.e. about 1500 ton/day. In 1995-96, slimes from the sewage-treatment were also disposed there. Łubna landfill is planned to be closed in 2000 year. Since 1996 remediation works have been carried out on the landfill. They include: bentonite barrier wall around the landfill (constructed in 1998), leachate drainage system (constructed in 1997 and 1998), treatment plant (under start-up), berms (under construction), degassing and mineral (or HDPE geomembrane) capping system (Fig.5).

INVESTIGATION ON OLD LANDFILLS
The *in situ* tests were performed in 1993-99 for Radiowo and in 1996-99 for Łubna landfill. They were to determine mechanical parameters of wastes for stability analysis, settlement prediction and estimation of bearing capacity for road foundation. The main purpose of the tests is to utilise the existing landfills entirely, i.e. to determine of shear parameters in order to assure safe slope inclination [Koda, 1998; 1999]. Field investigations consisting of settlement measurements, WST and CPT soundings, back analysis (as well as slope failure tests) were carrierd out. In the case of Radiowo landfill, morphological composition of wastes creates an additional factor influencing mechanical parameters. Organic matter content for non-composted wastes is ca. 4% [Koda, 1997]. For non-composted wastes, strength characteristics were confirmed based on slope failure test, and for old wastes, these characteristics were derived from the back analysis of 1991 landslide on the eastern slope of Radiowo landfill (Fig.4). The characteristics of fresh wastes were verified based on back analysis of 1995 landslide on the northern slope of Łubna landfill (Fig.5). This analysis also included test results presented in other references [Jessberger and Kockel, 1993; Sanches-Alciturri et all, 1993; Manassero et all, 1996]. Shear strength parameters for municipal solid wastes used in the stability analysis are shown in Table 2.

Table 2. Shear strength parameters for municipal solid wastes [Koda, 1998].

Category	Site	Unit weight γ [kN/m^3]	Normal stress σ [kPa]	Shear angle of friction ϕ [°]	Intercept cohesion c [kPa]	The tests methods
Non-composted wastes	Radiowo	9.0	35	20	25	slope failure tests, CPT, WST
Non-composted wastes with sand	Radiowo	12.0	50	25	23	slope failure tests, CPT, WST
Old municipal wastes	Radiowo	14.0	65	26	20	back-analysis of landslide, CPT, WST
Fresh municipal wastes	Łubna	11.0	125	21	15	back-analysis of landslide, WST

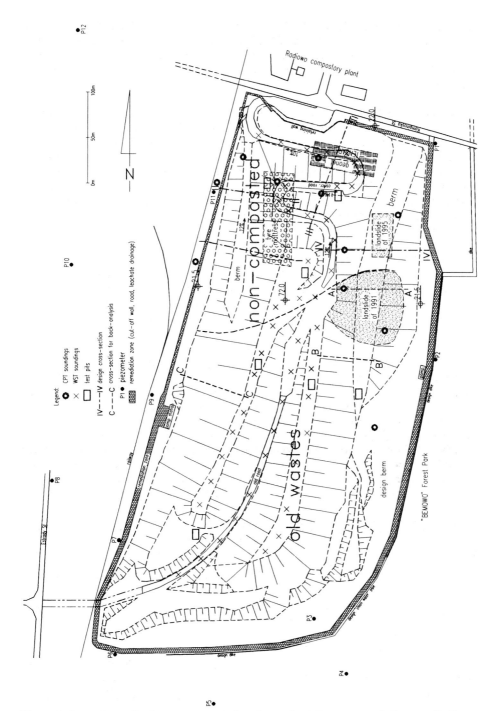

Figure 4. Layout, monitoring system, tests location and reinforcement solutions on Radiowo landfill.

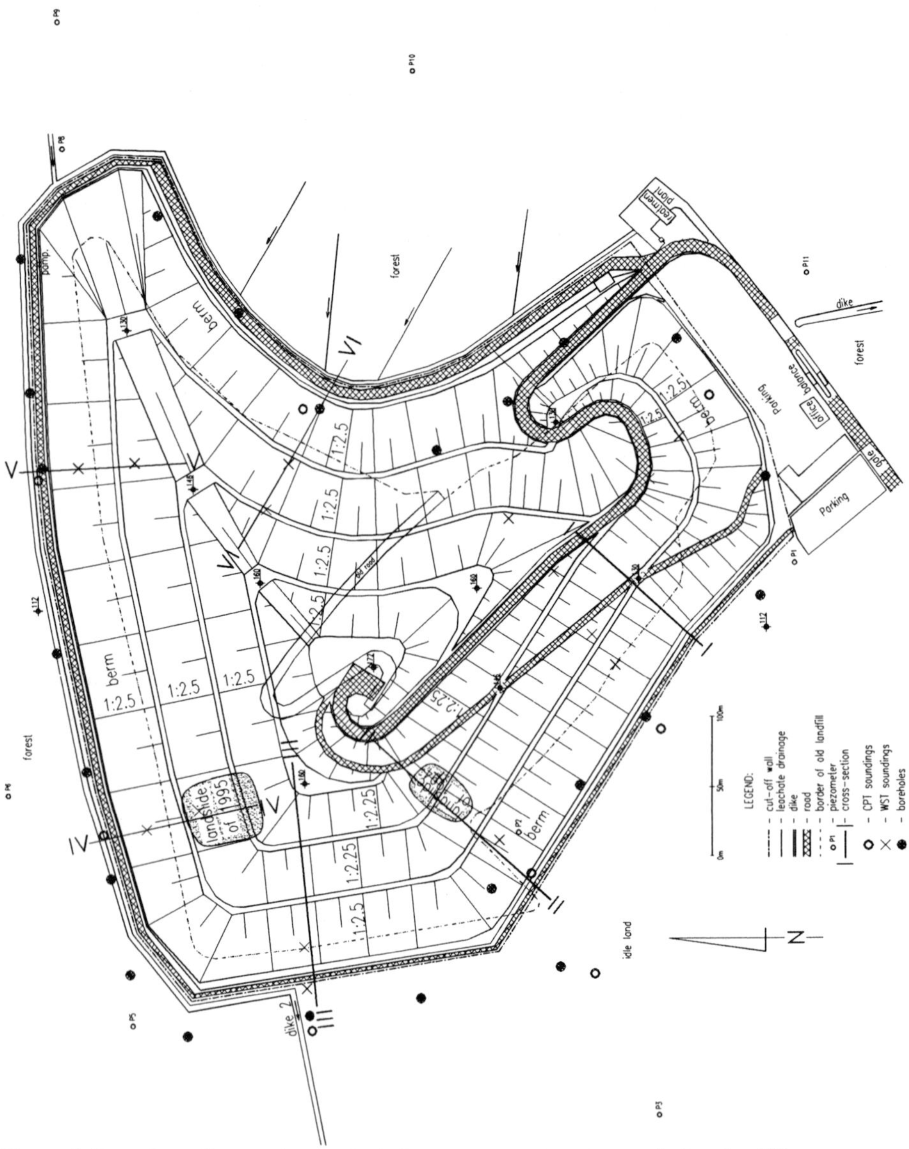

Figure 5. Location of test points and reinforcement construction on Łubna landfill.

Permanent surveys of the horizontal slope movements and settlements of the landfills body have been carried out since 1993 on Radiowo landfill and since 1996 on Łubna landfill. Moreover, permanent points for measuring settlements were fixed at the distance of 20 m along the alignment of the main access roads. Values of the initially evaluated modulus of compressibility ranged from 300 kPa (for fresh non-composted wastes) to 2000 kPa (for strongly compacted wastes with sandwiched sand layers). The test results are used to forecast settlement of the access roads on further sections, design the solid shape and assess the capacity of the landfills [Koda and Fołtyn, 1998].

REMEDIATION SOLUTIONS ON OLD EMBANKMENT LANDFILLS

In order to improve stability conditions of Radiowo landfill slopes the following have been done: retaining wall, moderate slope inclination, replacement of non-composted waste in the road foundation, lateral reinforcements (by geogrid and tyre mattress) and berms (Fig. 6). All factors of safety for reinforced slopes are higher than 1.3 [Koda, 1999].

In the case of Łubna landfill, when the slopes are high and of considerable inclination, the berms seem to be the most effective solutions for the slope stability reinforcement. The berm enables the achievement of additional capacity for waste disposal (Fig. 7). The surface of the berm was made of cohesive soil and compost.

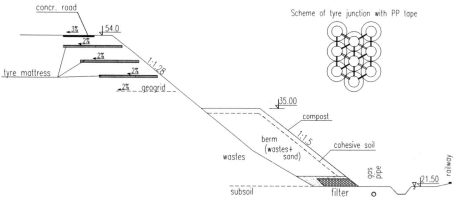

Figure 6. Cross-section II-II and reinforcements of the western slope of Radiowo landfill.

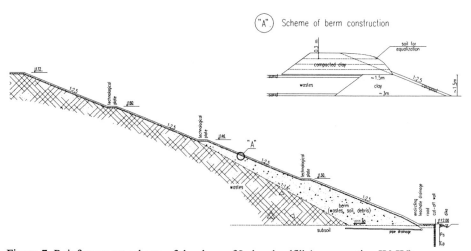

Figure 7. Reinforcement scheme of the slope of Łubna landfill (cross-section IV-IV).

Łubna landfill is located on the wet area. The upper part (first aquifer) of the subsoil consists of sandy soils and muds (thickness of 2 to 15 m). This layer is polluted by the leachate from the landfill. Deeper part of subsoil consists of low permeable boulder and varved clays (thickness of 25 to 40 m, average permeability coefficient of 10^{-6} m/s). Set of clay layers makes the protection of the main Quaternary aquifer possible (Fig. 8).

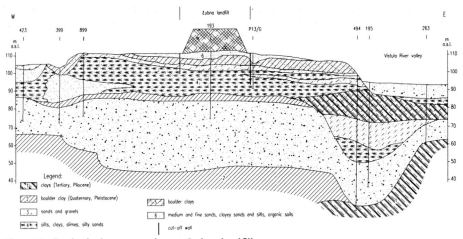

Figure 8. Geological cross-section on Łubna landfill area.

The bentonite cut-off wall was constructed in order to protect the first aquifer against the transport of pollutants. The permeability coefficient for the barrier wall is below 10^{-9} m/s (from the BAT *in situ* and laboratory tests). The initial hydraulic gradient for the bentonite is ca. $i_0=50$, while existing gradient in the field is $i=2-3$. In fact, the constructed cut-off wall is practically impermeable. The depth of the barrier wall, depending on subsoil conditions, was 5.5 to 17 m. Additional protection against the migration of pollution is assured by the flow direction from the vicinity to the leachate drainage (Fig. 9). It will also protect the bentonite material against degradation caused by the leachate.

In the framework of the remediation design, the pheripherical drainage around the landfill was constructed. Moreover, at the bottom of berms pipe drains were also installed so that to direct the leachate to the pheripherical drainage.

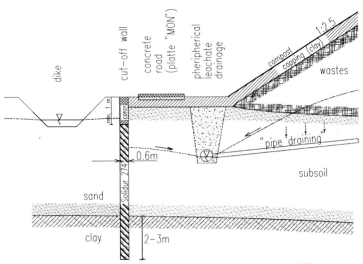

Figure 9. Scheme of remediation solutions around Łubna landfill.

The analogical protection system was designed for Radiowo landfill, where similar subsoil conditions were observed. Presently, remediation works are under construction on this site (Fig. 10).

The leachate can be partly utilised on the landfill (Radiowo) or discharged to the treatment plant and then to the surface streams (Łubna). The investigation of the field capacity of wastes, i.e. maximum water content without the leachate, was also carried out on Radiowo landfill. The test results were used in the analysis of the water balance for the waste body on the landfills [Koda and Żakowicz, 1998].

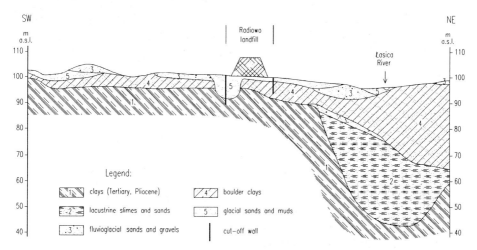

Figure 10. Geological cross-section on Radiowo landfill area.

CONCLUSIONS

Investigation of the old sanitary landfills refers mainly to the determination of the kind and age of wastes, hydrogeological conditions, waste mechanical characteristics as well as pollution of soil and groundwater in the subsoil and vicinity.

Transport of pollutants from old landfills depends mainly on hydrogeological conditions. Installation of the local monitoring system is important for the control of this process.

For calculation of landfill slope stability and settlement prediction of the waste mechanical characteristics should be determined by using *in situ* test methods, taking into account the composition and age of wastes. These characteristics should be verified by slope failure tests, settlement measurements and back analyses or laboratory tests.

Berm construction seems to be the most effective method for slope stability improvement of old landfills for further remediation works. Berm construction forces the landfill to be extended.

Radiowo and Łubna landfills have no protection systems. Hydrogeological conditions enable the construction of cut-off wall to protect against groundwater pollution. The leachate can be partly utilised on the landfill (Radiowo) or discharged to the treatment plant and then to the surface streams.

Preliminary investigation of gas volume carried out on both landfills, shows a possibility of its utilisation for energetic purposes.

Compost from municipal wastes (produced in Radiowo compostory plant) can be used for the covering layer as a substitute of humus.

Groundwater protection system, based on cut-off wall and drainage system installed below the water table, is safe and easy to build for the old sanitary landfills remediation. This system is the most effective and very economic, if only it is applied to large areas.

When all remediation works are completed, in the framework of remediation design, the landfill surface can be adopted for recreation purposes (cycleways, ski-lift, scenic points).

REFERENCES

Altlasten-Handbuch, Teil I: Altlasten-Bewertung und Teil II: Untersuchungsgrundlanden. (1987). *Ministerium für Umwelt Baden-Württemberg,* Karlsruhe.

Daniel, D.E. ed. (1993). Geotechnical practice for waste disposal. *Chapman and Hall,* London.

Jessberger, H.L. and Kockel, R. (1993). Determination and assessment of the mechanical properties of waste materials. *Proc. of the 1st Intern. Symp. on Geotechnics Related to the Environment,* 313-320, Bolton.

ISSMFE/ETC5. (1993). Geotechnics of Landfill Design and Remedial Works. Technical Recommendations - GLR. 2nd ed. *Ernst & Sohn.* Berlin.

Koda, E. (1997). In situ tests of MSW geotechnical properties. *Proc. of the 2nd Intern. Symp. on Geotechnics Related to the Environment, Thomas Telford ed.,* 247-254, Kraków.

Koda, E. (1998). Stability conditions improvement of the old sanitary landfills. *Proc. of the 3rd Intern. Congress on Environmental Geotechnics,* Vol.I, 223-228, Lisboa.

Koda, E. and Żakowicz S. (1998). Physical and hydraulic properties of the MSW for water balance of the landfill. *Proc. of the 3rd Intern. Congress on Environmental Geotechnics,* Vol.I, 217-222, Lisboa.

Koda, E., Król, P. and Żakowicz S. (1998). Utylization system of leachate from landfill and technological water from Radiowo compostory. *Proc. of the Intern. Conf. on Environmental and Technical Problems of Water Management for Sustainable Development of Rural Areas,* Vol.II, 281-290, Warsaw.

Koda, E. and Fołtyn, P. (1998). Experience in road construction on sanitary landfills. *Proc. of the 3rd Intern. Workshop on Secondary Rural Roads,* 143-156, Józefów.

Koda, E., (1999). Stability reinforcement of the old embankment sanitary landfills for remediation works. *Intern. Symposium on Slope Stability Engineering: Geotechnical and Geoenvironmental Aspects,* Shikoku (in print).

Manassero M., van Impe, W.F. and Bouazza A. (1996). Geotechnical properties of MSW. *Proc. of the Intern. Congress on Environmental Geotechnics,* Vol.3, 1425-1474, Osaka.

Schanches-Alciturri, J.M., Palma, J., Sagaseta, C. and Canizal, J. (1993). Mechanical properties of wastes in a sanitary landfills. *Proc. of the 1st Intern. Symp. on Geotechnics Related to the Environment, 357-*364, Bolton.

Design and Performance of a Compacted Clay Barrier for Heathrow Express Rail Link Tunnel

DR K. O'CONNOR, Formerly of Sir William Halcrow and Partners, London, UK, DR D.C. WIJEYESEKERA, Department of Civil Engineering, University of East London, UK and DR. D.E.SALMON, Sir William Halcrow and Partners, London, UK

ABSTRACT

This paper presents the construction and performance details of a compacted clay surround that was designed to prevent any leachate or methane penetrating a stretch of the cut and cover box section tunnel of the London Heathrow Express Rail Link. This stretch of the tunnel runs through a disused landfill site. The barrier was constructed of London clay excavated from concurrent bored tunnelling operations on adjacent sites. The acceptance criteria for the barrier design were based on permissible levels of permeability, swelling pressures and differential movements in the compacted London clay. This provided an economical and environmentally conducive solution compared with other alternatives considered.

Laboratory investigations carried out on the compacted London clay indicated that, despite the propensity of the clay surround to swell with access of ground water and inundation, it was possible to design the clay surround to meet the acceptability criteria. Results of the laboratory investigations carried out using a computer controlled consolidometer cell are discussed in this paper.

The associated construction features are described together with the stringent construction specifications for control of compaction quality. Monitoring of field instrumentation installed at the site, subsequent to construction of the barrier is also reported in the paper. Magnetic extensometer gauge measurements indicate that the settlements are lower than that predicted by the laboratory tests.

INTRODUCTION

A part of the construction of the new Heathrow Express rail link between the Airport and Paddington consisted of approximately 0.5km of cut and cover box tunnel through a landfill site. The transition from surface to underground rail occurred in this cut and cover section of the line coinciding with a disused landfill site. The dimensions of the box section tunnel were 12m x 7.5m and had a slope of 1 in 43.

The obvious design option of total relocation of the landfill proved to be unacceptable on both environmental and cost grounds. Consequently any engineering solution needed to accommodate the special design requirements to protect structures in the vicinity of landfill site. The option of using welded plastic membranes that would encapsulate the box tunnel as it passed through the site raised concerns regarding the long-term durability of such

Geoenvironmental engineering, Thomas Telford, London, 1999, 39–46

membranes. This led to the adoption of an alternative solution of a compacted clay surround to the box tunnel as being both more economical and environmentally advantageous.

The special design requirements of the compacted clay surround needed assurance that leachate or methane from the landfill will not penetrate into the tunnel. The surround comprised a barrier of London Clay conditioned to specific water contents and compacted around the box tunnel to form a flexible yet impermeable barrier. This design made use of the London Clay that was excavated from the nearby bored tunnel operations. This utilised in excess of 60,000 m^3 of the excavated London Clay, saving the costs that would have been otherwise incurred in transport and disposal.

Site Location and Geology

The landfill site is located north of Shepiston Lane, which is north of junction 4 on the M4 motorway. Published geological information indicates that Brickearth and Taplow Gravel overlie the London Clay at the site (British Geological Survey, 1981). The London Clay dips gently from north to south at an approximate angle of 3 degrees. This site was worked in the past for the extraction of gravel up to depths of 8m below the ground level. On exhaustion of the gravel, the site was used as a landfill facility, accepting a combination of domestic and light to medium industrial waste. The facility stopped receiving refuse during the 1960s. There are no known records of the manner and nature of these wastes. Figure 1 shows a plan of the site with a longitudinal profile along the line of the tunnel illustrating the existing geology, the construction section and locations of extensometers for post construction monitoring. The tunnel can be seen to achieve full penetration of the in situ London Clay at the southern end of the site, at which point the construction method changes from cut and cover to a bored tunnel.

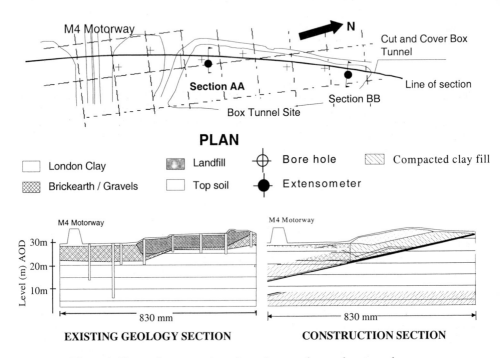

Figure 1. Plan and cross sections through cut and cover box tunnel.

DESIGN & CONSTRUCTION

The compacted clay surround was designed as a barrier to prevent methane penetrating the box tunnel in the form of gas or as gas dissolved in the leachate from the landfill. The three main acceptance criteria for the compacted clay surround were:

- A low permeability in the clay fill of less than 10^{-9} m/s was required.
- The swelling pressures in the clay should be relatively small in comparison with the horizontal stress caused by the overburden.
- The clay fill should be flexible in accepting a degree of differential movement without fissuring or cracking.

Compaction and Permeability Criteria

For a given compactive effort, permeability of the clay soil is lower when it is compacted at water contents wetter than the optimum. For clay compacted in this condition the advantage of low permeability is offset by increased settlements which could lead to fissuring and cracking. Day (1997) suggests that landfill covers should not be compacted at a water content dry of optimum or at high compactive effort as it can lead to significant swelling. The excavated London Clay comprised of varying sizes and shapes of "clay clods" that significantly influenced the proctor compaction (Wijeyesekera & O'Connor 1997). The permeability of compacted soils is further affected by the nominal size of these clay clods, their water content and the specified density. A soil conditioning machine was used to pulverise the excavated clay to an acceptable clod size of 50mm, whilst adding pre-determined quantities of water to produce three required classes of clay at specified water contents. The conditioned clay was stockpiled to allow the water within the clay to equilibrate resulting in a homogeneous material. The water content of the London Clay at its plastic limit (PL = 28%) was used as the parameter of the soil that defined the water contents for placement. The ranges of clay class, including their working tolerances, were as follows:

- 1.1PL Class Clay - From the base of the box tunnel to mid height (0 to 3.75m), specified water content range of between 26 and 30 percent.
- 1.2PL Class Clay - From mid-height to the top of the box tunnel (3.75 to 7.5m), specified water content range of between 29 and 32 per cent.
- 1.3PL Class Clay - From the top of the box tunnel to 4m above the box tunnel (7.5 to 11.5m), specified water content range of between 31 and 35 percent.

Placement moisture contents greater than the plastic limit ensured that any soil suctions developed will be negligible (Wijeyesekera & O'Connor 1997). An average placement moisture content of 1.2PL was appropriate for London Clay, to develop the required minimum shear strength of 50kPa for acceptable trafficability in highway embankment fills (Jones and Greenwood, 1993). The fill near the top of the tunnel was of 1.3PL clay to give flexibility and reduce any cracking as a result of any differential settlement (see Figure 2).

Compaction and Swelling Pressure Criterion

Compacted clay in a confined state will swell. In general, greater the change in water content combined with confinement, greater is the swelling pressures that can develop (O'Connor 1994). Also, for a given air voids, the swelling pressure increases as the dry density increases and the water content decreases (Wijeyesekera & O'Connor 1997). The specifications for the placement of the clay required the compaction process to achieve a maximum air void ratio of 5% with all three water content levels (1.1PL, 1.2PL and 1.3PL).

Settlement Criterion

To accommodate any settlements and maintain the integrity of the clay surround the uppermost layer of clay was of a high water content (1.3PL) to be more flexible. The clay fill could deform plastically when subjected to consolidation due to overburden pressure. Since vertical stress increases with depth, the water content of the clay surround with depth needed to decrease with depth to minimise the overall settlement.

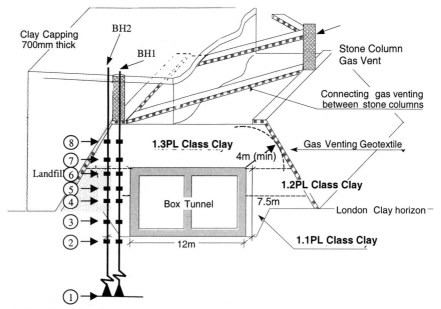

Figure 2. Details of the clay surround

BOX TUNNEL & CLAY SURROUND

The box tunnel is a twin cell reinforced concrete box having external dimensions of 7.5m by 12m. The external walls are 0.5m thick with a central dividing wall 0.3m wide at the northern portal. At the southern end, the box tunnel was connected with the two separate bored tunnels. The two cells of the box tunnel gradually diverged with the central wall becoming increasingly wide leading to an overall box tunnel width of 13m. The completed clay surround has a minimum thickness of 4m from any point of the tunnel. Figure 2 shows details of a typical cross section through the box tunnel showing the disposition and water content ranges in the three zones of the clay surround. This figure also shows the positioning of gas venting geotextiles used to ensure that any methane in a gaseous state will be collected at the boundary of the landfill and the clay surround and conducted to vertical stone column vents. A capping layer greater than 700mm thick was used to cover the landfill on the site.

LABORATORY INVESTIGATIONS

Undisturbed samples of the compacted clay surround were taken and tests carried out to estimate both the volumetric and stress changes that would occur. A computer controlled consolidometer cell (Figure 3) which is described in Ting, Sills and Wijeyesekera , 1994 was used to apply a stress path to the soil that will simulate the stresses acting on samples of clay fill adjacent to the side wall of the box tunnel whilst measuring any changes in the horizontal stress and volume.

Consolidometer results

The vertical stress applied to the test samples were the calculated overburden pressures. Horizontal stress observed was equivalent to the horizontal pressures on the box tunnel. Figure 4 illustrates the stress and strain vs. time response for tests performed on samples from section AA.

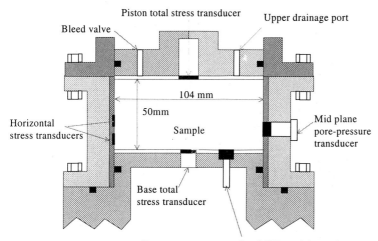

Figure 3. Details of consolidometer cell.

The horizontal stress variation with vertical stress at three distinct stages is shown in Figure 5. The first point refers to the stage of maximum vertical stress in the test. At this stage the full vertical stress was applied but was prior to the consolidation of the sample under constant vertical load. The second point shown is the horizontal stress just prior to the inundation of the sample. The difference between this and the maximum stress is attributed to the dissipation of the pore pressures during consolidation. The final point refers to the horizontal stress after the sample was inundated and allowed to reach equilibrium. These values are summarised in Table 1 for both sections.

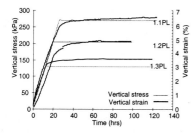 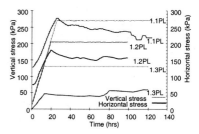

Figure 4. Development of stress and vertical strain in the clay surround at section AA.

Vertical strain of the sample was also monitored during the tests. Table 2 shows the respective vertical strain recorded at the same time as the horizontal stress values were measured. Figure 6 show the plots of vertical strain.

Table 1. Horizontal and vertical stresses

	Section AA			Section BB	
	1.1PL	1.2PL	1.3PL	1.1PL	1.2PL
Vertical confining stress during test (kPa)	270	200	130	120	57
Increase in horizontal stress after inundation kPa)	6	3	2	10*	1

* Sample underwent further compression and horizontal stress reduced

Table 2. Vertical strains

	Section AA			Section BB	
	1.1PL	1.2PL	1.3PL	1.1PL	1.2PL
Vertical strain at maximum vertical stress (%)	5.9	3.7	3.1	0.7	1.7
Vertical strain after consolidation (%)	6.5	4.7	3.5	0.9	1.8
Vertical strain after inundation (%)	6.5	4.7	3.5	0.8	1.5

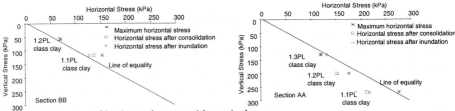

Figure 5. Variation of horizontal stress with vertical stress

Undrained Triaxial Compression Tests

The failure envelope derived from the triaxial tests on specimens from each of the samples of clay classes are shown in Figure 7. Also indicated are total and effective stress states both after consolidation and after inundation. As can be seen the movement of the stress states in terms of both total and effective stress lie well within the failure envelope.

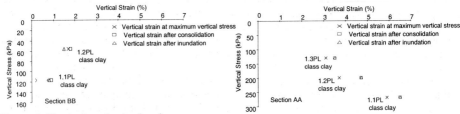

Figure 6. Vertical strains in the clay

Oedometer Permeability Evaluation

The permeability results from a series of one – dimensional consolidation tests are shown in Figure 8. It shows that even at low vertical stress, the compacted clay has a permeability of less than 10^{-9} m/s and with increasing vertical stress the permeability decreases to 10^{-11} m/s.

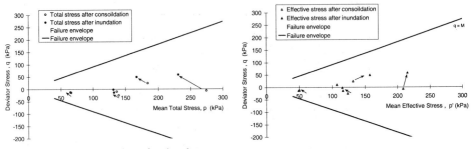

Figure 7. Strength envelope for the clay

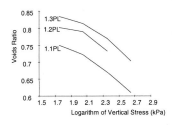 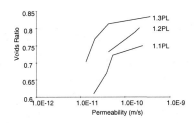

Figure 8. Voids ratio - permeability - vertical stress results from oedometer tests.

FIELD INSTRUMENTATION AND MONITORING

Settlement of the clay surround to the tunnel box were monitored using four magnetic extensometer gauges installed at the two sections AA and BB. The boreholes for the extensometers 1 and 2 at AA are indicated in Figure 1. At each section, two magnetic extensometers were installed in both the compacted clay and the in situ London Clay. Each extensometer installation comprised a base magnet set in the in situ London Clay and seven spider magnets positioned within the clay surround (see Figure 2). Monitoring began in September 1996. With the extensometers being read weekly for the first ten weeks, every two weeks for four months, and monthly thereafter. The movement of the compacted clay barrier at section AA is shown in Figure 9. Here the movement of the compacted clay is plotted relative to the base magnet that was inserted in the in situ clay below the foundation level of the box tunnel. By measuring the movement relative to the base magnet, it was possible to measure any changes in the compacted clay surround whilst excluding any other movements arising from changes that may have occurred within the in situ clay during construction. At each of the two instrumented locations it was possible examine the way in which the compacted soil behaved as the distance from the tunnel increased. It can be seen from Figure 10 that at the deepest section (Section AA) the compacted clay nearer the base of the box compressed while the upper layers of the compacted clay expanded. Secondly that there was little difference in the settlement behaviour as the distance from the box increased. In contrast at the northern section (section BB) of the tunnel there was a notable difference in the settlement observations. The predominant movement observed at this section was swelling at all layers and also the soil further from the box was seen to expand by twice the value of the column of soil nearest to the box. The reason for this different behaviour lies in the fact that the construction of the compacted clay barrier remained constant along the entire length of the box tunnel. In the upper reaches of the tunnel (boreholes 3 and 4) the clay surround with

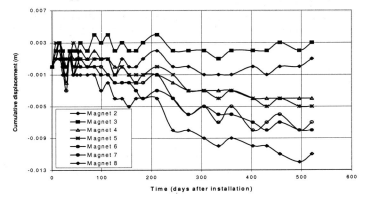

Figure 9. Magnetic Extensometer Data : cumulative displacement vs time - borehole 1

lower water content (1.1PL) had less vertical stress to counter balance the swelling pressures and therefore swell. However, in the deeper section (boreholes 1 and 2), where the lower levels had a greater vertical stress imposed restricted swelling. A further reason for the difference is due to the fact that at section 1&2 the water table is above the box and clay surround whereas at section 3&4 the water table is nearly half way up the clay surround.

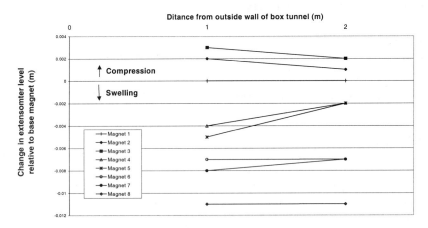

Figure 10. Differential settlement between boreholes 1 and 2

CONCLUSIONS

- The permeability values evaluated for the compacted clay demonstrate that it has a very low permeability and is practically impervious.
- The consolidometer results demonstrate the maximum horizontal swelling pressures and vertical strains exhibited by the compacted London Clay fill specimens for each class of clay surround and following inundation, are small in comparison with the measured horizontal stress developed after consolidation to the in situ stress level at each horizon.
- The vertical strains developed in the test specimens that modelled field conditions indicate that the estimated strains producing differential settlement at the uppermost corners of the box tunnel are significantly less than the strains that would lead to a shear failure within the surround. Magnetic extensometer gauge measurements indicate lower settlements than that predicted by the laboratory tests.

REFERENCES
British Geological Survey 1981. Sheet 269,Windsor, Solid and Drift edition.
Day, R.W. 1997. "Discussion paper on unsaturated hydraulic conductivity of two compacted barrier soils." *J. Geotech. & Geoenv. Engrg.*, ASCE, 122(12),1186.
Jones R.H. and Greenwood J.R., 1993, Relationship testing for acceptability assessment of cohesive soils, Proceedings of the Conference of Engineered Fills, Newcastle Upon Tyne, UK. Thomas Telford Ltd., pp 302-311
O' Connor K 1994, Swelling behaviour of unsaturated fine grained soils. PhD thesis, City University, London pp270.
Ting C.M.R, G.C. Sills & D.C.Wijeyesekera 1994, Development of K_0 in soft soils. *Géotechnique* 44 (1); 101-109.
Wijeyesekera, D.C. & K. O'Connor, 1997. Compacted clayey fill: an assessment of suction, swell pressures, shrinkage and cracking characteristics. *Proc. Int. Conf. Ground Improvement Techniques, Macau, 7-8 May 1997:* :621-630

First results about the influence of leachate on the properties of Boom Clay

A-S. OURTH
Faculté Universitaire des Sciences Agronomiques de Gembloux, Unité de Résistance des Matériaux, Belgium
J-C. VERBRUGGE
Universite Libre de Bruxelles, Laboratoire de Mécanique des Sols, Belgium

ABSTRACT

The design and carrying out of landfill bottom liner is a hot problem because it concerns the environment. Modern sanitary disposal sites are based on the concept of ensuring minimum environmental impact. The good and well known properties of clay makes it nearly unavoidable in the design of landfill bottom liner. Though very little knowledge exists about behaviour of clay permeated with a leachate. In this study, various tests have been carried out with and without leachate to investigate how clay reacts when in contact with a pollutant. In order to obtain sensible results, the leachate used has a composition that is representative of such met in general landfill sites. The initial tests concern the changes in intrinsic clay properties such as suction, Atterberg limits and permeability. Standard oedometer test were used to study the influence on the consolidation of clay. This paper will present the first result gained about the way that leachate influence the properties of Boom clay.

INTRODUCTION

The time when the garbage dump was seen as a pile of wastes is over. New solid waste management involves conception and design of bottom liner in sanitary disposal sites. Our concern is to improve the environmental quality of these sites.

The aim of the investigation described here is to compare the behaviour of clay - the mechanical properties of which as landfill bottom liner has long been proved - wet by water with that wet by municipal solid waste leachate. The effects on the intrinsic properties (Atterberg limits, suction, permeability) of the clay has currently been studied, as well as some geomechanical properties, particularly the settlement characteristics using a standard oedometer test.

The clay chosen here is Boom clay, well known in Belgium. The natural material is pure, without sand, and remoulded samples were prepared in the laboratory at the appropriate moisture content.

The leachate used is a mixture of leachate from different cells of a municipal solid waste landfill close to Brussels. Two of the cells are old ones, and the third one is still in an active state. Therefore, the leachate is one of 'middle age', coming from a regular landfill. All the liquid has been pumped in one step and subsequently frozen.

As the short term results obtained from clays prepared separately with distilled water and leachate fail to show any significant behaviour difference, it seemed necessary to let the clay be in contact with the pollutant for a long time. That way, modification of chemical properties of clay can take place. So, a special apparatus was used to leave the clay 'soaking' in leachate, safe from any further oxidation, for a period of about three month.

TEST MATERIAL : CLAY AND LEACHATE
The clay used in the investigation is Boom clay, coming from a well known deposit in Belgium. The benefit of this choice was that this needs no mix with sand. That way, there were no further parameters necessary to be taken into account arising from the nature and quantity of sand. The samples were obtained in an air-dried condition and then pulverised to a particle size smaller than 0.5 mm. The dry soil was mixed with the appropriate liquid to the desired moisture content and was left to stand for a minimum of two days period to produce an uniform and homogeneous mixture.

It is difficult to define a standard for leachate as it differs from one dump to another and it varies also with time according to the age of the waste. The liquid was taken from a municipal solid waste landfill site. It is a mixture of liquid from three cells of different age. The mixture is then like a 'middle-aged' leachate.

As leachates evolve naturally, due to organic matter and micro-organisms, it was necessary to find a conservation method preserving it's composition :

• Desiccation seemed a good solution, but it was impossible to desiccate humic and fulvic acid. The precipitation of those matters needed support of chemicals, so the composition was necessarily altered.

• It was also impossible to concentrate them in order to freeze them separately.

• Finally, the least solution was to freeze the whole quantity (about 1 m^3) without separation of components.

To study how possible modification in clay behaviour will evolve, the program involved tests carried out on clay contaminated for about three months with the leachate. This 'soaking' took place in a can where one part of the soil was kept in an anaerobic ambience mixed to five parts of pollutant. The liquid was first deaired, and any remaining air was replaced by nitrogen. Afterwards, the receptacle was closed and a water trap used to avoid any form of over-pressurising of the bottle. The apparatus (figure 1) was kept in a place away from light.

Figure 1. 'soaking' apparatus

TEST PROGRAM AND FIRST RESULTS
Several kinds of tests are to be used in this investigation. The research is now at its early stage and the results given here are those from the first investigations.

Material identification
The studied properties of clay are Atterberg limits, particle size distribution, soil micro structure and suction. To satisfy the aims of the investigation the behaviour of clay with water and with leachate (after soaking) are to be evaluated. For the leachate, the characterisation requires the knowledge of chemical composition, osmotic pressure and pH. These are also required for the liquid expelled during testing.

Leachate
• The chemical analysis of leachate (before freezing) is given in table 1:

N_{kjel}	(mg N/L)	258	K	(mg/L)	653
$N_{NH_4^+}$	(mg N/L)	248	Na	(mg/L)	2000
$N_{NO_2^-}$	(mg N/L)	<0.02	Fe	(mg/L)	4.73
$N_{NO_3^-}$	(mg N/L)	2.98	Mn	(mg/L)	0.459
N_{org}	(mg N/L)	10	Pb	(µg/L)	90
SO_4^{--}	(mg/L)	217	Cd	(µg/L)	1.1
Cl^-	(mg/L)	273	Zn	(µg/L)	258
P_2O_5	(mg/L)	8.13	Hg	(µg/L)	0.13
COT	(mg C/L)	316	Cu	(µg/L)	48
Ca	(mg/L)	208	Ni	(µg/L)	140
Mg	(mg/L)	131	Cr	(µg/L)	48

Table 1. Chemical analysis of leachate.

• The osmotic pressure of leachate, using a Wescor thermocouple psychrometer (CRR, 1975) was found close to-5 bars.

Clay
• The Atterberg limits of the Boom Clay are : Liquid limit: 67 %
                                                Plastic limit: 28 %
                                                Shrinkage limit: 19 %

• The Proctor test using the CBR mould and the low compaction energy presented an optimum value for clay remoulded with water at $\gamma_d = 14.2$ kN/m^3 and $w_{opt} = 28\%$.

• The suction curve measured (using a Wescor thermocouple psychrometer) on the clay prepared with water and with leachate (without 'soaking') are showed in figure 2.

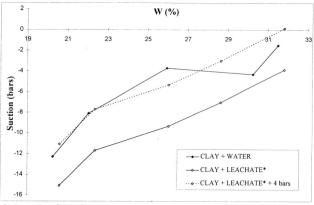

Figure 2. Result of suction test

Geomechanical properties
The program involves triaxial on saturated and unsaturated samples, oedometer and permeability tests.

The principal problem encountered for these tests is the evolution of the leachate when in contact with air and its corrosive effect on the testing apparatus. Some modifications have been done to prevent this corrosion and the evolution of leachate.

For the tests, there are four possible combinations for the pore fluid used to remoulded and to permeate the clay. The table below summarises the different possibilities.

Test series	Code	Remoulding Fluid	Permeating Fluid
1	W - W	Water	Water
2	W - L	Water	Leachate
3	L - W	Leachate	Water
4	L - L	Leachate	Leachate

Table 2. Combination of pore fluid at remoulding and permeating pore fluid

Moreover, the symbol L* in the text below, means that the test has been carried out on clay material remoulded with the leachate over a short term period of two days and not on the 'soaked' one with longer period for contamination.

Test series 1 is the reference, as it corresponds to the normal experimental conditions encountered in soil mechanics. Test series 2 corresponds to reality, clay remoulded with water before being placed as a liner for the landfill, and then subsequently permeated through by the leachate. Third and fourth series are fictitious situations of academic interest, to show the importance of the osmotic phenomenon. When clay is prepared with leachate, the soil solution contains more salt than in a sample that is remoulded with water. This will influence the interaction of solid and liquid phases. These four different combinations will enable to investigate this hypothesis, and present an explanation to some of the phenomenon that occurs.

At present time only oedometer and permeability tests have been performed.

Oedometer test

• Testing device :

Oedometer tests for W - W and L* - W were conducted in a standard 50 mm diameter oedometer. Tests with leachate as a permeant used an apparatus where the liquid was out of contact with the atmospheric air. So, a cell combining the principles of the triaxial cell with regard to watertightness / airtightness and the oedometer was constructed for the test.

• Results :

Oedometer test data presented are from the series W - W and L* - W, results observed for these were nearly the same for samples prepared either with water or with leachate.

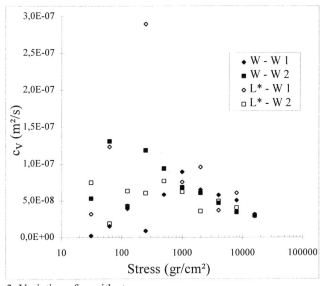

Figure 3. Variation of c_v with stress

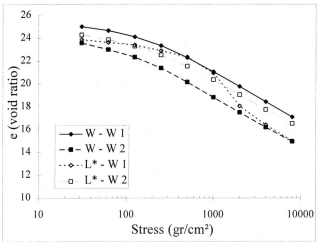

Figure 4. Compressibility curve : Variation of void ratio with stress

Permeability
• Testing device :
Permeameters used have rigid walls. The samples tested were of similar geometry (50 mm diameter and 16 mm height) as that used in the standard oedometer. These were compacted directly in the apparatus to avoid leakage along the wall. The tests were carried out over many weeks, with a measurement being taken daily to observe the evolution of the leachate with time.

• Results :
The permeability tests were performed on samples compacted at the normal optimum Proctor density and then saturated with water. At first, water was used as the permeant. When the value of permeability was well stabilised, leachate was used as the permeant. There was no appreciable difference of permeability coefficient observed over a period of three weeks after the change of permeant. The average value of permeability observed was 3 E-10 m/s. The standard deviation is about 5 E-11 m/s.

CONCLUSIONS
This paper summarises the first results about the influences of leachate on the properties of Boom clay. This search being at its early stage, the results concern clay moistened with water or with leachate only for two days before testing.

At present time, no significant difference between clay with or without leachate has been observed, except in the suction test. The osmotic pressure value of the leachate is about - 5 bars and the suction curve of clay wet with leachate is shifted about 4 bars under these (see Figure 2). The presence of leachate does therefore influence the amount of suction.

However, a longer time of contact between the clay and the pollutant would possibly result in more significant changes in the properties. For the future, it is planed to investigate the effect of leachate when in contact with the clay for at least three month before testing. These studies are currently being undertaken as an ongoing research programme.

ACKNOWLEDGEMENTS

The research work described in this paper is supported by a grant from Belgian National Fund for Scientific Research (FNRS). This support is gratefully acknowledged.

REFERENCES

BOYTON, S.S. and DANIEL, D.E. (1985) Hydraulic conductivity tests on compacted clay, *Journal of Geotechnical Engineering,* ASCE, Vol 111, No 4, p 465-478

CRR. (1975) Mode opératoire de la succion dans les sols au moyen du psychromètre a effet Peltier, *Méthode de mesure C.R.R.*, Bruxelles, Belgium, 36p.

DANIEL D.E., SHACKELFORD C.D. (1989) Containment of Landfill Leacheate with Clay Liners. In : *Sanitary landfilling : Process, Technology and Environmental Impact.* Academic Press, Harcourt Brace Jovanovich. San Diego. pp. 323-341

EKLUND, A.G. (1985) A Laboratory Comparisson of the Effects of Water and Waste Leachate on the Performance of Soil Liners. In : *Hydraulic Barriers in Soil and Rock, ASTM STP 874,* Philadelphia, USA, p 188-202.

ELSBURY B.R., DANIEL D.E., SRADERS G.A., et al. (1990) Lesson Learned from Compacted Clay Liner. *Journal of Geotechnical Engineering, ASCE.* Vol. 116. No 11. November 1990. pp. 1641-1660.

GIANNOPOULOU, D.M.(1995) The geomechanical characterisation of bentonite/soil mixes for use in the construction of barriers for pollution containment. In : *IX th Young Geotechnical Engineer's Conference,* Ghent, Belgium.

OURTH, A.-S. (1998) *Influence des lixiviats sur les propriétés géomécaniques, en conditions saturée ou non, de l'argile constituant des barrières d'étanchéité pour centre d'enfouissement technique et sites contaminés,* Mémoire de fin d'études, Diplôme d'études approfondies en sciences agronomiques et ingénierie génétique, Faculté Universitaire des Sciences Agronomiques de Gembloux, Belgium, 74p.

Investigating the relative importance of advective and diffusive fluxes through cement-bentonite slurry walls

LUCY PHILIP, School of Earth Sciences, University of Leeds, Leeds, UK

ABSTRACT

Cement- bentonite walls are commonly used for remediation by containment or pollution prevention, in addition to their use as a barrier to water flow in more conventional construction. This paper details results from an investigation of an in-situ single-phase cement bentonite slurry wall with particular attention paid to the role of advective and diffusive fluxes in contaminant transport.

The wall under investigation was installed, using standard methods and with a normal mix design, in January 1996 in order to contain leachates arising from 'piggy-backing' of an existing 'dilute and disperse' landfill site. In September 1998 an investigation of the wall was undertaken during which the top of the wall was exposed at adjacent locations allowing the drilling of three boreholes using rotary coring techniques. Core recovery and quality was high and samples obtained during coring were retained for testing in the laboratory.

Laboratory testing proved slightly difficult due to some initial problems obtaining high-quality sub-samples from the cores. The material is initially quite strong and not treatable by conventional soil testing and preparation techniques. In addition the material oxidises and dries upon exposure to the atmosphere with an accompanying loss of structure. In spite of this some meaningful laboratory tests were undertaken. Hydraulic conductivity of samples was measured using constant flow-rate methods and the relative importance of the diffusive flux was investigated using a variety of simple techniques. The results of the laboratory test programme are presented and placed in the context of long-term performance of cement-bentonite slurry walls.

INTRODUCTION

Cement-bentonite slurry walls are in common use as vertical cut-off barriers to allow containment of liquids and gases in the ground. Cement-bentonite slurries support the trench during excavation and then self harden in-situ. The constituents of these walls are:
- cementitious materials (cement and ground granulated blast furnace slag-ggbs or pulverised fuel ash-PFA)
- bentonite clay (usually sodium exchanged calcium montmorillinite)
- water

It is important that the long-term behaviour of these walls is understood before heavy reliance is placed upon their long-term integrity as pollution containment barriers. Although a reasonable knowledge base has been built up on the performance of freshly-mixed samples and laboratory mixes, relatively little work has been undertaken to give a realistic insight to the true condition of slurry walls in-situ (Tedd et al 1995, Jefferis 1993 and ICE et al, 1996).

Geoenvironmental engineering, Thomas Telford, London, 1999, 54–61

In addition the design requirement of these walls tends to be hydraulic conductivity $<10^{-9}$m/s and therefore at these low permeabilities the advective flux may be less important than the diffusive flux may play an important role in contaminant transport. Little attention has been given to the role of the diffusive flux in these walls.

SITE
It is rare that barrier wall owners will allow exhumation and/or sampling of existing walls, however this paper describes field investigation and follow-up laboratory work of a wall at an active landfill site. Details of the wall can be found in Table 1 below. The wall was installed as a remedial measure and therefore it is possible that it has been exposed to contamination on either side.

Table 1. Wall constituents, details and age at the site.

Site	Year of wall installation	Natural Geology	Depth of wall	Mix proportions Bentonite	OPC	GGBS	Water
Yorkshire	1996	Glacial clays, sands and gravels	1-6 mbgl	40kg	40kg	150kg	920kg

FIELDWORK
Fieldwork was undertaken in Yorkshire in September 1998 and a summary of the work is displayed in Table 2 and Figure 1 below.

Table 2. Summary of field work and associated findings.

Site	Extent	Equipment	Problems	Findings
Yorks.	3 Exposures of top of wall	Hand equipment	Management problems due to environmental issues	Wall seen to be in good condition with no oxidation visible
	3 Boreholes	Track mounted wireline rotary rig	Access difficulties	Excellent core recovery[3]

[3] Wall material became more stony near the base of the hole due to intermixing with the surrounding material.

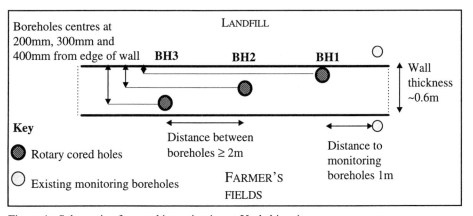

Figure 1. Schematic of ground investigation at Yorkshire site

Fresh slurry wall material is a rich blue colour due to the impurities originating from the GGBS. Upon exposure and oxidation the material turns beige-white allowing easy visual assessment of defects and cracks. The oxidising reaction causes a slight expansion and oxygen influx prevents the wall from remaining saturated. Any such 'oxidation' zones can be interpreted as a breakdown in the long-term integrity.

The wall investigated was found to be in excellent condition with no cracks or defects visible and only a very thin oxidation layer (~1mm) encapsulating the wall.

INDEX TESTING

A programme of simple laboratory testing was undertaken on samples collected during the fieldwork. The results are summarised in Table 3 below. As shown, A relationship between depth and moisture content was not found. SEM work showed the specimen structure to have large open pore spaces with a skeleton of cement minerals.

Table 3. Summary table of results of index testing

Wall	Moisture content % min/max./mean	Bulk density mean	Dry density
Yorkshire Site	212/307/249	1.23 Mg/m³	0.35 Mg/m³

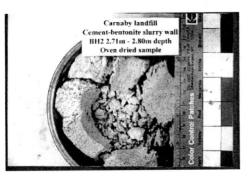

Upon drying samples collapsed with an associated loss of structure, as seen in Plate 1, adjacent. Some structure could be preserved by confining specimens during drying.

Plate 1. Total collapse of samples upon drying.

HYDRAULIC CONDUCTIVITY TESTING

Permeability testing was undertaken in a triaxial cell using constant flow rate techniques. The experimental set-up has been well used at Leeds and has been found to be ideal with a main advantage being the flexibility in test regimes and the ease of diagnosis of errors, equipment failure and leaks. A schematic of the experimental set-up can be seen in Figure 2 overleaf.

Tests were undertaken on 50mm diameter sub-cores taken from the rotary cores. Tests were routinely undertaken with cell pressures of 600 kPa and back pressures of 500 kPa and tests were computor logged for ease of determination of acheievment of stady-state conditions. Hydraulic gradient averaged about 27 and varied between 3 and 70.

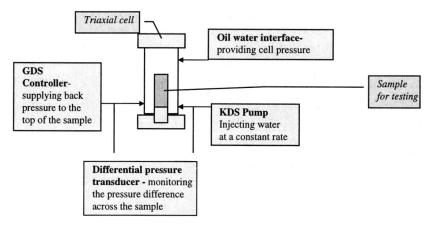

Figure 2. Equipment configuration for permeability in triaxial cell experiments.

Results from the permeability testing showed the permeability to vary and, except for the top of BH2, gave results of the order 10^{-8} - 10^{-9} m/s. The top of the Yorkshire Site wall had a slightly increased permeability, thought to be associated with leachate degradation of the wall but no other conclusive trend in permeability with depth was observed at lower depths.

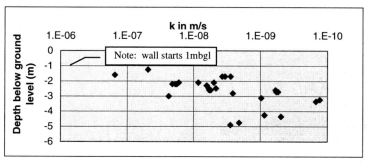

Graph 1. Variation in k with depth- Yorkshire Site BH2
         mean - 1.5×10^{-8} m/s
         mean excluding high result at 1.25 and 1.6m depth - 4.81×10^{-9} m/s

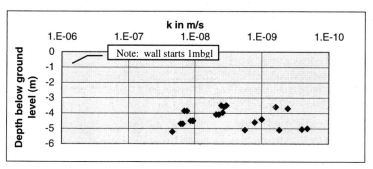

Graph 2. Variation in k with depth Yorkshire Site- BH3
         mean - 4.2×10^{-9} m/s

Hydraulic conductivity was found to be insensitive to hydraulic gradient but fairly sensitive to effective stress (see Graph 3 below).

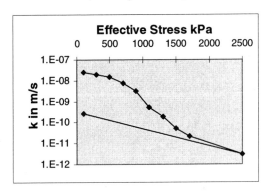

Graph 3. Variation in k with changes in effective stress
             Yorkshire Site Landfill BH2 3.5-3.6m.

DIFFUSION TESTING
Laboratory Testing

Two types of simple diffusion experiments were undertaken in an attempt to quantify diffusion parameters and the relative importance of diffusion as a means of contaminant transport through these walls. In both types of test the chloride ion was chosen as the diffusing ion as it is generally thought to be non-retardant and therefore conservative. A double reservoir diffusion cell was designed and constructed (see Figure 3 below) and a double reservoir tank with sample tubes was used (see Figure 4 overleaf). The diffusion experiments were conducted at constant temperature of 22 ^{0}C±1.

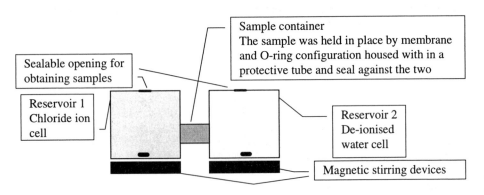

Figure 3. Double reservoir diffusion cell

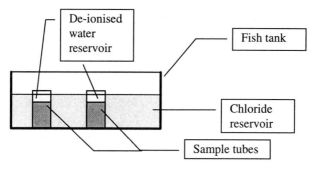

Figure 3. Double reservoir tank

Exclusion of leakage pathways was the main problem encountered and this was due to the nature of the material, which was strong enough to core like a rock but not strong enough to achieve completely perfect cores of consistent size. The material was too strong to be moulded into place.

At termination of each experiment the samples were divided into sub-samples along the length, which were subsequently crushed and diluted with de-ionised water in order to extract the pore fluids. Reservoir and pore fluids were analysed using ion chromatography (IC). Analysis of the results indicates the tortuosity factor ω_T to be between 0.0015 and 0.013. Some partitioning referred to by Rowe et al, 1995, was found and so both modelling and mass balance techniques were used in deriving ω_T. These values of tortuosity are very low, however any error in the tests, such as leakage paths etc. would tend to increase the values of ω_T obtained. The samples tested were from depth and were likely to have low 'natural' Cl levels, as indicated in the section below.

Although a conservative ion was used in these experiments it is possible that the slurry wall material has considerable retardation potential through ion exchange with CSH minerals and ettringite (Gougar et al 1996) and that the results were affected by retardation.

Field Profiles
Pore fluids were analysed using IC and Inductively Coupled Plasma techniques (ICP) to see if a contaminant plume due to leachate presence at the Yorkshire Site could be identified. ICP results were fairly inconclusive and indicated wide variation, thought to be due to micro-scale variations in wall chemistry, for many ions (Sr, Na, K) across short distances. However Lithium, not thought to be present in the original slurry mix, but likely to be present in any leachates showed a trend of decreasing concentration with depth indicating a possible advection/diffusion front. A front was successfully found (see Graph 4 overleaf) for chloride ion with significant Cl concentrations found at shallow depths (at least 1.7mbgl or 0.7m into wall) dropping off to virtually zero by 2.5mbgl.

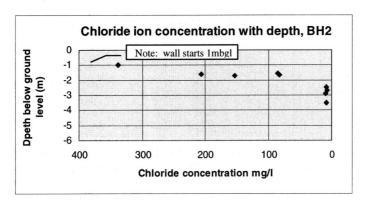

Graph 4. Variation in chloride ion concentration with depth, BH2

Due to the inherant unknowns assessment of transport parameters from field profiles is difficult.

SUMMARY OF FINDINGS
Advection versus Diffusion.
The advective flux has been shown as the dominant process for contaminant transport across this slurry wall. At the Yorkshire Site within 2.5yrs of construction an advective diffusive profile was seen to penetrate at least 300mm across and 700mm down into the wall. Models for diffusion run using the worst case ω_T and assuming a steady leachate concentration find that for a 600mm thick slurry wall it will be 20 years before breakthrough at 1/1000th of the source concentration and 32 years before breakthrough at 1/100th source concentration.. Assuming an hydraulic conductivity of 4.81x 10^{-9} m/s (average from BH2) and a low hydraulic gradient of 1, the time for leachate breakthrough could be ~ 3.5yrs. This is supported by field evidence at the Yorkshire Site. From the models of advective and diffusive transport it appears that the advective flux is dominant in contaminant transport - even if lower values of k are assumed over reduced thicknesses of wall. All these breakthrough times assume the barrier retains its chemical integrity with time which is a precondition outside of the scope of this work but has been investigated by others (Ahtchi-Ali & Casper, 1997 and Garvin & Hayes, 1998).

Long-term performance
The wall looked visually excellent when exposed and recovered in the field but the long-term k values and associated breakthrough times were disappointingly high.

Further work is continuing at Leeds to enable the long-term parameters governing contaminant transport through cement-bentonite slurry walls to be more fully understood and therefore predicted.

ACKNOWLEDGEMENTS
This paper reports part of a wider programme of research into the long-term performance of single-phase cement bentonite slurry walls. The programme was supported by EPSRC through grant number GR/L90828 and they are thanked for their support. Wastewise, in particular Simon Jones, are thanked for their collaboration and in addition a number of people

at the University of Leeds are also thanked for their support and/or chemical testing, Dr. D. Banks, Dr. J. West, Dr. R. Mortimer and Mr. K. Handley.

REFERENCES

AHTCHI-ALI, F. & CASPER, M.F. 1997. Effect of acidic leachate on material degradation of slurry trench cut-off walls. *Proceedings of International Containment Technology Conference, Florida*

GARVIN S. L.& HAYES, C.S. 1998. The chemical compatibility of cement-bentonite cut-off wall material. As yet unpublished

GOUGAR, M.L.D, ROY, DM AND SCHEETZ, B.E, 1996. Ettringite and C-S-H Portland Cement Phases for Waste Ion Immobilisation: A Review. *Waste Management.*, Vol 16, No.4, pp 295-303

ICE, CIRIA & BRE, 1996. *Draft notes for guidance for slurry trench cut-off walls (as barriers to Pollution Migration).*

JEFFRIS, S.A. 1993. Chapter 6, In-ground barriers. *Contaminated Land - problems and solutions,* ed. T. Cairney

ROWE, R.K, QUIGLEY, R. M & BOOKER, J.R , 1995. *Clayey Barrier Systems for Waste Disposal Facilities.* E&FN Spon.

TEDD, P., PAUL, V. & LOMAX, C., 1993. Investigation of an eight year old slurry trench cut-off wall. *Waste Disposal by Landfill- Green 1993,* p 581-590.

Pozzolanic Fly Ash as Hydraulic Barrier in Landfills

J P PRASHANTH[1], P V SIVAPULLAIAH[2] AND A SRIDHARAN[3]
[1] Lecturer, Department of Civil Engineering, UBDT College of Engineering,
Davangere - 577 004, Karnataka, India
[2] Assistant Professor and [3] Honorary Professor, Department of Civil Engineering,
Indian Institute of Science, Bangalore - 560 012, Karnataka, India

INTRODUCTION

Landfill liner acts as a barrier to minimize the migration of leachates. The liner system may utilize natural materials such as compacted clay or shale, bitumen, soil sealants, synthetic membranes otherwise known as geomembranes. The main requirements of liners are minimization of pollutant migration, proper swelling and shrinkage and resistance to erosion (Brandl, 1992). Compacted clay liners are widely used because of their cost effectiveness, large attenuative capacity and resistance to damage and puncture. Clay liners reduce the rate of contaminant migration by advection through their low permeability and through molecular diffusion by their sorption capacity. Bentonite because of its low permeability and high cation exchange capacity, is the most widely used mineral for construction of liners. But, because of its high swelling potential, bentonite cracks under drying condition and its hydraulic conductivity gets increased. Also, with some leachates it possesses high permeability. This paper examines the suitability of abundantly available fly ash, which is a waste product in the generation of power by burning of coal, for construction of hydraulic barrier in land fills. Under high alkaline conditions, which prevail in fly ashes, most of the toxic elements that are present in leachates precipitate and are not allowed to migrate (Pandain et al. 1995). The suitability of fly ashes as hydraulic barrier needs to be established from detailed geotechnical investigations.

FLY ASHES USED

Three different types of fly ashes have been obtained for the study. The fly ashes were obtained from Gulbarga, Neyveli and Vijayawada thermal power plants, India and hence the name Gulbarga fly ash (GFA), Neyveli fly ash (NFA) and Vijayawada fly ash (VFA) respectively. While Gulbarga and Neyveli fly ashes are collected directly from dry dumps, Vijayawada fly ash is collected from abandoned ash disposal pond.

PHYSICAL PROPERTIES

Specific gravity of the fly ashes used varied considerably and were found to be 2.58, 2.67 and 2.03 for Gulbarga, Neyveli and Vijayawada fly ashes respectively. Like most fly ashes, the particles of the fly ashes used are of silt size. Relatively, Gulbarga fly ash is finer than the other two. Neyveli and Vijayawada fly ashes possess almost the same gradation.

CHEMICAL COMPOSITION

Table 1 shows the chemical composition of the fly ashes expressed in percentage with respect to weight of fly ash. Like any fly ash, these fly ashes also found to contain large amount of silica and alumina. Reactive silica (part of total silica) of Gulbarga, Neyveli and Vijayawada fly ashes was found to be 5.30, 4.98 and 1.84% respectively. The free lime content (part of total calcium) was found to be 0.5, 3.92 and 0.86% for Gulbarga, Neyveli and Vijayawada fly ashes respectively (Sivapullaiah et al. 1998).

Table 1 Chemical composition of the fly ashes

Constituents	Fly ash		
	GFA	NFA	VFA
Silica (SiO_2)	51.06	50.40	58.88
Alumina (Al_2O_3)	20.29	18.81	29.67
Ferric oxide (Fe_2O_3)	10.82	16.61	5.87
Calcium as CaO	7.11	9.00	3.03
Magnesium as MgO	2.32	1.41	0.24
Titanium as TiO_2	0.26	0.28	0.27
Potassium as K_2O	0.25	0.23	0.28
Sodium as Na_2O	0.25	0.18	0.21
Loss on ignition	7.19	2.60	1.41

WATER CONTENT ON VOLUME BASIS

For soils, water content is usually expressed on weight basis (weight of water to weight of solids). Water content on volume basis is defined as the volume of water to the volume of solids which is conventionally called as water void ratio. In the study it has been termed as water volume content (Prashanth et al. 1998). It can be obtained just by multiplying the water content on weight basis by the specific gravity of the solids. Comparison of water content on weight basis for the materials of different specific gravity can be misleading. Specific gravity of the fly ashes used varied from 2.03 to 2.67, which is considerable. As such, in the present study all the water contents have been expressed on volume basis.

VOLUME CHANGE BEHAVIOR

The shrinkage limit (SL) test was conducted for the fly ashes at liquid limit (LL) initial water content. The liquid limit as determined by cone penetrometer, shrinkage limit and shrinkage index (SI) of the fly ashes are shown in Table 2. Difference between liquid limit and shrinkage limit is termed as shrinkage index (Ranganatham and Satyanarayana, 1965). As seen in Table 2, shrinkage of fly ashes is significantly lower.

Table 2 Liquid limit, shrinkage limit and shrinkage index of the fly ashes

Fly ash	Weight basis			Volume basis		
	LL	SL	SI	LL	SL	SI
GFA	62	52	10	160	134	26
NFA	44	38	6	117	101	16
VFA	49	42	7	99	85	14

The compressibility behavior of fly ashes is assessed from one dimensional consolidation test. Fly ash sample was mixed with required amount of water to bring it slightly above the liquid limit consistency. Then the sample was poured into consolidation ring and the test was carried out using load increment ratio of unity. The void ratio - pressure relationships for the three fly ashes are shown in Fig. 1. It can be seen that the compressibility of all the fly ashes are very less and its comparable to that of silts. The change in void ratio of Gulbarga, Neyveli and Vijayawada fly ashes for change in pressure of 6.25 kPa to 800 kPa, is 0.286, 0.119 and 0.113 respectively and lie in the same order of their liquid limit (volume basis). The compressibility of fly ashes is much less than expected from their liquid limit. As seen from Fig.1 there is no straight line portion in void ratio - pressure curves. Hence compression index of fly ashes are calculated for successive pressures of 400 to 800 kPa. These values of Gulbarga, Neyveli and Vijayawada fly ashes are 0.242, 0.133 and 0.116 respectively. They are low compared to those of soils generally.

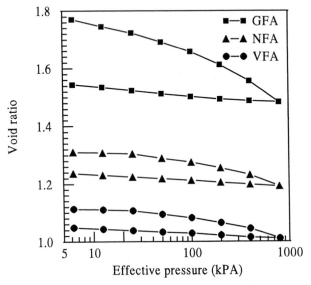

Fig. 1 Void ratio - pressure relationship of the fly ashes

Coefficient of consolidation of the fly ashes for various pressure ranges (between 6.25 to 800 kPa) was found to be varied from 100 m^2/year to 3000 m^2/year. The high rates of consolidation suggest that in most cases primary consolidation of the fly ash should be practically complete when the construction work is over.

COMPACTION AND STRENGTH CHARACTERISTICS

The pozzolanic reactivity of the fly ashes which is responsible for development of good strength can vary depending upon their chemical constituents namely reactive silica, carbon and iron contents; and physical properties like fineness etc. But now it is well understood that pozzolanic strength of fly ash mainly depends on the reaction between reactive silica and free lime content. Thus addition of lime may be beneficial to enhance the pozzolanic strength of fly ash. Hence the effect of lime on the compaction and strength characteristics of fly ash has been studied.

Expressing the compaction curves of fly ashes, having different specific gravity, for comparing their degree of packing and their optimum water content is misleading. For comparison the compaction curves have to be expressed on volume basis (Prashanth et al. 1998) namely solid volume occupation versus water volume content. Solid volume occupation is defined as the ratio of volume of solids to total volume of compacted soil. It can be obtained just by multiplying dry density by the specific gravity of the solids. Additional advantage of volume based expression is that any degree of saturation line or any percentage air voids line is unique irrespective of specific gravity. Proctor compaction curves of the fly ashes in terms of solid volume occupation and water volume content are shown in Fig. 2. It

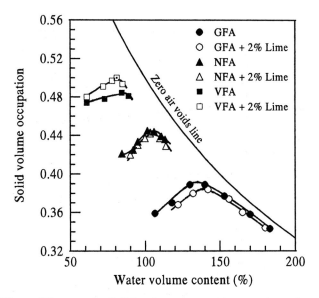

Fig. 2 Effect of lime on the Solid volume occupation - water volume content relationship of the fly ashes

can be seen that the shape of the compaction curve of the fly ashes is generally similar to that for fine grained soils. Figure 2 also shows the effect of lime on the solid volume occupation and water volume content relationship of the fly ashes. As seen effect of lime is marginal on the solid volume occupation - water volume content relationship of the fly ashes. Unlike soils, fly ashes does not contain any charges on their surface. As such addition of lime does not cause any fabric change (flocculation).

For strength test, dry fly ash samples were mixed with various quantities of lime. Then the required quantity of water is added and thoroughly mixed. The samples were statically compacted from either side of a static compaction mould to get specimen of standard size 7.62 cm height and 3.81 cm diameter. The dry density and water content of specimen compacted were equal to Proctor maximum dry density and optimum water content of the fly ashes. The unconfined compressive strength test was carried out after 28 days of curing. The specimens were cured in a desiccator at 100% humidity for 27 days. Then the specimen were soaked in water for 1 day before testing.

The unconfined compressive strength of the fly ashes with different percent of lime is shown in Fig.3. Addition of excess lime to fly ash is uneconomical and hence the lime content is limited to maximum of 15%. Neyveli fly ash has given an unconfined compressive strength of 1000 kPa even without addition of lime. Neyveli fly ash possesses sufficient amount of both reactive silica (4.98%) and free lime (3.92%). Reactive silica chemically reacts with free lime to produce cementitious compounds. These cementitious compounds bind the fly ash particles there by giving strength to fly ash. Addition of lime has further increased the

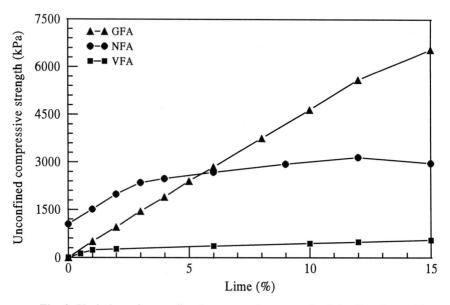

Fig. 3 Variation of unconfined compressive strength of the fly ashes with lime content after 28 days of curing

strength of the Neyveli fly ash. This indicates that the free lime present in fly ash is deficient for the reactive silica to react completely. Unlike Neyveli fly ash, Vijayawada fly ash by itself has no strength. Vijayawada fly ash contains very low reactive silica (1.84%) and free lime (0.86%). On addition of lime, Vijayawada fly ash has given some strength. Like Vijayawada fly ash, Gulbarga fly ash specimens without any added lime collapsed immediately after soaking. This is because Gulbarga fly ash though contain high reactive silica, contains very low free lime content (0.5%). However, due its high reactive silica content (6.42%), the strength of Gulbarga fly ash increased drastically on addition of lime. Neyveli fly ash has given higher strength than Gulbarga fly ash at lower free lime content. But at higher free lime contents, Gulbarga fly ash has shown higher strength than Neyveli fly ash. At lower free lime contents, as Gulbarga fly ash contains by itself very low free lime content, lime is not sufficient to react with reactive silica completely. But Neyveli fly ash by itself has significant amount of free lime content. The strength of the fly ashes without added lime lies in the order Neyveli fly ash > Vijayawada fly ash = Gulbarga fly ash = 0. This is because both Gulbarga and Vijayawada fly ashes contain very low lime contents. The strength of the fly ashes with added lime lies in the order Gulbarga fly ash > Neyveli fly ash >> Vijayawada fly ash. This is because reactive silica of Gulbarga fly ash > Neyveli fly ash > Vijayawada fly ash. As Neyveli fly ash given sufficient strength even without addition of lime it can be considered as self pozzolanic fly ash. Gulbarga fly ash as it has given high strength on addition of lime can be considered as pozzolanic fly ash. Vijayawada fly ash which has given only a small amount of strength even on addition of lime can be considered as non pozzolanic fly ash.

PERMEABILITY CHARACTERISTICS

For fly ash to be hydraulic barrier it should have low permeability. Coefficient of permeability of most of the fly ashes is in the order of 10^{-5} cm/s and change in density can cause only 5 fold change in coefficient of permeability (Raymond, 1961; PFA data books, 1967; Dayal et al., 1988; Dayal et al., 1989; Martin et al., 1990; Chen et al., 1992; Singh, 1994 and; Singh and Panda, 1996). This order of coefficient of permeability of fly ashes is because they are essentially of silt size particles and inhert in nature. This value of coefficient of permeability is too high for fly ashes to be used as hydraulic barrier. It is known that, hardened concrete is an efficient hydraulic barrier. It is seen from previous section that fly ashes develops considerable amount of pozzolanic strength on addition of lime. Hence an attempt is made to decrease the coefficient of permeability of the fly ashes by adding lime.

Dry fly ash samples were mixed with various quantities of lime. Then the required quantity of water was added to have Proctor optimum water content. Then the mixture was statically compacted in a permeability mould of internal diameter 8.0 cm and 6.0 cm height, to Proctor maximum dry density. The compacted specimen if of 8.0 cm and 5.0 cm height. The compacted specimen along with the permeability mould is kept in the desiccator of 100% humidity for 28 days. After 28 days of curing variable head permeability test was conducted on the hardened fly ash specimen. Readings were taken when the flow reached a steady state. Hydraulic gradient was varied from 10 to 50. Variation of the coefficient of permeability with the hydraulic gradient, in the range of 10 to 50, was found to be marginal. Figure 4 shows the variation of coefficient of permeability of fly ashes with percent lime, at an hydraulic gradient of 40. As seen in Fig. 4 coefficient of permeability of Gulbarga, Neyveli and Vijayawada fly ashes were respectively 6.5×10^{-5}, 3.2×10^{-5} and 7×10^{-5} cm/s respectively. Like other fly ashes (reported in literature) the coefficient of permeability of these fly ashes are also in the same order of 10^{-5} cm/s. Though solid volume occupation of Gulbarga fly ash

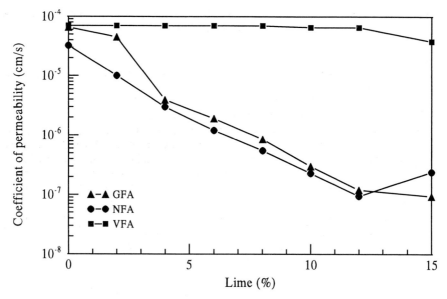

Fig. 4 Variation of coefficient of permeability of the fly ashes with
lime content after 28 days of curing

is considerably lower than the Vijayawada fly ash, both the fly ashes have almost the same permeability. Gulbarga fly ash possesses fine particles. As such size of the pores are also smaller.Thus the effect of lower solid volume occupation has been compensated by the effect of smaller pore sizes of Gulbarga fly ash. Addition of lime decreased the coefficient of permeability of Gulbarga and Neyveli fly ashes upto a value of 10^{-7} cm/s. But the effect of lime in decreasing the coefficient of permeability of Vijayawada fly ash is marginal. Both Gulbarga and Neyveli fly ashes have considerable amount of reactive silica. Thus they produce large amount of gelatinous cementitious compounds on addition of lime. These cementitious compounds fill the voids between the fly ash particles and reduce the coefficient of permeability. The reduction would be even more effective if the pore fluid contain any ions that can be precipitated under high alkaline condition. Vijayawada fly ash which is having low reactive silica can only produce small amount of cementitious compounds to fill the pore spaces. Amount of cementitious compounds formed in the fly ashes is also reflected from the unconfined compressive strength of the fly ashes. Without any added lime, though the solid volume occupation is lower and particle size distribution almost being the same, Neyveli fly ash has given lower coefficient of permeability than Vijayawada fly ash. This is because even without addition of lime, Neyveli fly ash is capable of producing some cementitious compounds as it contains considerable amount of both reactive silica and free lime content.

CONCLUSIONS

From the detailed investigations it has been shown that fly ashes possess suitable geotechnical properties for construction of liners. However, the permeability of fly ashes are generally higher. With lime addition the permeability of pozzolanic fly ashes can be lowered to the required level of about 10^{-7} cm/s. Lime addition which precipitates metal ions that are present in the leachate can further reduce the permeability.

REFERENCES

Brandl, H. (1992). "Mineral Liners for Hazardous Waste Containment," Geotechnique, Vol. 42, pp. 57-65

Chen, Y.J., Zhu, Y. and Shi, F.X. (1992). "Air Pollution Prevention During Fly Ash Disposal," *Geotechnical Engineering*, Vol. 23, pp. 1-10.

Dayal U., Chandra S. and Bohra N.C. (1989). "Geotechnical Investigation of Ash Properties for Dyke Construction at Ramagundam Super Thermal Power Project," IIT, Kanpur Report prepared for National Thermal Power Corporation Limited, New Delhi.

Dayal U., Jain S.K. and Srivastava A.K. (1988). "Geotechnical Investigation of Ash Properties for Dyke Construction at Kobra Super Thermal Power Project," IIT, Kanpur Report for National Thermal Power Corporation Limited, New Delhi.

Martin, J.P., Collins, R.A., Browing, J.S. and Biehl, F.J. (1990). "Properties and Use of Fly Ashes for Embankments," *Journal of Energy Engineering*, Vol. 116, No. 2, PP. 71-85.

Pandain, N.S., Rajasekhar, C. and Sridharan, A. (1995). "Fly ash - lime systems for the retention of lead ions," *Proceedings of Indian Geotechnical Conference*, Bangalore, Vol. 1, pp. 219-222.

PFA Data Books (1967). *Load Bearing Fill - Part I and Stabilization - Part II*, Central Electricity House, Manchester 20, England.

Prashanth, J.P., Sivapullaiah, P.V. and Sridharan, A. (1998). "Compaction Curves on Volume Basis," *Geotechnical Testing Journal*, American Society for Testing and Materials, Vol. 21, No. 1, pp. 58-65.

Ranganatham, B.V. & Satyanarayana, B. (1965). "A Rational Method of Predicting Swelling Potential for Compacted Expansive Clays," *Proceedings of Sixth International Conference on Soil Mechanics and Foundation Engineering*, Montrial,Vol.1, pp.92-96.

Raymond, S. (1961). "Pulverized Fuel Ash as Embankment Material," *Proceedings of Institution of Civil Engineers*, Vol.19, pp. 515-536.

Singh, D.N. (1994). "Engineering Properties of Compacted Panki Fly Ash," *Proceedings of Indian Geotechnical Conference*, Warangal, Vol. 1, pp. 202-206.

Singh, S.R. and Panda, A.P. (1996). "Utilization of Fly Ash in Geotechnical Construction," *Proceedings of Indian Geotechnical Conference*, Madras, Vol.1, pp. 547-550.

Sivapullaiah, P.V., Prashanth, J.P., Sridharan, A. and Narayana, B.V. (1998). "Reactive Silica and Strength of Fly Ashes," *Geotechnical and Geological Engineering*, Vol. 18, pp. 239-250.

Construction Considerations in the Choice and Design of Engineered Barriers

K. J. Read
Hyder Consulting Limited, Plymouth, UK

ABSTRACT

The choice of an engineered barrier on a construction or landfill site is often made after a process of assessment and design involving the desired performance of the barrier, its ease of construction, quality control and longevity. This choice can on occasion be greatly influenced by engineering and construction factors relating to the site and any proposed engineering works.

Vertical cut-off barriers are commonly used to prevent lateral migration of contaminants, leachates and gases. Where development is sought over contaminated soils or wastes horizontal barriers can be installed to isolate the development from the underlying materials.

For both barrier types there are a variety forms of material suitable and available for use. These range from mineral barriers such a clay capping, granular capillary break layers and bentonite slurry trenches to synthetic barriers such as HDPE membranes and factory prepared clay impregnated geofabrics.

Each form of barrier has differing engineering properties and these can become significant if the barrier is required to perform a temporary or permanent engineering role. In the case of vertical barriers they can be required to act as a foundation element to a structure or as a retaining wall where excavation adjacent to the barrier is required. Horizontal barriers may be required to support shallow foundations or allow construction of deep foundations through them without any loss of integrity, such a increased permeability, or any adverse affect on the structures above such as causing excessive settlements or heave.

This paper will illustrate some of the considerations necessary by summarising construction considerations and examining the process followed in two case studies. In these examples both mineral and synthetic barriers were used on the same sites to suit differing engineering requirements and local site specific needs.

INTRODUCTION

This paper is intended to illustrate the engineering and construction factors that require consideration when selecting and designing an engineered barrier to contaminants. It will also illustrate the influence of site specific engineering and construction factors may have on the selection and installation of engineered barriers.

With growing economic, moral and political pressure to develop so called "brown field" sites engineers have been called on to provide efficient and economic means of protecting the public from the effects of ground and groundwater contamination.

Geoenvironmental engineering, Thomas Telford, London, 1999, 70–77

Until comparatively recently development of contaminated sites was dominated by what is now known as a "dig and dump" philosophy, involving the excavation and removal to landfill of contaminated soils.   Increasing landfill costs together with growing pressure for sustainable development policies have resulted in alternative means of allowing development to proceed being sought.   Despite the development of increasingly sophisticated and effective methods of remediating contamination in situ, many such sites are now developed only after installation of engineered barriers to prevent contaminant migration and human contact. This is primarily due to timescale and cost.

Over the last 25 years management and control of landfilling has changed radically.   Former policies of allowing leachate to enter the environment in order for it to "dilute and disperse" have been largely abandoned and landfill sites are now primarily designed to contain leachates, and in some cases limit leachate generation.   Similarly landfill gas is now recognised both as a hazard, and if properly managed, as a resource.   In consequence engineered barriers to contain and control leachate, contaminants and landfill gas are routinely installed in new landfill sites and in the remediation of former landfills.

GENERAL CONSTRUCTION CONSIDERATIONS : VERTICAL BARRIERS
Vertical barriers are installed to prevent lateral contaminant transport and migration. These can be divided into two principle types, "trench" and "drain" type installations.

"Trench" type barriers usually comprise a very low permeability medium being installed into a prepared excavation on the margins of a development site or landfill.

"Drain" type barriers comprise drains or abstraction wells installed so as to intercept leachate, contaminant groundwater plumes or gas flow and lead the resulting flows to locations where the effluent or gas can be dealt with in a controlled manner.

Tables 1 and 2 summarise engineering and construction considerations requiring attention in the selection and installation of the vertical barrier types in common usage.

GENERAL CONSTRUCTION CONSIDERATIONS : HORIZONTAL BARRIERS
Horizontal barriers are installed to prevent vertical contaminant transport and migration. These can be divided into two principle types, "capping" type barriers and "sub-surface" type barriers.

"Capping" type barriers are typically used to provide a physical barrier between contaminated materials and the ground surface thus preventing human contact with the materials. they also act as low permeability barriers preventing groundwater ingress, thereby preventing or limiting leachate generation.

"Sub-surface" barriers are typically used on landfill sites to prevent downward migration of leachates from the fill materials.

Table 3 summarises engineering and construction considerations requiring attention in the selection and installation of the horizontal barrier types in common usage.

Barrier Type	Common Usage	Engineering and Construction Considerations
Gravel filled trench	Landfill gas venting/ migration interception.	Depth can be limited to approximately 6m depth using conventional plant. Depth may also be limited by trench side stability and groundwater level. "Drain effect" on local groundwater regime
Gravel filled trench with low permeability membrane	Landfill gas venting/ migration interception.	As above with following additional considerations : Increasing difficulty of installing the membrane with increasing depth. Trench support required below 1.20m if man entry required to install/weld membrane. "Dam effect" on local groundwater regime
Low permeability membrane	Aqueous contaminant transportation and gas flow prevention	Entry into confined space with potentially hazardous atmosphere may be required. Risk of damage to membrane by gravel and sharp edges in adjacent soils. Suitable source of clay.
Compacted Clay margin to landfill	Aqueous contaminant transportation and gas flow prevention	Compaction properties of clay, permeability, moisture content and plasticity. Shrinkage properties. Quality control and testing of compacted clay. "Dam effect" on local groundwater regime
Engineered Clay filled trench.	Aqueous contaminant transportation and gas flow prevention	Usually installed using bentonite clay mixes with cement and or other additives to aid contaminant adsorption. Clay can be designed to support the excavation. Depths in excess of 30m possible using diaphragm walling plant. Requires large plant and site for mixing equipment. "Dam effect" on local groundwater regime
Engineered Clay filled trench with low permeability membrane.	Aqueous contaminant transportation and gas flow prevention	As above with the following additional considerations:- Requires large plant and site for mixing equipment. Membrane on prefabricated frames, requires craneage.

Table 1 : Summary of General Engineering and Construction Considerations for Common "Trench Type" Vertical Barriers.

Barrier Type	Common Usage	Engineering and Construction Considerations
Gravel filled trench with low permeability membrane	Groundwater / leachate flow interception.	Increasing difficulty of installing the membrane with increasing depth. Trench support required below 1.20m if man entry required to install/weld membrane. Entry into confined space with potentially hazardous fluids present may be required. Risk of damage to membrane by trench fill gravel and sharp edges in adjacent soils. Need for topographic low point or possible deepening excavation to maintain gradient in drain. Possible need for long term pumping. Disposal / treatment of effluent required. Effect on regional groundwater regime / ground settlements.
Closely spaced gravel filled vertical columns	Gas vent barrier	Borehole or vibro-column installation dependant on soil conditions. Access/height restrictions on plant. Possible "do-nut" effect of reduced gas permeability around vibrated columns.
Interceptor well field	Groundwater / leachate flow interception	Requires good knowledge of regional groundwater conditions. Pumping trials required to confirm assumptions. Disposal / treatment of effluent required. Long-term running costs. Effect on regional groundwater regime / ground settlements.

Table 2 : Summary of General Engineering and Construction Considerations for Common "Drain Type" Vertical Barriers.

Barrier Type	Common Usage	Engineering and Construction Considerations
Natural Clay Capping.	Low permeability cap or basal layer to prevent/limit leachate generation and/or physically isolate contaminated soils	Post compaction mass permeability of clay material. Clay properties, moisture content and plasticity. Compaction quality control and testing regime. Clay layer thickness. Shrinkage properties. Surface and under drainage, Landfill gas venting measures. Gradient limitations
Low permeability Geo-Membrane. (HDPE or equivalent)	Low permeability cap or basal layer to prevent/limit leachate generation.	Strength and tear resistance Gas and moisture permeability Welding quality control and testing Installation supervision Surface and under drainage, Landfill gas venting measures. Surface protection. Gradient limitations
Clay impregnated Geo-synthetics	Low permeability cap to prevent/limit leachate generation	Strength and tear resistance Gas and moisture permeability Installation supervision. Surface and under drainage, Landfill gas venting measures. Surface protection.
Granular blanket	Capillary break layer	Grading design Layer thickness Provision of filter layers/geotextile to prevent siltation Installation supervision. Drainage provision.

Table 3 : Summary of General Engineering and Construction Considerations for Common Horizontal Barriers.

SITE SPECIFIC CONSTRUCTION CONSIDERATIONS

The comments given in Tables 1 to 3 all assume the installation is required to perform no other function than act as a barrier. On many developments however the installation may be required to provide an engineering service in addition to this.

Vertical barriers may be called upon to provide a permanent or temporary structural role in a development, such as a retaining wall or foundation element. If called upon to meet such a role the following engineering properties of the barrier may require assessment in addition to those listed previously in Tables 1 and 2 :-

- structural shear or compressive strength of barrier.
- vertical compression of barrier.
- long term structural durability of barrier.
- long term resistance to chemical attack.
- generation of stress cracks in barrier.

Inspection of Table 1 shows that very few of the common "trench" type barriers listed have any significant structural shear or compressive strength and it is quickly apparent that these forms of barrier are not suitable to act as supporting structures. A consequence of this is the limited use of such barriers as structural elements.

One barrier that may be used in this way however is the "engineered clay" type, when designed to act as a diaphragm wall. It should be noted however that the inclusion of reinforcement and concrete may greatly reduce the effectiveness of such a structure as barrier to fluid transport, concrete having a low permeability.

Capping barriers are frequently required to support development over the contaminated materials beneath. If so the following engineering properties of the barrier may require assessment in addition to the engineering properties of the underlying soils and those factors described on Table 3 :-

- thickness of barrier with respect to possible service and footing excavations.
- settlement properties of barrier under foundation loading.
- possible heave pressures due to clay mineral capping under foundations and floor slabs.
- integrity of the barrier if penetrated by piled foundations.
- long term CBR for pavement construction.

EXAMPLE SELECTION PROCESS : GAS WORKS SITE REMEDIATION

In 1995 Hyder were requested to assist with the development and remediation of a former gas works site in Stirling, Scotland. Site investigations showed the site to be underlain by up to 11m of contaminated soils which were found to extend beyond the site boundary beneath an adjacent, busy, duel carriageway.

Due to programme constraints, the low cost of local landfill disposal and a cheap supply of suitable fill materials excavation and removal of contaminated soils was selected as the most practical means of meeting the Clients brief.

The initial development brief required complete removal of all contaminated soils from beneath the site. This required consideration be given to the installation of a vertical barrier to prevent contaminants re-entering the site from beyond the site boundaries.

It soon became apparent that any such barrier would be required to support up to 11m of soil and that none of the conventional trench type barriers would be capable of achieving this. Consideration was given to use of a diaphragm wall, a contiguous bored pile wall, and soldier piles with close boarding as means of both supporting the excavation, and preventing migration of contaminants back onto the site. All were ruled out on grounds of cost and/or poor qualities as a contaminant barrier.

An agreed amendment to the brief resulted in just the upper 3.5m of contaminated soils being removed. This allowed safe installation of services and minor strip footings in "clean" soils and kept excavations above the water table. A horizontal migration barrier was installed to separate remaining contaminated soils from clean imported fill, and this was tied into the vertical barrier along the perimeter of the site.

The horizontal barrier was a graded granular blanket acting as a capillary break layer with geotextile filter layers above and below the blanket to prevent siltation. Drainage pipes were installed within the blanket to prevent it becoming saturated in the event of a rise in groundwater levels. A gradient was imposed in the blanket so that any flows were lead to an oil interceptor, which was in turn connected to the public sewer.

An HDPE geo-membrane was selected as the vertical barrier. The initial vertical cut was supported by sheet piling, the membrane was then draped over the piles with a protective boarding separating the membrane from the piles. Sheet piles were then withdrawn as filling proceeded. Figure 1 illustrates the nature of junction between the barriers.

Selection of 3.5m as the limiting depth of excavation allowed the geo-membrane, 4.5m wide, to be turned through 90° and unrolled on its side. This resulted in only 1 weld being required along the entire length of the barrier.

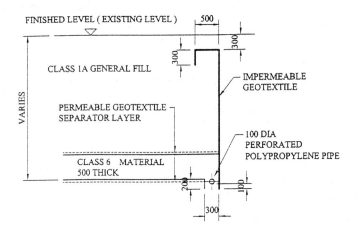

FIGURE 1 : VERTICAL AND HORIZONTAL MIGRATION BARRIER DETAILS

EXAMPLE SELECTION PROCESS : LANDFILL CAPPING

In 1998 Hyder were requested to assist with the design and selection of a landfill capping installation at a landfill site in mid Devon.

Both clay mineral and geomembrane capping designs were assessed and costed.

One of the main engineering considerations was a source of clay with which to form a capping layer. The need for a regular and uniform supply was identified in order to keep compliance testing for compaction and permeability to practical levels.

Significant deposits of readily available clay soils in Devon are however limited to Tertiary aged "ball clays" of the Bovey Tracy and Newton Abbot areas. In general these clays are of high plasticity with a tendency to significant volume change and cracking on drying. Additional consideration therefore has to be given to the layer thickness in order to ensure a mass permeability below the limit of 10^{-9}m/s is maintained within the barrier.

Capping of the landfill was scheduled for the late winter and spring of 1998 at it was recognised that there was a significant weather risk associated with using a moisture susceptible clay to form the capping.

In addition to the weather risk, the as dug condition of much of the Tertiary clays is dry of the required moisture contents for compaction. In consequence conditioning of the clay by adding moisture prior to placing can be required increasing placing costs.

Clay rich soils from obtained from a number of development sites had been used during an initial phase of capping. However to complete the capping operation a comparison of the following options was carried out :

1. Mineral capping using "ball clays".
2. Mineral capping using local clay rich soils,
3. Geo-membrane capping

Option (2) was the cheapest however timescale, lack of a reliable source and increased site testing/supervision precluded this option. Use of a geo-membrane (Option 3) was selected on the basis of price and offering the lowest risk.

CONCLUSIONS

Selection, design and construction of appropriate engineered barriers is a procedure requiring assessment of contaminant type, the means of contaminant transport and the existing ground and groundwater conditions together with any changes made to ground and groundwater conditions by the barrier installation.

Any engineering requirements, temporary or permanent, require to be assessed early in the selection process as these may have an over riding influence on the choice of barrier. Other factors such as local availability of materials may also have a major influence on the choice and construction of barriers.

ACKNOWLEDGEMENTS

The Author would like to thank Stirling City Council and the St Regis Paper Company for thier permission to use their projects as case histories in this paper.

Hydraulic Conductivity And Pore Fluid Chemistry In Artificially Weathered Plastic Clay

KELLY YUEN*, JIM GRAHAM**, TEE BOON GOH*
*Research Student, **Professor University of Manitoba, Winnipeg, Canada
PAUL JANZEN,
Engineer, Manitoba Hydro, Winnipeg, Canada
VINAYAGAMOORTHY SIVAKUMAR,
Lecturer, The Queen's University of Belfast, Belfast UK

INTRODUCTION

Compacted clays are used as liners and caps for landfills, retention ponds and waste water lagoons, and for remediation of contaminated sites. Their integrity can be altered by physical, chemical and biological processes, especially when they have low plasticity (Johnston and Haug 1986). Two common processes leading to inadequate performance are seasonal freezing-thawing and drying-wetting, both of which produce fracturing and higher hydraulic conductivity (Wong and Haug 1991, Yuen 1995). Clay liners that meet environmental requirements when newly constructed and intact, can subsequently experience unfavourable increases in their hydraulic conductivity following seasonal weathering

There are significant deposits of expansive proglacial clays in Western Canada (Quigley 1980, Graham and Shields 1985). If these exhibit the self-sealing properties (decreases in hydraulic conductivity) that have been observed in bentonite and sand-bentonite mixtures (Dixon et al.1993), there would be less concern about their long-term performance as seals and covers. Improvements (decreases) in hydraulic conductivity can also be envisaged following carefully managed changes in pore fluid chemistry. Knowledge of the mechanisms involved in these processes could help in remediating facilities that currently do not satisfy regulatory guidelines.

This paper deals with the hydraulic conductivity of a natural plastic clay from Eastern Manitoba. Initial hydraulic conductivities K_i were first measured using water as permeant. Specimens were then subjected to cycles of freezing-thawing or drying-wetting, and their 'weathered' hydraulic conductivity K_{weath} measured. Finally, some results are shown for hydraulic conductivities K_{perm} measured after permeation with sodium carbonate or dimethyl sulfoxide.

2. TEST MATERIAL

Tube samples were taken from 0.3 m - 10 m depth in proglacial Lake Agassiz deposits near Lac du Bonnet MB. The material is generally intermediate- to highly-plastic clay containing significant proportions of smectites (Graham and Shields 1985). Yuen et al. (1999) show that water contents near the ground surface are about 40%, increasing approximately linearly to about 70% at 9 m depth. Liquid limits range from 50 near the surface to over 100 at 8.5 m depth. Plasticity and liquidity indices in 'intact' clay below 3 m are 53.4±8.7% and 0.58±0.15 respectively. The average specific weight of the material is 2.76±0.06. The clay fraction varies somewhat irregularly from about 60% to 90%. Saturated bulk densities were 1.68±0.07 Mg/m^3. While the deposit shows some variation with depth, in geological terms it must be considered uniform. Average cation exchange capacity (CEC) for drying-wetting specimens (from 6.34 m - 6.60 m) was 43.6 ± 2.4 cmol(+).kg^{-1}. For specimens subjected to freezing-thawing (from 4.17 m - 4.42 m) the CEC was 22.0 ± 0.4 cmol(+).kg^{-1}. The differences in CEC are believed to reflect natural variability in the deposit (Baracos 1977) with the smectite component dominant and increasing with depth. Natural weathering at the site has produced a current 'crust' about 3 m deep, though it may have extended to perhaps 6 m during earlier periods of colder or dryer climate. To achieve consistent specimens for examining the effects of weathering, testing was done on initially intact specimens from 3 m to 10 m depth.

Figures 1a and 1b show mercury intrusion porosimetry (MIP) results from samples from 8.8 m and 4.2 m. 'Undisturbed' specimens (shown by dashed lines) have a unimodal pore size distribution with pore sizes concentrated at about 0.01 μm. Drying-wetting and freezing-thawing produced bimodal distributions with more pores in the range 10 μm to 100 μm. These correspond with 'inter-ped pores' or 'macropores'.

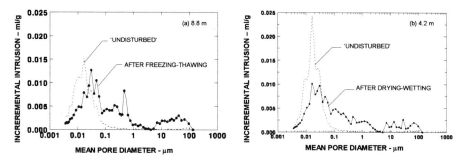

Figure 1 MIP data before and after (a) freezing-thawing and (b) drying-wetting

TEST PROCEDURES
Hydraulic conductivity tests used flexible wall (triaxial) permeameters (ASTM D5084-90) and (some) fixed-ring oedometers. The triaxial tests used (a) 71.1 mm diameter × 76.2 mm long specimens, (b) an effective consolidation pressure of 35 kPa, and (c) constant hydraulic gradients of about 25. More details of the test arrangements are given in Yuen *et al.* (1999). Liners and covers seldom experience overburden pressures greater than 30-40 kPa or gradients greater than 10 - 20 (Houston and Randeni 1992).

The cells shown in Fig.2 have stainless steel top caps and pedestals that penetrate through the cell base (Yuen 1995). This allows easy manufacture of parts of the cell in contact with permeant. The pressure vessel is made from lucite. Double burettes are attached at the top and at the bottom of the specimen to permit flushing of gas from drainage lines and back-pressuring to enhance saturation. A differential pressure regulator controls water pressure differences between the top and bottom of the specimen.

As mentioned earlier, specimens were first tested to determine the initial hydraulic conductivity K_i in their 'unweathered' state. It is appreciated that the deposit has experienced various forms of weathering in the 10,000 years or so since deposition. Here, 'unweathered' is used to distinguish the initial hydraulic conductivity of an 'undisturbed' specimen, from the 'weathered' value obtained after 3-6 cycles of freezing-thawing or drying-wetting in the laboratory. An isotropic effective mean stress of 35 kPa (with back pressure 172 kPa) was used for consolidation. It was held constant until the volume change rate decreased to 0.15%/day, usually after about one week. Figure 3a shows results for a specimen from 8.6 m depth which swelled by about 2% of its volume after 6 days. Following consolidation, the average B-value was 0.97, representing a saturation close to 100%.

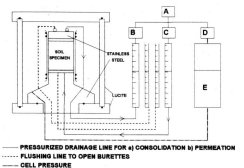

A = regulated pressure supply, B = top pressure regulator, C = bottom pressure regulator, D = cell pressure regulator, E = cell fluid reservoir

———— PRESSURIZED DRAINAGE LINE FOR a) CONSOLIDATION b) PERMEATION
- - - - - FLUSHING LINE TO OPEN BURETTES
— - — CELL PRESSURE

Figure 2. Triaxial hydraulic conductivity test arrangement:

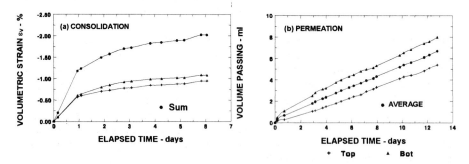

Figure 3 Typical initial conductivity test results on "unweathered" specimen

Tap water was used as permeant for measuring K_i. Figure 4 shows cation concentrations measured by flame absorption spectrophotometer for the water before and after permeation. Cation concentrations after permeation were higher than for tap water, indicating the presence of soluble cations in the clay. Figure 3b shows permeation results from the same specimen with an initial hydraulic conductivity $K_i = 4.6 \times 10^{-11}$ m/s. Initial hydraulic conductivities in the deposit average about 1×10^{-10} m/s, with some tendency to decrease with depth (Fig. 5a).

After K_i values were established, specimens were subjected to drying-wetting (DW) cycles or freezing-drying (FT) cycles before the hydraulic conductivity was again measured. Drying at room temperature to constant mass produced only a limited number of microcracks or micro-fissures in DW specimens. To exaggerate the effects of drying, dried specimens were sawn in half along their vertical axis and then reassembled in the test cell with wire spacers forming an artificial 2 mm wide fissure from top to bottom. The same pressures as before (207 kPa cell pressure and 172 kPa back pressure) were applied for approximately 2 weeks. Figure 6 shows typical consolidation and permeation results before and after drying-wetting. Significant amounts of water were absorbed during rewetting, almost 40% of the dry volume. When permeation was begun, water flow was initially rapid. The flux rate gradually slowed as additional swelling occurred. After about 6 days, permeation became constant with a conductivity K_{weath} value of 2.0×10^{-10} m/s. This is somewhat larger than the initial value of $K_i = 0.61 \times 10^{-10}$ m/s and gives $K_{rel} = 3.3$. Visual inspection showed that clay laminations had coalesced and the spacers had become embedded in clay.

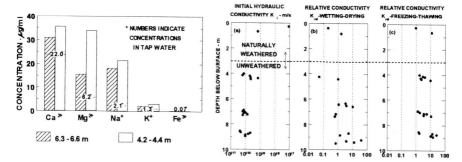

Figure 4. Cation concentration of permeant collected after initial permeation with tap water

Figure 5. Variation of hydraulic conductivity with depth. (a) Initial hydraulic conductivity K_i, (b) relative conductivity K_{weath}/K_i after drying-wetting (c) relative conductivity K_{weath}/K_i after freezing-thawing.

Freeze-thaw (FT) specimens were frozen one-dimensionally to -20°C in an insulated box with access to unfrozen water from below. Freezing rates (10°C/day, 120 mm/day in 1-D open-system tests) were greater than those observed in the field which are generally not more than 0.1°C/day, 30 mm/day (Yuen *et al.* 1999). Laboratory freezing produced reticulate ice lenses 2 - 3 mm thick and 10 - 20 mm long. The general trend after each freezing-thawing cycle was for specimens to become smaller and their masses to decrease through water expulsion from pedal aggregates. After 6 cycles, mass loss was about 5%. Despite the water loss, specimens became softer on thawing, possibly due to destruction of cementation bonds (Graham and Shields 1985).

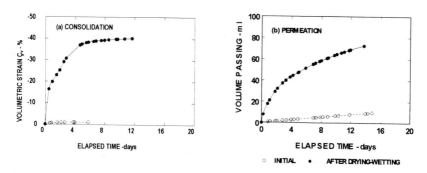

Figure 6 Effect of drying-wetting on (a) consolidation and (b) permeation data

After drying-wetting or freezing-thawing, specimens were subjected to the same test procedures as before and the hydraulic conductivity K_{weath} measured. Swelling and self-sealing returned specimens to 93±7% saturation. Figures 5b and 5c show measured values of relative conductivity $K_{rel} = K_{weath}/K_i$ for drying-wetting and freezing-thawing respectively (see also Yuen *et al.* 1999). Results indicate that weathering increased the hydraulic conductivity by one order of magnitude or less. These increases are smaller than those found in other less plastic clays, for example by Boynton and Daniel 1985, Johnston and Haug 1992,

Daniel and Wu 1993. They may be influenced by the effective confining pressure of 35 kPa which is rather higher than the value of 10 – 20 kPa recommended for compacted clays. The higher pressures may have closed some fissures and reduced the hydraulic conductivities by about one-half order of magnitude (Boynton and Daniel 1985).

SODIUM CARBONATE PERMEATION
Remedial measures may be required if weathering increases the hydraulic conductivity of a barrier above permissible values. One approach suggested by McConkey et al. (1990) is to add sodium carbonate Na_2CO_3. Tests were therefore conducted to determine whether changes in pore fluid chemistry would change the hydraulic conductivity of this plastic clay. Results are shown in Table 1.

Drying and Wetting.
Four specimens were examined from 6.34 m to 6.60 m depth (Table 1). Their average initial hydraulic conductivity K_i was $0.76 \pm 0.24 \times 10^{-10}$ m/s. Two of these specimens were permeated with tap water during the entire conductivity test. The other two specimens were permeated with 0.05 M and 0.10 M Na_2CO_3 respectively. Figure 7 shows results from a DW specimen with initial conductivity $K_i = 1.1 \times 10^{-10}$ m/s and $K_{weath} = 28 \times 10^{-10}$ m/s representing a 25-fold increase in conductivity. At this stage 0.10 M Na_2CO_3 solution was substituted for the tap water. At once, the permeation rate under the same gradient decreased sharply and the hydraulic conductivity dropped to $K_{perm} = 3.5 \times 10^{-10}$ m/s, roughly three times K_i and eight time less than K_{weath}. Similar results were obtained with 0.05 M Na_2CO_3, though the decrease in conductivity was less and more gradual (Table 1).

Table 1. Effect of sodium carbonate permeation on weathered hydraulic conductivity

Depth (m)	Init. Water content %	Process*	Permeant	$K_i \times 10^{-10}$ m/s	$K_{weath} \times 10^{-10}$ m/s	$K_{perm} \times 10^{-10}$ m/s
6.34	62.0	DW	tap water	0.61	2.0	-
6.42	62.0	DW	tap water	0.74	0.8	-
6.60	62.3	DW	0.05 M	0.57	5.9	2.4
6.52	67.3	DW	0.10 M	1.1	28.0	3.5
4.17	47.2	FT	tap water	1.9	7.3	-
4.27	48.2	FT	tap water	2.1	4.2	-
4.42	39.6	FT	0.05 M	2.2	21.0	7.9
4.34	40.0	FT	0.10 M	6.8	15.0	6.6

* DW = drying-wetting ; FT = freezing-thawing

Figure 8a shows cation concentrations in water samples collected from the DW specimen in Fig.7. During tap water permeation (1.5 pore volumes), the water samples displayed essentially constant or slowly decreasing cation concentrations. However, when the permeant was changed to Na_2CO_3, the Na^+ concentration increased rapidly and the Ca^{2+} increased more slowly. Nevertheless, the increase in calcium concentration indicates that calcium ions were being liberated from exchange sites of the clay particles. This specimen had the highest weathered conductivity which allowed cations to move rapidly through the specimen and be available for collection. If the 0.1 M Na_2CO_3 solution had passed through unchanged, the Na^+ concentration would register 46,000 $\mu g.ml^{-1}$. The relatively small value of 65 $\mu g/ml$ that was observed indicates that the interstitial pore fluid was being replaced by the Na_2CO_3 solution. Figure 8 shows that 10 days of permeation with 0.10 M Na_2CO_3 (0.27 pore volumes) was not enough to fully exchange and saturate the exchange complex with Na^+.

Freezing and Thawing.
Table 1 also shows results from FT specimens. After three freezing cycles, specimens increased in volume by 4 - 8% above initial values, while masses were 0.5-2.0% higher. Suctions at ice-water interfaces during freezing attracted water that was not released during subsequent thawing. Specimens from 4.17 m and 4.27 m produced relative conductivities 3.8 and 2.0 respectively, while less plastic specimens from 4.42 m and 4.34 m had relative conductivities of 9.5 and 2.2. Subsequent permeation of these latter specimens with 0.05 and 0.10 M Na_2CO_3 produced relative conductivities of 3.6 and 0.97. Figure 8b shows cation concentrations from collected water samples from one of the FT specimens. After 6 days and 0.72 pore volumes of tap water permeation, an additional 0.60 pore volumes were exchanged in 11 days of permeation with 0.10 M Na_2CO_3. The Na^+ concentration began to rise within a few days after the permeant was changed. Ca^{2+} and Mg^{2+} began to increase later. It is again noted that insufficient time (and pore volume exchange) was allowed for chemical equilibrium to be reached. Compromises had to be made between hydraulic gradient, chemical exchange processes (pore volumes of permeant), and test duration.

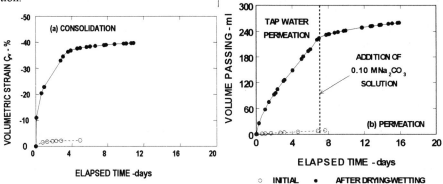

Figure 7. Effect of adding 0.1 M Na_2CO_3 after drying-wetting and tap water permeation

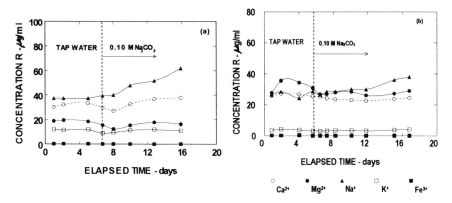

Figure 8 Cation concentration of collected water samples following addition of 0.1 M Na_2CO_3: (a) drying-wetting specimen, (b) freezing-thawing specimen,

DMSO PERMEATION
The second chemical selected for study was dimethyl sulfoxide (DMSO). This has the capacity to bind water molecules and reduce ice crystal formation during freezing (Keng et al. 1984). This property has led to its use as a cryoprotective agent for both plant and animal

tissues. It is commonly used to prevent the damage caused by freezing-thawing of soil specimens being prepared for mercury intrusion porosimetry or scanning electron microscopy. It has also been used in agriculture for the addition of nutrients and/or disease controlling chemicals in plants. Basic chemical properties and health hazard data can be found in Merck Index (1989).

The objective was to determine whether DMSO would inhibit ice penetration into clay microstructure during freezing-thawing and thereby limit increases in hydraulic conductivity. When specimens were initially built into the triaxial cell, a solution containing either 0%, 5%, 10% or 20% DMSO by volume was used for measuring the initial hydraulic conductivity K_i. Water, and not DMSO, was used in the fluid reservoir during subsequent freezing-thawing or when the weathered hydraulic conductivity K_{weath} was measured.

Table 2. Effect of DMSO on inhibiting changes in hydraulic conductivity due to freezing-thawing

Depth m	Init. Water content %	Process*	Initial permeant	K_i 10^{-10} m/s	K_{weath} *10^{-10}m/s†	K_{weath}/K_i
7.09	64.3	FT	tap water	0.97	3.2	3.3
7.22	50.9	FT	5% DMSO	1.3	4.2	3.2
6.99	64.6	FT	10% DMSO	0.63	1.1	1.8
6.89	63.0	FT	20% DMSO	0.74	0.95	1.3

* FT = freezing-thawing † Distilled deionized water as permeant

Table 2 summarizes measurements made on DMSO specimens. Despite some differences in water content and density (which are not large in terms of what is frequently observed in natural clays) the initial hydraulic conductivities after DMSO permeation were reasonably constant, ranging from 0.63 to 1.3 $\times 10^{-10}$ m/s. In general, freezing-thawing produced volume increases (2.0 - 2.7%) and mass decreases (1.1 – 2.3%), with the magnitudes of the changes depending on the concentration of DMSO used earlier (Yuen 1995). After freezing-thawing, specimens were returned to the cell for consolidation and permeation with distilled deionized water. Compression strains during consolidation ranged from 3% (5% DMSO) to 7% (20%DMSO). The corresponding relative conductivities were 3.2 and 0.95.

The results show that DMSO was partly effective in inhibiting changes increases in conductivity caused by freezing-thawing. It probably did not prevent ice forming in macropores and fissures but may have prevented ice-water interfaces from penetrating aggregates and increasing the micropores. This possibility could be explored using MIP testing.

DISCUSSION
Because they are cheaper and easier to perform than flexible-wall triaxial tests, rigid-wall oedometers sometimes preferred for measuring hydraulic conductivities. They may use pseudo-constant head conditions, or load increments and time-compression curves. Rigid-wall testing might be expected to produce less fissure sealing but this may be incorrect. Stresses are commonly higher than in flexible-wall tests and there are quite strong shear fields in which lateral stresses are generally unknown.

Opportunity was taken to add to the limited data base that compares hydraulic conductivities in the two test types. Specimens that had been permeated with Na_2CO_3 or DMSO were subsequently trimmed for oedometer tests. In each case, the fluid in the oedometer cell was the same as the original permeant. Figure 9 shows hydraulic conductivities deduced from compression-time curves in three oedometer tests on FT specimens. It also shows values from the corresponding triaxial tests. (The loops in the hydraulic conductivity data correspond with unload-reload cycles. They represent the influence of decreased compressibility during unloading on the algorithm $c_v = k/m_v\gamma_w$.

Hydraulic conductivities measured by oedometer were consistently lower than those from triaxial tests. They decreased strongly with pressure, probably because of the nonlinear relationship between compression and pressure. If oedometer tests are to be used, it is necessary to note the pressure level at which hydraulic conductivity has been interpreted. Closure of fissures and cracks by high vertical stresses and lateral confinement appear to produce hydraulic conductivities that are lower than field values.

Figure 1 shows that both freezing-thawing and drying-wetting alter the microstructure of this clay-rich soil. Permeation behaviour like that shown Figs.6,7 indicates that expansion on wetting in plastic clay can produce some self-sealing of fissures and joints. However, the self-sealing was not sufficient to return the hydraulic conductivity to its original 'undisturbed' or 'unweathered' value, even after 25 days of testing (Table 1).

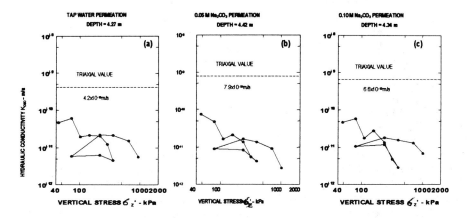

Figure 9 Comparison of hydraulic conductivities measured in triaxial and oedometer tests after freezing, thawing and permeation with (a) tap water (b) 0.05 M Na_2CO_3 and (c) 0.1 M Na_2CO_3

Lake Agassiz clay typically contains more than 55% montmorillonite and illite, with >25% kaolinite. The montmorillonite is often calcium saturated (Baracos 1977). Normally, sodium in low concentrations would not replace the calcium ions due to the cation selectivity of the clay (Tan 1993). However in this program, sodium is able to replace the adsorbed cations through a mass action effect. Replacing Ca^{2+} on exchange sites with Na^+ increases the hydration of the clay, and hence increases its swelling. This is believed to have produced the reductions in hydraulic conductivity that accompany Na permeation (Fig.7) (Kosmos and Moustakas 1990).

For calcium rich soils, introduction of exchangeable Na^+ for Ca^{2+} ions occurs mainly on the external surface of the clay domains (Keren and Singer 1990). Swelling is primarily due to hydration, with double-layer repulsion playing a lesser role. In studying the effects of sodium-calcium changes on the hydraulic conductivity of Wyoming bentonite, Shainberg and Caiserman (1970) observed a decrease in conductivity of an order of magnitude when the exchangeable sodium percentage (ESP) reached 50%. The decrease may result from 1) reduction in macropores caused by the swelling of the clay, or 2) clogging of pores by dispersed clay particles. The dominant mechanism will depend on the clay mineralogy, exchanging cation, and the pore fluid chemistry (Dane and Klute 1977). Another mechanism that contributes to the reduction of hydraulic conductivity by addition of sodium carbonate is precipitation of insoluble compounds. Sodium carbonate was selected for use instead of sodium chloride because the calcium-rich clay that was being studied would precipitate calcium carbonate. It may be argued that the addition of a salt solution should flocculate clay particles and cause the hydraulic conductivity to increase. The thickness of diffuse double layers (DDL) is given by

$$L = x + \left(\frac{DkT}{8\pi n_0 e^2 v^2} \right)^{0.5}$$

where L = thickness of the DDL measured from the clay surface, x = thickness of the Stern layer, D = dielectric constant of the medium, k = Boltzman's constant, T = temperature (Kelvin), n_0 = number of cations per ml of solution, e = electron charge, and v = valence of cation in solution (Mitchell 1976). Equation [1] indicates that increasing the concentration of the pore fluid, should compress the diffuse double layer DDL and bring the clay particles closer together. this will increase the macropores and hence the hydraulic conductivity. Such increases have been observed by Pupisky and Shainberg (1979) and Abu-Sharar *et al.* (1986) among others.

The effect of electrolytes on swelling may be caused by other factors as well as repression of the DDL. The ions of a dissolved electrolyte (in this study sodium and carbonate ions) interact with water molecules, thereby influencing their arrangement. Hence, they would be expected to affect the integrity and extent of hydration shells surrounding clay particles. Depending on the ions in the permeant and the cations adsorbed on the clay surface, flocculation may not occur, and the hydraulic conductivity will decrease, not increase. This is the case when comparing the flocculating effects of Na^+, Ca^{2+} and Mg^{2+}. Due to their lower charge, Na^+ ions in solution have a lower ionic potential (charge/size ratio) and usually act as a dispersant rather than a flocculant like Ca^{2+} or Mg^{2+} (Fuller *et al.* 1995). Flocculation would not have occurred in this program until the molarity of the Na_2CO_3 solution reached 0.23 M (instead of the maximum value of 0.10 that was actually used

CONCLUSIONS

Triaxial testing allowed good control of pressures and hydraulic gradients at values similar to those encountered in field applications. Oedometer tests produced hydraulic conductivities that were at least one order of magnitude lower than the triaxial results, and decreased strongly with pressure. Weathering by drying-wetting and freezing-thawing can cause strongly developed fissure structures in high-plastic clay. These fissures could be expected to cause large increases in hydraulic conductivity. However, expansivity of the clay allowed some self-sealing to occur. Typically, the increases in hydraulic conductivity following weathering averaged one order of magnitude, from 1×10^{-10} m/s to 10×10^{-10} m/s. The increases were significantly less if the permeant was changed to sodium carbonate or if FT specimens were pre-treated with dimethyl sulfoxide (DMSO). It must be kept in mind that the chemicals proposed for this purpose may themselves be contaminants and potentially toxic.

ACKNOWLEDGEMENTS
Funding was provided by Manitoba Hydro, Manitoba Environment, and the Natural Sciences and Engineering Research Council of Canada. We are grateful to Daryl McCartney, Narong Piamsalee, and B.Wiebe for their interest and support.

REFERENCES

Abu-Sharar, T.M., Bingham, F.T., and Rhoades, J.D. 1987. Reduction in hydraulic conductivity in relation to clay dispersion and disaggregation. J.Soil Science Soc.America **51**, 342-345.

Baracos, A. 1977. Compositional and structural anisotropy of Winnipeg soils - a study based on scanning electron microscopy and X-ray diffraction analyses. Can.Geotech.J. **14**, 125-137.

Boynton,S.S. and Daniel, D.E. 1985. Hydraulic conductivity tests on compacted clay. J.Geotech. Eng. ASCE **111**, 465-478.

Dane,J.H. and Klute, A., 1977. Salt effects on the hydraulic properties of a swelling soil. Proc. Soil Science Soc. America **41**, 1043-1049.

Daniel, D.E. and Wu, Y.-K. 1993. Compaced clay liners and covers for arid sites. J.Geotech. Eng. ASCE **119**, 223-237.

Dixon, D.A., Graham, J., and Gray, M.N. 1999. Hydraulic conductivity of clays under low hydraulic gradients. Accepted for publication, Can. Geot.J. **36.**

Fuller, L.G., Tee Boon Goh, Oscarson, D.W., and Bilianderis, C. 1995. Flocculation and coagulation of Ca- and Mg-saturated montmorillonite in the presence of a neutral polysaccharide. Clays and Clay Minerals **43**, 533-539.

Graham, J., and Shields, D.H. 1985. Influence of geology and geologicl processes on the geotechnical properties of a plastic clay. Eng. Geology **22**, 109-126.

Houston, S.L. and Randeni, J.S. 1992. Effect of clod size on hydraulic conductivity of compacted clay.. Geotech. Testing J. **15**, 123-128.

Johnston, K., and Haug, M. 1986. Impact of wet-dry freeze-thaw cycles on the hydraulic conductivity of glacial till. Proceedings, 45th Can. Geotech. Conference, Toronto ON.

Keng, J.C.W., Morita, H., and Ramia, N.T. 1985. Cryptoprotective effect of dimethyl sulfoxide on soil structure during freeze-drying. J. Soil Science Soc. America **49**, 289-293.

Keren, R. and Singer, M.J., 1990. Effect of pH on permeability of clay-sand mixture containing hydroxy polymers. J. Soil Science Soc. America **54**, 1310-1315.

Kosmas, C., and Moustakas, J., 1990. Hydraulic conductivity and leaching of an organic saline-sodic soil. Geoderma **46**, 363-370.

McConkey, B.G., Reimer, C.D., and Nicholaichuk, W. 1990. Sealing earthen hydrualic structures with enhanced gleization and sodium carbonate. I. Laboratory study of the effect of a freeze-thaw cycle and a drying interval. Can. Agric. Eng. **32,** 163-170.

Merck Index 1989. 11th edition, publ. Merck and Co. Inc., Rahway NJ, USA.

Mitchell,J.K.1976. Fundamentals of soil behaviour. John Wiley and Sons Inc.,New York, NY.

Pupisky, H., and Shainberg, I. 1979. Salt effects on the hydraulic conductivity of a sandy soil. J. Soil Science Soc. America **43**, 429-433

Quigley, R.M. 1980. Geology, mineralogy, and geochemistry of Canadian soft soils: a geotechnical perspective. Can.Geotech.J. **17**, 261-285.

Shainberg, I., and Caiserman, A. 1971. Studies on Na/Ca montmorillonite systems 2. The hydraulic conductivity. Soil Science **111**, 276-281.

Tan, K.H., 1993. Principles of soil chemistry. 2nd edition. Marcel Dekker (Publishers) Inc. New York, NY.

Wong L.C. and Haug, M.D. 1991. Cyclical closed-system freeze-thaw permeability testng of soil liner and cover materials. Can.Geotech.J. **28**, 784-793.

Yuen, K. 1995. The effects of freezing-thawing weathering, drying-wetting weathering, and changes in pore fluid chemistry on the hydraulic conductivity of Lake Agassiz clay. MSc thesis, Univ. of Manitoba, Winnipeg MB.

Yuen, K., Graham, J., and Janzen, P. 1999. Weathering-induced fissuring and hydraulic conductivity in a natural plastic clay. Accepted for publication, Can.Geotech.J. **36**.

Analysis of Settlement Characteristics of Municipal Solid Waste Landfills

G. L. SIVAKUMAR BABU, Assistant Professor, Department of Civil Engineering, Indian Institute of Science, Bangalore, INDIA

INTRODUCTION

Knowledge of settlement behaviour of municipal solid waste landfills is critical to drainage and contour requirements   besides the ultimate development and utilization of landfill sites. Excessive settlements, both differential and total, significantly affect the function and integrity of structures built after closure following mandatory regulations. An understanding of settlement characteristics of municipal landfills also helps in using the additional space available for filling due to settlement so that higher land filling capacity can be achieved. In this paper, a technique for data analysis to examine the time-settlement behavior of landfills is presented. It is shown that time-settlement behavior at a point in a landfills under a given set of loading and environmental conditions can be represented by two coefficients, representing the initial deformation in reference time and, the other, the variation of subsequent deformation.

LITERATURE REVIEW

The importance of understanding the settlement characteristics of MSW landfills can be realized from the volume of literature that exists today on this topic (Merz and Stone, 1962, Sowers, 1973), Yen and Scalon, 1975, Edil et. al, 1990, Coduto and Huitric, 1990, Mitchell et.al, 1995, Wall and Zeiss,1995, Ling et.al 1998). Edil et. al (1990) examined the applicability of two models, Gibson and Lo model and power creep law to study landfill settlements and showed that power creep law gives a better representation of settlement than Gibson and Lo model. Ling et. al (1998) provide a critical study of Yen and Scalon approach and use of power law in the study of landfill settlements and proposes the use of rectangular hyperbola model. These studies suggest that consideration of total settlements or strains in relation to time is appropriate. This is also substantiated as follows. The conventional examination of time-dependent behavior in terms of coefficient of secondary compression has difficulties (Sivakumar Babu and Fox, 1997). The slope of the time-compression response (C_α) in terms of differentials is given as

$$C_\alpha = de/d \log t = 2.3\ t\ de/dt \qquad (1)$$

The equation (1) shows that, C_α depends on time (t) and change in void ratio with time (de/dt) and suggests that the time-compression at any instance of time is a function of time as well as the rate of compression up to that point. It implies that if the changes in void ratio over time (de/dt) are commensurate with changes in time, C_α is constant. If changes in (de/dt) are different, , C_α is not constant with time. It also suggests that both time and (de/dt) or settlement/time could be considered together to examine the time-settlement behavior of soils. For linearisation, the relationship can be expressed as,

$$\ln(S/t) = -(a + b \ln t) \qquad (2)$$

Geoenvironmental engineering, Thomas Telford, London, 1999, 88–94

The term S/t represents the magnitude of total percent settlement or settlement up to time t, and a & b are constants. The coefficient "a" denotes the initial percent settlement or settlement in a reference time such as month, and "b" represents subsequent variation of time-deformation response. When the effects of stress history such as those resulting from excavation exist, recompression response increases with time till, a state on compression path is reached. Subsequent compression response decreases with time. Increase in compression response with time could also occur, when the state is on the verge of large deformations leading to failure. To examine these propositions and the above relationship, of the data available on compression behavior of MSW landfills, the data of Merz and Stone (1962), Sowers (1973), Coduto and Huitric (1990), and Sheurs and Khera (1980) are considered.

Analysis
Data of Merz and Stone (1962)

Merz and Stone (1962) presented settlement behavior of landfill cells treated with different stabilization methods. The details of cell construction are given in Table 1.The observations are examined in terms of settlement/time versus time and shown in Fig. 1. The deformation response under a given set of loading and environmental conditions is high initially and reduces with time. The variation in deformation response with time could be linearized using equation (2) as shown in Fig.2 indicating that the relationship between the variables is tenable. Statistical analysis shows that the relationship has very high degrees of correlation in the range of 0.99. The relationship is extended to cover a period of 10 years (120 months) and is shown in Fig.3. It suggests that   i) compared with other cells, the cell 1, in which the refuse was placed continuously with depth, watered and subjected to maximum compaction, exhibited minimum rate of settlement   immediately and ii) during and beyond one year of observations, cell 1 exhibited faster rate of settlement than other cells over the same duration of interest, presumably due to the presence of moisture (the settlement between 1 to 12 months is maximum for cell 1 and minimum for cell 4, and iii) comparison of time-settlement behavior of different cells enables the evaluation of differential settlements at any time, and extendable beyond one year of observations.

Table.1. Landfill Cell construction methods of Merz and Stone (1962)

Cell no	Method of construction
1	Refuse was placed continuously till full depth and watered, cell moisture content on dry weight basis 80.1% , subjected to maximum compaction
2	Refuse was placed in 4 ft thick layers and one ft soil cover and water content 43.5% refuse was subjected to standard compaction
3	Refuse was placed as in cell 2 but no water was added, standard compaction was achieved by bull dozers
4	Refuse was placed continuously. overall moisture content was 51.9% minimum compaction
5	Soil was admixed with refuse in the ratio of 1:2 by volume, cell moisture was 34.8%, standard compaction with bull dozers

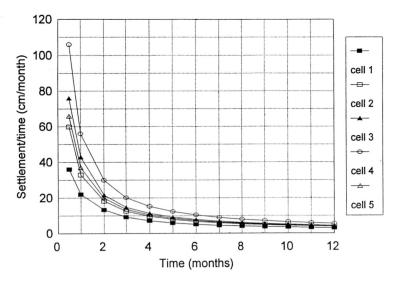

Fig.1 Relationship between settlement/time (S/t) and time (t) for the data of Merz and Stone (1962)

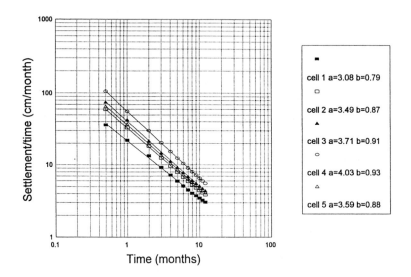

Fig.2 Relationship between settlement/time (S/t) and time (t)(ln(x) and ln(y) form)

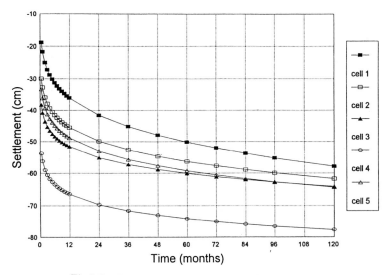

Fig.3 Prediction of long term settlements

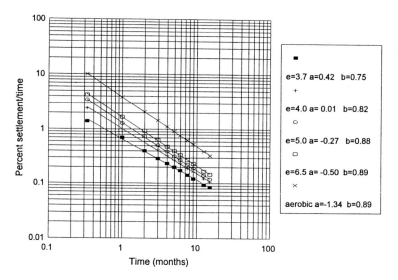

Fig.4 Analysis of data in ln (%S/t) and ln t form for the data of Sowers (1973)

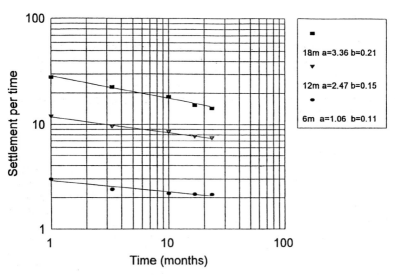

Fig.5 Time-settlement behaviour for the data of Coduto and Huitric (1990)

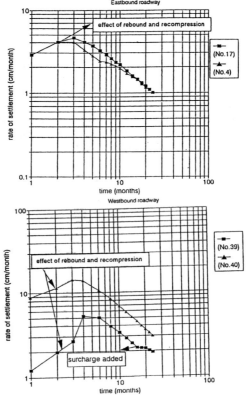

Fig.6 Time-settlement behaviour of landfill wastes at different locations along eastbound
roadway and westbound roadway (from Sheurs and Khera, 1980)

Data of Sowers (1973)

Sowers (1973) presented perhaps, the most widely quoted study on landfill settlements. Results of Sowers' (1973) study, are examined in terms of eq.2 and the resulting relationships are shown in Fig. 4. It suggests that i) under aerobic conditions, settlement in the initial stages is more compared with anaerobic conditions, and ii) under anaerobic conditions, the higher the initial void ratio, higher is the percent settlement immediately and higher is the rate of settlement subsequently. The approach or technique for reduction is also validated with reference to the data of Data of Coduto and Huitric (1990) and shown inFig.5.

Data of Sheurs and Khera (1980)

Sheurs and Khera (1980) studied settlement behavior of experimental roadways constructed on sanitary landfills. The roadways were directly constructed on landfills that contained partially decomposed garbage. The experimental roadways, I-85 eastbound and I-85 westbound have a total length of 975 m. To have proper leveling, material was excavated from I-85 eastbound and in both the cases, the material was compacted. The compaction of material results in prestress effects and, involve rebound after the removal of overburden pressure or compaction pressure. Once the effect was nullified, the settlement/time begins to decrease with time as shown in Fig.6. It also shows that the compression behavior beyond the zone of excavation and recompression could be linearized as before and long term compression behavior predicted. The hyperbolic model approach can not address some of these aspects. The above analysis separates the areas where monotonic compression is possible from recompression and rebound zones suggests the use of the approach proposed for prediction of long term settlements.

Concluding remarks

The paper presents an approach to examine the compression behavior of landfills based on periodic measurements of settlements. The relationship is valid for a particular location in a landfill and forms datum line with respect to which any measured changes at that location could be examined. More information on the mechanisms controlling landfill settlement and the relationship between gas and leachate generation and observed settlement is necessary and is not available so far. Nevertheless, reliable field data on landfill settlements extending to long durations at different locations with due consideration to the initial conditions of applied stress and environment are necessary for the development and verification of settlement models as is demonstrated herein.

ACKNOELEDGEMENTS

The author thanks Late Prof. G. A. Leonards and Prof. P. J. Fox of Civil Engineering Department, Purdue University, West Lafayette, Indiana for review and helpful suggestions during authors stay at Purdue University, USA.

REFERENCES

Coduto, D. P. and Huitric, R. (1990) Monitoring of landfill movements using precise instruments Geotechnics of waste fills-theory and practice, ASTM STP 1070, 358-370.

Edil, T. B., Ranguette, V. J. and Wuellner W. W. (1990) Settlement of municipal refuse, Geotechnics of waste fills-theory and practice, ASTM STP 1070, 225-239.

Ling, H.I, Leshchinsky, D, Mohir, Y and Kawabata, T (1998) Estimation of municipal soild waste landfill settlement, Journal of ASCE, Geotechnical and Geoenvironmental Engineering, vol.124, No.1,21-28.

Merz, R. C. and Stone. R. (1962) Landfill settlement rates, Journal of public works 93(9), N Y, 103-106.

Mitchell, J. K., Bray, J. D. and Mitchell, R. A. (1995) Material interactions in landfills, Geoenvironment 2000, ASCE, Geotechnical Special Publication No. 46, 568-590.

Sheurs, R. E and R. P.Khera (1980) Stabilization of sanitary landfill to support a highway, Transportation Research Record 754, 46-53.

Sivakumar Babu, G.L. and P.J. Fox (1995) Discussion on Municipal landfill biodegradation and settlement, Journal of Environmental Engineering, Vol.123, No.5, ASCE, 521.

Sowers, G. F. (1973) Settlement of waste disposal fills, Proc. of eighth ICSMFE, Moscow, 2:207-210

Wall, D. K. and Zeiss, C (1995) Municipal landfill biodegradation and settlement, Journal of Environmental Engineering, Vol.121, No.3, ASCE, 214-224.

Yen, B. C. and Scalon, B. (1975) Sanitary landfill settlement rates, Journal of Geotechnical Engineering, ASCE, 101:GT5:475-487.

The Influence of Physicochemical Stresses on Sorption and Volume Change Behaviours of Bentonite

C.C. SMITH and C.L. PEARCE
Department of Civil and Structural Engineering, University of Sheffield, UK

ABSTRACT

The volume, permeability and sorption behaviours of bentonite under varying mechanical and osmotic pressures were investigated. Modified flexible wall permeameter cells and oedometers were used to apply effective stresses. Osmotic pressures were applied by exposing the bentonite to varying concentrations of zinc chloride solution. From initial tests it was found that above a threshold concentration the bentonite particles aggregate in batch tests. Stiffness and permeability behaviour is distinctly differentiated by whether the bentonite was aggregated or non-aggregated at the outset of a consolidation test. Initial sorption test results indicate that high confining pressures reduce the sorption capacity of bentonite.

INTRODUCTION

Bentonite is a commercially available clay which is widely used in the construction of barriers for waste containment due to its low hydraulic conductivity and high cation exchange (attenuation) capacity. As a barrier to contaminants the bentonite is exposed to mechanical and chemical stresses. Numerous researchers have found that the pore fluid has a strong influence on the clay particle interactions. An increase in the pore fluid concentration may cause shrinkage of aggregates of clay particles, causing the opening of macropores and an increase in permeability (Barbour and Fredlund, 1989). It has also been demonstrated in tests with bentonite and caesium (Oscarson et al., 1994) that density has a significant effect on sorption characteristics. Sorption capacities for soil are generally measured in batch tests where all the soil surfaces are assumed to be available for adsorption. Oscarson et al. (1994) postulate that dense clay has small pores that caesium cannot access, which reduces the sorption capacity by up to one-half.

The current work follows up a previous study by Hird et al. (1997) who mapped the volume and permeability changes of a bentonite–sodium chloride system using two stress state variables (effective stress and osmotic pressure). In this paper, a bentonite–zinc chloride system is investigated. Specimens of bentonite are prepared at a range of concentrations of zinc chloride and consolidated in stages in modified flexible wall permeameter and oedometer cells to develop constitutive relationships between permeability, void ratio, effective stress and the osmotic pressure of the pore fluid. Further tests are being conducted to examine the sorption capacity of the bentonite at differing confining stresses by changing the pore fluid of the specimen by diffusion at selected stages of consolidation. The experimental set-up and testing methodology for these consolidation and sorption tests are described and the results so far available are presented.

Geoenvironmental engineering, Thomas Telford, London, 1999, 95–101

MATERIALS AND METHODS

The Wyoming bentonite used in this study consists mainly of the clay mineral montmorillonite and the principal exchangeable ion is sodium. The material was used in the previous study by Hird et al. (1997), who established liquid and plastic limits of 416% and 47% respectively in distilled water and a specific gravity of 2.58. Zinc chloride solutions of various concentrations from 0.001M to 0.5M were prepared in distilled water.

Batch test characterisation

Initial batch tests of 1:10 mixtures of bentonite and a range of zinc chloride solution concentrations and distilled water were conducted. Samples of the fluid were taken after 24 hours, 1 week and 4 weeks and analysed by inductively coupled plasma atomic emission spectrometry (ICP-AES). Analysis showed an average variation of approximately 20% in concentrations over this period. The results from the 4 week samples are shown in Figure 1 and indicate that as zinc is adsorbed sodium (and some other ions, principally calcium) are desorbed. It is assumed that the base load of sodium detected in the water batch tests was originally present as a soluble salt rather than as an exchangeable cation. The total desorbed data is corrected to remove this. There is reasonable comparability between the milliequivalents of ions sorbed and desorbed except at the highest initial concentration of zinc chloride. Below a concentration of approx. 0.014M virtually all the added zinc was removed from solution, however, at concentrations greater than this, significant quantities of zinc remain in solution. It was noted that a while a significant quantity of other ions appeared in solution in the bentonite-water batch test (including aluminium and iron), these did not appear in any of the bentonite–zinc chloride tests.

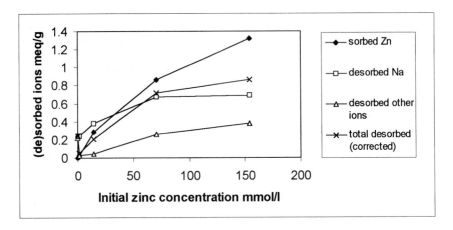

Figure 1: Adsorption/desorption behaviour from bentonite-zinc chloride batch tests.

Oedometer and permeameter test methodology

Following the procedure of Hird et al. (1997), samples of saturated and homogeneous bentonite in blocks that could be handled were produced from consolidated slurry. A ratio of 1:10 bentonite powder to water or zinc chloride solution was utilised to create a thin slurry. After maturing for two days, to allow ion exchange and clay dispersion, the slurry was de-aired in a vacuum desiccator for several hours. The slurry was then poured into a 100 mm diameter consolidation pot, and consolidated under a load of 60 kPa. The expelled solution was retained and sampled to establish initial conditions within the sample.

Modified oedometers and flexible wall permeameter cells were employed to apply mechanical stress to the samples. The flexible wall permeameter cells provide better control over the experiments, while the oedometers provide a more straightforward test method. The flexible wall permeameter was modified to allow fluid circulation around the top and bottom of the sample. Figure 2 shows the arrangement of top and bottom caps linked to a peristaltic pump which continuously circulates fluid from a reservoir. The fluid is pumped into the top cap where parallel spiral grooves allow the fluid to come into contact with the sample through a porous disc. The fluid then moves to circulate around the base, which is also provided with spiral grooves, before returning to the reservoir.

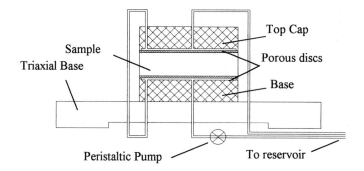

Figure 2: Schematic diagram of flexible wall permeameter base and sample arrangement.

The oedometers were also modified to permit the circulation of fluid around the base and top of the sample through the porous discs in a similar way to the permeameter cell. As the oedometers are not sealed at the top the fluid was allowed to overflow through a tube back to the reservoir. Layers of cling film were used to prevent evaporation from the top reservoir.
Continuous circulation of fluid from a reservoir around the edges of the specimens allows continuous monitoring of changes in the pore fluid concentration (assuming that the fluid external to the clay is in equilibrium with that internal to the clay).

Bentonite samples were extruded from the consolidation pot and trimmed to provide a sample of a suitable thickness, not more than 20 mm. Due to the low permeability of bentonite thicker samples required a considerable length of time to consolidate in the flexible wall permeameter or oedometer cell. The porous discs were saturated in the same concentration of fluid as that in the specimen. To maintain a bentonite to fluid ratio of 1:10 the moisture content of the sample was ascertained from the trimmings and the required volume of fluid was added to the reservoir. In initial experiments new fluid was used for circulating from the reservoir, in later experiments, where possible the fluid expelled from the consolidation pot was used for circulating as this maintained the same distribution of desorbed ions in the system as were in the original slurry. At a load of 60 kPa in the test apparatus, the sample should be in chemical equilibrium with the circulating solution, with no ion exchange occurring. In the oedometer, consolidation was monitored by measuring the specimen height using a LVDT. In the flexible wall permeameter apparatus, the expelled fluid that flowed into the reservoir was weighed periodically. The change of fluid volume in the flexible wall permeameter cell was also monitored.

Samples were consolidated in stages up to a selected pressure, at which point the concentration of fluid in the reservoir was increased to observe the effect of changing the osmotic pressure, and to monitor the influence of pressure on zinc sorption. The pressure was then removed in increments. Samples of fluid were taken from the reservoir at the end of each loading and unloading stage for analysis. When the pressure had been fully released half the bentonite sample was placed in half the fluid from the reservoir to investigate the complete removal of mechanical stress on zinc sorption. A batch test was run in parallel with the permeameter and oedometer tests at a 1:10 ratio of bentonite to zinc chloride solution as a control to eliminate time effects and to determine the influence of stress and/or density.

RESULTS AND DISCUSSION

Batch tests
The initial batch tests indicated that the concentration of the zinc chloride influenced the swelling behaviour of the bentonite. The results showed that up to a concentration of approximately 0.01M zinc chloride the majority of the zinc ions were adsorbed. The bentonite mixed with the concentrations below and at this threshold swelled to form a dense gel. At the chosen concentrations above this level the bentonite formed a thin layer on the base of the container. At the lower concentrations, it is assumed that the bentonite has thick diffuse double layers, which force the particles apart and cause swelling. Where the adsorption capacity is exceeded the diffuse double layers are suppressed and an aggregation of the particles occurs. Similar behaviour was observed by Hird et al. (1997) with a threshold concentration of 0.1N sodium chloride. This alteration in fabric may influence the permeability of the clay, which was investigated in the flexible wall permeameter and oedometer tests.

Volume change behaviour
Figure 3 shows the variation of void ratio with applied effective stress for various pore fluid concentrations from the current oedometer study together with results from Hird et al. (1997) for comparison. The behaviour of the 0.1M zinc chloride sample correlates closely with the behaviour of the 1N sodium chloride sample reported by Hird et al. and was attributed to the initial aggregated structure of the bentonite. The 0.01M zinc chloride sample behaviour does not show the same degree of correlation with the sodium chloride pattern of behaviour, though close correlation between a multiionic and homoionic system may not be expected. A detailed comparison will be undertaken when further test data becomes available. It should be noted that all data presented from the current study is based on void ratios forward calculated from the moisture content of the consolidation pot sample trimmings. More precise computations of void ratio are generally based on backward calculations from final moisture contents. These are unavailable at present since the experiments have yet to complete their unloading stages.

Permeability
Figure 4 shows the variation of permeability (indirectly derived from consolidation data) with vertical effective stress (taken as the value at the end of each loading stage). The sodium chloride results are reproduced from Hird et al. (1997) for comparison (the data for 0.01N and 0.1N NaCl is not reproduced since it is almost identical to that for the 0.001N NaCl). The specimens prepared in zinc chloride solutions seem to demonstrate the same behaviour as was noted by Hird et al. For the same vertical effective stress the permeabilities of the non-aggregated specimens (0.1N, 0.01N, 0.001N sodium chloride, 0.01M zinc chloride and water) vary within a narrow range and are significantly lower than the aggregated specimens (1N sodium chloride and 0.1M zinc chloride) which again vary within a narrow range. The

anomolous water result at 100kPa is attributed to a large change in void ratio during the consolidation phase from which it was derived. The implication, tentatively proposed by Hird et al., is that for non-aggregated bentonite clay, the effective stress controls the permeability of the material, which appears little influnced by the osmotic pressure.

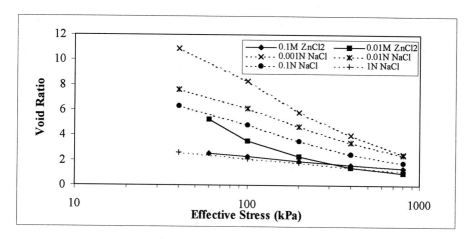

Figure 3: Relationship between void ratio and effective stress for Wyoming bentonite in different pore fluids

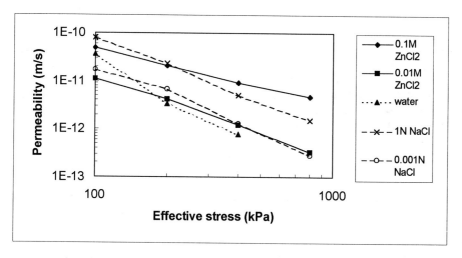

Figure 4: Relationship between permeability and effective stress for Wyoming bentonite in different pore fluids

Sorption

Results from a preliminary sorption test comparing a bentonite sample in a flexible wall permeameter with that in a batch test are presented in Figure 5. The permeameter sample was initially prepared in 0.01M zinc chloride, and consolidated to 800kPa. The data appears to indicate that sorption is suppressed at higher confining stress levels. The two sets of data are not exactly comparable because of differing concentrations of other ions (particularly sodium) in the fluid. However the concentration of sodium at the highest zinc concentrations is only 2.5 –5 % that of the zinc and so is unlikely to have a significant influence. The pH levels were also very similar (5.4 in the batch test and 5.3 in the permeameter test). Experiments currently in progress are designed to maintain greater comparability in pore fluid composition at all stages and to independently compare the roles of stress and density on the sorption capacity.

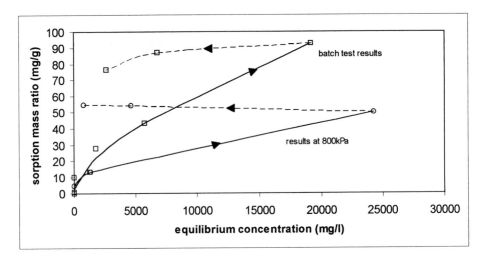

Figure 5: Comparison of zinc sorption for a flexible wall permeameter test at 800 kPa and a batch test (dashed lines denote desorption).

CONCLUSIONS

Test procedures involving modified flexible wall permeameters and oedometers for investigating the influence of mechanical and osmotic stress on void ratio, permeability and sorption in Wyoming bentonite have been described. The volume change behaviour of bentonite is significantly affected by the concentration of zinc chloride in the pore fluid. From initial tests it is tentatively concluded that for samples prepared in a concentration greater than 0.01M of zinc chloride the clay particles are aggregated. In common with results reported by Hird et al (1997) for bentonite prepared with sodium chloride solutions, this aggregation significantly increases the permeability and stiffness of the bentonite. It appears that effective stress is the dominant control on permeability and that permeability is not a direct function of solute concentration – its only relevance is whether it renders the initial fabric aggregated or non-aggregated.

Preliminary results comparing zinc sorption at a confining effective stress of 800 kPa with that deduced from batch tests indicate that application of pressure does reduce the adsorption of zinc by bentonite. However, this conclusion is tentative at present and is being investigated in further tests.

ACKNOWLEDGEMENTS

Financial support by the EPSRC and technical assistance by Mr P.L. Osborne and Mr M. Foster is gratefully acknowledged.

REFERENCES

Barbour, S.L. and Fredlund, D.G., 1989. Mechanisms of osmotic flow and volume change in clay soils. Canadian Geotechnical Journal, Vol. 26, pp. 551-562.

Hird, C.C., Smith, C.C.,and Prakash, V.J., 1997. The influence of physico-chemical stresses on permeability and volume change behaviour of bentonite. Proceedings, 1st BGS GeoEnvironmental Engineering Conference, Cardiff, pp. 331-336.

Mesri, G. and Olson, R.E., 1971. Mechanisms controlling the permeability of clays. Clays and Clay Minerals, Vol. 19, pp. 151-158.

Oscarson, D.W., Hume, H.B., and King, F., 1994. Sorption of cesium on compacted bentonite. Clays and Clay Minerals, Vol. 42, pp. 731-736.

Shrinkage and Desiccation Cracking in Bentonite-Sand Landfill Liners

Y.Y. Tay, D.I. Stewart, T.W. Cousens
School of Civil Engineering, University of Leeds, Leeds LS2 9JT, UK

ABSTRACT

Data are reported on the shrinkage and desiccation cracking exhibited by bentonite-enhanced sand mixtures (BES) upon air-drying. Mixtures containing 10% and 20% bentonite by dry weight, compacted at moisture contents ranging from 8 to 30%, were investigated. Hydraulic conductivity data on BES specimens saturated and tested immediately after compaction and on similar specimens that had no visible damage after air-drying are also presented.

All the mixtures exhibit volumetric shrinkage upon air-drying, with the amount of shrinkage increasing with increasing moisture content during compaction. At any initial moisture content mixtures containing 20% bentonite shrink more than those containing 10% bentonite, but the shrinkage is insensitive to the compactive effort. Compacted beds of BES containing 10% and 20% bentonite exhibit no visible desiccation cracking as the top surface is dried when compacted at 15 and 14% moisture content, respectively, and only minor cracking when compacted at initial moisture contents of 20% and 15%, respectively. At initial moisture contents of 20% and 15%, 10 and 20% mixtures both undergo 2.7% volumetric strain upon air-drying. Compacted beds of both these mixtures suffered severe desiccation cracking when compacted at initial moisture contents of 30% and 20%, respectively, when both mixtures undergo around 4% volumetric strain when air-dried. The saturated hydraulic conductivity of intact BES specimens is unaffected by a drying episode prior to testing.

INTRODUCTION

Bentonite-enhanced sand mixtures (BES) are increasingly being used as landfill liners because they are an economic alternative to compacted clay when suitable clay is not locally available. As the main function of the liner is to prevent, or at least minimise, the movement of water out of the waste disposal facility, BES should satisfy three performance criteria if it is to perform satisfactorily as a barrier. It should have low hydraulic conductivity (typically $<1 \times 10^{-9}$ m/s), sufficient strength for stability during construction and operation, and not be susceptible to excessive shrinkage cracking due to any moisture content change that occurs during the lifetime of the landfill (cracks may close on rewetting, but do not completely heal and cause a permanent increase in hydraulic conductivity; Phifer et al., 1995).

This paper is on shrinkage cracking of BES. It presents data on the shrinkage of compacted specimens of BES upon air-drying. It reports observations of the desiccation cracking exhibited by large (800x800mm) beds of compacted BES upon air-drying. It also presents measurements of the saturated hydraulic conductivity of intact BES specimens that were, (i) tested after compaction, and (ii) subjected to a drying episode prior to saturation and testing.

Geoenvironmental engineering, Thomas Telford, London, 1999, 102–109

Table 1. Properties[1] of SPV 200 Wyoming bentonite and Sherburn yellow building sand

SPV 200 Wyoming Bentonite	Sherburn Yellow Building Sand
Average particle size = 2 μm [2]	Effective size (D_{10}) = 212μm
Specific gravity = 2.751 [2]	Fines = 0.36%
Liquid limit = 354% [2]	Coefficient of Uniformity = 2
Plastic limit = 27% [2]	Specific gravity = 2.68
Moisture content = 13% (as supplied)	Max & min void ratio [3] = 0.767 & 0.375
	Moisture content ~ 4% (as supplied)

(1) Tests performed in accordance with B.S. 1377: Part2: 1990 unless otherwise indicated
(2) After Studds et al. (1998)
(3) Determined by methods described in Head (1980)

MATERIALS

The materials used in this study were SPV 200 and Premium Gel Wyoming bentonites supplied by Cetco Europe Ltd (formerly Volclay), and Sherburn yellow building sand supplied by Tom Langton and Sons. SPV 200 bentonite originates from Lovell, Wyoming and is a well ordered sodium montmorillonite with minor quartz and cristobolite impurities (after Studds et al., 1998). Premium Gel (which was used for about half the liner desiccation tests) is the same material further processed to remove most of the coarse impurities. Sherburn yellow building sand is predominantly a quartz sand (suppliers specification). Other properties of these materials are given in Table 1.

METHODS
Shrinkage tests
BES specimens containing either 10 or 20% bentonite (by dry weight) were prepared at moisture contents in the range 8 to 30% by light and heavy manual compaction in accordance with B.S.1377: Part 4: 1990. After extrusion from the compaction mould, the specimens were air-dried in a well ventilated room until they exhibited no further volume change (at which point the moisture contents were on average 1.2 and 2.3% for the 10 and 20% mixtures respectively). After air-drying the specimen were visually inspected for cracks and the external dimensions were measured using a Vernier calliper. All samples were then oven dried at 105^{0}C to determine the moisture content after air-drying.

Liner Desiccation Tests
Large BES specimens containing either 10 or 20% bentonite (by dry weight) were prepared in 800mm square, by 250mm deep wooden boxes using a hand-held vibrating Kango pneumatic hammer with a 80x100mm rectangular plate fitted to the vibrating probe. BES mixtures at varying moisture contents were compacted in 5 layers, with each layer being compacted for 13.3 minutes, to achieve a final specimen thickness of 200mm. The Kango hammer was applied in a series of two second bursts, and moved systematically over the surface of each layer between bursts. The top of the specimens was then subjected to evaporative drying by passing warm air (at 30°C) across the exposed surface.

After one week, the specimen was photographed, any visible cracks were recorded, and the moisture content-depth profile was measured. Most of the liner desiccation tests ceased at this point. However one specimen (containing 10% bentonite prepared at 30% moisture content) was inundated with water, and then subjected to a second drying episode.

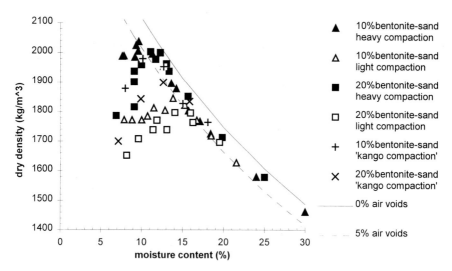

Figure 1. Compaction curves for mixtures containing 10% and 20% bentonite

Small samples were also prepared by "Kango compaction" in a 200mm square, 200 deep mould. Compaction was performed in 5 layers, with the Kango hammer being applied for a total of 54 seconds per layer (again in a series of two second bursts, and moved systematically over the surface between bursts), to determine the compaction curve for BES mixtures using this method.

Hydraulic Conductivity tests
BES containing 10% bentonite (by dry weight) were prepared at various moisture contents by heavy manual compaction in accordance with B.S.1377: Part 4: 1990. After extrusion from the compaction mould, two 12-18mm thick specimens were trimmed from the compacted mixture and weighed. One was tested immediately, while the other was air-dried prior to testing. Thin specimens were used to minimise the saturation time, and these were inspected to ensure they were not cracked. Testing was conducted in triaxial permeameters using an effective cell pressure in the range 50 to 100kPa, and a back-pressure of 300kPa to improve specimen saturation. The back-pressure alone was applied for two weeks, then a hydraulic gradient was superimposed and a constant head test performed. Hydraulic conductivity was calculated when the in-flow and out-flow rates were equal.

RESULTS
Figure 1 shows the compaction curves for mixtures containing 10 and 20% bentonite for both light and heavy manual compaction. The compaction curves for mixtures containing 10 and 20% bentonite subjected to "Kango compaction" are also shown. For light manual compaction the maximum dry density achieved with the 10% mixture was 1840 kg/m^3 at an optimum moisture content of 14.0 %, and with the 20% mixture it was 1800 kg/m^3 at an optimum moisture content of 15.1%. For heavy manual compaction the maximum dry density for the 10% mixture was 2010 kg/m^3 at an optimum moisture content of 9.5 %, and for the 20% mixture it was 2000 kg/m^3 at an optimum moisture content of 11.5%.

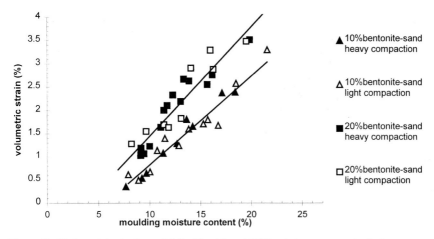

Figure 2. Volumetric strains exhibited by 10 and 20% mixtures upon air-drying

The maximum dry density achieved by "Kango compaction" was 1980 kg/m^3 at an optimum moisture content of 10.5% for the 10% mixture, and 1900 kg/m^3 at an optimum moisture content of 13.0% for the 20% mixture. The compactive effort of "Kango compaction" was therefore intermediate between that of light and heavy manual compaction.

Shrinkage tests
Figure 2 shows the volumetric strain upon drying exhibited by the 10 and 20% BES mixtures prepared by both light and heavy manual compaction. Only data for specimens where no cracks were observed are plotted in Figure 2, as cracks result in an underestimation of the volumetric strain. All the specimens exhibited shrinkage on drying, with the amount of shrinkage of a particular mixture increasing with increasing moisture content during compaction. At any initial moisture content the 20% bentonite mixtures shrunk more than the 10% bentonite mixtures, but shrinkage was insensitive to compactive effort. Kleppe and Olson (1985) found similar trends, reporting that shrinkage of a particular soil depends on moulding moisture content but is insensitive to compactive effort.

Liner Desiccation Tests
Figure 3 (a)→(d) shows liner specimens containing 10% bentonite compacted over a range of moisture contents by "Kango compaction" after the top surface was subjected to evaporative drying. Similarly, Figure 4 (a)→(d) shows liner specimens containing 20% bentonite prepared and tested in the same manner. In all liner specimen tests there was moisture content decrease to a depth of about 80mm below the exposed top surface. Liner specimens containing 10% and 20% bentonite exhibited no visible desiccation cracking when compacted at 15 and 14% moisture content (no photo for the latter), respectively. Mixtures containing 10% bentonite exhibited minor cracking (crack width <1mm) upon evaporative drying when compacted at 20% moisture content (Figure 3b), whereas mixtures containing 20% bentonite exhibited minor cracking when compacted at only 15% moisture content (Figure 4a).

Mixtures containing both 10 and 20% bentonite exhibited increased amounts of cracking upon surface drying as the moisture content during compaction increased (see figure 3 and 4,

(a) Compacted at 15% moisture content

(b) Compacted at 20% moisture content

(c) Compacted at 23% moisture content

(d) Compacted at 30% moisture content

Figure 3. BES mixtures containing 10% bentonite after air-drying.

respectively). Mixtures containing 10% bentonite exhibited severe cracking (crack widths up to 4.5mm) upon evaporative drying when compacted at 30% moisture content (Figure 3d), whereas mixtures containing 20% bentonite exhibited severe cracking when compacted at only 20% moisture content (Figure 4c).

After drying, the specimen containing 10% bentonite and prepared at 30% moisture content was inundated with water, and then subjected to a second episode of evaporative drying. Inundation resulted in the apparent closure of most of the cracks visible after the first drying episode, but the second drying episode resulted both in the re-opening of the large cracks caused by the first drying episode, and the formation of additional, closely spaced, fine cracks between the original cracks (photographs omitted for reasons of space). During inundation there was moisture content increase to a depth of about 40mm, with the near surface soil

(a) Compacted at 15% moisture content

(b) Compacted at 18% moisture content

(c) Compacted at 20% moisture content   (d) Compacted at 30% moisture content

Figure 4. BES mixtures containing 20% bentonite after air-drying.

reaching about 40% moisture content. It is therefore likely that the additional closely spaced, fine cracks were the result of the thin near surface zone drying out from this higher moisture content.

Hydraulic Conductivity tests

Table 2 reports the saturated hydraulic conductivities of compacted BES mixtures containing 10% bentonite. The average hydraulic conductivity of the five specimens that were saturated and tested after compaction was 5.0×10^{-12} m/s, and the average hydraulic conductivity of the four specimens that were air-dried prior to saturation and testing was 7×10^{-12} m/s. The measurement of very low hydraulic conductivities is difficult and scatter in the data is normal. Therefore, the difference between the average value for specimens tested immediately and that for air-dried specimens is not thought to be significant, particularly as the highest and lowest measurements (which differ a factor of 4) were both made with specimens that underwent an air-dying episode.

DISCUSSION

During the liner desiccation tests, drying occurred in the top 80mm of the 200mm deep beds, with the moisture content of the deeper soil remaining essentially unchanged. Therefore the zone of drying was 10 times wider that it was deep. Such a large aspect ratio indicates that the liner specimens were representative of real liners, which have very large horizontal extent, provided the desiccated layer is similarly restrained at its the base (i.e. mineral liners compacted onto geomembranes may behave differently).

Liner specimens containing 10% and 20% bentonite exhibited no visible desiccation cracking when compacted at 15 and 14% moisture content, respectively. The data in Figure 2 therefore suggest that compacted beds of BES can undergo more than 2% volumetric strain without cracking (the volume change must occur one-dimensionally). BES mixtures containing 10% and 20% bentonite exhibited only minor cracking when compacted at initial moisture contents of 20% and 15%, respectively, when both would have undergo 2.7% volumetric strain upon drying. However, these mixtures suffered severe desiccation cracking when compacted at initial moisture contents of 30% and 20%, respectively, when both would have undergo about 4% volumetric strain when air-dried (determined by extrapolation of the trends in Figure 2). Kleppe and Olson (1985), who established that it required 4-5% volumetric shrinkage to cause significant cracking of mineral liners, found a similar result.

The liner specimen that was inundated with water highlighted two important points. Firstly, on inundation BES that is insufficiently surcharged can swell to moisture contents greater than that during compaction, and on subsequent drying will suffer more cracking than a specimen that has not had access to additional water. Secondly, the close spacing of the additional cracks after the second drying episode are thought to indicate the sensitivity of crack spacing to the thickness of layer undergoing volume change, and that crack spacing in a liner is related to depth of drying at the instance of cracking. If correct, then larger but more widely spaced cracks would be anticipated in mixtures where drying penetrates further below surface before the desiccated zone suffers enough volumetric strain to cause cracking.

Measurements show that specimens that suffered a drying episode after compaction have the same hydraulic conductivity as specimens that did not suffer such an episode, apparently contradicting the results of Phifer et al (1995) who found that a decrease in void ratio during drying resulted in a decrease in hydraulic conductivity for intact specimens. This may be solely because most of the air-dried specimens were compacted at 9.5% moisture content, and would only have undergone about 0.7% volumetric strain. However it may also indicate that the internal structure of BES differs from that of conventional compacted clay soils. Compacted clay soils are thought to be made up of intact aggregates separated by discontinuities, and Phifer et al. (1995) suggest that desiccation causes shrinkage of the aggregates, but can open up the discontinuities (they found that the hydraulic conductivity of intact specimens decreased upon desiccation, whereas specimens with hydraulic defects showed a marked increase in hydraulic conductivity). A better idealisation of BES may be that proposed by Mollins et al. (1996), who studied the swelling of BES against a surcharge stress. They suggested that BES at low moisture contents can be idealised as a continuous sand matrix with partially swollen bentonite occupying the pores. At moisture contents

Table 2. Hydraulic conductivity of compacted mixtures containing 10% bentonite

Test ID	Moulding moisture content (%)		Effective cell pressure (kPa)	Back pressure (kPa)	Hydraulic gradient	Hydraulic conductivity (m/s)
1w	9.5	tested immediately	50	300	96	7.6×10^{-12}
2w	9.5	tested immediately	100	300	183	4.7×10^{-12}
3w	9.5	tested immediately	100	300	209	4.0×10^{-12}
4w	13.5	tested immediately	100	300	149	3.8×10^{-12}
5w	19.1	tested immediately	100	300	147	5.1×10^{-12}
1a	9.5	air-dried	50	300	46	1.2×10^{-11}
2a	9.5	air-dried	100	300	112	3.5×10^{-12}
3a	9.5	air-dried	100	300	40	9.6×10^{-12}
5a	19.1	air-dried	100	300	204	2.9×10^{-12}

where there is a continuous sand matrix, desiccation would cause shrinkage of the bentonite within the sand pores, but little change in sand porosity. On rewetting the bentonite can swell to fill the sand pores again, and thus the state of the mixture is not greatly effected by drying.

Compacted liners are usually placed at moisture contents close to the optimum (typically optimum + 2%) for the compactive effort being employed. The optimum moisture contents of BES for light manual compaction were 14% and 15%; for heavy manual compaction 9.5% and 11.5%; and for "Kango compaction" 10.5% and 11.5%; for the 10% and 20% mixtures, respectively. Therefore a BES mixture similar to those tested is unlikely to suffer significant damage due to desiccation cracking if it is placed by heavy compaction, unless the moisture content is allowed to increase after liner placement.

CONCLUSIONS
- The volumetric shrinkage of compacted BES mixtures upon air-drying increases with increasing moulding moisture content, with mixtures containing 20% bentonite shrinking more than those containing 10% bentonite, but it is insensitive to compactive effort.
- Compacted beds of BES containing 10% and 20% bentonite do not exhibit desiccation cracking if the volumetric strain during drying is less than about 2%.
- Desiccation induced volumetric strain greater than about 4% causes severe cracking in compacted beds of BES.
- The saturated hydraulic conductivity of BES is unaffected by a drying episode prior to measurement provided drying does not cause desiccation cracking.

REFERENCES
British Standards Institute, BS 1377 (1990) Methods of test for soils for civil engineering purposes. HMSO Stationary, London.

Head K.H.(1980) Manual of Soil laboratory testing., 1, Pentech Press, London

Kleppe, J.H. and Olson, R.E. (1985) Desiccation cracking of soil barriers. ASTM STP 874, 263-275.

Mollins L.H., Stewart D.I. and Cousens T.W. (1996) Predicting the properties of bentonite-sand mixtures. Clay Minerals, 31, 243-252.

Phifer, M., Boles, D., Drumm, E. and Wilson, G.V. (1995). Comparative response of two barrier soils to post compaction water content variations. ASCE GSP 46, 591-608.

Studds, P.G., Stewart, D.I. & Cousens, T.W. (1998). The Effects of Salt Solutions on the Properties of Bentonite-Sand Mixtures. Clay Minerals, 33, 651-660.

Geosynthetic Clay Liners and long-term slope stability

KENT P. VON MAUBEUGE, Naue Fasertechnik GmbH & Co. KG, Luebbecke, Germany

CHRISTOPHER MICHAEL QUIRK, Naue Fasertechnik GmbH & Co. KG, Manchester, UK

ABSTRACT

Geosynthetic clay liners (GCLs) are relatively thin composite materials combining bentonite clay and synthetics (usually geotextiles). GCLs have been employed by the waste industry for well over a decade now, and their level of usage is rapidly increasing both in the British Isles and world wide. In landfill facilities, GCLs are generally used to replace or augment compacted clay liners. Until recently, the decision to do so has primarily been based on the availability of clay material on site (i. e., economic considerations). However, the advantages in using a GCL over other sealing elements such as compacted clay are not only economic but technically based, and the economic benefits extend beyond the construction phase, as a thin GCL can increase the revenue earning potential of a facility. This paper will highlight the shear behaviour of GCLs and demonstrate their long-term stability.

INTRODUCTION

Geosynthetic Clay Liners (GCLs), also called bentonite liners, are industrially manufactured, whereby a set quantity of natural sodium bentonite is confined between two geotextiles (Fig. 1).The carrier layer is either a woven or a combination woven/nonwoven geotextile which allows for good anchorage of fibres. The geotextiles are then needle-punched together through the intermediate bentonite layer, securing the bentonite in place and reinforcing the otherwise weak layer of clay (when hydrated).

Figure 1: Schematic drawing of the manufacturing process of a needle-punched GCL

The low midplane friction angle of bentonite (in hydrated condition peak approx. 9°, residual about 4° to 5°) is overcome by direction independent needle-punching or partly by direction dependent stitch-bonding (Fig. 2) [1].Geosynthetic Clay Liners achieve hydraulic conductivities in the range of approx. 1×10^{-11} m/s at high confining stresses. At low confining stresses hydraulic conductivities of $< 5 \times 10^{-11}$ m/s may be expected for needle-

Geoenvironmental engineering, Thomas Telford, London, 1999, 110–117

punched products. Higher hydraulic conductivities may occur with other GCLs, depending on the type of manufacturing process (e. g. 9 x 10^{-11} m/s) [2].

The focus on hydraulic conductivity becomes less insignificant since all manufacturers mainly use natural sodium bentonite. "Polymer additives for bentonite" is no longer a

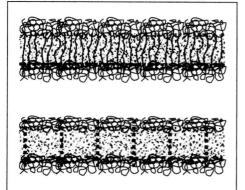

relevant topic since the long-term efficiency could not be proven in practice. Egloffstein [3] has questioned the long-term effects and has proven that the amounts of polymer had to be much higher than the amount of bentonite used (approx. 4 – 5 kg/m²) to ensure a long-term efficiency. The Environment Agency are currently drafting guidelines for the use of GCLs and are taking the position that the use of polymers is not to be encouraged. More important is the question of the long-term shear strength transfer.

Fig. 2: Schematic cross-section of a needle-punched GCL (top) and a stitch-bonded GCL (bottom).

THEORETICAL BACKGROUND

Heerten et al. [4] published a theoretical approach which describes the shear behaviour for needle-punched GCLs, and a design diagram that shows the conditions under which the critical shear plane is not midplane but at the GCL interface (fig. 3). The main parameters are the confining stress, the slope inclination, and the GCL peel strength. The peel strength is an index parameter where one geotextile layer of the GCL is fixed in the upper clamp and one in the lower clamp of the tensile force testing machine and the needle-punching which holds the GCL together is pulled apart. The peel strength achieved is reported in N/10 cm and is brought into relation to the normal stress and to the slope inclination (fig. 3). This design basis could be considered conservative as the needle-punched GCLs tested were pre-hydrated for 24 hours without any confining stress and the cohesion intercept determined was not taken into consideration.

A 24-hour-hydration was selected as according to the installation instructions for GCLs, a cover soil should be placed within 24 hours after deployment [13]. Heyer et al. [12] further propose that GCLs with a moisture content prior to covering of more than 50 % should be replaced, as the risk of bentonite thinning of the GCL after placement rises with an increasing moisture content.

The shear plane is assumed to be outside the needle-punched GCL if the determined value in the chart (fig. 2) lies above the chosen slope inclination; thus e. g. a slope inclination of 1.5 (h) : 1(v) could be achieved at a peel strength of 69 N/10 cm and a cover thickness of 4 m provided that the interface friction angles as well as the midplane friction angles of the adjacent layers achieve at least the same value. At a peel strength of 29 N/10 cm and 1.5 m cover, a slope with an inclination of approx. 33° is possible (dot A in fig. 3) but with 10 N/10 cm a internal failure would occur (dot B).

One commercially available needle-punched GCL has, according to von Maubeuge [6], approx. 2.5 million fibres per m² with a minimum strength of 40 cN per fibre. The result of this is the theoretical short-term shear strength of 1,000 kN/m² assuming that all fibres are locked and tear at the same time. In the shear test, however, lower shear strengths (load dependent) than the theoretical 1,000 kN/m² are achieved (fig. 4) so that it can be concluded – as recognised in the shear tests - that not the tearing of the fibre, but the pull-out resistance is relevant.

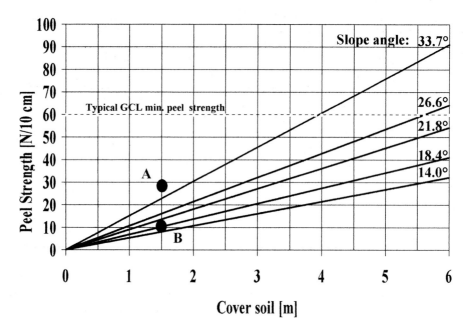

Fig. 3: Design diagram for the determination of the max. permissible slope angle as a function of cover soil (γ = 20 kN/m³) and GCL peel strength. If the point lies above the line, the GCL should not fail internally. Shear tests are required with typical soils from the site.

Once the fibres are activated in the field by the shear transfer of the actual load, a long-term pull-out (sometimes referred to as pull-out creep) is not of concern because the 2.5 million needle-punched fibres are not just simply anchored. Due to their crimped shape they are curled together three-dimensionally like knots. The staple fibres are also thermally fixed and this acts to prevent releasing of the knots. Thus a long-term pull-out of the anchored fibres is not to be expected and has been proven in long-term tests.

GENERAL SHEAR BEHAVIOUR OF GCLs
In order to examine the general shear behaviour of geosynthetic clay liners in hydrated condition, tests with different GCL types were conducted in the laboratories of Naue Fasertechnik GmbH & Co. KG, Germany. On an automatic tilt-table (1 m x 1 m) a textured geomembrane (GM) was fixed upon which each GCL type was placed. After 24 hours of free hydration (horizontal position) again a textured geomembrane was placed on the hydrated GCL. The setup was then loaded with a 30 cm thick gravel layer (approx. 6 kN/m²). The box was constructed in such a way that the shear plane could only occur between one of the

geomembranes and the GCL or in the GCL itself. After a consolidation (short term) time of 0.5 hours, the tilt-table was raised at a speed of 1°/min. Some representative results from the conducted tests are shown in Table 1.

Table 1: Tilt-table resultss for the determination of the general shear behaviour of GCL types

#	Structure	shear plane	shear angle
1	5 kg bentonite between two geotextiles without additional reinforcement	bentonite	8°
2	5 kg bentonite between two geotextiles, fixed with water-soluble glue	bentonite	8°
3	4.5 kg bentonite between two geotextiles, needle-punched with 8 N/10 cm peel strength	bentonite	18°
4	4.5 kg bentonite, stitch-bonded between two nonwovens (200 g/m²)	between upper GM and GCL	29°
5	4.5 kg bentonite between two nonwovens 300 g/m², needle-punched with a peel strength of 30 N/10 cm	between upper GM and GCL	33°
6	4.5 kg bentonite between a 200 g/m² nonwoven and a 100 g/m² woven, needle-punched with a peel strength of 65 N/10 cm	between GM and woven of GCL	22°

The conducted tests highlight two significant factors in the behaviour of GCLs:

a) The peel strength between the single geotextile layers has a decisive influence on the shear behaviour of GCLs. The results are in agreement with the design diagram (fig. 3).

b) At a sufficient midplane shear force transfer the selection of the adjacent geosynthetics is significant for interface shear. Light needlepunched nonwovens (~ 200 g/m²) and wovens show lower shear angles than thicker needlepunched nonwovens (~ 300 g/m²).

In addition, mechanically bonded nonwovens with a higher mass per unit area (~ 300 g/m²) provide a better protection for the encapsulated bentonite and thus are more resistant against construction conditions than thinner geotextile components.

Upon selecting a GCL, particularly needle-punched GCLs, not only the reinforcing (midplane) shear strength is relevant. Interface shear behaviour (interface friction angle) is just as important. Nonwovens achieve higher shear values against adjacent soils than wovens, due to the three-dimensional structure. Without considering the influence of possible extruding bentonite (see [7] or section 6), the following relationships ($\tan\varphi'$ / $\tan\psi'$) for interface friction angles of geotextiles can be assumed according to Grett [11] (ψ' = soil friction angle, φ' = interface friction angle of soil / geotextile):

	needle-punched nonwoven	woven
clay	~ 0.92	~ 0.84
fine sand	~ 0.92	~ 0.80
coarse sand	~ 0.95	~ 0.83

LONG-TERM SHEAR BEHAVIOUR

In order to prove the design diagram described in section 2 and the long-term durability of the fibre composite as function of the peel strength, Naue Fasertechnik GmbH & Co. KG constructed long-term shear boxes and examined the long-term shear behaviour and possible creep deformations on Bentofix type B (300 g/m² needle-punched cover nonwoven – 3500 g/m² natural sodium bentonite powder – 350 g/m² woven-reinforced needle-punched nonwoven).

Fig. 5: Long-term shear boxes with Bentofix® type B

On slope inclinations of 2.1 : 1 (25°) the GCL was covered with crushed gravel (2/8 mm) in a thickness of 30 cm (approx. 6 kN/m²) and the gravel was loaded with steel plates (25 kN/m²). In order to keep the bentonite permanently hydrated, a watering device was installed that waters the GCL daily with 10 litres water. Only the bottom (carrier) geotextile was fixed at the upper edge of the long-term shear box. This ensures that the shear force is actually transferred by the fibre reinforcement. A displacement of the upper geotextile to the lower geotextile layer could be observed at the edges during the entire testing period.

In the first test, a needle-punched GCL with a peel strength of only 10 N/10 cm was selected. It can be seen from Figure 3 (dot B) that a sliding within the GCL had to be expected since the minimum peel strength of the GCL at a normal stress of 31 kN/m² (approx. 1.5 m cover) and at a slope inclination of 1:2.1 should have been at least 18 N/10 cm.

In the first 10 days a displacement of approx. 2 cm between the upper and the lower geotextile layer occurred. After 23 days a midplane slide – as anticipated - occurred. It was recognised that the anchoring fibres had been pulled out of the lower geotextile layer.

In a second test, a needle-punched GCL with a peel strength of 29 N/10 cm was selected. According to fig. 2 (dot A) with this peel strength on a slope inclination of 1:2.1 a normal stress of approx. 52 kN/m² (approx. 2.5 m soil) could be safely applied without having to expect that the shear plane would occur within the GCL.

This set-up was installed on October 3, 1993. As in the first test, a displacement of 2 cm of the upper to the lower geotextile layer occurred after 2 days. Since that time no further displacement has been encountered, not even after more than 40,000 hours (4 ½ years on April 1, 1998).

The design diagram was verified in the long-term test as well as the fact that a detectable creeping (see also section 5) or pulling out of the needle-punched fibres did not occur.

CREEPING

Since all geosynthetics tend to creep (longitudinal movement under permanent load), a safety factor is used for designs using geosynthetic reinforcement. Normally this value is supposed to be 4 so that the possibility of creeping of the Bentofix fibre reinforcement would only occur at > 250 kN/m² according to the theoretical approach of section 1. Hewitt et al [5] carried out various shear tests with GCLs and it was recognised that such shear stresses occurred in needle-punched GCLs at normal stresses of > 400 kN/m² (approx. 20 m cover). However, creep testing was carried out on Bentofix® GCLs [8] even though such high shear stresses are rarely expected in GCLs (e.g. in piggy-back landfills, high dam constructions)

In shear boxes with a size of 30 cm x 30 cm (lower box 30 cm x 35 cm) the creep behaviour of the Bentofix® types B 4000 (NW) and NSP 4900 (NS) was examined. The saturation, consolidation, and shear stages were carried out with loads upto 630 kN/m².

After applying 50 % of the normal stress as creep stress (shear stress) no significant displacements between the cover nonwoven and the carrier geotextile had been recognised – at low normal stresses (21 kN/m²) as well as at high normal stresses (630 kN/m²). The duration of each test was > 500 hours.

Thus the long-term shear tests in the 1 m x 1 m boxes were not only confirmed for low normal stresses, but also for normal stresses of up to 630 kN/m²; a subject matter which may be relevant for intermediate sealings and base sealings where shear stresses in this range have to be transferred.

LARGE-SCALE TEST PLOTS
In 1994 the US American Environmental Protection Agency (EPA) initiated large shear tests in Cincinnati to verify the midplane shear strength of GCLs. These tests were supervised by the University of Texas at Austin, Geosynthetic Research Institute in Philadelphia and Geosyntec Consultants in Atlanta. For the first time the results were presented by Koerner [9] on the VDI (Association of German Civil Engineers) seminar in Karlsruhe on October 9 and 10, 1996. Latest results were presented at the Geo-Bento conference in Paris [10].

The aim was to prove that the midplane friction angle of GCLs is sufficiently high on slopes (length between 30 and 20 m) with inclinations between 3:1 and 2:1 so that it may be assumed that a safety of 1.5 is given on slopes with an inclination of 3:1.

The GCLs were installed in November 1994 on a subgrade of silty sand and were covered with a 1.5 mm thick textured geomembrane, a geonet composite (with mechanically bonded nonwoven on both sides) and a 1 m thick soil layer. To avoid passive forces, a toe support was not installed. After one and two months the first two slides occurred on the 2:1 slopes. In both cases the failure surface was the woven side of the GCL against the geomembrane. Afterwards it was discovered that the interfacial friction angles from shear tests had values between 20° and 23° and more, thus significantly lower than the actual slope inclination. Additionally it was found that bentonite had partly extruded from the woven side and further decreased the interface friction between geoemembrane and woven. This bentonite extruding has been reported for thin mechanically bonded (< 220 g/m²) nonwovens by Gilbert [7]:

"Second, for GCLs with bentonite encased between geotextiles, the bentonite may extrude through the geotextiles into adjacent interfaces and affect the interface strength. Bentonite extrusion is normally associated with woven geotextiles, although it has been observed for thin (i. e. mass per unit area less than 220 g/m²) nonwoven as well."

Meanwhile Koerner [10] reported that plot F had an internal failure. Bentonite between two geomembranes hydrated and the wet bentonite sheared as it was not reinforced.

The peel strength of the two needle-punched GCLs in the plots which did not fail was tested > 90 N/10 cm and thus they remained stable. According to Fig. 2 at least 6 m of cover soil can be transmitted with this peel value.

In the meantime deformation has been recognised in the plots I, J. K, and L due to occurring subsoil failure. Koerner [10] reported: ".. *the general indication was that the entire lower half of these four test plots were deforming within the subgrade soils beneath the GCLs.*"

A conclusion after the test trial is:

♦ needle-punched nonwovens give a better interface friction contact than wovens
♦ bentonite can extrude through wovens and nonwovens ($\leq$ 200 g/m²)
♦ thicker (> 270 g/m²) nonwovens are safer solutions for GCLs due to better interface contact and bentonite retention in the GCL
♦ needle-punched nonwovens should be on both sides of the GCL where shear stress is applied to the GCL.

GLOBAL TRENDS

As a result of the type of work carried out above and the importance of long term slope stability and the confidence that must be placed in the materials used, needle-punched shear transferring GCLs account for the majority of world wide sales.

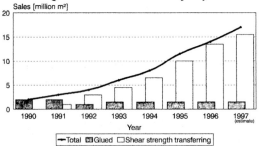

Fig. 7: Estimated GCL sales world-wide

SUMMARY

Needle-punched geosynthetic clay liners show a lot of technical advantages. Beside the low hydraulic conductivity, the self-sealing capability, and the elongation capability, the shear strength is an important criterion for the long-term efficiency of this product group.

The requirement for a minimum peel strength is necessary for steep slope applications. It is important that the proof of long-term stability is conducted by means of field studies. In order to achieve an interface friction angle as high as possible against the adjacent interfaces (e. g. textured geomembrane or soil), mechanically bonded nonwovens are especially suited. With a mass per unit area > 220 g/m² not only a good bond is achieved, but also a possible extrusion of bentonite is prevented.

Also, substantial costs can be saved by using geosynthetic clay liners. This cost-saving sealing layer can be installed more easily and faster and thus also more economically than compacted clay liners, capillary barriers or concrete sealings.

The existing examinations on the long-term performance exactly show that needle-punched geosynthetic clay liners are a predictable and long-term stable sealing element.

It is important that as the United Kingdom uses more and more of these materials that designs are carried out properly and that the shear transferring properties of the GCL are taken into consideration. The potential for creep particularly on landfill caps also needs to be addressed, whilst the specification for caps can be lower, the GCL must still perform adequately, ie that creep does not occur.

REFERENCES.

[1] Stewart, D., von Maubeuge, K. P.; "Cost-effective and efficient solutions with GCLs for sealing operations in the mining industry", Tailings and Mine Waste '97, Denver 1997
[2] ASTM round robin "Precision Statement for ASTM D5887", April 1997 (not published)
[3] Egloffstein, T. A..; "Geosynthetic Clay Liners, part six: ion exchange"; GFR, Vol. 15, No. 5, June/July 1997
[4] Heerten, G.; Saathoff, F.; Scheu, C.; von Maubeuge, K. P.;"On the long-term shear behavior of geosynthetic clay liners (GCLs) in capping sealing systems"; Proceedings "Geosynthetic Clay Liners"; pages 141 – 150, Nuremberg, April 1995
[5] Hewitt, R. D.; Saydemir, C.; Stulgis, R. P.; Coombs, M. T.; Effect of Normal Stress during Hydration and Shear on the Shear Strength of GCL / Textured Geomembrane Interfaces"; ASTM Symposium "Testing and Acceptance Criteria for Geosynthetic Clay Liners", pages 65 – 70, Atlanta, January 1996
[6] Von Maubeuge, K. P., Eberle, M. "The use of GCLs for sealing applications in the waste industry", Geoenvironment '97, Melbourne, November 1997
[7] Gilbert, R. B., Scranton, H. B., Daniel, D. E., "Shear strength testing for Geosynthetic Clay Liners", ASTM Symposium "Testing and Acceptance Criteria for Geosynthetic Clay Liners", pages 65 – 70, Atlanta, January 1996
[8] Seibken, J., Swan, R. H., Yuan, Z., "Short-term and creep shear characteristics of a needlepunched thermally locked Geosynthetic Clay Liner", ASTM Symposium "Testing and Acceptance Criteria for Geosynthetic Clay Liners", pages 65 – 70, Atlanta, January 1996
[9] Koerner, R. M.; Carson, D. A.; Daniel, D. E., Bonaparte, R.; "Current Status of the Cincinnati GCL Test Plots"; VDI seminar "Oberflächenabdichtung oder -abdeckung? Regelwerke oder alternative Systeme?; pages 3-1 – 3-25, Karlsruhe, Germany, October 1996
[10] Koerner, R. M., Carson, D. A., Daniel, D. E., Bonaparte, R.; "Update of the Cincinnati test plots", Geo-Bento '98, Paris, February 1998
[11] Grett, H. D., "Das Reibungsverhalten von Geotextilien in bindigem und nichtbindigem Boden", Heft 59, Mitteilungen des Franzius-Instituts für Wasserbau und Küsteningenieurwesen, Hanover, 1984
[12] Heyer, D.; Ascherl, R.; "Design and construction of sealing systems with geosynthetic clay liners (GCL)", Proceedings of the First European Geosynthetics Conference EUROGEO 1, Maastricht, Netherlands, September/October 1996
[13] Bentofix Installation Guidelines, Naue Fasertechnik GmbH & Co. KG, Luebbecke, Germany, September 1996
[14] Berard, J. F., "Evaluation of needle-punched GCL's internal friction", Geosynthetics '97, Long Beach, March 1997

Selective Sequential Extraction analysis (SSE) on Estuarine Alluvium Soils

R.N. YONG*, S.P. BENTLEY*, C. HARRIS**, and W.Z.W.YAACOB**/*
* Geoenvironmental Engineering Research Centre, Cardiff School of Engineering, Cardiff University, Cardiff, UK.
** Department of Earth Sciences, Cardiff University P.O. Box 914, Cardiff CF1 3YE, UK.

ABSTRACT

Selective Sequential Extractions (SSE) were used to study the retention mechanisms of heavy metals on three estuarine alluvium soil columns obtained from leaching experiment for up to 5 pore volumes (PV). Leaching results indicate that almost 99 % of heavy metals (Pb^{2+}, Cu^{2+}, and Zn^{2+}) were retained in the soils with the C_e/C_o values in the order of 10^{-3} to 10^{-4}. This is in accord with the results obtained in respect to the resultant pH of the effluents and pore waters. The heavy metal extraction profiles from SSE show very similar trends with the retention profiles from the leaching experiment, where heavy metals were retained mainly at the top part (Entry of leachate) of the column. SSE indicates qualitatively that heavy metals precipitated with carbonates and amorphous materials (oxides/hydroxides) are higher than heavy metal retention via exchangeable mechanisms. The mass balance calculation gives range of deviation of 35-80% of the total soil extractions. The retention mechanisms of the heavy metals in soil solids are ranked in the following order: Carbonates>Amorphous>Organics>Exchangeable phases.

INTRODUCTION

The use of clay soils as impermeable or attenuating barriers is becoming more and more popular as the "material of choice" in landfill liner systems. Many researchers (Griffin et al., 1976; Yanful et al. 1988; Yong et al. 1992, 1993, etc.) have discussed the different aspects and potential use of soil material not only for liners, but also as substrate material under landfills. Heavy metals (H.M.) such as Pb, Cu and Zn that are commonly found in leachates from landfills can be effectively attenuated by such soils. The amounts of heavy metal retained depends on the pH of the soil-water system and the soil buffer capacity (Yong and Phadungchewit, 1993). Heavy metals are retained in soils by hydroxide and carbonate when the pH of the soil solution is higher than 4. The primary mechanism for H.M. retention in clay soils is through precipitation of the metal ions with carbonates and amorphous oxides or hydroxides (Griffin et al. 1977). Yong and Phandungchewit (1993) have shown that the presence of carbonates in a soil contributes significantly to the retention capability of the soil.

This study was undertaken to determine the retention capability of the heavy metals Pb, Cu and Zn by soils in South Wales using leaching column experiments and selective sequential extraction analyses of the contaminated column samples. The role of the various soil fractions in sorbing the contaminants is examined in relation to the capacity of the soil for H.M. retention.

Geoenvironmental engineering, Thomas Telford, London, 1999, 118–126

MATERIALS AND METHODS
The soil samples used in the study were collected from three different sites adjacent to an active landfill sites around South Wales, United Kingdom (Figure 1, Table 1). Five samples from each site were randomly taken from shallow trenches and pits. All samples were air-dried and sieved through 20mm sieve in the laboratory, to remove all coarse pebbles.

Four separate sets of tasks were undertaken during the course of the study:
- Basic characterisation of all soils using physical and chemical tests.
- Determination of the attenuation characteristics of these soils via leaching column tests - conducted on a selected sample from each group (NEA4 for NEA samples, PEA3 for PEA samples and CEA3 for CEA samples).
- Acid digestion on slice soil samples obtained from leaching column tests to determine the "gross" amount of heavy metals retained.
- Confirmation studies of retention mechanism by determine the heavy metals extracted from different soil solids via SSE.

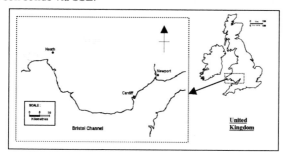

Figure 1. Study area and sampling location

Table 1. Samples descriptions

Group	Samples	Description
NEA	NEA1-NEA5	Estuarine Alluvium from Neath
PEA	PEA1-PEA5	Estuarine Alluvium from Newport
CEA	CEA1-CEA5	Estuarine Alluvium from Cardiff

Physico-chemical tests
Physical tests of all samples were conducted according to British Standard, BS 1377 (1990). These included natural moisture content, specific gravity test, particle size distribution, compaction test and permeability test. Physico-chemical tests performed according to procedures detailed in the Laboratory Manual of the Geotechnical Research Centre of McGill University, included organic content determination, specific surface area (SSA), cation exchange capacity (CEC) and pore fluid chemistry. Mineralogy of the soil samples was determined using x-ray diffractometry. Carbonate content was determined using the titration method by Hesse (1972), and amorphous oxides/hydroxides content was obtained using the procedure described by Segalen (1968). Leachate solutions and pore water were analysed using atomic absorption spectrometry (AAS) for Si, Al, and Fe.

Leaching column test
The leaching cell used consisted of a Plexiglas cylinder with diameter 115 mm and height 125 mm. The soil samples were compacted at maximum dry density and optimum moisture

content into triplicate soil columns i.e., soil series as illustrated in Figure 2. The leaching experiments were conducted under air pressure at 10psi (68Kpa) to reduce the time factor. The cells were leached with a test leachate that was obtained from a Municipal Solid Waste Landfill (MSWL) and spiked with heavy metals Pb, Cu and Zn. The pH of the test leachate was reduced up to 1.45 to promote increased mobility of the heavy metals in the soil columns.

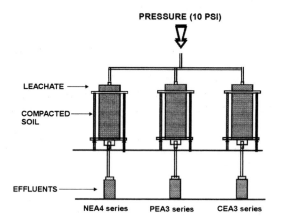

Figure 2.   Three series of leaching column experiments

The effluents (leachate discharge) were collected and analysed at every 0.5 PV using an ICP-Mass Spectrometer (ICP-MS). At the end of the experiments the columns were extruded and the columns were sliced into six equal slices (~20 mm each). The soil slices were tested for exchangeable cations using ammonium acetate (pH 7.0) and pore water analysis. All solutions were analysed using the ICPMS.

Acid digestion
About 200 mg of dry sample was placed into a clean dry savillex vial. The vial was placed in a fume cupboard before addition of 5 ml of Romil HF to the vial. It was then left overnight to digest on the hotplate. The solution obtained after digested was reduced by partial evaporation, following which about 5 ml of aqua regia was add to the reduced solution. The aqua regia solution was prepared by mixing 60 ml of conc. HNO_3 and 180 ml of conc. HCl. The sample was then digested for another 24 hours and subsequently evaporated to obtain a dry sample. Following this, 5-6 ml of 5M primar HCl was added to the sample which was left on a hotplate for several hours. The solution obtained therefrom was diluted using 100ml of plastic volumetric flask prior to analysis using the ICP-MS.

Selective sequential extraction
Methods of extraction have been adopted from Yong and Phandungchewit (1993), Yanful *et al.* (1988), Tessier *et al.* (1979) and Gupta and Chen (1975). The basic utility of selective sequential extraction is its use of appropriate reagents to release different heavy metal fractions from soil solids by destroying the binding agents between the metals and the soil solids and permitting the metal species to be detected (Yong and Phandungchewit, 1993).

The soil samples from the leaching column tests which were used for the SSE analysis were sliced into six equal slices. For the SSE analysis, 1 g of the soil slice was weighed into the 50ml polypropylene centrifuge tubes to avoid any loss of soil sediment during the extraction stages. Only 1 g of soil was used in order to comply with the *stage 5* requirement using HF. On the fifth stage, the teflon containers (Savillex vials) can only contain a small amount of sample (about 1 g). The procedure for H.M.extraction for the different sorption mechanisms were as follow:-

- Stage 1-Exchangeable Cations: 8 ml of 1M Potassium Nitrate, KNO3. The pH was adjusted to the natural soil pH and shaken continuously for 1 hour. This release all metals retained via cation exchange.
- Stages 2-Carbonate: 8 ml of 1M Sodium Acetate (NaOAc) adjusted to pH 5.0 by acetic acid (HOAc) and agitated continuously for 5 hours to release metals which precipitated with carbonates.
- Stages 3-Oxides and hydroxides: 20 ml of 0.04M of Hydroxylamine hydrochloride ($NH_2OH.HCl$) in 25%(v/v) Acetic Acid (HOAc), agitated occasionally at $96\pm3°C$ for 6 hours. This will extract all metals precipitated as hydroxides and/or adsorbed on the oxides or the amorphous hydroxides.
- Stage 4-Organics: Heavy metals bound to organic constituents in soil via complexation, adsorption and chelation. This extraction is divided into three phases: - Phase i: 3 ml of 0.02M Nitric Acid (HNO_3) and 5 ml of 30% H_2O_2 adjusted to pH 2.0 with HNO_3 occasionally agitated for 2 hours at temperature $85\pm2°C$. Phase ii: 3 ml of 30% H_2O_2 (at pH 2.0), intermittently agitated for 3 hours at the temperature $85\pm2°C$. Phase iii: 5 ml of 3.2M Ammonium Acetate (NH_4OAc) in 20%(v/v) HNO_3, diluted to 20 ml, and continuously agitated at room temperature for 30 minutes.
- Stage 5-Residual Fractions: Metals that are bound to the soil solids via specific adsorption in soil mineral lattice will be extracted by digestion. The residue (after the previous 4 stages) was digested with 5 ml of hydrofluoric acid (HF), 5 ml of aqua regia i.e., nitric acid (HNO_3) mixed with hydrochloric acid (HCl) and finally with 5 ml of hydrochloric acid (5M HCl).

In between the stages, solids-liquid separation was obtained by centrifugation. The supernatant collected and analysed, and the residue left was washed, centrifuged again, and the second supernatant discarded.

3. RESULTS AND DISCUSSION
Physico-chemical properties
The physical and chemical properties of soils used in the leaching experiments are listed in Tables 2 and 3 below. The CEA3 sample can be classified as a silty clay (CL) with intermediate to high plasticity, whilst the PEA3 sample is classified as a silty clay (CH) with high plasticity, and the NEA4 sample classed as a clayey silt (MH) with high plasticity.

The coefficient of permeability for all samples was found to be around 10^{-10} cm/sec. This appears to be consistent with the clay percentage of from 50-57%. The chemical properties of the samples in Table 3 show that sample PEA3 has a better potential to retain heavy metals via cation exchange because of the high CEC and SSA values. The clay minerals content in all three soils are similar. Both PEA3 and CEA3 have high carbonate contents and low organic matter content--in comparison to NEA4. The percentage of oxides/hydroxides material in NEA4 is higher than PEA3 and CEA3.

Table 2. Physical properties of the estuarine alluvia

Property	NEA4	PEA3	CEA3
$\omega_o\%$ initial	51.8	46.9	27.4
LL %	65.8	56.2	47.3
PI %	30.0	29.6	25.8
ω_{opt}	30.0	23.0	19.2
γ_{dry} Mg/m^3	1.36	1.57	1.7
Gs	2.49	2.65	2.52
Clay %	50	52	57
K x 10^{-10} (m/sec)	2.2	0.6	1.9
Class	MH	CH	CL

$\omega_o\%$= initial water content; LL % = Liquid limit; PI % = Plastic limit; ω_{opt} = optimum water content; γ_{dry} = dry density; k = permeability falling head.

Table 3. Chemical properties of the estuarine alluvia

Property	NEA4	PEA3	CEA3
Carbonates (%)	3	16.5	18
Amorphous oxides/hydroxides (%)	9.9	13.4	8.5
Organics (%)	5.11	3.78	3.63
SSA (m^2/g)	73.34	74.97	66.62
CEC (meq/100g)	14.84	39.43	38.5
Clay mineralogy	K, I, C	K, I, C	K, I, C

SSA = specific surface area; CEC = cation exchange capacity; K = Kaolinite;I = Illite; C = Chlorite

Retention Assessment via leaching column

Results of Pb concentration in the effluents (C_e) relative to the influent concentration (C_o) are shown as breakthrough curves for NEA4, PEA3 and CEA3 in Figure 3. The Ce values obtained for the first 2 pore volumes are somewhat puzzling, and should be discounted at this stage of the analysis.

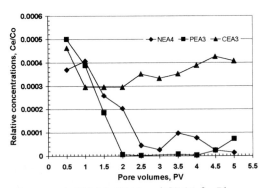

Figure 3. Breakthrough curves in NEA4, PEA3 and CEA3 for Pb.

The high Ce values in the first 2 pore volumes suggest that improper sealing of the samples in the columns allowed for quick passage of the leachate during the initial stages of the experiment. This appears to be self-correcting due to accumulation of Pb at the top of the sample (as seen in Figure 4). Since the pH values for the pore water showed a range of 6 to 8, one could conclude that all the heavy metals were precipitated in the soil columns -- as confirmed earlier by Yong and Phandungchewit (1993). Considering the results after the first 2 pore volumes in Figure 3, it is seen that almost 99% of heavy metals are retained in the columns. The relative concentrations, C_e/C_o recorded are in the order of 10^{-4}.

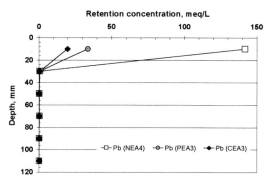

Figure 4. Retention profiles of Pb in NEA4, PEA3, and CEA3

Extraction profiles and retention mechanisms

H.M. extraction results obtained via SSE have been used to plot the "extraction profiles" of various soil solids in sample NEA4, -- as shown in Figure 5 for Pb retention.

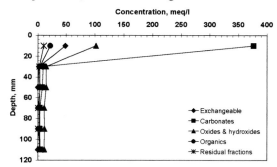

Figure 5. Extraction profiles of Pb in sample NEA4

The results from SSE analyses for all the samples are shown in the form of bar plots -- as seen in Figure 6, 7, and 8. These plots indicate that the H.Ms were retained primarily through mechanisms associated with precipitation of the metals with carbonates and hydroxides. This is consistent with the high pH recorded in the pore water of the soil columns.

Total heavy metals extracted from the various soil fractions are shown in Figure 9 below. As indicated in the figure, the carbonates phase of PEA3 shows the highest retention of heavy metals as compared to NEA4 and CEA3. PEA3 which appeared to retain about 17% carbonates almost similar to CEA3 about 18% (from Table 2). NEA4 shows the lowest content of carbonates -- about 3%. The H.M. extraction values do not correspond very well with the carbonate content in these soils except for PEA3. Apparently, precipitation of the H.M. as separate phases could account this discrepancy; i.e. co-precipitation with the carbonates and hydroxides was not a significant factor. This is further reinforced as observation when the results of the NEA4 soil are studied. Thus whilst the NEA4 soil shows the highest extraction compare to PEA3 and CEA3 in hydroxides phase, this does not accord directly with the hydroxides content, since PEA3 (13.4%) has the highest hydroxides content compare with NEA4 (9.9%) and CEA3 (8.5%).

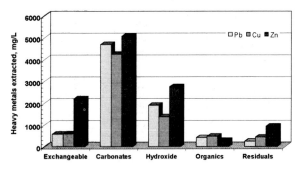

Figure 6.   Heavy metals extracted in NEA4

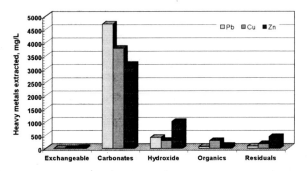

Figure 7.   Heavy metals extracted in PEA3

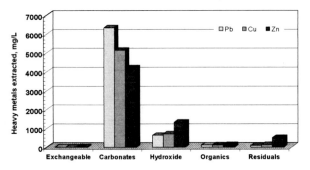

Figure 8.   Heavy metals extracted in CEA3

The retention of H.M. by the soil organics appears to correspond with the organic content in the soils. In this case, NEA4 (5.1% of organic content) shows the highest H.M. retention as compared with PEA3 and CEA3. The correlation between H.M. retention through exchange mechanisms cannot be directly sought through association with the SSA and CEC of the soils principally because studies on the manner of distribution of the various soil fractions have not been conducted. Since particle coatings and aggregations of particles can be significant factors when hydrous oxides and organics are present in the soil, no easy correlation can be made between the SSA and CEC with cation exchange mechanisms for H.M. retention.

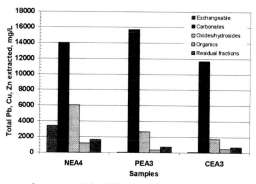

Figure 9. Total heavy metals extracted in different soil solids (mg/L)

Mass balance calculations

Mass balance calculation have been performed by calculating the deviation of the total heavy metals extracted from SSE and gross heavy metal extracted from acid digestion in NEA4, PEA3 and CEA3 soils. The deviation of SSE obtained from Table 4 ranges from 35-80%. The highest deviation was 80.9% in the case of Cu in sample PEA3. The percentages of deviation in Cu were high in comparison with Pb and Zn. This corresponds with the results reported by Gupta and Chen (1975). They discovered that in some of their results, the deviations exceeded 30%, and in Cu it was found to deviate as much as 80%.

The effectiveness and usefulness of SSE in extracting the heavy metals vary according to researchers. Because the results obtained are at best operationally defined, Yong and Phandungchewit (1993), stated that the SSE method of analysis provides one with a qualitative appreciation of the capability of the various soil constituents to accumulate heavy metals.

Table 4. Mass balance calculations

Elements/Soil slices	Total extracted via SSE (mg/L)	Gross concentration via acid digestion (mg/L)	Deviation (%)
Pb			
NEA4	7866.89	5791.47	35.8
PEA3	7197.97	4567.30	57.6
CEA3	5170.81	3390.19	55.5
Cu			
NEA4	7159.42	4419.84	62.0
PEA3	6180.84	3395.31	80.9
CEA3	4569.65	2755.92	65.8
Zn			
NEA4	11265.84	7610.32	48.0
PEA3	6212.04	4194.33	48.1
CEA3	4802.45	3244.33	48.0

CONCLUDING REMARKS

Heavy metals in acidic leachate (pH~1.4) enter the top of the column when they encounter the highly pH of the pore water in the soil (pH 6-8). The heavy metals are retained almost instantly by the top part of the soil solids via precipitation mechanisms either as separate phases or in association with the carbonates and oxides/hydroxides. From the SSE analyses, the retention mechanisms of the heavy metals in soil solids can be ranked in the following order--Carbonates>Amorphous>Organic>Exchangeable.

ACKNOWLEDGEMENTS

The authors acknowledge Prof. B. K. Tan from University Kebangsaan Malaysia for his contribution during preliminary work of the study. Prof. H. R. Thomas and The Division of Civil Engineering for their continuing support. Also to the technical assistance from Mr. L. Czekaj, Mr. M Griffiths and Ms. S. Goldsmith.

REFERENCES

British Standard Institution, BS1377. 1990. *Method of Test for Soils for Civil Engineering Purposes.*

Griffin, R.A., Shimp, J.D., Steele, J.D. Ruch, R.R., White, W.A., and Hughes, G.M. (1976). Attenuation of Pollutants in Municipal Landfill leachate by Passage through Clay. *Environmental Science and Technology.* **Vol. 10** pp. 1262-1268.

Gupta, S. K., and Chen, K. Y. 1975. Partitioning of Trace metals in Selective Chemical Fractions of Near Shore Sediments. *Envir. Lett.*, Vol. 10, pp. 129-158.

Hesse, P.R. 1972. A Textbook of Soil Chemical Analysis. *Chemical Publishing Co., Inc.* New York.

Segalen , P., 1968. Note Sur Une Methode de Determination des Produits Mineraux Amorphes dans certains sols a Hydroxides Tropicaux. *Cah, Orstom Ser. Pedol.*, 6 : pp. 105-126.

Tessier, A., Campbell, P.G.C., and Bison, M. 1979. Sequential Extraction Procedure for the Speciation of Particulate Trace Metals. *Analyt. Chem.* Vol. 51, pp. 844-850.

Yanful, E.K., Quigley, R.M., Nesbitt, H.W., 1988. Heavy Metal Migration at a Landfill site, Sarnia, Ontario, Canada—2: Metal Partitioning and Geotechnical Implications. *Applied Geochemistry*, Vol. 3, pp. 623-629.

Yong, R.N., Mohamed, A.M.O & Warkentin, B.P. 1992. *Principles of Contaminant Transport in Soils.* Elsevier, New York.

Yong, R.N. & Y. Phandungchewit, 1993. PH Influence on Selectivity and Retention of Heavy Metals in Some Clay Soil. *Can. Geotech. J.* 30: pp. 821-833.

Contaminated Ground and Constructed Facilities

Land Disposal of Dredged Sediments from the Jacarepaguá Basin

M S S Almeida
COPPE, Graduate School of Engineering, Federal University of Rio de Janeiro, Brazil

L S Borma
CETEM, Centre for Mineral Technology, Rio de Janeiro, Brazil

M C Barbosa
COPPE, Graduate School of Engineering, Federal University of Rio de Janeiro, Brazil

INTRODUCTION

The Jacarepaguá Lagoon Complex, shown in Figure 1, which comprises the Jacarepaguá, Camorim, Tijuca, and Marapendi lagoons, is located west of the city of Rio de Janeiro, in a region of intense population and economic growth. Hydrodynamic studies have recommended dredging these lagoons to restore their circulation capacity and revitalize the water system. Laboratory and field studies (Almeida et al. 1998; Alves et al. 1998; Borma et al. 1998) began in 1996. Camorim Lagoon was dredged that same year, following heavy rains. Shaped much like a canal, this water body receives some 70% of the flow of rivers belonging to the Jacarepaguá Basin.

More recently, the Rio de Janeiro Mayor's Office defined dredging of Tijuca Lagoon and this hydrographic basin's main rivers as a priority within the environmental recovery of the Jacarepaguá Lowlands. Ocean disposal was deemed prohibitive not only because of the large volume to be dredged and the transportation distance involved but also because an area already available is intended for future use in commerce and urbanization. Therefore, land disposal was considered the most attractive alternative for the final destination of dredged sediments.

Field and laboratory geo-environmental studies were conducted in order to characterize river sediments as regards granulometry, mineralogy, and concentration of heavy metals and, further, to evaluate potential contamination of the selected disposal site. This article reports on the results of these studies.

SEDIMENT CHARACTERIZATION

Characterization of lagoon sediments (Almeida et al. 1998) revealed the predominance of clayey material with a high organic matter content (between 10% and 30% in weight). Particle size variation was detected in vertical profiles, with deepest layers composed of quite sandy material. Since hydrodynamic studies have defined dredging works to involve a depth of 2.0 meters, most of the sediment to be dredged will consist of sludgy material with a high percentage of fines and a high organic matter content.

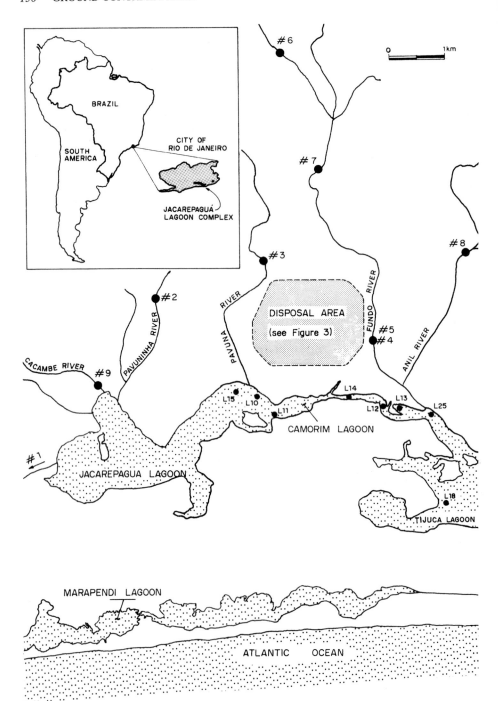

Figure 1. – Location of the sampling points in rivers and lagoons of the Jacarepaguá basin

In terms of mineralogical composition, the lagoon sediments are composed basically of kaolinite clay, filled in with partially degraded micas and large quantities of NaCl crystals. The presence of iron sulfides (FeS_2) is also significant (Borma et al., 1998). A smaller quantity of quartz and carbonates was detected, originating from the remains of mollusk shells and feldspars.

Characterization of river sediments entailed a sampling campaign involving the collection and analysis of 10 samples distributed across the main river basins (location shown in Fig. 1). Bottom material is predominantly sandy because of greater flow velocity. Organic matter content, cation exchange capacity, and plasticity indexes for river sediments are thus lower than for lagoon sediments. The ion Na^+ is less abundant, reflecting lower salinity in the river environment. As regards mineralogical composition, quartz, micas, kaolinites, and potassium feldspar predominate. Iron sulfides were detected in some of the more clayey samples, although in much smaller quantities than in samples from the lagoon environment.

CONTAMINATION OF DREDGED SEDIMENTS

The presence of heavy metals and their release to the foundation soil
The spatial distribution of heavy metals within bottom sediments in the Jacarepaguá Lowlands water system is associated not only with the discharge of industrial liquid effluents but also with seasonal variations in the flow regime for rivers and lagoons. Geochemical studies (Azevedo et al. 1987; Fernandes et al. 1994) provided evidence that metals are conveyed to the lagoon system primarily through the discharge of industrial effluents from rivers that drain areas of greater urban and industrial concentration, that is, Arroio Pavuna, Arroio Fundo, and Rio do Anil.

Chemical and metallurgical industries have been indicated as the main source of discharge of heavy metals of anthropic origin into the system (Azevedo et al. 1987; Fernandes et al. 1994). These studies show that Zinc (Zn), Copper (Cu), and Lead (Pb) are the most available metals in the environment under study. For the purposes of this paper, the following elements have also been investigated: Arsenium (As), Barium (Ba), Cadmium (Cd), Chromium (Cr), Cobalt (Co), Mercury (Hg), Molybdenum (Mo), Nickel (Ni), and Selenium (Se). The latter elements were included because the literature (IADC/CEDA, 1997) offers reference values that can serve in comparing and evaluating the contamination status of Jacarepaguá's river-lagoon environment.

Results on the presence of these 12 heavy metals in the sediments in question reveal a tendency for greater contamination of the lagoon environment than of the river environment. Figure 2 shows detected behavior for Cu, Ni, Pb, and Zn. Despite this tendency, concentrations do not differ greatly across environments. For the purpose of comparison with available criteria (presented in the next item), the maximum value obtained for each element was therefore used, whether it represented river or lagoon sediments.

Comparison with reference values
Because Brazil is still in the process of defining a national criterion for evaluating the quality of dredged sediments vis-à-vis contaminant concentrations, sediments under study were classified according to values proposed by the United Kingdom, the Netherlands, and Canada. Table 1 shows classification of each analyzed element in comparison with these reference values. Based on this evaluation, it can be concluded that detected values lie between the ranges defined as not contaminated (A) and slightly contaminated (B).

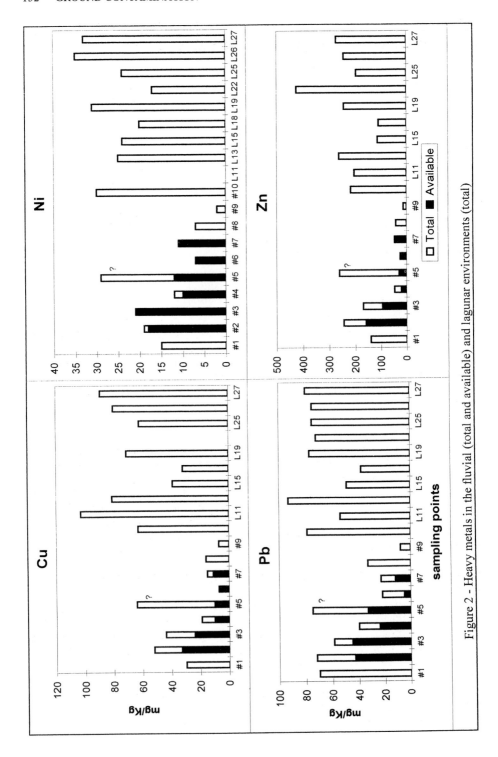

Figure 2 - Heavy metals in the fluvial (total and available) and lagunar environments (total)

Table 1 – Comparison with criteria from three countries (IADC/CEDA (1997)

Element	United Kingdom	Netherlands	Canada
Ar	< A	<A	<B
Ba	B	>C	-
Cd	A	A	B
Cr	< A*	<A	B - C**
Co	-	<A	-
Cu	A	<B	B - C**
Hg	A	A***	B***
Mo	-	A - B	-
Ni	A – B	<A	B - C**
Pb	< A	A - B	B - C**
Se	A– B	-	-
Zn	B*	B	B - C**

A: not contaminated; B: slightly contaminated; C: highly contaminated
* Reference values established for the potentially mobile concentration (mg/Kg);
** Reference values established for total concentration divided by total organic carbon concentration (μg/g of organic carbon);
*** Values determined in present study lie below the detection level of atomic absorption equipment (1 mg/Kg)

Comparison with background values

Another way of evaluating the contamination status of a given lagoon environment is by reference to background values. Although research of background values for the Jacarepaguá Lowlands water bodies lies outside the scope of the present study, an analysis of Figures 1 and 2 lends some idea of these levels. These figures show that more distant points on the rivers (#6 through #9) display much lower values of Cu, Ni, Pb, and Zn than values detected at other points in river and lagoon environments. This distribution offers clear indications that heavy metals are being released from anthropic sources.

Total concentration versus available concentration

Another aspect that should be kept in mind regarding heavy metals concentration is the difference between total concentration and available concentration. The available fraction, which is also considered that fraction originating from anthropic sources, is more easily released upon disposal. This value is usually lower than total concentration, since the fraction of metals originating from the rock-producing sediment is usually not released through simple leaching with weak acids. Total concentration and available concentration were compared for the samples #2 to #7. As can be seen in Figure 2, most results show a tendency for available concentration values to be lower than total concentration, as expected.

EVALUATION OF SEDIMENT QUALITY FOR THE PURPOSES OF LAND DISPOSAL

Based on the results above and the criteria applied, although detected contamination levels are higher than local background values, this level is not high enough to preclude land disposal. Furthermore, sediments display a high percentage of fines and a high organic matter content, lending them a good capacity to retain metallic ions. In addition, adopting a conservative approach, the following two aspects were not taken into consideration during comparison with reference values:
a) the mixture, which takes place during dredging and disposal, between more and less contaminated sediments leading to the dilution of observed concentrations;
b) the fact that total concentration was used instead of available concentration.

The presence of iron sulfides
One aspect inherent to the lagoon sediment under study, which does not favor land disposal, is the significant presence of iron sulfides. These substances have been reported as the main culprit in acid drainage of mining residues. In such situations, it is not rare to observe acidification of surface or underground waterways at extremely low pH levels (around 3.0 or even lower) along with a high concentration of heavy metals. In a recent study, Borma (1998) and Borma et al. (1998) showed that in the case of Jacarepaguá lagoon sediments, oxidation of iron sulfides arising from land disposal may prompt acidification of sediments due to simple exposure to the oxygenated atmosphere and may consequently release heavy metals.

It should be stressed that in the present case, this concern with acidification is not related merely to the release of heavy metals since, as mentioned earlier, critical levels of the latter are not detected. The major problem lies in the possibility of altering the pH of surface waterways lying adjacent to the area, especially Arroio Fundo and the lagoons themselves.

In relation to the formation of acid drainage, the process of oxidation of iron sulfides and its effects depend greatly on environmental conditions at the disposal site. The possibility that the produced acidity will reach surface or underground waterways depends upon the characteristics of the foundation soil and upon the flow tendency of the water that percolates through the oxidized sediment. Given existing uncertainties, plans foresee construction of an experimental cell for in situ monitoring of these processes and evaluation of the efficiency of any ameliorative measures that may prove necessary.

GEOTECHNICAL STUDIES REGARDING LAND DISPOSAL OF SEDIMENTS
The site selected for disposal is shown in Figure 1 and in greater detail in Figure 3. In accordance with the local urbanization plan, this area lies about 3.0m below the minimum elevation, which means that land owners need a considerable amount of fill to bring their sites up to minimum occupation standards. Considering that river sediment tends to be sandier than lagoon sediment, in order to deposit these dredged sediments on land, plans call for the construction of dikes built of river sediment to contain lagoon sediments. The distribution of these dikes has been devised so as to coincide with the street layout designed for the location. Figure 3 shows the distribution of dikes and basins for lagoon sediment disposal.

Of the total available area (i.e., some 375 ha), only 70% will be effectively available for disposal of dredged material. The remaining 30% include areas that present outcroppings of sandy soil, which displays a low capacity to retain contaminants. The effective disposal area, shown in Figure 3, will be divided into cells. The lagoon sediments will be discharged directly into each cell via a movable, flexible pipeline. After sedimentation, floating water will be directed back to the lagoons via the Arroio Fundo.

Based on analysis of borings, most of the chosen disposal site displays layers of soft organic clay with thickness varying from 2m to 12m, most being around 4m thick. Conventional geotechnical stability and settlement studies indicate the need for dikes of up to 1.8m in height. The soft foundation clay will settle under the load of dikes and sediments, with the magnitude of this settlement depending upon the thickness of the clay layer. Suitably placed settlement plates will indicate when it is necessary to correct dike height over time so as to maintain design level and thus hold in lagoon sediments.

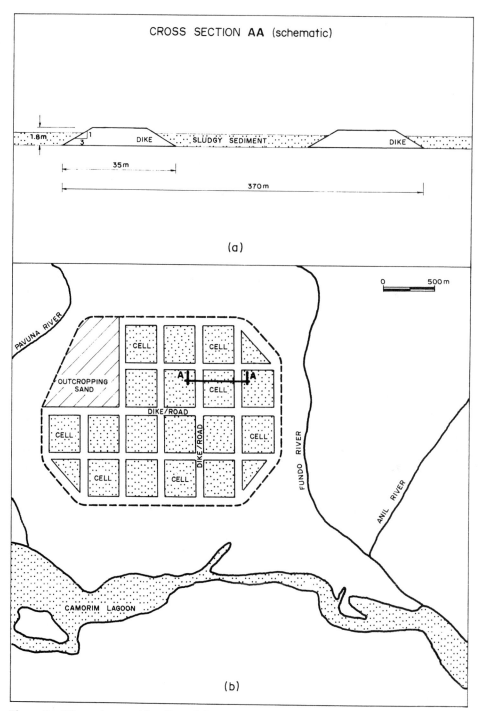

Figure 3 – Disposal area with cells and dikes

CONCLUSIONS

The following are the main conclusions regarding the disposal of lagoon and river sediments to be dredged from the Tijuca Lagoon and neighboring rivers:

- Due to a high percentage of organic matter, material to be dredged from the lagoon environment does not differ substantially from the foundation material at the disposal site, composed basically of peat and soft clay with a high organic matter content and low support capacity;

- As regards the concentration of heavy metals, although some of these substances are discharged into the river-lagoon environment from anthropic sources, the level of concentration does not render land disposal infeasible;

- One aspect of land disposal that merits further research is the fact that when seawater penetrates an aquatic system and oxygenation conditions are low, this system is vulnerable to the formation of iron sulfides. When exposed to the oxygenated atmosphere, these substances tend to acidify their surrounding environment, increasing the likelihood that heavy metals will be released into neighboring aquatic environments. In this case, the problem affects lagoon sediments more than river sediments.

REFERENCES

Almeida, M. S. S.; Barbosa, M.C.; Castro, F.J.C.O.; and Alves, M.C. (1998) "Characterization of the sediments of Jacarepaguá Lagoon for dredging purposes," Int. Congress on Environmental Geotechnics, Vol. 1, p. 69-76, Lisbon, September 1998.

Alves, M. C. M., F. J. C. O Castro, Santos, A. C. B. 1998, A physico-chemical process for the improvement of the sedimentation behaviour of a dredged sediment Int. Congress on Environmental Geotechnics, Vol. 2, p. 455-459, Lisbon, September 1998.

Azevedo, H.L.P; Monken, H.R.; and Melo, V.P. (1987) "Study of heavy metal pollution in tributary rivers of Jacarepaguá Lagoon, Rio de Janeiro State, Brazil, through sediment analysis," in *Heavy Metals in Coastal Environments of Latin America* (Berlin: Springer-Verlag, pp.21-29).

Borma, L.S. (1998) Acidification and release of heavy metals in dredged iron sulfide rich sediments (in portuguese), DSc thesis, COPPE/UFRJ, Rio de Janeiro.

Borma, L.S.; Ehrlich, M.; Barbosa, M.C.; and Cabral, A. (1998) "Land disposal of dredged iron sulfide rich sediments," Int. Congress on Environmental Geotechnics, Vol. p. , Lisbon, September 1998.

Fernandes, H.M.; Conti, L.F.C.; and Patchineelam, S.R. (1994) "An assessment of the pollution of heavy metals in Jacarepaguá Basin, Rio de Janeiro, Brazil: a statistical approach," Environmental Technology 15: 87-94.

IADC/CEDA (1997) Environmental aspects of dredging – Conventions, Codes and Conditions: Land Disposal, Guide 2B, The Netherlands, March, 1997.

Case study of heavy metals in estarreja (Portugal)

Dr. Carlos Costa
University New of Lisbon
Campus Faculdade Ciências e Tecnologia
2825 Monte de Caparica- Portugal
Tel. 351-1-2948580
Fax. 351-1-2948398
e-mail: cnc@mail.fct.unl.pt

Eng. Cláudia Jesus
University New of Lisbon
Campus Faculdade Ciências e Tecnologia
2825 Monte de Caparica-Portugal
Tel. 351-1-2948580
Fax. 351-1-2948398
e-mail: caj@mail.fct.unl.pt

ABSTRACT:

Due to five decades tradition on chemical industry, Estarreja is one of the most important industrial areas in Portugal. Intensive industrial activity along with both direct discharges of the effluents into natural water-streams and waste disposal on the ground has, throughout the years, had strong impact on health and welfare.

Recently it was created an association between industry and local authorities – ERASE. The main objective of this association is to find, in co-operation with the Portuguese Environmental Ministry, a cost-effective solution to solve the environmental problem.

Site investigation carried out during September/November 1998 took place both within the industrial area and along the natural water-streams. Pollution inside the industrial area is manly due to uncontrolled disposal of industrial wastes, containing high quantities of pyrite ashes and dust and sludge with mercury.

The results revealed high concentrations of heavy metals in the soil of the Chemical Complex (namely As and Hg, also Pb and Zn) but mainly in sediments of the water-streams. Here, the quantities of pollutants are far above the ones found within the industrial area.

Site investigation included geotechnical laboratory and in situ testes on the sediments in order to evaluate the suitability of the sediments to compaction. Studies carried out didn't validate the proposed ERASE solution to dispose the contaminated soil together with the sediments from the water-streams into a landfill.

It was concluded that, at the present state of knowledge, it doesn't seem advisable to remove the sediments from the main water-stream. Further studies should be undertaken. Meanwhile landfill should only confine pyrite wastes, sludge with mercury and sediments from the smaller ditches.

1. INTRODUCTION

Awareness of the contaminated land and groundwater problem has been enhanced in Portugal for almost twenty years mainly due to Estarreja chemical industrial activity. Nevertheless we dare say that, as far as Estarreja contamination problem is concerned, only now, and after a long period of tackling hesitation, the subject is starting to approach its growth of maturity.

Like in many other soil and groundwater contamination problems, the first basic question to be answered for Estarreja is "what contamination is present? how far? how deep?", which gives rise to the site investigation issue. This paper deals mainly with this question but it also bares in mind that, after site investigation problem, three other problems remain to be solved: the migration problem "where are the contaminants going to?", the impact problem "is the impact significant?" and the remediation problem "what can be done about it?". All of them should be seen in the context of the "source-path-target" methodology risk analysis (Petts et al., 1997; Loxham, 1998).

2. THE REGION
Unlike the majority of the rivers that flow into the Atlantic Ocean, the river Vouga meets the sea by way of great expanses of marshes and lagoons. It's estuary is more like a delta, ranging from Ovar, to the North, till Costa Nova and Vagos, to the South, 30 km far. Aveiro, the district capital, lies SE, some 5 km away from the coast, while Estarreja, the industrial regional pole, is located 10 km to the North of Aveiro and 15 km distant of Vouga mouth, artificially opened almost every year.

This large region of more than 500 km^2 is known to be one of the most beautiful natural regions of Portugal – the Ria de Aveiro region, famed for its wildlife. It is also very rich in terms of fishery and agricultural resources. Together with industry, these activities are able to support nearly a quarter of a million people, with one of the lowest ratios of unemployment in this country. However, health and welfare are now clearly threatened: urbanisation and industrialisation of the last 50 years imposed unbearable stresses over the lagoons and the wetlands, with both direct discharges of domestic and industrial effluents into natural water-streams and toxic waste disposal in the ground.

3. THE INDUSTRY
Most part of heavy industry is located inside the "Complexo Químico de Estarreja – CQE" (*Estarreja Chemical Complex*). This industrial area has 2 km^2 and is 1 km away from the town of Estarreja. The most significant industrial units, working for many decades, are:

- *Quimigal*, installed in 1952, produced ammonium sulphate, from sulphuric acid and ammonia, since the beginning, and nitric acid and ammonium nitrate since 1974. These productions stopped in the early nineties. In 1978 started a new unit (*Anilina de Portugal*) for the production of nitric acid and aniline and mononitrobenzene, which is still running.
- *Uniteca*, working since 1956 for sodium and chlorate compounds from rock salt through electrolytic cells, using graphite and mercury cathodes.
- *Cires,* installed in 1963 to produce synthetic resins, mainly PVC (polyvinyl chloride) from vinyl chloride monomer (VCM). This raw material was also produced in this plant till 1986.
- *Dow Portugal*, producing since 1978 isocyanide polymers of aromatic base.

These industrial units are greatly interdependent for the exchange of raw materials and subproducts (Fig. 1.)

Over the working decades these industrial units rejected huge quantities of waste, both solid waste, which was stocked directly on the soil, and liquid effluents, to ditches and water streams.

a) **Solid waste production** - according to last estimates (WS Atkins, 1997) the following quantities are stocked inside the Chemical Complex:
- 150,000 tonnes of pyrite wastes, ashes, dusts and sludge from gaseous effluents treatment, containing heavy metals (As, Pb, Zn, Cu and others), from Quimigal production;
- 60,000 tonnes of sludge containing Hg, from Uniteca;
- 300,000 tonnes of sludge with calcium hydroxide, thought to be inert, from CIRES.
-

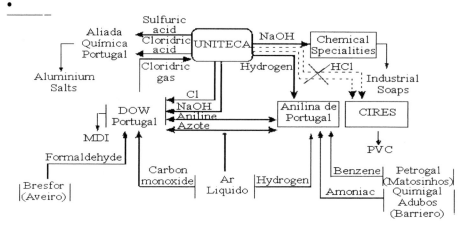

Figure 1 – Industrial Units at CQE and their products (modified after WS/Atkins Portugal, in Validação da solução proposta pela ERASE para a recuperação e regeneração de águas e solos contaminados, 1997)

b) **Liquid Effluents** - Until 1975 the liquid effluents were discharged directly to non impervious ditches (Vala de S. Filipe, Vala da Breja and Vala do Canedo), transporting for several kilometres through the agricultural fields a number of contaminants to the main water streams, especially to the river branch named Esteiro de Estarreja.

After 1975 an emissary pipe was constructed for Quimigal/Uniteca effluents; Dow Portugal has its own pipe, both discharging in Esteiro de Estarreja. Cires unit maintains discharges to Vala da Breja. The main liquid effluents should contain: aniline, benzene, monochlorobenzene, mononitrobenzene, arsenic, mercury, zinc and lead, among others. Figure 2 shows the location of CQE and the solid waste deposits and also the pathway of the ditches and pipes.

4. THE ENVIRONMENT
4.1Geomorphology and Hydrology
CQE lies on a plain with slopes less than 1%. The more affected water streams, all of them draining to Ria de Aveiro, are:
- river Antuã, Esteiro de Estarreja and Vala de S. Filipe, at South, flowing from NE to SW, till Ria de Aveiro, both at Largo do Laranjo (5 km SW of CQE);
- rio Fontela and Vala da Breja, at North, draining to Largo da Coroa (5km NW of CQE)

There is also Vala de Canedo, draining to a inner basin, the Veiros lagoon, an environmentally interesting natural water unit.

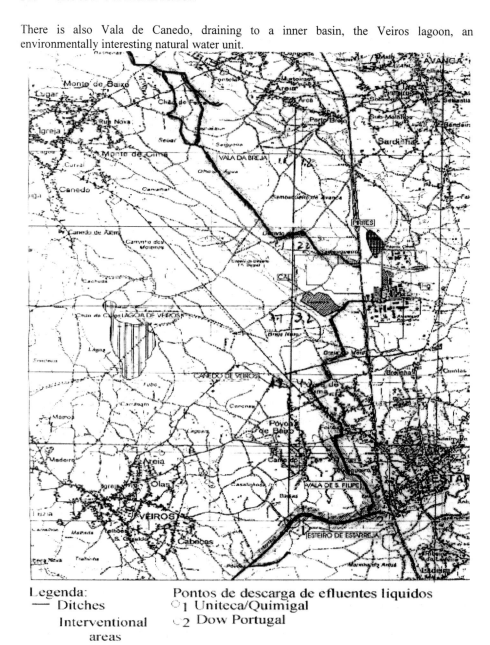

Legenda:
— Ditches
   Interventional
   areas

Pontos de descarga de efluentes líquidos
○1 Uniteca/Quimigal
○2 Dow Portugal

Figure 2 – Identification of the industrial units, solid waste deposits and pathway of the ditches and pipes.

4.2 Geology
CQE is located near the northeastern edge of the Aveiro sedimentary basin, were quaternary deposits are no more than a dozen meters thick, covering the Precambrian schistous bedrock.

As the structure is plunging to the West, these sediments may attain several tens of meters, above the Cretacic formation, composed by interstratified conglomerates, sandstones, shales and marlstones.

In the Quaternary formation three lithologic complexes may be individualised:
• The upper complex, composed by dune sands and alluvium silty-clayey sands, covered by organic topsoil;
• The intermediate complex, very heterogeneous, with irregular geometry, due to sudden vertical and horizontal geologic variations;
• The lower complex, formed by loose conglomerates.

4.3 Hydrogeology
Following the geological structure described above three Hydrogeological Units can be defined:
• The Quaternary Unit, installed in sandy loose formations,
• The Cretacic Unit, composed mainly by sandy and carbonated consolidated formations (linked to the Jurassic Unit, below, absent in the vicinity);
• The Precambrian Unit, unproductive and practically impervious.

The Quaternary and the Cretacic units form together an important aquifer system, which can by divided into two subsystems, the Quaternary subsystem and the Cretacic subsystem, with a rather independent functioning. The Quaternary subsystem occupies a total area of 650 km^2, with an average thickness of 40m, thickening from E to W and N to S.

Between the sandy formations at the top and the conglomerates at the base (see geological description) some interstratified beds of mud may function as aquitards, lowering the hydraulic conductivity and making difficult the connection between the confining units. This may give place to the subdivision of the Quaternary subsystem: the upper unit, superficial, hydraulically free, and the lower unit, semi-confined. However, due to the marginal position of CQE in the hydrogeological Quaternary subsystem context, it is possible that, inside the CQE area, this division does not exist.

According to studies performed by different authors (Cristo, 1985; LNEC, 1992) the direction of regional groundwater flow, at least for the superficial aquifer, seems to be E-W and NE-SW, but significant variations may occur locally, which call for an extended groundwater monitoring programme, yet to be implemented.

4.4 Agricultural uses
In Ria de Aveiro region every household used to have a well for its own drinking and agricultural supply. These wells rarely exceeded 20 to 25 m deep, exploiting the superficial aquifer. High permeability of quaternary deposits and shallow groundwater table made the aquifer extremely vulnerable to agronomic, urban and industrial pollution. Eventually, the presence of nitrates, chlorates and heavy metals made inadvisable the domestic supply from this source. Nevertheless, and in spite of some casualties, namely among cattle, it is not uncommon the use of water from shallow wells for agricultural and even drinking supply.

5. SOIL AND WATER CONTAMINATION STUDIES. THE *ERASE* SOLUTION
There is a strong conscience among the scientific community, at least since the early eighties, that the Ria de Aveiro ecosystem is heavily polluted, both in terms of soil and sediments as well as water streams and groundwater.

Most of the previous studies were focused on mercury in water (Hall *et al.*, 1987, 1988), following the "Minimata Disease" investigation phenomena around the world. Even more recently (Pereira, 1997) very high mercury concentrations were found in Esteiro de Estarreja waters dissolved fraction: up to 8 000 ng/dm^3. Mercury in soil, sediments and groundwater was detected by Barradas (1992); Ferreira (1993) studied mercury in soils inside CQE. High levels of arsenic were also found soil and sediments around Veiros lagoon (Fonseca et al., 1995).

The need for a systematic approach of contaminated soil and groundwater aiming the treatment and clean-up of land, water streams and the aquifer affected by CQE industrial pollution, became clear, particularly after the LNEC 1992-1994 studies. An association between industry and local authorities – ERASE, was then created with a main objective: to find, in co-operation with the Portuguese Environmental Ministry, a cost-effective solution for this environmental problem.

The solution proposed by ERASE in 1997was to design and built a landfill in the existing disposal area of pyrite wastes, ashes and dusts from Quimigal. Together with this material it would be confined the sludge containing Hg, from Uniteca and all the contaminated soils lying underneath. Soil and sediments from ditches and *esteiro*, would also be disposed and sealed in the landfill - Fig. 3

To validate the solution a site investigation programme was designed in order to eliminate the uncertainties, namely:
- Exact volume of existing wastes and underlying contaminated soils to be excavated and confined inside CQE;
- Exact volume of contaminated soils and sediments to be removed from ditches (8.5 km extension) and *esteiro* (only the first 2 km) and confined together with the wastes;
- Physic and chemical characteristics of contaminated soils and sediments, including workability in terms of disposal in landfill.

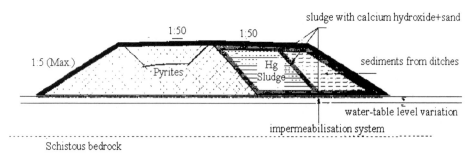

Figure 3 – Sketch of the of the possible landfill (modified after WS/Atkins Portugal, in Validação da solução proposta pela ERASE, 1997)

6. SITE INVESTIGATION
6.1 The site investigation programme
The site investigation programme was carried out from September to November 1998, as follows (Quimitécnica/Engil, 1998):

a) Area of pyrite wastes
- 25 exploratory trial pits above groundwater level were undertaken. In each pit it was collected 2 soil samples (bellow the pyrite wastes) at different levels. All the soil samples were submitted to the following chemical analyses: determination of As, Zn, Pb, Cd and Cu content.
- Additionally3 exploratory trial pits were made, to investigate contamination "hot spots".

b) Area of sludge piles with mercury
- In this area 20 exploratory trial pits above groundwater level were undertaken. In each pit 3 soil sample (bellow the wastes) at different levels were collected and submitted to Hg and NaCl chemical analysis.
- 5 drillings were also made for the geological understanding of the underlying strata.

c) Esteiro de Estarreja
Along the upstream 2 km *esteiro* 10 transversal section were designed; in each 7 exploratory trial pits were undertaken in the following manner:
- 3 pits located in the water stream bed;
- 4 pits located 2 in each water stream margin;

In each water-stream bed pit 3 soil samples at different levels were collected. In each water stream margin pit only 2 soil samples at different levels were collected. A total of 170 soil samples were submitted to the following chemical analysis: As, Hg, Pb, Zn. The particle size distribution, organic content and pH were also determined.

Sediments excavated from 3 transversal sections representative of the different particle size distribution (200 m^3 in total) were subjected to geotechnical laboratory and *in situ* tests in order to investigate the suitability of the sediments to compaction in the landfill.

d) Ditches (Vala de S. Filipe/Vala da Breja/Vala do Canedo)
A total of 15 transversal sections with 5 exploratory trials pits each gave a total of 170 soil samples were As, Hg, Pb and Zn determinations were carried out.

6.2 Some comments on the results
Site investigation and chemical analysis results revealed, with the exception of Uniteca area, a much rather pessimistic scenario than it was foreseen.

a) Area of pyrite wastes
Severe and generalised contamination with As (several samples above 1 g/kg) was detected, increasing with depth and reaching the groundwater table. Values increase from NE to SW, following the direction of groundwater flow. Pb contamination is also severe but follows an erratic pattern (hot spots?).

b) Area of sludge piles with mercury
Contamination with mercury seems to be confined to the waste piles and underlying soil, with a maximum of 30 mg/kg.

c) Esteiro de Estarreja
- Extremely severe and generalised contamination with As, with several samples above 10 g/kg. 2/3 of the As analysis was above 50 mg/kg. Also extremely severe but not so generalised contamination with Hg, with several samples above 1 g/kg. More than half of the

Hg analysis was above 10 mg/kg. Important contamination levels with Pb, also, and less important contamination with Zn.

• The more contaminated was the sediment with a heavy metal (for example As) the more probable is the occurrence of other heavy metals with significant high concentrations. The sequence for the occurrence of high concentration levels of heavy metals in the sediments seems to be: As, Hg, Pb, Zn.

• At some sections the pattern of contamination, unequal from right to left side of the section, seems to be controlled by erosion/depositional phenomena.

• In terms of spatial horizontal distribution, contamination in the bed of the *esteiro* seems to increase from section 1 (upstream) to section 4; is more or less constant (at higher levels) between section 4 to section 7; then apparently decreases from section 7 to section 10 (downstream). The existence of a "sandy channel" inside the sediments between sections 1 to 4 may partially justify this pattern. Then the sandy channel closes and gives place to a finer fraction (silty-clayey) after section 6, preventing contamination from progressing downstream. Moreover, the lower gradients at the central sections may be responsible for lower flux velocity, dominance of the finer fraction of the sediments and higher concentration of heavy metals.

• Contamination seems also to increase with depth in the more severely contaminated sections, which contradicts Hall et al. (1987) and confirms Pereira et al. (1995). Great variability in nature and content of contamination in depth should also be seen as a result of constant changes at the source in nature, quantities and treatment effectiveness of the pollutants. The fact that the lower strata are more polluted may indicate that treatments are more effective in recent times that they were before.

• At the margins contamination doesn't tend to decrease with distance but it does with depth. The pattern of contamination seems to be controlled by occasional dredging and spreading on the fields.

• Geotechnical laboratory and *in situ* tests showed that the sediments are not suitable to dispose in a landfill, mainly due to the high values of water content, organic matter and plasticity of the finer fraction.

c) Ditches (Vala de S. Filipe/Vala da Breja/Vala do Canedo)
Contamination is not relevant except in the 1.5 to 2 km downstream S. Filipe ditch. Here the pattern is similar to Esteiro de Estarreja.

7. CONCLUSIONS
Site investigation greatest impact on the previous solution proposed by ERASE results from the fact that, neither horizontally nor vertically, could reasonably low levels of contamination be spatially defined at Esteiro de Estarreja. Here, contamination is enough high below 1.3m depth, in the water-stream bed, as well as outside a zone of 6m width, at the margins.

Esteiro de Estarreja has levels of contamination in As and Hg sometimes one order of magnitude and more, greater than those found near the source, some 5 km upstream. Excavation of contaminated sediments at Esteiro de Estarreja without reaching reasonable low levels of contamination in depth means to uncover highly polluted sediments and allow them to be removed by the tides in an newly unbalanced environment. Further studies reaching deeper and wider zones must be made in order to define a more accurate scenario for the *esteiro*.

At the present state of knowledge it doesn't seem advisable to remove the sediments from *esteiro* and the last 2 km of S. Filipe ditch. Landfill should only confine pyrite wastes, sludge with mercury and sediments from the smaller ditches. It is also of doubt to remove soils below pyrite wastes before the construction of the landfill. Total impermeabilisation of the foundation may be technically ineffective and/or economically unacceptable. In order to prevent continuous flux of water through the contaminated soils a cement-bentonite cut-off wall combined with migration interception techniques such as drainage trenches might be a good solution to minimise impacts on groundwater.

8. REFERENCES

Barradas, J. (1992) – Geoquímica de Elementos Maiores e Vestigiais em Amostras de Solo, Sedimentos de Vala e Águas Subterrâneas. Contribuição para a Caracterização Ambiental da Zona Envolvente ao Complexo de Estarreja, PhD Thesis, Univ. Aveiro.

Chambino, M. T. (1998) - Groundwater Contamination in Portugal: Overview of the Main Problems. Nato/CCMS Pilot Study. Evaluation of Demonstrated and Emerging Technologies for the Treatment and Clean-Up of Contaminated Land and Groundwater. EPA 542-R-98-001a. June 1998.

Cristo, F. P. (1985) – Estudo Hidrogeológico do Sistema Aquífero do Baixo Vouga, Min. Equip. Soc., DGRAH.

Ferreira, M. M. (1993) – Mercúrio em Solos da Àrea do Complexo Químico de Estareja. MSc Thesis, Univ. Aveiro.

Fonseca, Cardoso E., Barradas, J. M.; Ferreira da Silva, E.; Reis, A. P (1995) – Distribuição de Arsénio em Solos e Sedimentos da Vala na Envolvente do Complexo de Estarreja. Implicações de Ordem Ambiental, Geociências, Rev. Univ. Aveiro.

Hall. A., Duarte, A. C.; Caldeira, M. T.; Lucas, M. F. (1987) – Sources and Sinks of Mercury in the Coastal Lagoon of Aveiro, Portugal. Sci. Total Environ. 64 (1-2), 75 –87.

Hall. A., Duarte, A. C.; Oliveira, J. P.; Lucas, M. F. (1988) – Particulate Mercury in the Coastal Area off Aveiro, Portugal. In: Heavy Metals in the Hidrological Cycle, M. Astruc and J. N. Lester (Eds.), Selper Ltd, London, 507-512.

LNEC (1992) – Metodologias para a Recuperação de Àguas Subterrâneas e Solos Contaminados. Partes A e B.

Loxham, M. (1998) – Methodologies for development Perspectives in Contaminated Land. NECER Meeting at Lisbon, LNEC, May 1998.

Pereira, M. E.; Duarte, A. C.; Millword, G. E. (1995) – Seasonal Variability in Mercury Inputs into the Ria de Aveiro, Portugal. Netherlands Journal of Aquatic Ecology.

Pereira, M. E.; Duarte, A. C. (1997) – Contaminação da Ria de Aveiro com Mercúrio. Indústria da Água nº 24, Julho-Setembro 1997, pg. 47-57.

Petts, J.; Cairney, T.; Smith, M. (1997) – Risk-Based Contamination Land Investigation and Assessment. John Wiley & Sons.

Quimitécnica/Engil (1998) – Trabalhos de Levantamento Preparatórios na Zona do Complexo Químico de estarreja. Relatório Final.

WS Atkins (1997) – Validação da Solução Proposta pela ERASE para a Recuperação e Regeneração de Águas e Solos Contaminados. Relatório Final.

Trials for the construction of a cement solidified retaining structure in a domestic landfill site using deep soil mixing

M.S. DE SILVA,* N.J. O'RIORDAN** & L.N. PARRY*
* Bechtel Ltd., London. **Ove Arup & Partners, London
The authors undertook this work as part of the CTRL design and construction within the Rail Link Engineering consortium. This paper is published with the permission of Union Railways.

INTRODUCTION

This paper describes the use of a cement stabilization and solidification process in a 10 year old domestic refuse waste disposal site in Kent. Although there is now a considerable body of literature on cement stabilization and solidification, much of it originating in the USA, its use in the UK has been limited (Al-Tabbaa and Evans, 1998a). A review of the literature and consultation within the industry has established that the technique has not been used in material of this nature previously although widely used in the stabilization and solidification of contaminated soils (Al Tabbaa and Rose, 1996).

The authors have recently undertaken trials on a variant of the process known as Deep Soil Mixing within a domestic landfill site which lies along the alignment of the Channel Tunnel Rail Link (CTRL) as it passes through the Parish of Harrietsham adjacent to the M20 in Kent, UK (see Figure 1). The technique if successful will be used to construct a retaining structure approximately 200m long, 15m high and 10m deep of cemented domestic waste. The CTRL main line will run a few metres from the foot of the retaining structure. The site trial was undertaken by Keller Ground Engineering.

Prior to trialing the technique a number of options were evaluated for the construction of the CTRL through this site. The alignment is such that CTRL track elevation is at approximately 12m below ground level as it passes through the site, with the depth of waste extending for up to 2m below track level. Construction of an open cut through the site would result in significant disturbance to the site, which combined with land ownership issues would require the purchase of the site and excavation and removal of the entire waste body to an alternative landfill, whilst construction of a more conventional retaining structure would require a high cantilevered structure necessitating deep piling into an unprotected aquifer.   Deep soil mixing, if successful, would offer considerable cost advantages over these other techniques.

This paper gives details of the landfill site and the local geology and hydrogeology, and the construction of trial blocks and columns within the site using several different cement mixes. Details will also be given of the testing regime and the technical constraints of using this technique in these difficult ground conditions and how they might be overcome.

This section of the CTRL route is underlain by strata belonging to the Lower Cretaceous system of Mesozoic Age and the following formations outcrop. The strata generally dip gently to the north-east.

Formation	Description	Thickness	NRA Aquifer Classification
Folkestone Beds	Fine to medium sands with some clay	50 - 55m	Major aquifer
Sandgate Beds	Sands, silts, and clay	5 - 20m	Minor aquifer

Ground conditions on the site were proved by cable percussion boreholes and trial pits (Figure 1), and can be described as follows:-

Capping Layer. The capping layer is slightly domed and was found to be of variable thickness from 1.4 - 4.8m thick (averaging 2.7m from thirteen locations) and generally thinner on the western edge of the site. The material is generally described as soft to firm clay or sandy clay.

Domestic Refuse. The sandpit, now backfilled with domestic waste and topped by a capping layer, covers the majority of the site. In the south-west of the site where an attenuation or buffer layer of sand and clay has been placed between the M20 cutting and the landfill, the domestic waste is not present (Boreholes SA1651 and SA1652). At borehole SA1651, the made ground comprised a 6.3m layer of generally medium dense sand and SA1652 records a 6.5m layer of soft to firm clay.

Putrescible domestic refuse between 4.8 - 9.5m thick (where penetrated) was identified beneath the capping layer. The base of the refuse is recorded in three locations at 95.5, 94.1 and 92.9m AOD and appears to be sloping down from west to east. The refuse is variable, but contains a considerable quantity of undecomposed to slightly decomposed wood and putrescible materials.

Methane is being generated and the gas is being drawn off through a system of pipes to a gas flare located on the site.

Attenuation Layer. Below the domestic refuse over most of the site, there appears to be a basal attenuation layer separating the waste from the natural ground. The thickness of this layer was proved in two locations (SA1654 and SA1655) to be 2.3 and 3.2m thick respectively. The material is generally described as sand or gravel, and occasionally as a clay.

Folkestone Beds. The top of the Folkestone Beds, described generally as a fine to medium sand, was encountered in five locations, two of which are directly below the domestic refuse. The interface appears to slope from about 93.5m AOD in the west to about 92m AOD on the east side of the pit with local variations, in particular the level drops to around 89m AOD in the western corner of the site. – see cross section Figure 2.

One borehole (SA1683) was located at the edge of the tip. The upper 1.7m of the hole showed the interface between natural ground and the tip material. This hole showed the natural succession to be 1.7m of silty sand (possibly Head deposits) over Folkestone Beds sand which was proved to a depth of 21.0m bgl (81.2m AOD).

Sandgate Beds. The Folkestone Beds are underlain by the Sandgate Beds, generally described as laminated very silty clay and were encountered by a single borehole (SA1652) at a depth of 14m bgl (84.05m AOD). Boreholes drilled under other contracts (Refs. R15 and R16) to the west, south and east of the site, indicate that the surface of the Sandgate Beds dips at approximately $1°$ to the west.

Groundwater. Groundwater was encountered as a seepage during the boring of SA1654 within the made ground, suggesting localised perched water. However there is no evidence to suggest that it extends across the site.

Piezometers installed in four boreholes monitored between 1994 and 1996 show the standing water level is below the base of the made ground at between 87.79 and 88.69m AOD. A borehole terminating within the made ground indicated a water level at 97.08m AOD in December 1996.

Four variable head permeability tests were carried out in the Folkestone Beds. The results ranged from 1.3×10^{-6} to 7.0×10^{-6} m/s, with an average 3.5×10^{-6} m/s. The results of one of the tests has been discounted as the test section included an interface between adjacent strata.

Gas Monitoring. Gases were recorded from six boreholes. In general the boreholes which penetrated the domestic refuse showed high levels of flammable gases (measured as equivalent methane) even after being allowed to settle for several days (29 and 37% by volume for SA1655 and SA1654 respectively). A level of 61% by volume of flammable gases was recorded in AC1656 during boring; the hole was abandoned when this level did not significantly reduce after one hours venting. These levels were recorded accompanied by readings of low oxygen and high carbon dioxide concentrations.

Borehole SA1686 did not exhibit elevated flammable gas levels despite its central position adjacent to AC1656.

The two boreholes on the south-western boundary of the site, adjacent to the M20, have not indicated the presence of any landfill gases.

The Mixing Trials

The trial is required to investigate the practicality of installing soil mix grout columns, an established treatment for conventional soils, within domestic landfill material. The domestic landfill is producing gas that is flared off approximately 100m north east of the trial area. The latter is in an area where no gas abstraction wells are located.

Soil Column Installation

The objectives of the operation were as follows:-

- That the columns can be installed to the maximum depth anticipated i.e., 17m.
- That an interlocking structure of columns can be successfully constructed within the domestic waste.
- To experiment with different grout mixes with a specified UCS of $10N/m^2$ at 28 days and a permeability of $<10^{-9}$ m/s.

To achieve the above the following was proposed:-

1. Install six number columns of 0.9m diameter through the full depth of the waste and 1m into the underlying natural strata at various locations across the site.
2. Construct a trial block measuring approximately 2.4x2.4m of 25 interlocking columns of to 10m depth using a variety of grout mixes (Block A).
3. As Block A above but using 20 piles clustered around an untreated core (Block B).

The configuration of these blocks is depicted in Figures 3 and 4.

Measurements of grout take for all soil-grout columns was recorded and onsite testing was undertaken using a Marsh cone and samples taken for cube strength which was recorded after 7 days and 28 days.

Method of Work

The work was undertaken using a Soilmec R16 set up for the following parameters:-

Soil mix tool diameter 0.9m
Rotation rate 2-20rpm

Insertion/extraction rate 0.25-0.5m per min.
Grout flow 20-100 litres per min.

Specific records were kept for the installation of each column to enable
repeatability of a successful permutation of parameters.

The soil-mixing tool consisted of a 0.9m diameter cutting tool comprising two
shallow angled blades mounted at 180 degrees which upon rotation cut into the
ground at an insertion rate controlled by the drilling rig. Grout injection ports
are located at the trailing edge of the blades. Below the mixing-tool was a 1m
length of auger to help direction and pull down of the mixing tool.

The mixing tool was mounted on a modified Kelly bar which was attached to a
crawler mounted drill rig, with the rotation rate and insertion/lift rate controlled
via cab mounted controls on the drilling rig.

The mixing tool was positioned over the centre of the designated column and
the tool rotated into the ground whilst grout was injected at the tool. Mixing
continued until the required depth was achieved or refusal was met.

Grout flow was adjusted to suit the insertion rate achieved and therefore reduce
the level of surface run-off. Due to the slow insertion rate generally the
minimum pumping rate was used. Upon reaching the required depth the tool
was withdrawn with the rotation in the same direction as during insertion. The
design of the system is such that both during insertion and withdrawal of the
tool, the material through which it is cutting remains very much in place
although it is mixed with considerable energy. The effect is somewhat like a
blender, and the angle of the tool is such that on withdrawal there is no
extraction of material at the surface and on this site, with comparatively large
grout-take, immeasurably low heave.

Construction of Trial Columns

Test Columns. The objective of the test columns was to try several grout mixes
whilst obtaining penetration of the waste and key in the columns 1m into the
underlying Folkestone Bed Sands. An initial test column using the 0.9m
diameter mixing tool and 0.45m auger achieved only 7.4m penetration within the
waste and was unable to proceed further due to ground resistance. A second test
column using a pre-drilled hole (created by the rigging of the Kelly bar) initially
appeared unsuccessful in penetrating the waste, although on withdrawal sands
were found on the edges of the tool, suggesting that it had in fact penetrated
through a shallower part of the waste mass into the underlying natural strata. A
third test column was attempted using the 0.9m mixing tool but with a 0.75m

diameter 1m auger. Only 3m penetration was achieved, and on withdrawal the tool was seen to be clogged with clay from the landfill capping layer.

Trial Columns. Six trial columns were attempted, the intention being to penetrate fully through the waste and key in at approximately 1m into the underlying Folkestone Beds. The grout mixes used in each of the columns is given in Table 2.

Trial Column 1 achieved a depth of 15m keying in as intended into the Folkestone Beds. The rate of insertion was, however, considerably slower than anticipated resulting in significant grout loss to start with at the surface. Flow was therefore reduced during the insertion and compensated for by increased flow if necessary during withdrawal.

Trial column 2 achieved a depth of 13.3m, which from previously drilled pilot boreholes indicated that although penetrating the waste the auger seemed unable to key into the Folkestone Beds.

Trial column3 achieved a depth of 11.2m although the target depth to key in was 13.8m. Again from pilot borehole information this indicated that the auger was unable to penetrate the Folkestone beds.

Trial Column 4 achieved the target depth of 12.m , but had to be withdrawn and cleaned of clay from the capping layer at 6.6m. On re-entry, target depth was achieved.

Trial column 5 achieved a depth of 11.3 m compared to a target depth of 13.8m. Problems were encountered with downward pull on the rig at 9.5m, and the column was therefore re-mixed using a different mixing tool which achieved the 11.3m penetration which again was indicative of an inability to penetrate the Folkestone Beds.

Trial column 6 utilized a 0.9m mixing tool The target depth for this column was 10.8m, but there was a catastrophic failure with the tool snapping off at 9.3m, indicative of a large dense obstruction, probably metal. The Kelly bar was lifted out of the hole, and it was evident that as mixing has only occurred during insertion, the column was less well mixed and clearing the top of the column post construction revealed voids to 2m depth.

Two trial blocks were attempted the configurations for which are given in figures 2 & 3. For Trial Block A, a construction sequence was adopted for this block using primary and secondary columns with a target depth of 10m for all. The columns were begun with a 450mm auger. The production rate was generally good, and although grout takes were high (typically 70 to 80% of the theoretical volume of the column), they were seen to be consistent. Back spinning of the tool had to be employed as the column was progressed in order to clear the auger and tool of clay, but this presented a risk of blocking the grout injection ports. To mitigate against this the tool was modified by welding on a metal plate to protect the ports during backspinning.

The insertion of the secondary columns proved problematic, as by the time they were inserted the grout had begun to cure and the mixing tool was having to cut its way through the primary columns. This resulted in very high wear, which was partially alleviated by the welding of additional teeth on the outside edges in order to take the brunt of the cutting action. The auger was increased to 750mm in order to disturb a larger area of soil and waste material and therefore reduce wear and tear on the mixing tool. To reduce the risk of clogging the mixing tool and auger, as experienced in trial Column 3 in particular, the top section of the flight was removed to allow the tool and auger to act separately.

Target depth for secondary columns was not achieved on columns 9 (8.9m), 15 (8.1m) and 13 (9.0m). During the installation of column 13, the tool seized, unable to move in a forward or reverse direction, before finally snapping. This appears consistent with the presence of bulky, probably metal objects., which have been grouted into the edges of the primary columns and cannot be spun out of the way.

Trial Block B was originally scheduled to comprise 25 columns, but due to difficulties only 19 columns were completed. Due to the difficulties encountered in Block A, soft and hard grout mixes were used, whereby the primary columns were constructed of a 10:1 cement/bentonite mix. Columns 17,19,23 and 25 having 50% less bentonite and column 21 no bentonite. All primary columns reached the target depth, although columns 7, 11 and 15 proved problematic and needed several attempts. This involved the withdrawal of the auger and tool and cleaning off clay build up. Problems were encountered once more with secondary columns, but it was apparent that problems with these columns occurred next to locations where problems had been encountered with the primary columns. This supported the hypothesis that bulky objects within the waste were hindering the progress of the auger. Grout mixes for the columns are given in Table 1.

Mix No.	W/c ratio	OPC, kg	PFA, kg	GGBFS, kg	Bentonite, kg	Water, l
1	0.6	667	-	-	-	400
2	1	400	-	-	-	400
3	0.6	333	333*	-	-	400
4	1	200	200*	-	-	400
5	0.6	333	-	333	-	400
6	1	200	-	200	-	400
7	2	250	-	-	25	500
8	2	250	-	-	12.5	500
9	2	250	-	-	-	500

Table 1 : Grout Mixes

Post Installation Investigation

The objectives of post installation investigation were to:-

- Establish the condition of the external face of the treated blocks.
- Investigate the nature, strength and permeability of the treated material.
- Assess the degree of bonding between adjacent columns.

This was achieved by:-

- Excavation of a trial trench to 3m depth along one face of each treated block.
- Excavation of a pit to 7m depth through the untreated core of Block A using an auger tool.
- Rotary coring of holes to produce cores of nominal 100mm diameter to full depth through Block B and testing of core for density, strength, permeability,

chemical composition, leachability under pH of 6.5 and freeze thaw, immersion in distilled water, leachate and acetic acid.

Visual Observations and Laboratory Test Data

Excavation of waste around the trial blocks exposed a distinct solidified face of material, the curvature of each column forming the outer face of which could be readily identified. Close inspection clearly indicated a shredding of the waste which was embedded in the mass. There was little evidence of over break, indicating the dense nature of the waste and evidently little void space for the grout to flow into. This was supported by the tendency for grout to flow out readily from the top of the column during installation at times when penetration rates were low.

Excavation of the untreated core of Block A was not entirely successful due to steam obscuring views but it did provide further evidence that grout had not spread from the columns as the core remained essentially clear of obstructions. The material recovered on the auger flights also gave an indication of the degree of mixing achieved by the auger in advance of the mixing tool. The material recovered was a fibrous, shredded mix of paper, string and plastic in a black, silty, clay matrix of an appearance similar to compost.

Finally, block B was slowly dismantled using the excavator, first deepening the excavation around the columns and then using the machine to prise apart the columns. This was only done once each of the columns in the block had been identified from its location within the block. It was noted that a number of columns were significantly softer than others and broke apart readily, others were solidly held together. Of particular note was the lack of bonding between adjacent columns, such that with fairly gentle leverage from the excavator arm individual columns could be prised apart.

Examination of the core samples proved particularly interesting, as it was clear that obtaining good quality core was difficult. Often, section of core showed little or no recovery, and sections where recovery consisted only of shredded waste and up to cobble size stones and occasional pieces of metal. These poorly grouted up sections of core did not always correspond with observations of column integrity at comparable column depth in the field. It would appear that a combination of the flush and trapped revolving fragments had washed and stripped away the soft grout.

Grout cube strength has been measured at 7 and 28 days and shows considerable difference in strength with mixes simultaneously and significant change with time. After 7 days variations between mixes were from 1.0 N/mm2 for Mix 6 and 7 to 17 N/mm2 for Mix 1. After 28 days cube strength was between 1.0 N/mm2 for Mix 7 to 29.5 N/mm2 for Mix 1. Notably, measurement of cube strength has shown very little strength and very little improvement with time in the mixes containing bentonite with 2:1 water:cement ratio and the mixes containing GGBS at 1:1 water:cement ratio. The greatest cube strengths were consistently recorded in the straight OPC at 0.6 water:cement ratio.

A programme of laboratory testing is underway and includes strength, permeability (using a variety of liquids; site leachate, water, acetic acid pH2.9), long term immersion and freeze thaw (Head, 1992, ASTM. 1995a, 1995b). Most of the testing is ongoing but at the time of preparing this paper approximately half of the strength test results are available and these are summarised below.

Host Material	No. of Tests	Unconfined Compressive Strength (Range) MN/m^2	Unconfined Compressive Strength (Average) MN/m^2	Comments
Capping	4	3.08 – 4.98	3.91	
Domestic Waste	11	3.12 – 5.40	4.04	Excluding two value of 1.0 and 1.6 MN/m^2
Basal Layer	6	1.97 – 3.48	2.86	Excluding two values of 0.7 and 7.15 MN/m^2

Table 2 : Results of Unconfined Compressive Strength Testing (OPC mix at 0.6 w c ratio). Samples between 2-4 months old.

Host Material	No. of Tests	Unconfined Compressive Strength (Range) MN/m^2	Unconfined Compressive Strength (Average) MN/m^2	Comments
Capping	1	14.3	-	
Domestic Waste	5	1.97-7.13 (4.89 to 7.13)*	4.93 (5.67)*	*Excluding value of 1.97 MN/m^2
Basal Layer	2	1.94 and 7.96	-	

Table 3 : Results of Unconfined Compressive Strength Testing (OPC + GGBS mix at 0.6 w c ratio). Samples between 2-4 months old.

The results to date indicate similar strengths for the capping and waste material and slightly lower for the basal material. It is thought that the latter may be due to the difficulty of mixing at the greater depth that the basal material was encountered. There are also indications that addition of the GGBS admixture has a beneficial effect on strength. Testing on other grout mixes to those summarised in Tables 2 and 3 is still underway.

To date a total of seven laboratory permeability tests have been carried out using triaxial apparatus. The majority of these are on samples from columns constructed using the standard 0.6 w/c OPC mix and excluding one value of 1.5E-09 m/sec all others were between 1.4E-07 and 8.4E-8 m/sec.

Freeze/thaw testing of Mix 1, in accordance with ASTM D4842-90 showed a small loss in mass, in the order of 0.0 to 0.5%, a similar range to that reported by Al-Tabbaa et al (1998).

Long term immersion tests (Tedd et al, 1993; Tedd, pers. Comm., 1998) are underway using the same liquids as that in the permeability testing. In the short term, after one months immersion observations do not indicate any deterioration of cores.

Discussion and Conclusions

The trials have met with only moderate success. Although the technique has been successfully utilised elsewhere in the world, it has been done so in more natural and homogenous materials generally to shallower depth (Al-Tabbaa and Evans 1998a , 1998b). These trials indicated that a combination of depth and a difficult substrate as is likely to be encountered in older landfills will severely hinder the process of column installation. If the auger is able to reach the base of the waste then a rig with the power of even a Soilmec R16 is insufficient to drive through dense underlying strata as typified in the Folkestone Bed Sands. In this case having penetrated through 15m of waste the rig did not appear to have the capability to penetrate into dense Folkestone Beds. This could be a limiting factor in the technique, but does not preclude the use of the technique where the waste to be grouted is shallow.

In a large structure such as that envisaged in these works, failed columns (incomplete depth or mixing) are likely to be less critical. Therefore such a structure may be successful in this material providing that enough columns reach full depth. The problem would appear to be however, that in such an old site tipped under different licensing conditions, no amount of previous characterisation can determine exactly what would be found at the particular column location. This is less likely to be a problem in a modern landfill where many of these bulky metal objects would not be tipped. It is recommended that intensive site investigation using auguring techniques is used to characterise the waste mass. This should be supplemented with trial pitting where sufficient depth can be achieved. It is suggested that subsequent calculation of likely success rate for column installation can then be worked on the basis that if failure rate of cable percussion is greater than 15 % the technique is unlikely to be viable. In the Runham Lane site, failure was at 10%, whilst 35% of holes required chiselling.

Adopting a primary and secondary column approach for the construction has not provided a bond between the columns. An alternative sequencing of construction employing a cellular overlapping or touching layout would probably be more effective, possibly with post installation grouting.

During installation of the secondary columns it was also apparent that the mixing tool was relied on heavily to act as a cutting tool, resulting in severe wear, especially on the end plates.   This was helped by re-positioning of the outer teeth. It is also clear that the configuration of the tool is important to form a column with volume of consistent quality. This was best achieved with a combination of 900mm mixing tool and 750mm auger.

A high rate of failure in the installation of secondary columns could be attributed to the inability of the mixing tool to push aside obstructions that were held in place by primary columns. Sequencing of works to avoid creating pockets for obstructions to get caught may reduce this failure. Alternatively, use of a retardant in all columns would assist in this, allowing for obstructions to move with the rotation of the auger and mixing tool. The obstruction would become embedded in the structure, but the integrity of the wall would be maintained by its bulk, whereby the majority of columns were complete and would bind together as a contiguous structure.

Clogging with clay was also a regular problem. Excavation of the clay cap prior to column installation may be beneficial. Health & Safety concerns can be addressed by loosely backfilling around the excavated hole.

It is also necessary to consider environmental protection in undertaking such work. In the case of this particular site, it is located on a major aquifer approximately 1.0km from a (currently) disused Public Water Supply. The method had to be able to provide a safe technique for penetrating the base of the landfill whilst not allowing the escape of leachate from the site. It is believed that the technique itself provides significant protection, as it does not produce voids or temporary conduits for liquid due to its basic design which breaks up material whilst pumping and stirring in grout. A permanent "substrate/grout" plug is therefore provided during the work. Extra safeguards were employed during the works and consistent of down gradient groundwater quality monitoring and similar precautions are recommended in work in sensitive groundwater areas.

Long term immersion tests are in progress and will provide valuable information on the durability of the column mixes. It is also proposed that the remaining columns will be entirely exhumed in 2000, an exercise which is eagerly awaited as it should provide invaluable *in situ* durability information. It is the authors' intention to cover the results of the long term immersion test and UCS testing on other mixes in a subsequent paper.

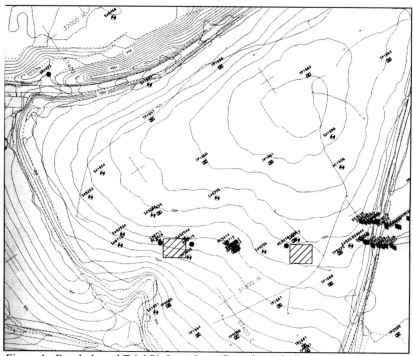

Figure 1. Borehole and Trial Pit Locations (Cross hatch indicates trial blocks).

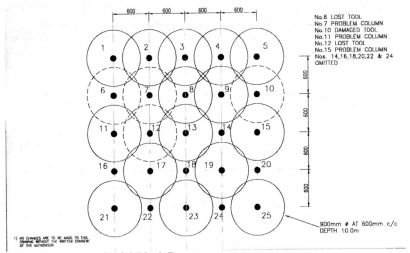

No.6 LOST TOOL
No.7 PROBLEM COLUMN
No.10 DAMAGED TOOL
No.11 PROBLEM COLUMN
No.12 LOST TOOL
No.15 PROBLEM COLUMN
Nos. 14,16,18,20,22 & 24 OMITTED

900mm ⌀ AT 600mm c/c
DEPTH 10.0m

1) NO CHANGES ARE TO BE MADE TO THIS
DRAWING WITHOUT THE WRITTEN CONSENT
OF THE AUTHORISER.

Figure 2. Layout of Trial Block B

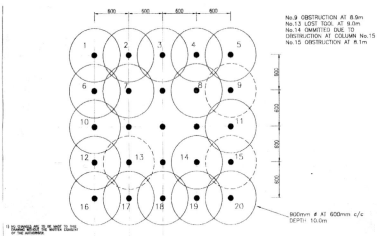

No.9 OBSTRUCTION AT 8.9m
No.13 LOST TOOL AT 9.0m
No.14 OMMITTED DUE TO
OBSTRUCTION AT COLUMN No.15
No.15 OBSTRUCTION AT 8.1m

900mm ⌀ AT 600mm c/c
DEPTH 10.0m

1) NO CHANGES ARE TO BE MADE TO THIS
DRAWING WITHOUT THE WRITTEN CONSENT
OF THE AUTHORISER.

Figure 3. Layout of Trial Block A.

Figure 4. Mixing tool emerging from grouted column.

References

Al-Tabbaa A., Evans C.W. (1998) Pilot in situ auger mixing treatment of a contaminated site Part 1: Treatability study.   Proc. Instn Civ. Engrs Geotech. Engng, 131, Jan., 52-59.

Al-Tabbaa A., Evans C.W. (1998) Pilot in situ auger mixing treatment of a contaminated site Part 2: Site Trial.   Proc. Instn Civ. Engrs Geotech. Engng, 131, Apr., 89-95.

Al-Tabbaa A., Rose S.P. (1996). Treatability study of ins-situ stabilisation/solidification of soil contaminated with methylene blue. Environmental Technology, 17, 191-197.

ASTM (1995a). Determining the resistance of solid wastes to freezing and thawing. Vol. 11.04, D4842-90, 148-151.

ASTM (1995b). Compressive Strength of moulded soil-cement cylinders. Vol. 04.08, Test D1633, 254-256.

Head, K.H. (1992). Manual of soil laboratory testing, volume 3, chapter 20: Triaxial consolidation and permeability tests. 1001-1027.

Tedd, P., Paul, V. and Lomax, C. (1993). Investigation of an eight year old slurry trench wall. Green '93. Int. Symp. On Waste Disposal by Landfill. Bolton Institute. Balkema 1995. 581-590.

Tedd, P. (Pers. Com.), (1998).

The contamination of soils and changes of Geotechnical conditions

Dr, Asocc. Professor K.J.DUNDULIS, Vilnius university, Lithuania, and V.IGNATAVIÈIUS, DR, Asocc. Professor, Vilnius technical university, Lithuania

INTRODUCTION

At the moment the pollution of geotechnical environment is one of the most important problems of environment engineering. Following the publications of scientific research of recent years the main focus is the solution of hydrogeological and ecological problems. Building and soil background interaction circumstances, change of soils geotechnical characteristics (Basalyk et. al, 1997, Dashko, 1984, Svirastava, 1997) are investigated not so often.

During recent years in conducting reasarch of former military sites, in oil storages places and terminal territories and in their surroundings there were noticed cases of pollution of geological environment. Also there were traced cases of technological solutions leakage in manufacturing enterprises. On the basis of committed research it is possible to determine typical cases of pollution influence on geotechnical conditions and the change of soil characteristics.

TYPICAL CASES OF POLLUTION AND GEOTECHNICAL

Geotechnical investigation on various areas of spread of pollution substances allows us to determine typical cases of geotechnical pollution:
- the building is designed and built in the area which was formerly polluted;
- the building is built, and the geological environment under it has a source of pollution;
- the building is polluting its environment itself.

According the pollution character the following cases are determined most often - oil products and inorganic chemical solutions. From soft hydrocarbonate - diesel fuel, hard hydrocarbonate - oil fuel. From technological liquids most often occurring pollution substances are - metal salts solutions, acids and alkaline solutions.

On the other hand the size of pollutions influencing change of circumstances is determined by the structure of pollution area. In the territory of Lithuania the fundament for buildings in most cases is till clay and sand soils of various origin. Lying condition of ground water have a big influence on the thickness of areas. The thickness of areas and the fluctuation of ground water level determine the thickness of soils polluted by oil products. The direction of flow of ground water and the parameters of soil permeability determine the area where pollution substances spread and the character of dynamics.

From Geotechnical stand point it is important to evaluate interaction of pollution substances and the soil. In Lithuania the following two cases of such interaction are determined:

Geoenvironmental engineering, Thomas Telford, London, 1999, 163–169

– polluting sand soils with oil products, these products cover particles of sand by pellices, partially fill pores and function as some kind as lubricant;
– soil polluting by chemical agents, which can have reactions with mineral part of the soil, the solution of a certain minerals occur and the changes of minerals, the amount of contact changes, compressibility increase, the strength of soil decrease.

The validity of these provisions was checked in the laboratory by using method of physical modelling, also research was committed in specific sites.

THE RESULTS OF PHYSICAL MODELLING

Following the investigation of changes of geomechanical characteristics of soils under the influence of pollution materials the following two cycles of research were committed.

During the first cycle of research formed fine sand mixtures were saturated by liquids of various viscosity - ethyl alcohol, water, burning oil, diesel fuel, black mineral oil. The mixtures of sand and the mentioned liquids were prepared in order to reach uniform porosity (e-0.7). The level of degree of saturation equals to 0.4. The shear strength of the soils was investigated using the method of direct shear in vertical load 0.1, 0.2 and 0.3 MPa. The results show that angle of internal friction of the sand is closely connected to the coefficient of dynamic viscosity of liquids (Fig. 1). The correlation coefficient of this interaction is 0.971.

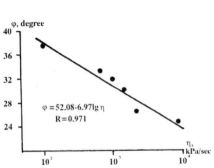
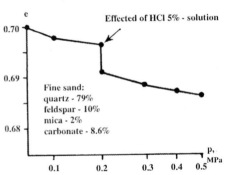

Fig. 1. Correlation between angle of internal friction and dynamic coefficient of viscosity

Fig. 2. Changes in sand compressibility under the effect of HCl 5% - solution

The second cycle of research was committed in saturating the same porosity sand by HCl 5% solution and in investigating compressibility by oedometers (Fig. 2). The tests show, that after saturating loaded sand soils with HCl solution in several hours collapse effect was noticed. The void ratio of soil at the effect of constant load of 0.2 MPa decreased from 0.06 to 0.1 points.

These tests of physical modelling prove the influence of pollution materials on Geotechnical characteristics of sand.

RESEARCH ON SITES
The district of dwelling houses
The district was filled by dwelling houses 40 years ago. 15 years ago it was noticed, that oil station in the neighbourhood pollutes the soil and soil water of this district. The research of the territory showed, that oil products covered approximately 200 thousand sq. m territory in the area of fluctuation of ground waters (Fig. 3).

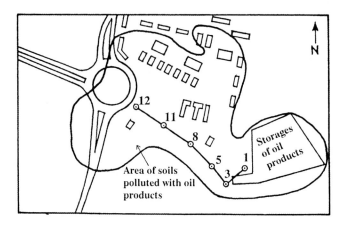

Fig. 3. The contamination situation in the housing area

The level of pollution substances fluctuates from 2 to 6 m (Fig. 4). Ten years ago special bores were established in the area of reservoirs of oil products, they were used to soak liquid oil products from underground. Collection dress of oil products and ground water was additionally designed and laid in the front part of pollution area. At the moment approximately 3 thousand cu. m of oil products are soaked (Marcinonis and Sherys, 1997).

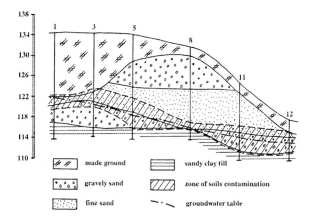

Fig. 4. Geotechnical section in the housing area

Additionally it was investigated if soil pollution did not change geotechnical properties of soils under buildings. The research of strength and compressibility of polluted soils (table 1) showed, that angle of internal friction of the sand has diminished approximately 5°, and modulus of deformation E from 35 MPa to 26 MPa. These changes are not so major, that they could harm the stability of the buildings. The limit of bearing capacity is sufficient, as loading from the buildings is only approximately 0.2 MPa.

Table 1. Geotechnical properties of natural and oil contaminated Vilnius sands

Soil type	Density, Mg/m^3	Moisture content, %	Void ratio	Cohesion, KPa	Angle of inter. friction, degree	Modulus of deformation, MPa
fine sand	1.89	17.4	0.65	0.09	36	35
poluted fine sand	1.89	18.6	0.66	0.08	31	26

Reconstruction of oil terminal

In carrying reconstruction of oil terminal in the building area of subsidiary buildings near natural soft soils there were determined soils polluted by oil products (Fig. 5). The main polluting substance is diesel fuel. Hard oil products are determined only in the upper part of geological section. The concentration of oil products was from 10 mg/kg to 63.3 g/kg.

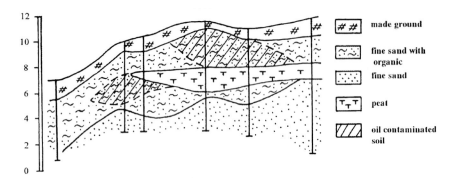

Fig. 5. Geotechnical section in the oil terminal (by Kuten & Shlauteris)

In accordance with N.Kuten et. al (1997) data, in committing CPT research on polluted soils cone resistence is 20-30 percent less than in non-polluted soils. Laboratory tests of strength characteristics of samples of disturbed structure of polluted and cleaned soils (table 2) shows considerable changes of characteristics. Medium course sand polluted oil products angle of internal friction is 8°, and cohesion 26 kPa less in comparison with cleaned soils. The angle of internal friction of fine sand depending on the character of pollution substances is from 2° to 17° less than cleaned soils. Cohesion differs 4-11 kPa respectively.

Table 2. Shear test results of the oil contaminated and refined Klaipëda sands

Soil type	Oil contaminated soil		Refined soil	
	Cohesion, kPa	Angle of inter friction, degree	Cohesion, kPa	Angle of inter. friction, degree
Medium coarse sand	4	27	30	35
Fine sand	0-7	18-34	26-30	35-36

Galvanic department of plant

In the galvanic department of one plant of Western Lithuania, which was built 30 years ago deformations of building constructions were traced. During investigation of causes of deformations it was determined leakage of technogenical solutions from technical canals. Before building the department it was determined, that the base for foundation will be till sandy clay with microlayers of silt and sand. Till clay fraction was formed of hydromica (75-85%), kaolinite (10-30%), chlorite (1-5%), montmorillonite (1-5%) and calcite (Fig. 6). Physical characteristics of soils, provided in table 3 shows that till was of high density and was stiff. The research of this time shows, that in the surroundings of this department there was fresh (mineralization 0.3-0.8 g/l) hydrocarbonic water.

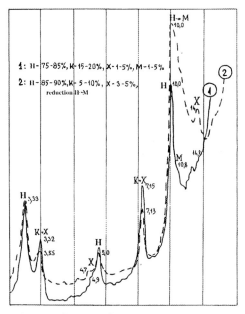

Fig. 6. Diffractografs of minerals of clay particles.
    1 - natural till; 2 - polluted till
    H - hydromica, K - kaolin, M - montmorillonite, X - chlorite, H→M - reduction

After several decades of exploitation technogenical water was formed - mineralised (19.3 g/l), having 9.3 g/l of chlorine, 2.4 g/l of sulphate, 6.8 g/l natrium and potassium. This water is aggressive not only towards concrete, but also to till, as no carbonate is traced in it, calcium is substituted by natrium. Montmorillonition process takes place in soil. The research on site as well as soil tests by saturating them in the laboratory (after 3 days and after 350 days of saturation) showed, that the characteristics of soil change (table 3).

Table 3. Avarege values of till's Geotechnical properties

Properties	Natural soil	Polluted in-situ soil	Saturated after 3 days	Saturated after 350 days
Moisture content, %	12.6	15.4	16.5	26.4
Void ratio	0.40	0.43	0.50	0.65
Density, Mg/m^3	2.17	2.07	2.06	2.09
Angle of inter. friction, degree	28	30	30	21
Cohesion, kPa	50	20	7	18
Modulus of deformation, MPa	35	24	13	9

Moisture and porosity increases in soil. Still soils become firm plastic and even soft plastic. Mechanical characteristics of soils change accordingly - angle of internal friction decreases from 28° to 15° (Fig. 7). Cohesion decreases from 50 kPa to 18 kPa. As a result of influence of technogenical water the compressibility of soils during exploitation period increased 1.5 times, from samples soaked for one year - 3.5 times, and forecasts to compressibility (after 10 years) - 4.5 times (Fig. 8).

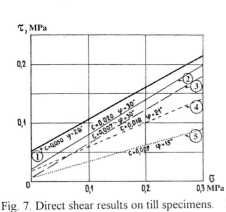

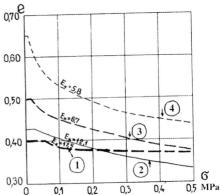

Fig. 7. Direct shear results on till specimens.
1 - natural till; 2 - polluted in-situ;
3 - after saturated 3 days; 4 - after saturated 350 days; 5 - prognosis after 10 years

Fig. 8. Oedometer test results on till specimens (explanation on Fig. 7)

CONCLUSIONS

1. The main polluting substances of geoenviroment are oil products and technogenical solutions.
2. Oil products forming pellices on sand particles diminish strength of soils and increase their compressibility. The changes of soil strength depend on viscosity of pollution substances.
3. Active chemical solutions change chemical structure of clay minerals, cause changes of their moisture and porosity, diminish strength and increase compressibility.
4. Geotechnical investigations must be carried out considering possible secondary and direct changes of polluting substances towards geoenvironment. It is necessary to evaluate

changes of soil characteristics depending on character of polluting substances and their possible intensity.

REFERENCES

Basalyk R., Frankovska J., Janotka I. (1997). Sealing mineral materials for landfill liners. Proc. Int. Symp. "Engineering Geology and the Environment". Balkema, Rotterdam, vol. 2. pp 1593-1597.

Dashko R. (1984). Naturale of saturated clay deformation under structures. Jour. "Notes of Leningrad Mining Institute", vol. 100. pp 40-47. (In Russian).

Kuten N., Shlauteris A., Dundulis K. (1997). Oil product pollution impact on physical mechanical sand properties. Proc. Int. Symp. "Engineering Geology and the Environment". Balkema, Rotterdam, vol. 2. pp 1955-1961.

Marcinonis A., Sheirys N. (1997). A Tragedy at Vilnius Gelezinkelieciu Water Intake. Jour. "Geologijos akiraciai". Vilnius, No. 2 (26), pp 16-24. (In Lithuanian).

Srivistava R.K. (1997). Geotechnical aspects of soils - Industrial waste interaction bechavior. Proc. Int. Symp. "Engineering Geology and the Environment". Balkema, Rotterdam, vol. 2. pp 2179-2184.

The Effective Integration Of Remediation Into The Redevelopment Of Brownfield Sites

C. ROUND, J. STRANGE & S. RHODES
WSP Environmental Ltd, Basingstoke, Hampshire, UK

ABSTRACT

Government policy actively encourages the redevelopment of 'brownfield sites' in its aim to achieve sustainable development; this is reinforced by planning policy guidance, such as that set down in PPG 23. This encourages the redevelopment of disused and contaminated industrial/commercial sites and promotes remediation through development. Although such sites are often a valuable resource, their redevelopment is often perceived as being problematic due to the time and costs involved in implementing the remediation works, and the constraints associated with the subsequent development, operation and financing of the scheme. These problems often result from the remediation being considered in isolation from the rest of the redevelopment works. However, by effectively integrating the remediation into the rest of the works, site redevelopment can be made more cost effective. The resulting environmental benefit gained from redevelopment can be used as a positive selling point with the planning authorities, investors, developers and purchasers. This paper draws on the experience of a leading consultancy in the industry. The paper outlines the major requirements of a potential developer, the requirements of the remediation process and how these two can be reconciled to achieve the greatest benefit to all parties. Case studies are used to illustrate how an integrated approach to development can turn a brownfield site from a liability into a positive asset.

INTRODUCTION

The re-development of sites with a previous use, the so called brownfield sites, is becoming increasingly more common as fewer suitable greenfield sites come on to the market. Furthermore, government policy on sustainable development promotes the use of brownfield land and the remediation of these sites through the development process. Recently, this emphasis has increased with the Government statement that 60% of new housing development are to be built on brownfield sites, whilst the introduction of sequential tests will make building on greenfield sites increasingly more difficult.

Many developers recognise that sites with previous usage represent a good development opportunity and actively seek brownfield land for their developments. Many other organisations still view these sites as representing too great a risk with the potentially prohibitive abnormal development costs, which can be involved. Successful development of a brownfield site can be achieved if all of the issues are addressed together and not in isolation and it is important for the project team to effectively integrate the remediation and construction works.

Geoenvironmental engineering, Thomas Telford, London, 1999, 170–181

REQUIREMENTS OF THE REMEDIATION AND DEVELOPMENT PROCESS
It is important to consider the remediation works as a necessary part of the overall development of the site so allowing their effective integration. There are several key items to be considered when designing and programming the development works. The relationship of these elements is presented in Figure 1 and are discussed below:

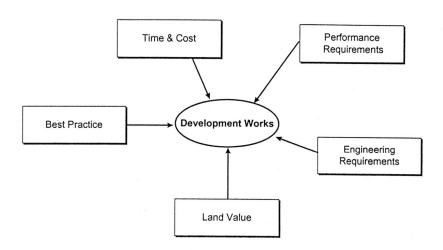

Figure 1. Key Issues Affecting Development Works

Performance Requirements
Whatever remediation method is to be adopted, it must be capable of cleaning the soils to an appropriate standard. This standard is frequently dictated, on development sites, by the requirements of conditions placed on the planning consent and will generally be to a level suitable for the proposed end use of the site. Traditionally, this has involved remediation to generic guideline values, such as the ICRCL levels, however, the assessment process has moved forward with the adoption of risk assessment methodologies by the Environment Agency. This allows targeted remediation to be designed taking account of the site specific conditions and the sensitivity of the proposed end use.

On some sites, it is possible that a two stage remediation process will be undertaken. The first stage may be the remediation to a statutory standard, i.e. a level at which the site is not causing a pollution risk and therefore is not in breach of the Environmental legislation. The second stage represents additional remediation works to bring the site to a standard suitable for its proposed end use.

Engineering Requirements

The remediation works are only one stage in the site development process and this needs to be recognised from the beginning. Problems can be encountered where the remediation works are undertaken in isolation. Typical problems which WSP Environmental have encountered include:

- Process based remediation methods can change the engineering properties of the soil making them unsuitable for re-use on site. This can occur due to the process increasing the moisture content of the soil above its optimum range, or the inclusion of organic matter which may represent a long term settlement risk.

- The use of capping materials can raise site levels, thereby incurring problems with access roads and service falls. In addition, the encapsulation barriers and any underlying aquitard, may be compromised when installing foundations and infrastructure.

- The remediation works also need to consider how to deal with other construction abnormals, such as underground obstructions.

Time and Costs

Many alternative remediation methods represent cost effective ways of treating a site, however, they may also take longer. Typically, options such as bio-remediation, have been shown to be effective ways of treating hydrocarbon contaminated soils, whilst being significantly cheaper than conventional excavate and removal methods. Bio-remediation, however, generally takes between 3 and 12 months for the remediation to be complete and may require large areas of land on which to construct the bio-treatment beds. Additional time delays may also occur from the requirements of the Waste Management Licensing Regulations, and the necessary field trials and feasibility studies.

These methods may be suitable for sites where the owner has time available prior to the construction works commencing, or where the works can be phased to allow the release of clean land for sale. However, where the developer has purchased the land, the cost savings associated with alternative remediation options have to be compared with the cash flow requirements of the developer. In many situations conventional excavate and remove methods may actually be more appropriate.

Value

The method and degree of remediation works which are required for a site are related directly to the sensitivity of the end use. The most sensitive end use is residential properties, whilst the least sensitive end use is generally employment type uses. However, land values tend to be much higher for residential properties and therefore there is a trend towards zoning a site for residential use, where possible, thereby significantly increasing the remediation costs.

The long term value of the site also has to be considered as funders may have reservations over certain remediation methods, thereby potentially de-valuing the land. Similarly, any lack of flexibility with the development could impact on its value.

Typical remediation options available are summarised on Figure 2, together with a several of the key issues related to their application.

REMEDIATION OPTION	APPLICATION	ADVANTAGES	POTENTIAL CONSTRAINTS	RELATIVE COST	RELATIVE TIME
Excavate and Remove	All Contaminants	Removes Contamination	Need to reinstate void. Not sustainable option. Traffic generation.	High	Short
Encapsulation	All contaminants	Uses well established principles	Contamination remains on site with potential liability. Limitation on Development	Low	Short
Bio-remediation	Hydrocarbons	Removes contamination No importation of spoil	Can change engineering properties of the soils. Licensing requirements	Medium	Medium to long
Soil Vapour Extraction	Volatile Compounds	Removes contamination No importation of spoil Undertaken in situ	Licensing requirements Sterilize site.	Medium	Medium to long
Solidification/ stabilization	Metals	Engineering benefits Limited importation of spoil Neutralizes contaminants	Not widely accepted in UK. Licensing requirements. Possible on-going liability.	Medium to high	Short to Medium
Soil washing/ flushing	Soluble Metals Light organics	Removes Contamination	Can change engineering properties of the soils. Not widely used	Medium	Short to medium
Incineration	Any contaminants	Removes Contamination	Produces residue Licensing requirements	High	Medium

Figure 2. Summary of Remediation Options

Best Practice

The approach of BATNEEC (Best Available Technique Not Exceeding Excessive Cost) promotes the use of the most appropriate technology. This can be used to portray the 'green image' of converting a brownfield liability into a positive asset. Similarly the use of on site techniques can reduce pressure on valuable resources, such as landfill void capacity, importation of clean fill and also reduced pressure on the environment through minimal vehicle movements etc.

The common approach employed in developing brownfield sites is to maintain separate environmental and engineering consultants and rely on a pro-active Project Manager to balance the requirements of each consultant. Where these projects are not effectively managed, this can lead to poor integration of the environmental and engineering requirements. An example of a project where the remediation could have been more effectively integrated is the redevelopment of a former railyard in East Anglia. The soils at the site were extensively contaminated with heavy oils from the railway yard. The chosen remediation method included the excavation of the contaminated soils for bio-treatment at the site, using the bio-pile system, mixing the contaminated soils with nutrients and substrate to promote microbial degradation of the hydrocarbons.

The site was being redeveloped for mixed retail and leisure use, and extensive preparation works were required as part of the redevelopment, including the raising of the site by up to

1.5 metres to bring the development above the flood plain. The large volume of remediated soils had to be retained on site, because of prohibitively expensive off-site disposal costs. Hence, the remediated soils were to be used as general backfill to raise the site.

A number of problems were encountered :

- *Time & cost* - because of unfavourable weather conditions the biological process was slower than expected, delaying the remediation by months. This eventually led to delays in the start of the site preparation works for the development.
- *Engineering* - the remediated soil met the clean-up requirements of the environmental consultant. However, the remediated soil was found to have a high moisture content, and a significant amount of organic matter such as straw, introduced as part of the substrate. This resulted in a remediated soil which was less suitable for use as backfill than the contaminated soil, from the geotechnical point-of-view. As a result, is was necessary to mix the remediated soil with imported clean sand to produce a more suitable fill.

INTEGRATING THE REMEDIATION INTO THE REDEVELOPMENT PROCESS
Problems may be encountered in matching the remediation requirements and the requirements of the development, however, most commonly it is the constraints of time and cost which conflict.

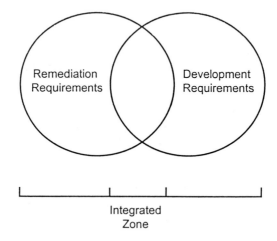

Figure 3. Overlap of remediation and development requirements

Time Constraints
It is often found that time constraints can limit the remediation approach taken, due to the need to finish the development to generate revenue - and often 'dig and dump' methods are adopted as the quickest and easiest option. Remediation works can often 'sterilise' the site for any other construction works for a considerable period of time, adding months or years to a construction programme. With careful planning and selection of appropriate remediation methods this can be overcome. WSP Environmental were consultants on a project in Croydon, where in situ bio-venting methods were used to remediate oil contamination. The process was designed to locate the wells away from the positions of the piles to the proposed structures, so allowing construction of the development to be undertaken while the remediation works were completed. This minimised adverse impact on cash flow for the scheme.

Risk Factors
The often inherently uncertain nature of remediation, in terms of time, cost and 'end product' presents a high risk scenario to developers. The technical and financial risks associated with remediation sets it apart from the rest of the project, where these factors are more easily quantified. This again can be overcome with careful planning and it is often possible to place time and clean up requirements on the remediation process through the Contract. In this way, if the remediation works take longer than planned, then the additional costs can be recovered via liquidated damages. Similarly, if the clean up standards are not met then the Contractor can be required to remove the materials at their own cost. If these type of conditions are placed on the contract, then it would be sensible to employ a company which can undertake both the investigation and remediation works.

Engineering Requirements
The appropriate selection of remediation technique to support engineering requirements can provide positive cost savings to the development. This was accomplished on a site in Oxford, by WSP where cement stabilisation was used to solidify contaminated made ground. The engineering advantage obtained was to improve the ground conditions such that 200 kN/m^2 foundation loads could be accommodated so removing the requirements for a piled foundation option.

Many remediation methods require considerable engineering works of their own, from major earthworks, to in-situ remediation techniques, which require considerable surface plant and equipment.

Potential conflicts may arise in a range of engineering factors :
- *Geotechnical* - for example, ground improvement measures such as stone columns, or piling may increase the potential for contamination migration.
- *Structural* - for example, the use of a capping layer may prevent the use of a ground-bearing slabs or piled foundation options.
- *Hydrogeological* - for example, the installation of a in-ground barrier wall may adversely affect the groundwater regime. WSP have been involved in a project where a substantial slurry wall was proposed, which potentially threatened groundwater flow within an aquifer and modelling of the hydrogeological conditions was undertaken to assess the risks.
- *Gas* - for example, the use of gas-venting trenches or similar may interfere with surface landscaping/features or surface drainage.

- *Site levels* - It is not unknown for an environmental consultant to recommend placing a clay cover to a site, in order to provide a protective barrier. However, such options can have an adverse impact on site levels rendering drainage falls and access road gradients difficult to achieve.

Design and implementation of the remediation works can often be carried out in isolation from the designers of the development - this poor interface can result in over-engineering by the main designers.

WSP Environmental have been involved in the investigation, assessment and remediation of a 20 hectare gas works site in Portsmouth since 1991. At the commencement of the site remediation works, the initial plots were treated using the traditional lift and shift methodologies. This was driven by the purchaser requirements of site suitable for industrial/commercial use within a set time scale and the Clients requirement for a minimal liability solution, which addressed their statutory obligations at least cost. In addition the requirements of the regulatory bodies with respect to clean up standards had to be incorporated into the remediation solution. Given the organic nature of the contamination, alternative techniques such as bio-remediation were addressed as the necessary open space was available, however lead in times for feasibility testing and trials made the purchasers programme impossible to achieve.

Integration of the developers requirements at an early stage allowed the remediation to be tailored to fulfil requirements. This was achieved on part of the site by advising on the locating of buildings in the most appropriate position to avoid major site constraints including the easement of a live gas pipeline and encapsulated hotspots of contamination. This enabled more efficient use of the site area and reduced foundation constraints on the building. Another area has been remediated by using lift and shift on the worst hotspot areas of contamination and the remainder of the site has been remediated for industrial re-use by applying a capillary break layer in conjunction with gas protection measures.

Early access to information on proposed development layouts enables the remediation to be targeted to reduce specific risks associated with that development, whereas traditionally a whole site is remediated to allow an overall use with specific associated risks, which in reality will not apply to all areas and a lesser scope of remediation may have sufficed.

On adjacent plots where timescales were not critical, it was possible to implement risk based solutions. This also involved liaison with the planning and regulatory bodies to obtain a consensus agreement to a lesser degree of clean up whilst minimising the most significant actual risks to the environment and public health. This also introduces the effects of implementation of the BATNEEC concept, as this type of remediation approach can reduce costs by enabling retention of greater levels of residual contamination but does not increase risk. This approach was successfully applied to remediation design to address an off-site contamination migration problem, where the revised remediation costs were reduced significantly over those for the original remediation requirements. A similar risk assessment approach to hydrocarbon contamination in natural soils, resulted in an agreed significantly increased limit for residual contamination so minimising the volumes of waste for removal and hence costs.

Where this approach was used, potential site purchasers have been provided with information on the risk based remediation design to enable them to make a decision with respect to development options and residual liabilities. If this liaison occurs at an early stage then the risk assessment can be further targeted to the specific proposed development.

The solutions for site remediation, in particular the risk assessment based options, require a degree of commercial awareness, as reducing the level of clean up may reduce the land values where residual *apparent* liabilities are higher and the planning and development constraints may be more restrictive. However, the reduced overall remediation costs may allow a site to be remediated and brought back into productive use, where a more traditional approach would result in the site having a negative value due to remediation costs and therefore not be feasible to remediate.

A knowledge of what potential purchasers are willing to accept and what they are willing to pay for land in a particular location is valuable information when trying to integrate the remediation options in to the land use master planning for the site.

IMPROVING THE INTEGRATION OF REMEDIATION INTO REDEVELOPMENT
Integration of the redevelopment and remediation processes has positive benefits to both site vendor and developer, but to achieve the full effects of these benefits, the integration has to occur at the earliest possible stage in the site reclamation process. The only way this can occur is improve communication between all members of the project team, such that all of the requirements can be identified and therefore accommodated. A commitment from the outset to clear and open discussion between the key players in a particular scheme is the only way that fully effective integration can be achieved.

An example of the benefits of early identification of all party requirements is taken from a site in Docklands. Contamination issues required the site to be capped in some way whilst the developers concerns related to minimising costs for pavement construction. Careful specification of the cover material to provide a layer which can act as both a capping to roadways and a capillary break can prevent accrual of abnormal road construction costs. In this instance, a suitable specification was developed by WSP Environmental, utilising a modified 6F2 material which had an appropriate California Bearing Ratio and was designed to have a limited capillary rise.

An example of a site where the remediation and development works were effectively integrated is a former cement works in Kent. The land was surplus to the requirements of the owner and was scheduled for re-development for employment type use. An investigation confirmed the presence of hydrocarbon contamination, resulting from fuel leakage and poor disposal of old lubricating oils. In addition, significant thickness of fill were present requiring the use of abnormal foundations.

The site is divided into a southern and northern area. The southern area was to be developed immediately as a distribution centre whilst the northern area was to be retained for a period of time, prior to eventual development.

Remediation of the southern area was designed and submitted to the planning authority as part of the overall development package and as a result all of the issues relating to remediation, site preparation and construction were undertaken concurrently thereby minimising delays to the start of works. The construction works involved:

- Demolition of existing structures, including underground tanks and pits.
- Remediation of hydrocarbon contamination.
- New highway access.
- On site infrastructure.
- Construction of the distribution building.

Due to the timescale for development an excavation and removal option was used for remediation of the southern area and this was incorporated into the contract for the site preparation works.   As a result, one contractor was responsible for undertaking the demolition, remediation, bulk earthworks, infrastructure and building works. In addition, one consultant was acting as the Engineer for the works. The use of a small core design team and site team meant that a holistic view could be taken on the development works and the issues of remediation and development could be easily integrated within the construction programme. Examples of this include:

- The remediation works were undertaken concurrently with the site preparation works and verified by the site engineer. This minimised claims over volumes and also allowed the excavation works to be sequenced such that down time on plant and equipment was minimised.
- Clean spoil was re-used on site and engineering properties were improved using vibro replacement techniques, so reducing abnormal foundation costs.

The northern area of the site was remediated under a separate contract. Due to the greater flexibility on time, an ex-situ bio-remediation option was used, so reducing the abnormal costs well below those which would have been incurred if a traditional 'dig & dump' solution had been implemented. Consideration was given to the need for re-using the materials to infill the voids and the remediation process was designed such that the treated soils could be used as an engineered fill.

These works are now complete and the site is awaiting development.

THE WAY FORWARD

Traditionally, sites have been remediated and then placed on the market for sale for a particular type of redevelopment.   This often results in duplication of environmental consultancy services and can result in an adversarial and prolonged land deal with the site developer trying to discredit pre-existing remediation works with a view to reducing the sale value of the land. The integrated approach deals with any conflicts of interests at the outset with all parties being aware of what they will get and what it will cost them, and the land sale can then be completed in the minimum timescale.

Currently it is common for the individual members of the development team to be appointed at separate times. Generally the first appointment is the Land Agent, followed by the Planning Consultant, Quantity Surveyor and Architect. These team members may plan out the site use in order to optimise the sale value of the land. It is only when a masterplan layout has been developed that the Environmental Consultant becomes involved. This approach is shown in Figure 4.

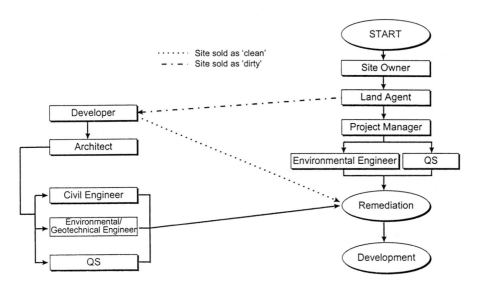

Figure 4. Traditional Procurement Route for Contaminated Site Development

However, the remediation works should be considered at the pre-planning stage in order that the site can be zoned for development potential. This can then be used as a tool in the development planning process to identify those areas which are more suitable for the high value, high sensitivity end use such as residential. The more difficult areas of the site can the be developed for less sensitive employment or leisure uses, so reducing the remediation costs, with minimal impact on the land value. This approach has been employed successfully on a forty hectare site in west London, were the significantly contaminated nature of parts of the site rendered them suitable for employment type uses only. The loss of land value was off set by the reduced remediation costs.

An approach to this is to undertake a risk assessment of the site and from this, zone the site for development potential. The high risks parts of the site become low development potential and their use is dictated by the costs and practicalities of developing these areas. This approach broadly coincides with the Risk Based Corrective Action (RBCA) method, developed in the United States and now advocated by current UK legislation. The RBCA approach is to targeted the remediation specifically at the risks within the site, thereby reducing un-necessary costs. Where substantial remediation costs are involved the use of more sophisticated quantitative risk assessment methods can be employed. This tiered approach is discussed in more detail in Boyd et al (1999).

On the larger sites, it is possible to develop this approach to programme the site development works to the benefit of cash flow. This may include identifying the low risk areas and developing these first, so releasing cash to fund the more problematic parts of the site.

WSP Environmental have been involved with several projects as a part of the development team at a pre-planning stage, which has allowed the site development to be tailored to the conditions present. This has saved significant development costs and the adoption of the RBCA approach has acted as a powerful tool in negotiations with the Regulatory bodies.

CONCLUSION
The constraints on development of greenfield sites will increasingly lead to a demand for development on brown field sites. Many of these developments will include a requirement for remediation of contamination. The integrated approach where the environmental consultant is appointed alongside the other members of the project team, has numerous benefits to both the site vendor and developer. This approach is shown in Figure 5.

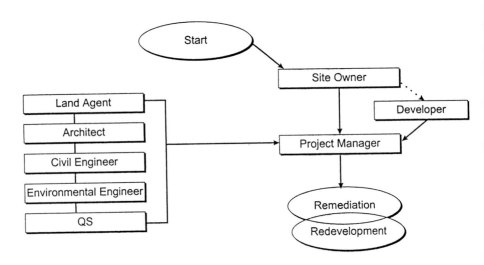

Figure 5. Optimised Procurement Route with Integrated Approach

Improved and early communication of the key issues and requirements allows the site redevelopment and remediation to be carried out in the most efficient, technically appropriate and cost effective manner with minimal conflict and with subsequent savings in both time and cost.

REFERENCES

Boyd et al (1999)
Boyd. M, Clifton A, O'Dea. A, Risk Assessment: Goldening the Brown, 2nd BGS Geo-Environmental Engineering Conference, 1999.

Effective Geoenvironmental Investigations on Riverside Sites

I M SUMMERSGILL, MA, MSc, MBA, CEng.
WS Atkins Ground Technology & Geoenvironment Division, Epsom, UK

INTRODUCTION

Riverside sites present some of the most complicated situations for investigating by engineers and scientists. They are 'mobile' sites, not only in respect of fluvial or tidal terms but also in their historical construction and usage.

Effective investigation can frequently read like a good detective story, with clues being obtained at intervals building up the entire picture. The impossibility of gaining all past information within a single intrusive site investigation needs careful explanation to site owners and purchasers. Cost-effective redevelopment of the site will almost certainly involve detailed desk study and a second or third stage (phased) investigation.

Examples of specific occurrences are given throughout the paper, as separate, brief case histories. Whilst most of the work herein relates to sites investigated on the tidal Thames and Medway, the concepts and basic geology apply to many riverine and estuarial conditions.

HISTORICAL ASPECTS

In most settlements built around a watercourse, wharves have been a feature since trading activities began. The prime site for landing or loading goods often determined the location for development of the industrial sectors of towns and cities, before docks and canals formalised the issue in the 18th and 19th Centuries.

A simple timber staging unit with piles driven into the river bed was soon superseded by larger timber wharves built over and out into the rivers, thus extending the area available to store materials off the firmer mainland (or onto previously unusable marshland). As time progressed, these wharves became more formalised and the river was excluded from underneath the platforms by timber walling and infilling with spoil. Gradually the land encroached on the river and outlines became more linearly defined.

Soon, greater loads were being stored on the wharf side and this would lead to settlement with time of the infilled ground, or eventual failure of the timber walls. New walls were inevitably driven in front of old ones, and sometimes tied into them for greater rigidity. Further infilling would raise the ground levels, in order to mitigate against the problem of frequent flooding.

During the last century, concrete and steel became useful and cost-effective materials for use in river walls. Various configurations of mass gravity walls or reinforced king-piles were developed, as was the use of tied anchor blocks and piled crane rail beams. On many river frontages, steel sheet-piling became the economic choice in the middle period of the 20th century. An example of a typical developing wharf frontage is shown in Figure 1.

Natural disasters, such as the 1953 tidal surge in the Thames Estuary, put paid to inadequate defences and led to more robust wharfages from the 1960's (Ogley, 1991). Now these sheet-piles are coming to the end of their design life and 'new' problems (such as accelerated low water corrosion (ALWC) on sheet piles) are becoming part of the picture, when assessing wall stability or remnant lifetime of the wall structure (Parker, 1998).

Late 20th Century economics is fast reducing the number of river wharves and warehouses, but regeneration of the river fronts is replacing this in the form of modern housing facing the water. Wharf areas are in demand for leisure activities such as mooring, parks and promenading. In the early part of next century, it is probable that the river will recover some of its margins from historical trespassers, as revetted and stepped embankments encourage environmental improvements (such as the Environment Agency have encouraged to be built at the Greenwich Millennium site).

DESK STUDY WORK

In investigating such riverside sites, the geoenvironmental professional needs to step back from immediate action and consider the layers of history. Many of the clues to the jigsaw beneath the wharf will lie in Museum and County archives, but much will not. However, the remnant information will allow excavations to take place at the appropriate locations rather than randomly. The site will then tend to reveal further clues upon careful excavation.

Archaeological matters cannot be dismissed during such investigations, especially given planning guidance PPG 16, and particularly in London Boroughs where there is specific guidance. Riverside sites have yielded much data on past generations, such as roman finds, early boats/preserved timbers, palaeo-environmental and waterlogged remains in the peat, or domestic debris from prehistoric settlements on gravel 'islands' in the former marshland. Combining the initial desk study and any geotechnical excavation work in trial pits with an archaeological watching brief may reduce the number of intrusive investigations eventually required for wharf developments. Certainly, the extent of city mapping that pre-dates the Ordnance Survey is frequently a surprising factor in archaeological studies; these can reveal invaluable clues to the piecemeal development by man.

Example 1

The geoenvironmental investigation of a large riverside site in West London included a study of potential cross-boundary contamination within/upon the gravel strata, at depth. Monitoring revealed a hydrogeological regime perpendicular to expected flowroutes towards the River Thames, 250m away. An archaeological desk study uncovered maps of this part of London dating back 400 years, indicating the tidal creek patterns on the marshland/pastures. It also uncovered details of the canalisation of a tributary creek in the late eighteenth century, before its infilling for railways. These factors indicated a main tributary crossing the site (along the boundary), enlarged by man and probably infilled with granular spoil for the railway. Hence the hydrogeological flow gradient along this 'buried channel' rather than across it.

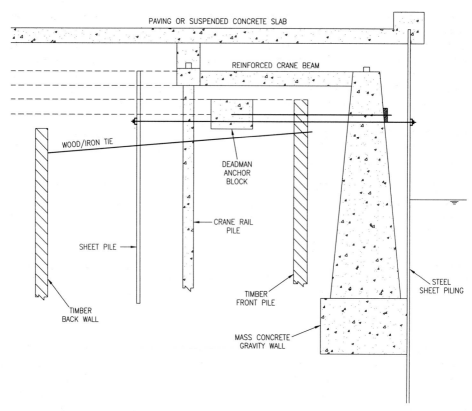

Figure 1 – Typical Wharf Frontage Succession

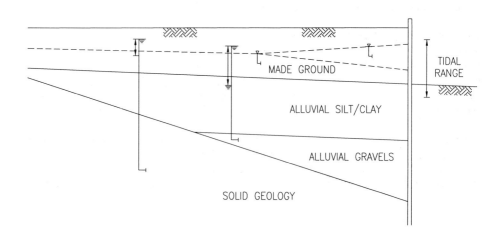

Figure 2 – Schematic – Typical Geological Sequence

The historical use of riverside wharves for industrial purposes encompasses many former potentially-contaminative uses - oil terminals, timber treatment, scrap metal yards, asphalt plants, to name but a few. In several cases, more than one usage has occurred and the actual activities carried out on site have varied as commercial practices changed. Useful information can be gleaned from local libraries, town archives and (Kelly's) commercial directories; however, much of the past activity on wharf areas was generally ad-hoc and unrestricted by formal inspection. A series of publications (DoE. series of Industry Profiles) will give guidance to the potential contaminants released by various industries, which assists in targeting investigation testing. Aerial photographic libraries are an important source of 20th century record of change of sites, and town centres have frequently been filmed both in wartime, and more recently for leisure publications.

Example 2

Developments along the river front in the Medway Towns had taken place since the Romans forded the river. A multiplicity of wharfs and industrial uses have occurred, including Napoleonic era fortifications/boat building, leading to Shorts' flying boat factories this century. When researching this history of an era of wharfing, the local library Archives produced several victorian Company publications alongside the usual town maps. The local Bridge Trust had records of the sequential construction of its bridges, indicating the adjacent wharf (under investigation), its names and uses at each stage. Thus a minor topographical variation in wharf edge on successive O.S. maps could be defined as a buried former divider wall. This was confirmed on site during initial remediation, and was not therefore an 'unexpected obstruction' causing delay or redesign.

Consideration also needs to be given to the evident point that the practices of past generations are not the same as the environmental protection standards of today. Soakaways and drainage systems may, in the past, have connected industrial washwaters to the underlying gravels or direct to the river. Tidal influences, and broken drains due to wharf movements, will have complicated the 'natural' percolation of such contaminants. Also 'obstructive' are the historical walls and buried construction, leading to a legacy of spilt chemicals which can have built-up a substantial depth of contamination in the ground below wharves; mostly in the made ground, but also in the top layers of the alluvial clay/silt beneath. Old drainage records and abandoned outfalls on the river wall are pertinent clues to be studied.

GEOTECHNICAL INVESTIGATIONS

There are two principal aspects here, the natural geology and man-made landraising. The former is generally predictable in an estuarine environment, albeit nature will have created its own buried channels and variable deposition. The latter is completely unpredictable, as man has probably had several attempts at infilling the land, using whatever material was handy (or economic) at that time, from river dredgings to chalk spoil to WWII bombing debris.

Typically, in the Medway and Thames, some 3-5m of made ground may be present over 3-5m of alluvial silts/clays then a similar thickness of alluvial gravels, before basal beds are encountered. In London's Docklands, the gravel thicknesses tend to predominate (Howland, 1991). Figure 2 indicates a rough schematic, showing some tidal influences on groundwater.

The made ground will have variability, both in extent and in vertical composition, related to what was deposited, how and when. Whilst it has frequently been compacted by transient loading, the fill will contain numerous local water bodies trapped by the use of cohesive material such as reconstituted chalk or silt dredgings. There will also be, in many cases, a perched water table upon the underlying Alluvium; this is particularly relevant to contamination aspects covered later.

Example 3
The same site described earlier comprised four distinct historical wharfs with dividing wet and dry docks. It was apparent, from old maps, the ages at which each dock was infilled to join the wharfs into a single waterfront unit by the 1970's. During initial investigation (by others), the different nature of the infill material was not highlighted and 'global' assumptions had been made about the materials' behaviour. In a second stage investigation, examination of each location was used to define whether the made ground and docks infill was river dredgings (silt-19th century), chalk spoil from adjacent quarrying (early 20th century), or surplus waste slag from a steelworks (mid-20th). This data then informed the analysis of geoenvironmental remediation studies, allowing more specific targeting of decontamination.

Adjacent to the wharf edge(s) and where piled structures were evident in the past, then the made ground may not have been loaded (or even been compacted when placed). Thus careful investigation of these upper layers, both visually and physically, might reveal the potential for significant settlements even under light, transient loadings. Sampling for oedometer tests needs to carefully take this into account. The natural Alluvium material is invariably very soft and of little use for load-bearing support. But, as well as comprising soft clay layers, it has the potential to contain much peaty material and also sandy horizons. Thus there are twin concerns over compressibility and downdrag on piles. Its impermeable nature is, however, of significant benefit to restricting downward passage of residual or mobile contaminants, if the clay is of sufficient consistent and cohesive thickness.

Example 4
Where major structures or crane rails are found on historical site maps, the ground alongside their (piled) foundations requires careful analysis. On one site, with very similar materials on both sides of the crane beam, it was only a comment by the borehole driller about the comparable ease of penetration of the riverside hole that alerted the site engineer to undertake a further review. A quick series of Macintosh probes along and across the site (after cover slab removal) confirmed a much less consolidated profile behind the river wall, beneath the crane area. Further soil sampling and analysis indicated that preload and active compression of the made ground had only been applied landward of the crane rails. The expectation of excessive settlements on the (uncompacted) infill beneath the crane rails and directly behind the river wall informed the structural designer's subsequent analysis.

The alluvial River Gravels are normally in a medium dense state rather than dense/very dense and occasionally loose where there has been minimal overlying clay/silt (or in 'scour hollow' deposition). Commonly, cobble-size material, whether river transported debris or of erosional origin, will occur randomly in the gravels (often affecting piledriving). The gravels are usually in hydrogeological continuity with the river, through the bed of its mainstream channel. Thus there will be significant tidal response, albeit with associated lag times.

In most built up areas, and particularly in London, the gravel aquifer is generally brackish/saline and only of a poor to adequate quality, as a result of commercial activity; thus it is not much used as a resource now. It is, however, often classified by the Environment Agency as a minor aquifer, which may require local remediation measures to protect existing abstraction licences. Where major aquifers such as the Thanet Beds or Chalk lie directly below the gravels, then appropriate precautions need to be taken (in liaison with the Environment Agency) to prevent cross-contamination from the perched water above, particularly when drilling boreholes or piling.

In geotechnical terms, the natural strata should provide very little in the way of atypical data for estuarial conditions. Note needs to be taken of the effects of tidal changes in groundwater levels and the potential for strata horizons to vary due to past geological/fluvial actions (especially scour at the front of walls). As well as the potential for obstructive cobbles in the gravel layers, there have been coastal locales where silt and sand particles have 'pumped' into the gravel interstices under tidal action, forming a virtually concrete-bound layer and making drilling (or pile-driving) progress very laborious.

CONTAMINATION INVESTIGATIONS
In respect of the contamination element of geoenvironmental investigations, comment has already been made concerning the leakage of drains/soakaways, sometimes related to transient movements of the wharf structure. The made ground will have been laid down by different methods, and most probably at different times using a variety of materials (some cohesive, some not). Thus, a mélange of ground layers may be present. In all such investigations, safe working parameters need to be followed (Steeds et al, 1996).

This makes the investigation of contaminated areas difficult to be precisely aimed; inevitably 'surprises' will occur during full-scale remediation. Relating discoveries to past practices is less rational than, say, on a factory site on land, where processes were more frequently fixed in one location through history (or, at least, were better documented). The small-scale nature of wharf businesses, and the different economic constraints of the past, make definitive conclusions from contaminated site investigations elusive creatures.

Example 5

One wharf area had a history running from timber treatment and storage in previous centuries, through asphalt, oils and concrete works to scrap metal car sorting post-WW2. An expanded suite of chemical testing was used in specific target areas to attempt to identify any residual spillage effects. This particularly involved an understanding of the 'natural' chemicals used before syntheticisation this century, and their potential to disperse within a buried environment or groundwater. Conclusions were indicative, but not definitive; they did, however, allow specific excavation targeting by the site environmental scientist during remediation excavations. For example, heavy oils from the scrapyard work appeared to 'fix' onto the upper 0.5m of chalk infill thus requiring removal of this layer but not the full depth of chalk (as originally envisaged after the first stage SI).

Apart from the 'traditional' chemical testing of soils and water garnered from intrusive excavations, leaching tests are necessary to define contaminant mobility (and also the suitability of 'inert' excavated material, for reuse in the works). Other techniques of potential value include interface metering of installations (especially over tidal ranges), bail-out tests for hydrocarbon concentrations (and recovery), dye tracing of leakage paths (combined with foreshore inspections), and down-the-hole soil vapour surveys. In most of these cases, the economy, speed and minimal disturbance provided by window sampling or mini-auger techniques is to be encouraged.

Example 6
A large wharf area had been constructed in the 1960's out into a tidal estuary, using a mixture of sheet metal piles and mass concrete panels within I-piles; infill was surplus motorway spoil, predominantly chalk. An initial investigation (by others) whilst in use as a scrap vehicle importer and breakers yard, indicated widespread hydrocarbon contamination in the soils beneath much of the wharf. After site closure and clearance, more detailed analysis of the waters in monitoring boreholes indicated (by interface meter) that there was now floating product in only one location. By sampling, baling tests for quantity/inflow, and GCMS typification, this turned out to be used engine oil (perhaps from a departing tenant). The effects of two years of tidal action appeared to have 'flushed out' earlier mobile hydrocarbon contaminants, and a more specific remediation programme for the wharf was designed after discussion with the owners/regulators.

Predictions of the hydrogeological regime affecting the site are also difficult, due to tidal influences and the numerous perched water tables. It is imperative that the movement of water is clearly understood, in order that design detail is fit for purpose and to inform discussions with the regulatory bodies. A sequence of tidal monitoring is required in order to understand the permeable and impermeable zones beneath the site. The appropriate location and screened zones of monitoring points is best decided after the first stage intrusive investigation, otherwise money will probably be wasted by 'random' location of wells.

Finally, the potential for gassing on riverside sites must be thoroughly investigated. Whilst there may be methane-generation from alluvial peat, this can frequently be proven to be of limited volume/flow (and hence of low risk) by appropriate investigation. More difficult to quantify is the vapour dispersal potential of contaminants, particularly certain hydrocarbons, in the made ground or within perched aquifer water. Adequate measurements will allow a detailed assessment of risk to be carried out, which will inform discussions with the local Council's environmental health officers.

STRUCTURAL INVESTIGATIONS
A sensibly-scoped site investigation (as recommended within SISG, 1993), incorporating both on-land drilling techniques and foreshore investigation/probing, will normally provide accurate geoenvironmental data for design. The determination of the extent of buried structures will be a major factor that informs subsequent design, and assists checks of the existing structural stability. In order to determine buried obstructions, both intrusive trial pits and non-intrusive ground radar/geophysical techniques are recommended. The latter can frequently be difficult to interpret, due to the inherent made ground variability and the masking effects of concrete (and steel) in the near-surface. However, one particular technique that has been used successfully is gamma-density 'sounding', within a borehole drilled directly behind (or in front of) the river wall.

Example 7
At a tidal river site, historical drawings existed which indicated two different lengths/types of sheet piling along one section of 1960's wharf. It was considered, by inspection, that these details probably represented (i) design and (ii) as built (differently). Designer's analyses indicated that the longer piles would be suitable for incorporation in the new development, without strengthening; the shorter ones would not. To provide confidence as to the length of sheet-piling, boreholes were drilled directly behind the piles and a specialist NDT firm carried out analyses using down-the-hole instrumentation. This confirmed a consistent, deeper toe to the piles. The extra cost was solely the NDT firm plus driller's standing time of two hours/holes; the location of boreholes having been moved to provide both geotechnical and NDT information in 'one hit'.

A useful reference document on the subject of older river walls is CIRIA Report B13 (Bray & Tatham, 1992). The Environment Agency, and other concerned river bodies such as Port Authorities, will have carried out audits of the wall conditions in major locations; they may also have records of previous remedial/construction works on file. Many new foundations come up against the buried history of the wharves in the form of obstructions. Of most concern is constructing basement floors or deep trenches close to the river edge, where old anchor ties may still be operational (albeit even when the old dock wall has been buried). Removal of these anchorages without consideration of load redistribution onto newer river walls has led to distortion in some cases, and considerable redesign costs. As a corollary to this, it perhaps over-conservative to design for conventional earth forces on a new sheet-pile wall when this is driven in front of an existing wall, and the gap structurally infilled (without removal of restraint to the old wall).

The condition of structures in marine environments is a subject in itself, and evaluation of their remnant lifetime will involve both intrusive and non-destructive methods (Summersgill, 1989). Of relevance to most wharf walls (and particularly sheet-piled ones) is the extent of 'hidden' corrosion on the buried face. Simply measuring the remnant steel thickness from the front face may not reveal the rate of unseen corrosion behind the pile; in fact, it may provide unsafe data if there are significant forces at work there.

For example, some low steel thickness readings have been found to be due to micro-laminations in the centre of the (rolled) sheet-pile, whilst thicknesses greater than the original rolling specification can be a warning sign of partial corrosion on the rear face. Obviously, if there is a voided area behind a wall, tidal fluctuations will occur there and aerobic corrosion activity could take place as swiftly as on the outside face. If possible, trial pits directly behind the wall and holes drilled through it (to confirm NDT thickness readings) are strongly recommended, as they will reveal more to the assessor than 'remote' testing.

Example 8

A substantial dock site had been the victim of a wall failure during the 1953 tidal surge in the Thames. One wall was rebuilt with steel sheet-piles, the other two being unaffected and of concrete construction with tie-backs. An investigation of the sheet-piles after 40 years in service indicated localised repairs and a requirement for cathodic protection to the faces. The tie-bolts appeared secure both in the face and behind the panels, where visually inspected (and accessed). It was not clear, from records, whether new anchorages were installed in 1954 or if the old concrete walls' anchorages were used. Selected tie-bars were thus exposed full length and the tie-backs inspected; there was found to be a secondary sheet-pile wall, equally in good condition. This assisted the designer's analysis; sealing/wrapping of the tie-bars at turnbuckles and corrosion pits on the bars was incorporated as part of the trial pit reinstatement work.

In many cases, corrosion behind the wall can be exacerbated by entrapped contaminants, subject to some tidal fluctuation but which cannot escape from their (hydraulically) confined location. In such an event, mobile hydrocarbons or acidic compounds may accumulate at the physical barrier of the wall over many years/decades. Tidal fluctuations merely spread the contaminants over a deeper zone of soil or concrete/steel in this zone of the wharf. Although frequently difficult and expensive, full-depth coring of concrete walls will allow detailed petrographic examination of both concrete faces, as a first step towards evaluation of potential attack on both the buried rear and visible front surfaces.

Where existing anchorages are considered to be unserviceable or of doubtful remnant lifetime, it may be more cost-effective to supplement rather than replace the units. One technique coming into usage is that of inclined anchor ties (Soudain, 1998), drilled from a platform over the riverfront. Relying on anchorage in the underlying solid geology (and/or gravels), an important element in their success is sufficient knowledge of past, buried walls that may need to be drilled through at depth and at angles.

CONCLUSIONS

Redevelopment cannot take place without adequate investigation of the aspects noted above. The cost of such investigations can be quite high, but the potential capital savings in, say, re-use of an existing river wall (through renovation) are substantial. Similarly, decontamination costs may be overestimated if the site is viewed in isolation, rather than in its local environmental context (particularly in groundwater remediation).

Redeveloping old wharf sites requires a detailed appreciation of historical, geotechnical, fluvial, structural and contamination characteristics. Each site tends to have its own unique problems and a widely-trained soils/water engineer is often better equipped to manage such investigation than a specialist in only one particular field. The input of specialist knowledge is, however, crucial to a full understanding of the site's history and its potential impact.

ACKNOWLEDGEMENTS
The Author wishes to acknowledge the many current and former clients on the Thames and Medway, who may recognise their individual sites reflected generally herein. Additionally, the site expertise of several specialist testing organisations is acknowledged. The contents of this Paper are solely the Author's personal views, not necessarily those of WS Atkins.

REFERENCES
Bray, R.N. & Tatham P.F.B. (1992) - *Old waterfront walls - management, maintenance and rehabilitation.* CIRIA Report B13. E&FN.Spon. London, 1992. ISBN 0-419-17640-3.

Howland, A.F. (1991) - "London's Docklands: engineering geology" in Part 1 of *Proceedings of Institutions of Civil Engineers.* London, December 1991 (Paper 9659 pp 1153-1178. ISSN 0307 8361.

Ogley, B (1991) - "Tidal Surge of 1953 - New Defences for Kent" in *The KENT Weather Book,* Froglets Publication, Westerham. October 1991, pp *68-70 & 85. ISBN 1-8723-87-35X.*

Parker, A (1998): "Warming to Limpets" - *New Civil Engineer.* London., 12/3/98, pp 14-16.

Site Investigation Steering Group (SISG) : *Site Investigation in Construction.* Part 2: Planning, procurement and quality management. Thomas Telford, London 1993. ISBN 0-7277 1983-1

Soudain, M (1998)- "Multiple Choice" - *Ground Engineering.* London, October 1998, pp 20/1

Steeds, J.E., Shepherd, E and Barry, D.L.(1996) - *A guide for safe working on contaminated sites.* CIRIA Report R132. London, 1996. ISBN 0-86017-4514.

Summersgill, I.M. (1989): "Remedial Works on Grovehurst Jetty, Kent" in *Proc .3rd Int.Conf Structural Faults & Repair* - London, July 1989. Vol 1, Engineering Technics Press, pp.347-353. ISBN 0-947-644-09.

Multifaceted Evaluation of the Ecological Risk of Maintenance of Oil and Gas Fields in Western Siberia

DR. E.A. VOZNESENSKY, Associate Professor, DR. E.N. SAMARIN, Associate Professor, Department of Engineering and Environmental Geology, Moscow State University, Russia
DR. V.V. VOZNESSENSKAYA, Senior Researcher
A.N. Severtzov Institute of Ecology and Evolution RAS, Moscow, Russia
O.YU. SAMOILOVA, B.Sc., Graduate Student
Faculty of Geology, Moscow State University, Russia

THE SCOPE AND THE MAIN TASKS OF THE STUDY

An inter-disciplinary methodology to investigate pollution from maintenance of oil and gas fields in Western Siberia and to explore potential impact of new ones has been developed and tested at Luginetzkoe oil/gas condensate field (OGCF) in Tomsk district in 1998. The field and laboratory investigations focused on the impact of oil, its derivatives and other chemicals on human health and biocenoses. Evaluation of ecological risk in such a system must be based on the study of mutual interrelationships of the various components of natural conditions. This caused the involvement of the specialists from different scientific fields (engineering geology, geophysics, biology, ecology, analytical chemistry) in the project.

This ongoing project is aimed at the solution of the following main tasks:
• to reveal the principal regularities of the oil pollutants migration in the environment from the various sources in the soils, surface and ground water, soil biota (with the due consideration of their sorption and transformation) and, finally, via the food chains to small mammals;
• to study the peculiarities of the oil derivatives migration in soils and to quantitatively estimate their sorption capacity with respect to different kinds of pollutants;
• to develop multi-species test-systems basing on the complexes of soil invertebrates allowing to evaluate in situ the degree of oil contamination of ecosystems;
• to determine concentration levels of pollutants having no evident influence on the biota;
• to estimate the indices of productivity and biodiversity of mammals in the areas with different level of contamination and to find the reliable parameters characterising the degree of deviation from the reference conditions and the influence of pollution on the frequency of chromosome distortions in small mammals.
The final purpose is to find the most informative and reliable criteria for evaluation of ecological risk of maintenance of oil and gas fields.

METHODOLOGY OF THE MULTIFACETED APPROACH

All investigations have been carried out in Luginetzkoe OGCF which is under development since only 1982, so the total industrial load on the environment at present is not high here. Moreover, part of the licensed territory of this field is almost intact and can be considered as the area with reference conditions.

Geoenvironmental engineering, Thomas Telford, London, 1999, 191–198

The research included combination of methods of environmental geology, geophysics, bioindication, cytogenetics and analytical chemistry and is based on the idea that reliable control of concentration, regularities of sorption and the complete tracing of oil pollution in the environment from the source of contamination through the soils, ground and surface water, soil biota and, finally, via the food chains and nets to small mammals and birds are needed for ecological risk assessment. The most dangerous toxic compounds in oil and gas fields are polyaromatic hydrocarbons, phenols and amines. Resulting from multiple spills and accidental leakage of oil light saturated hydrocarbons reach ground water table and spread laterally in the direction of water flux to the places of discharge, the heavy fraction is being sorbed on soil minerals. Only the direct determination of oil and its derivatives concentration in soil and ground water can provide a basis to understand the results of bioindication. Field investigations in Luginetzkoe OGCF were run within a series of "key sites", distinguished according to the lithology of the upper part of geological section and the distribution pattern of the potential sources of contamination. Luginetzkoe is exploited since 1987, now there are 480 productive wells. The primary sources of pollution are: productive oil and gas wells, casing-head gas burning facilities ("torches"), oil reservoirs, local and main pipelines, pump stations and roads. For 1998 studies 5 key sites 8,2-11,7 km^2 each were selected; they fell into 3 grades - heavy, medium and intact - according to their possible contamination and technological infrastructure.

Site investigations, sampling of water and soils with the determination of their density and moisture content, permeability tests, geophysical studies - georadiodetective sounding of soils in the frequency range 25-25000 MHz (ground penetrating radar - GPR), express-analysis of the quality of surface and ground waters (Eh, pH, conductivity, salinity), geobotanical profiling, bioindication studies (capturing and identification of rodents, study of soil mesofauna) were carried out simultaneously in all key sites. All the obtained information was used to develop the geographic information system (GIS) of the Luginetzkoe OGCF. This work included: a) study and processing of the available records, b) processing of the new data from the numerical cosmic photos by RESORCE-F in May 1993, c) investigation sites positioning by the global positioning system (GPS) Magellan 2000, d) field refinement of the indicative attributes for decoding of cosmic materials, e) input of site positioning data into computer and integration of all the numerical layers into the general GIS.

Methods of soils and waters investigation
Observation and sampling points were related to the groups of productive oil and gas wells of different age, casing-head gas-burning torches, places of oil spills and leakage. All these points were positioned with the help of GPS. The upper part of geological section has been studied by means of hand drilling: the mean depth of boreholes was 7-8 m (maximum 12,7 m) and in prospective pits 2-3 m deep (in technogenic, lacustrine-paludal and blanket deposits). In the every productive wells group one drift was located in the drilling slime sump, one - on the poured off soil (wells were drilled from the sand fills about 1 m thick), two boreholes were made in the direction of ground water flux from the wells. In cases of surface oil spills the upper part of aeration zone was studied in a pit 2-2,5 m deep. Soil samples were taken from all the drifts. Water probes were taken from the boreholes reached the ground water table, rivers and small creeks, and from all sources of long-term contamination. Additional soil samples were taken along the radii around the gas-burning torches from the depth of 5 and 40 cm. Soil samples were also obtained in the sites of bioindication studies from the depth of 5 cm. The subsequent laboratory tests included study of soils mineralogy, grain-size distribution, index properties and total carbon content (burning with potassium bichromate). Contamination has been evaluated using 4 most important from the environmental point of view components: to-

tal hydrocarbons content, concentration of polycyclic aromatic hydrocarbons (PAH), phenols and anionic surfactants. Routine techniques have been used in this part of the study: anionic surfactants were determined in the form of ionic associate with methylene blue, phenols concentration by reversed phase high performance liquid chromatography (HPLC) with electrochemical and fluorescent detectors and dynamically modified sorbents; hydrocarbons with concentration less than 500 mg per 100 ml or 1 g of soil were detected by thin-layer chromatography on «Syluphol» plates using hexane as a mobile phase, and with higher concentration - gravimetrically; PAH were estimated to naphthalene content using the normal phase HPLC.

Multipurpose device Zond-12c ("Radar Inc.", Riga) and data processing software (Department of Seismometry and Geoacoustics, Moscow State University) have been used in GPR studies. The reflecting layers were identified basing on the different wave types and correlated with the available geological and hydrogeological information.

Methods of bioindication
Bioindication of the environmental pollution was based on the evaluation of the present state of vegetation, soil invertebrates and rodents in the areas with different influence from the oil and gas production facilities and other sources of contamination.

Soil mesofauna was studied in the equivalent biotopes but with different pollution level near the old casing-head gas burning facility surrounded with oil spills. The relative number of different groups of soil invertebrates (segmented worms, earthworms, worms-nematodes, different groups of insects and their larvae, spiders, myriapods, mollusks, etc.) has been estimated according to the routine sampling technique: 16 specimens 0.25x0.25 cm each to the depth of the uninhabited mineral horizon were taken from each site. The number of samples in every series (16) is sufficient to obtain the reliable data. A model test has been also performed on the cultures of Enchytraeus crypticus and Eisenia foetida worms. These soil organisms are very sensitive to oil contamination and can be easily cultivated in laboratory conditions. The toxicity of soils from polluted and reference sites has been studied and the reproduction of worms have been compared with the one in control OECD mixture (10% crushed peat, 70% fine silica sand, 20% kaolinite with addition of calcite to get pH close to 6,0).

Small rodents were captured in animal traps 10 m from each other put into lines along the natural boundaries of the biotopes. Each line comprised of 25 traps. 6 lines were located close to the contamination sources and 5 - far from such places. All those biotopes were distinguished as «forest», «cuttings» and «intermediate» ones. Then the weight, length and mass of internal organs of every animal have been determined. The internal organs index is found as a ratio of their mass to the total weight of the animal. The state and morphometry of internal organs including liver, spleen and digestive tract have also been studied. As a whole over 150 animals of different species from different biotopes have been investigated. Samples of somatic chromosomes for the study of chromosome aberrations were prepared according to S. Ford and G. Hamerton modified technique from the marrow cells.

SITE DESCRIPTION
Geologically the territory of Luginetzkoe OGCF is located on a monocline with the gentle inclination to the north and north-west, complicated with a gentle Luginetzkaya brachyanticline. The sedimentary cover is composed of mostly terrestrial Jurassic-Neogenic rocks everywhere overlaid by the Early Pleistocene lacustrine-alluvial clayey sediments of Kelvatskaya Formation or sandy-clay deposits comprising three river terraces and well-developed floodplain.

Geomorphologically the studied territory is situated on the Early Pleistocene lacustrine-alluvial plain, composed of Kelvatskaya Formation and the enclosed into it complexes of terraces of Chizhapka river and its tributaries. The third terrace is a base one, it has a flat surface and within the field is composed mostly of clayey lacustrine-alluvial sediments. The second and the first terraces are sandy and also flat. The well-developed floodplain is 1-2 m high and 1-2 km wide with a crest-sink surface, pronounced swells along the river bed and oxbows. Among the geological processes the most important for this territory is the intensive development of swamps and peat accumulation. Neotectonic movements, rivers erosion and accumulation, influencing the intensity of undermining and related stability of slopes should also be considered.

Hydrogeologically the territory of Luginetzkoe OGCF belongs to the West-Siberian artesian basin having a ternary section: two hydrogeological levels and interstitial waters of Paleozoic foundation. Two complexes comprise the upper hydrogeological level, confined in the Eocene-Holocene sediments and separated from the lower level by thick impermeable Turonian-Eocene deposits with separate aquiferous layers. Basing on our field investigations in addition to this regional scheme we have distinguished two sporadic aquiferous beds: 1) in the technogenic deposits, and 2) in the lacustrine-alluvial soils of Kelvatskaya Formation. All the units of the first hydrogeological complex are closely related with each other. They recharge from the precipitation and outflow from the deeper layers, the significance of the latter decreasing downwards. All waters are insipid with mineralization up to 0.4 g/l and with high phenols content - up to 0.01 mg/l caused by the contamination during the drilling of wells. The lower hydrogeological level is characterised by a high mineralization - up to 60 g/l and a constrained water exchange with the overlaying aquiferous units and complexes.

Cryopedologically the territory is located in the subzone of the seasonally frozen soils and a permafrost potential development under an anthropogenic influence. The average annual soil temperatures are positive - mostly from 0 to 3^0C. The depth of seasonal soil freezing varies from 0.5-0.6 m on peat to 3-3.5 m near well drained sandy slopes of alluvial terraces. For the natural central taiga landscapes of the territory the thickness of the seasonally freezing layer in silty-clayey soils usually comprises 1.2-2 m. Frost heaving and cracking are wide-spread.

CONTAMINATION, ITS CONSEQUENCIES AND WAYS OF RISK EVALUATION
The obtained data revealed the main regularities of the pollutants migration in soils and ground waters, the most dangerous sources of their contamination, for the first time we can demonstrate the impact of gas torches on soils, vegetation and soil-dwelling animals using new reliable and independent indices. The GIS and attributed databases are used to retrieve, analyse and present the multifactorial information, necessary to evaluate the ecological risk in oil/gas fields. The synthetic electronic map of the environment stability and resistance to the impact of oil/gas mining complexes will be developed in this ongoing project.

Patterns of soil contamination
The comparison of the measured concentrations of hydrocarbons (HC), phenols, PAH and surfactants in the soil profiles near the casing-head gas burning facility (torch) #1, working since 1987, has revealed low HC (less than 1.8 mg/g) and, somewhere, surfactants (5-10 μg/g) contamination. The zone of a pronounced influence of torches on soils appeared to be confined within 300 m from the facility - this is the distance, where the stable decrease of pollutants concentration in soils begins. The concentrations measured in the soil samples from the profile near the 2 years old torch #2 gave evidence of a very low contamination level: phenols - <0.7 μg/g, PAH - <0.2 μg/g, HC - <0.3 mg/g and surfactants - 5-7 μg/g. The radius of con-

tamination zone around this torch can be estimated as 300 m as well. The most dangerous for soil biota compounds in all samples are phenols and PAH. A very informative index of contamination appeared to be the total organic carbon content, varying from 0.7% for podzolic soil (2.3% for a marshy one) at a depth of 5 cm to 9-10% 300 m from the torch due to sedimentation of gas black from the burning gas condensate.

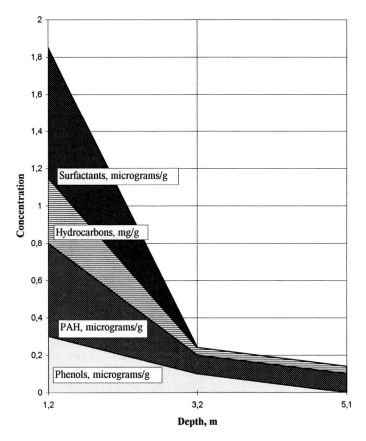

Fig.1. Decrease of concentrations of different pollutants in a borehole with depth to ground water table

Contamination of soils in aeration zone has been studied on all the geomorphic levels. It has been revealed at the sites with the maximum industrial load in soil massifs adjacent to the groups of productive wells and situated along the ground water flux from the latter - to the first (3-4) meters deep. Below the concentrations of all components decrease by the order of magnitude (Fig.1). The most fine soils - loams and clays were found to be more contaminated (the maximum concentrations of phenols - up to 0.8-1.3 µg/g, PAH - up to 30 µg/g, HC - up to 10 mg/g, surfactants - 14 µg/g). Besides, all lenses of sands and sandy silts near the wells were, as a rule, polluted at least 5-6 m deep. The low barrier properties of poorly decomposed peats with respect to oil and its derivatives were also detected. Moreover, if peats are located along the ground water flux, they tend to accumulate toxicants, especially HC, due to the high water-retaining capacity. Soils from the sites with reference conditions either contain practically no organic components under consideration or their concentration is below the maximum permissible level (MPL). It has been obtained that: a) the principal persisting source of soil contamination, especially with phenols and PAH, are drilling slime sumps; b) the duration (first years) of existence of these sumps and other pollution sources is very important; c) the most distinct barrier for oil migration from spills is the contact of sand fills and underlying clay soils.

Shallow ground water table and mostly clayey type of geological section sufficiently complicated the interpretation of GPR data and decreased its penetrating depth to only 1-2 m. The most informative for detecting water table level was a 300 MHz antenna. In relatively homogenous soil section the reflecting boundaries are associated with their different saturation. The best results were obtained for the soils with dielectric permittivity not exceeding 8-10.

Contamination of surface and ground waters

Two small rivers, crossing this territory in the latitudinal direction, have different level of contamination. In one of them no pollutants were detected as a result of the natural dilution during their transport from the local sources. However, in another one the HC concentration in separate specimens is almost by the order of magnitude higher than the MPL - up to 0.4 mg/l due to the location of productive wells practically at the boundary of the rather narrow water-protecting zone. The major source of surface and, probably, ground waters contamination are slime sumps, puddles around old torches and technological oil spills, altogether forming a technogenic aquiferous horizon in sand fills of productive wells sites. Similar to soils a relationship between the duration of a wells group operation and pollutants concentration in ground waters is observed, being the most distinct for HC and surfactants. In general, low but stable phenols contamination (0.03-0.05 mg/l) is observed in the aquiferous units of alluvial deposits of the third terrace and in Kelvatskaya Formation because of their mobility: unlike surfactants and PAH, phenols are water-soluble and they are practically not sorbed by soils.

Vegetation status

The natural vegetation of the studied territory is represented mostly by forest communities of south taiga with the prevalence of dark coniferous and small-leaved boreal forest with dominating cedar and spruce. Bogs, mostly raised ones, are concentrated together with the swampy forests in the central parts of interfluves. The diversity of plant communities is conditioned by the humidification, causing the development of an ecological series with the branches of circulating and stagnant humidification.

Under industrial influence phytocommunities are either destroyed or they gradually transform and degrade. The major disturbances of vegetation are caused by the total destruction of plant communities followed by a partial disturbance of soil cover during the construction of wells. The principal chemical stresses for vegetation - sewage and burning gas - result in destruction of arboreal species and development of cat-tail and reed grass scrubs, stable to contamination and simultaneously contributing to purification of waters. The structure of forest communities does not change with the distance from the source of pollution, and morphological disturbances are observed only in regrowth. Both thermal and chemical influence of burning gas results in stackheadedness and necrosis of coniferous species, morphological transformations of the branches of leafy ones and in early partial of complete coniferous needles necrosis, especially in regrowth, as far as 120 m from the torches.

It is fir that was found to be the most sensitive species to contamination. Since the fir prevails in regrowth, the reproduction of forests in contaminated sites is endangered. At present degradation of vegetation is observed only in the close vicinity to the groups of wells. However, considering the effect of time as a dose factor and a possible increase of oil and gas production, we can suppose a worse status and reproduction of forests in this territory with time.

Response of soil biota to pollution

The effect of oil pollution on the soil fauna has the following negative consequences (Fig.2): 1) disappearance of the soil-forming groups - segmented worms, earthworms (Lumbricidae) and Enchytraeidae; 2) two-fold decrease in number of the major soil-litter dominants - spiders and rove beetles (Staphylinidae) - the principal soil entomophages, causing the alteration of the trophic structure of the whole community; 3) abrupt decrease of biodiversity. The model tests have demonstrated the highest death rate for adult soil animals with the highest PAH, HC and surfactants content. Surprisingly, however, the second generation of Enchytraeidae appeared to be more sensitive to soil acidity than to oil contamination.

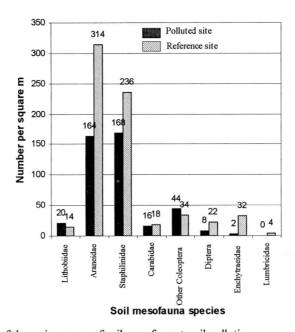

Fig.2. Response of the main groups of soil mesofauna to oil pollution

Status of small rodents

Large number of red voles, relatively small number of gray voles and the absence of field mice, observed in all studied "forest" biotopes, is typical for the forests of the south Siberian taiga. Field mice and gray voles, typical species of the succession biocenotic complexes, prevailed in the recently stressed biotopes (early succession stage). For the "intermediate" biotopes the number of different rodent species, as well as their sexual and age composition had the average parameters. Contamination of the territory has not influenced the sexual and age composition of voles populations. However, a relatively higher number of females and youngsters was observed in field mice from reference sites. The reproduction cycles of animals in polluted areas can also be somewhat (for 8-10 days) shifted with respect to those in reference biotopes, which is in agreement with investigation results. Any of these responses of field mice to oil pollution may be an indicator in the system of monitoring, but needs further clarification. Laboratory studies revealed abnormalities in the architectonics of the mucous of the digestive tract, sometimes even mucous hypertrophy or atrophy caused first of all by the disturbance of its microflora. Increase in weight of liver, spleen and kidneys was found in red voles from the contaminated biotopes. The highest ratio of animals with abnormally big liver and spleen - 28.6% was characteristic for animals captured close to the torch. Near the oil spills this ratio comprised 22.2%, and for reference sites - it was below 10%. Change of the spleen colour - its clarification, giving the evidence of the distortion of hemogenesis, as well as change of liver colour and distinctly visible vascular texture were found in red voles from the polluted areas.

GIS of the territory of Luginetzkoe OGCF

GIS technologies in such studies helps to create various the in-line maps, necessary for ecological risk evaluation. Basing on the results of field studies of 1998 we have prepared GIS layers which may be demonstrated. These layers characterise: 1) geographic situation, 2) industrial load on the environment, 3) geological conditions, 4) 1998 investigation data.

CONCLUSIONS

Field investigations, carried out in 1998 in the territory of Luginetzkoe oil and gas condensate field in Tomsk district, resulted in the development of a multifaceted methodology for evaluation of ecological risk. The main results of field and subsequent laboratory studies can be summarised as following:

1. The drilling slime sumps were found to be the major source of contamination of soils in aeration zone with oil and its derivatives. The main barrier for these pollutants near the productive wells is a contact between the sand fill and underlying, usually clayey, soils. Hydrocarbons, phenols, PAH and surfactants migrate downwards few meters deep.
2. Soils contamination with HC, PAH, surfactants and phenols around the torches is low and evident within approximately 300 m. The radius of this zone depends also on the "age" of gas-burning facility and the wind-rose.
3. Low but stable phenol contamination is observed in the aquiferous units of units of alluvial and lacustrine-alluvial deposits of the third terrace and interfluve plain because of the high mobility of this pollutant. A technogenic aquiferous unit in sand fills and slime sumps is the main source of surface and, probably, ground waters contamination.
4. Degradation of vegetation, caused by pollution, are rather local. Its main factors are the sewage and burning gas. Fir is the most sensitive species to their influence. Since the fir prevails in regrowth, the reproduction of forests in contaminated sites is endangered.
5. The soil mesofauna respond to oil contamination by the abrupt decrease in biodiversity, disappearance of worms, 2-fold decrease in the number of spiders and rove beetles, regulating the number of other Arthropoda.
6. Small rodents respond to oil contamination of soils by the proved increase of liver, spleen and kidneys sizes, as well as by their qualitative changes. The developed indices of liver and other internal organs may be used as a reliable indicator of environment pollution.
7. A good agreement among the results obtained by different methods of environmental geology, geophysics, analytical chemistry and biology provides a reliable basis for their integration in the multifaceted approach to the evaluation of ecological risk of the maintenance of oil and gas fields.

ACKNOWLEDGEMENT

The study is supported by the Russian Presidential Federal Program «Integratziya» under the project # K0151.

Detection and Monitoring

Soil Pollution Characterisation during Infiltration Tests

DR I. ALIMI-ICHOLA, L. GAIDI
Laboratoire URGC-Géotechnique, INSA-LYON, VILLEURBANNE, FRANCE

INTRODUCTION
Liner project efficiency (material choice and liner layer thickness) depends on soil - pollutant interaction. To carry out the pollutant transfer mode in unsaturated soil and pollutant action onto the soil, infiltration and standard laboratory hydraulic conductivity tests were performed on soil column. During infiltration tests, soil moisture and soil electrical conductivity were measured to observe moisture and leachate solute movement. Infiltration curves were used to describe leachate inflow and the steady infiltration rate. Moisture profiles allowed the observation of the soil saturation process and the determination of the leachate transit time in the soil column. Soil electrical conductivity profiles give the pollutant distribution in the soil column.

Studies have shown that the hydraulic conductivity of compacted soil permeated with leachate can be significantly higher than the hydraulic conductivity of the same soil permeated with water (Gipson 1985, Gleason et al. 1997). Laboratory permeability tests performed with different water head, carry out the influence of the leachate and the hydraulic gradient on the soil hydraulic conductivity.

MATERIALS AND METHOD
Pollution tests were performed with leachates produced by leaching waste clinker, lead slag and REFIOM. The percolation of these leachates through soil layer, was compared to water percolation. To achieve test results in a reasonable time, soil used must have a hydraulic conductivity close to 10^{-7} m/s after compaction.

Soil Geotechnical Characteristics
To characterise the tested soil, different geotechnical tests were performed, e.g. particle size distribution, Atterberg limits. Standard permeability tests were performed on compacted soil at different dry unit weight γ_d. These tests results were used to determine the soil unit dry weight to achieve hydraulic conductivity equal 10^{-7} m/s. It is noticed that this hydraulic conductivity value is obtained when the dry unit weight γ_d is equal to 16 kN/m^3. Particle-size distribution and Atterberg Limits tests showed that the used soil is clayey sand (SC). Proctor compaction test gave water content and dry unit weight at the optimum state. Geotechnical test results are summarised in table 1.

Table 1

w % natural	LL	IP	w_{opt}	$(\gamma_d/\gamma_w)_{opt}$
11	30	19	10	1.9

Soil suction for different volumetric water contents is measured to obtain soil water characteristic curve. Measured data and the Van Genuchten model curve are shown in figure 1. This model proves that the soil suction becomes zero when the volumetric water content reaches 33%.

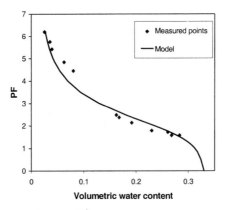

Figure 1. Tested soil water characteristic curve

Devices and Experimental procedure
For different tests, the dry soil was wetted at optimum water content (w_{opt} = 10%) and compacted to achieve a dry unit weight of $16 kN/m^3$. During infiltration tests, moisture movement was monitored with TDR probes. Soil water content and electrical conductivity were assessed by TDR pulse analysis.

TDR Method
Measuring the transit time and the amplitude of an electromagnetic pulse which propagated along a transmission line embedded in soil, soil dielectric constant and voltage reflection coefficient can be calculated. Topp et al.(1991) proposed the following relationship to determine the soil volumetric water content when the soil dielectric constant is known :

$$\theta = -0.053 + 0.29 K_a - 5.5 \ 10^{-4} K_a^2 + 4.3 \ 10^{-6} K_a^3$$

where K_a dielectric constant and θ volumetric water content

To improve volumetric water content determination, TDR probe was calibrated with the different leachates used for infiltration tests. The three used leachates have different pH, different electrical conductivity and different solute concentration. Chemical characteristics of these leachates are shown in table 2.

Table 2 Leachates chemical characteristics

Leachate	pH	Electrical conductivity (mS/cm)	Redox potential (mV(ENH)	Solute concentration (mg/l)
Waste clinker	7.9	12.63	317	7996
Lead slag	6.3	53.3	298	75446
REFIOM	7.6	11.2	312	6980

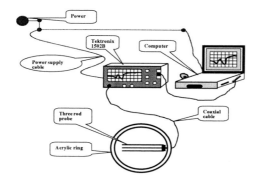

Figure 2 Schematic TDR method devices

Figure 2 shows different devices for calibration tests and for TDR measurement. Dry soil wetted at known water content is compacted in a ring equipped with a three rod TDR probe. Soil dielectric constant versus soil volumetric water content is evaluated with TDR curves. The calibration function of volumetric water content versus dielectric constant for soil used in infiltration tests is computed for the three leachates and water. The following relationships provide volumetric water content assessment during infiltration tests :

water $\qquad \theta = 8\ 10^{-5}Ka^3 - 0.0022Ka^2 + 0.0361Ka - 0.043$

waste clinker leachate : $\qquad \theta = -6\ 10^{-7}Ka^3 - 0.0006Ka^2 + 0.0256Ka - 0.0199$

lead slag leachate : $\qquad \theta = 4\ 10^{-5}Ka^3 - 0.0015Ka^2 + 0.0324Ka - 0.043$

REFIOM leachate : $\qquad \theta = 7\ 10^{-5}Ka^3 - 0.0033Ka^2 + 0.0541Ka - 0.0949$

Infiltration Test Device
The infiltration tests are carried out in PEHD column having 550mm in length, composed with rings of 50mm height and 106mm of diameter. The soil was wetted at desired water content and compacted at desired bulk density in each ring. Before soil compaction, five rings were equipped with three rod TDR probe of 80mm length. After the setting of the column, TDR probes are located at 75, 175, 275, 375, 475mm depth. The column is installed as it is shown in figure 3.

Table 3 Initial water content and unit dry weight during infiltration tests

	Water	Waste clinker leachate	Lead slag leachate	REFIOM leachate
Water content (%)	12.50	11.20	10.98	10.26
Bulk density γ_d/γ_w	1.6	1.6	1.6	1.6
h_0	0	0	0	0

The infiltration is performed at constant water head supply h_0. Table 3 presents the initial infiltration conditions for different liquids.

Cumulative infiltration versus time curves are determined from the variation of water level in Mariote burette. TDR curves obtained from Tektronix 1502B, are recorded in computer for further analysis. The records are made every 2 hours

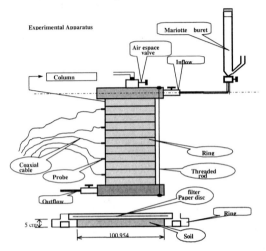

Figure 3 Schematic of soil column used for infiltration tests

RESULTS AND DISCUSSION

Experimental results concerned infiltration curves, moisture and bulk electrical conductivity distribution and the trends in the soil permeability coefficient. The use of different leachates must reveal the leachate action on the soil. The choice of a soil of a permeability equalises to 10^{-7} m/s, allows a percolation after one or two days. Moisture profile analysis must show the soil saturation degree when the liquid outflow starts. Soil bulk electrical conductivity profiles associated to water content profiles allow the comparison of moisture and solute migration.

Infiltration curves analysis

Figures 4 (a) and 5(a) show infiltration curves when water and lead slag leachate inflow the soil column. Water and lead slag leachate infiltration curves are represented in figures 4(b) and 5(b) by the Philip model :

$I = S\sqrt{t}$ where I is the cumulative infiltration and S the sorptivity.

It is noticed that the sorptivity is constant during water infiltration but its value changes during the lead slag leachate infiltration. Referring to Philip infiltration theory, the change in the sorptivity value corresponds to the change in the capacity of the soil to absorb or remove water.

Table 4 Soil hydrodynamic parameters

	v_0 (m/s)	S (cm/s$^{1/2}$)	D (m^2/s)
Water	2.01 10^{-7}	1.148 10^{-2}	1.60 10^{-7}
Waste clinker leachate	2.24 10^{-7}	2.891 10^{-2}	4.01 10^{-6}
Lead slag leachate	2.06 10^{-7}	1.017 10^{-2}	1.185 10^{-7}
REFIOM leachate	7.90 10^{-7}	4.673 10^{-2}	1.47 10^{-5}

It is shown in table 4, the steady infiltration rate v_o, the sorptivity and the water diffusivity for water and leachates. These hydrodynamic parameters depend on the liquid infiltrated. The steady infiltration rate and the sorptivity of the REFIOM leachate are four times higher than the water steady infiltration rate. The REFIOM leachate diffusivity is hundred times higher than the water diffusivity. These results prove that the solute contained in leachate change the soil structure. It increases the infiltration rate, the water diffusivty and the sorptivity.

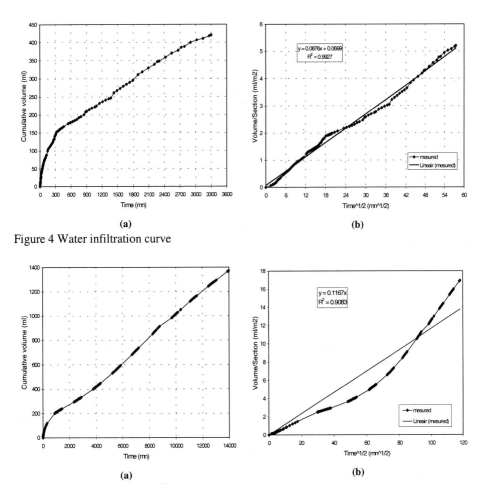

(a)   (b)

Figure 4 Water infiltration curve

(a)
Figure 5 Lead slag leachate infiltration curve

Soil permeability results
Standard laboratory hydraulic conductivity tests were performed with three water head pressure, 20, 50 and 100cm. Before permeability test, each sample is saturated under water head pressure equals to 50cm. Test results are represented by the average liquid flux q ($m^3/s/m^2$) versus hydraulic gradient i = dh/dL. In figure 6, are presented the results of the tests realised with water and REFIOM leachate. It is noticed that the average water flux is proportional to hydraulic gradient (fig.6 (a)) but REFIOM leachate flux is not proportional to

the hydraulic gradient. Hydraulic conductivity changes when hydraulic gradient applied to move leachate, changes (fig.6 (b)).

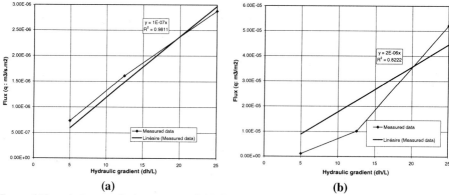

(a)   (b)

Figure 6 Darcy's law applied to water and REFIOM leachate permeation

It is shown in table 5, the average value computed with all measured data during the permeability tests, regardless of hydraulic gradient influence. It is noticed that the hydraulic conductivity hardly does not change, when the soil is permeated by water, waste clinker leachate or lead slag leachate. It becomes ten times higher when the soil is permeated by REFIOM leachate.

Table 5 Average hydraulic conductivity k for water and leachates

	water	waste clinker leachate	lead slag leachate	REFIOM leachate
k (m/s)	$1.64 \ 10^{-7}$	$1.43 \ 10^{-7}$	$1.46 \ 10^{-7}$	$1.61 \ 10^{-6}$

Moisture and Electrical Conductivity Profiles

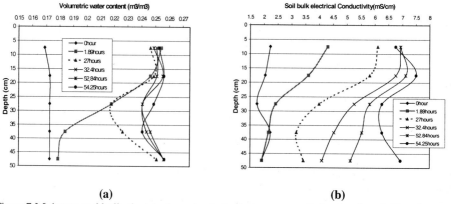

(a)   (b)

Figure7 Moisture and bulk electrical conductivity during waste clinker leachate infiltration

Figures 7 and 8 present moisture profiles and bulk electrical conductivity profiles in the column, during the permeation of waste clinker and lead slag leachates. Soil bulk electrical conductivity is determined with TDR curve, using the method described by Nader et al (1991) and Alimi (1998).

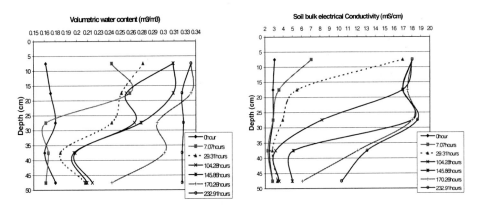

Figure 8 Moisture and bulk electrical conductivity during lead slag leachate infiltration

Profiles represented by dash line are the profiles when the leachates outflow the column. Leachate transit times in the soil column are 27 and 29 hours respectively for waste clinker leachate and lead slag leachate. The transit time is not really affected by the leachate quality but it is half time less than water transit time (52 hours) in the soil column. At the transit time, moisture is not uniform in the soil column but the reached saturation degree is higher in column infiltrated with waste clinker leachate than in column infiltrated with lead slag leachate. The leachate action on the soil depends on the leachate characteristic. It is noticed also that soil saturation with leachate continues during the outflow. The leachate flows through the soil column before water content becomes uniform in the soil.

The bulk electrical conductivity profiles shown in figures 7 (b) and 8 (b), prove that the upper layers retain more solute than the lower layer before leachate outflows the soil column. During the waste clinker percolation, pollutant retaining remains more important in upper layer although the water content along the soil column becomes uniform. The soil purifies the percolated leachate.

When the lead slag leachate percolates throughout the soil column, the column upper layer becomes saturated with solute before a uniform water content was obtained along the column. This upper part saturated with solute remains saturated when the uniform water content is reached along the column. The interaction between soil and lead slag leachate is more important than that of waste clinker leachate.

It is shown with these profiles that the soil volumetric water content does not exceed respectively, 26% and 33% when waste clinker and lead slag leachates are percolating. These water content values represent respectively 66% and 85% of the porosity of the soil compacted at the bulk density γ_d/γ_w equal to 1.6. The soil porosity occupied during water percolation reaches 92%. The soil water characteristic curve (fig.1) shows that soil pF is zero when the volumetric water content is over 33%. So the soil column cannot retain more moisture and the values of the waste clinker leachate sorptivity and diffusivity presented in table 4 explain the reduction of the moisture which is retained.

CONCLUSION

This study allows the evaluation of the soil pollution process. It is shown that the solute action on the soil can be described by the transit time, that to say the time for the leachate to percolate unsaturated soil. The reduction of this time for the leachate percolation means that liner layer thickness must be increased to prevent rapid percolation and to secure ground water.

This leachate action is also described by the soil hydraulic conductivity. Standard permeability tests show that the soil permeability coefficient can increase ten times when it is percolated by leachate. This result proves that the liner must be tested with leachate to determine its permeability and to project its thickness.

The use of TDR probes during infiltration tests allows to observe that the soil is not saturated before water or leachates flows out of the column. When uniform water content is reached in the column, the saturation degree remains less than 95% as it is recommended in hydraulic conductivity test. According to measured data in tables 4 and 5, steady infiltration rate values are close to hydraulic conductivity and are not influenced by this low saturation degree. It seems that the steady infiltration rate estimates the hydraulic conductivity when the infiltration tests are carried out under low water head. This estimation is not influenced by the leachates.

REFERENCES

ALIMI ICHOLA I., (1998) Use of Time Domain Reflectrometry Probes for monitoring unsaturated soil pollution Proc. of 3rd Int. Cong. On Environmental Geotechnics, Lisboa,Portugal 7-11 Sept. 1998

GIPSON A. H., Jr, (1985) Permeability Testing on Clayey Soil and Silty Sand Bentonite Mixture Using Acid Liquor, Hydraulic Barriers in Soil and Rock, ASTM STP 874, A. I. Johnson, R. K. Froebel, N. J. Cavalli and C. B. Petterson, Eds., ASTM, West Conshohocken, PA, pp. 140-154

GLEASON, M. H., DANIEL, D. E., and EYKHOLT, G. R. (1997) Calcium and Sodium Bentonite for Hydraulic Containment Applications, Journal of Geotechnical Environmental Engineering, ASCE, Vol. 123, No 5, pp. 438-445.

NADLER, A., DASBERG S. and LAPID I. (1991) Time Domain Reflectrometry Measurements of Water and Electrical Conductivity of Layered Soil Columns, Soil Sci. Soc. Am. J. vol. 55 : 938-943

RHOADES, J.D. and van SCHILFGAARDE (1976) An Electrical Conductivity Probe for Determining Soil Salinity. Soil Sci. Soc. Am. J. vol. 40 : 647-651

TOPP, G. C., DAVIS J. L. and ANNAN, A. P. (1980) Electromagnetic Determination of Soil Water Content : Measurement in Coaxial Transmission Lines. Water Resour. Res. 16 : 574-582

Carbon stable isotope fractionation of CO_2 flux through soil columns

J.A.C. BARTH, Å.H. OLSSON, R.M. KALIN

Environmental Engineering Research Centre & QUESTOR Centre
The Queen's University
David Keir Building, Stranmillis Rd
Belfast BT9 5AG
N. Ireland UK
fax: xx44 (0)1232 663754
phone: xx44 (0)1232 274438
e-mail: J.Barth@qub.ac.uk

ABSTRACT
Diffusional effects between 'soil respired CO_2' and 'soil CO_2' lead to an enrichment of ^{13}C at the bottom of the soil column. This effect has to be accounted for when using stable isotope compositions of CO_2 as an environmental monitoring tool. A laboratory experiment with a limited supply of CO_2 confirmed the ^{13}C enrichment of bottom CO_2. The maximum isotopic difference between bottom and top CO_2 was 5.2 ‰, slightly higher than the theoretically predicted value of 4.4 ‰. This discrepancy was likely caused by a higher flux rate in the experiment.

INTRODUCTION
A multitude of kinetic and equilibrium processes lead to fractionations of the stable isotopes of carbon (^{13}C and ^{12}C) and produce CO_2 with varying carbon isotopic compositions (e.g. Hoefs, 1997; Clark and Fritz, 1997). The determination of the stable isotope composition of CO_2 ($\delta^{13}C_{CO2}$) can therefore serve as a useful monitoring tool in many environmental applications. For instance, this technique can be used to establish carbon balances within soil and groundwater systems in order to understand global and local carbon budgets (e.g. Fritz et al., 1985, Amudson et al. 1998). Furthermore, the determination of the isotopic composition of CO_2 that results from aerobic microbial decomposition of contaminated systems may help to quantify the biogeochemical processes taking place (Hall et al., in press, a). Thus, a potentially useful application of stable isotope biogeochemistry is Monitored Natural Attenuation (MNA) (Aggarwal and Hinchee, 1991; Baedecker et al., 1988, Suchomel et al. 1990; Revesz et al. 1995). Often MNA may be the only cost-effective option if there are low levels of contamination or if very large areas are involved. In a first instance it is important to understand the biotransformation processes that take place during mineralisation of contaminants (Lovley and Lonergan, 1990; Landmeyer, et al., 1996) in order to optimise the design and monitoring of such engineered in situ bioremediation schemes.

However, secondary isotope effects that are associated with the migration of respired CO_2 in the soil column also have to be considered, if stable isotope methods are to be used as a monitoring tool. Cerling et al. (1991) showed that the isotopic composition of soil CO_2 is not constant with depth and predicted a minimal isotopic difference of 4.4 permil when comparing soil respired CO_2 to soil CO_2. This difference is attributed to diffusional effects during CO_2 migration in the soil. These findings are mainly based on theoretical considerations and have been tested at specific soil sites. To date, no systematic investigation of these isotope effects in relation to soil properties (i.e. grain size, porosity, water content...) and varying flux rates has been carried out. Our objective was to investigate if the predicted isotope effects do also exist in a controlled laboratory experiment. This paper presents initial results from an experiment with coarse sand in an artificial soil column and a limited supply of CO_2 at the lower column end.

METHODOLOGY
A glass column of 600 mm length and 78 mm diameter was equipped with four sampling ports at 70, 215, 313 and 526 mm above the inlet (Fig. 1). These ports were capped with butyl septa to allow sampling with a gas tight syringe. The column was filled with coarse, dried sand of an average grain size of 3 mm and a size range of 0.6 to 10.0 mm. The sand was washed and autoclaved in order to avoid any secondary CO_2 production within the column. The bottom of the column was then connected to a 5 L flask that was filled with pure CO_2 from a cylinder with a known isotopic composition. Each port was sampled every ~14 minutes at the beginning of the experiment, while at a later stage the sampling intervals were several hours.

The stable isotope measurements of the CO_2 were performed in the stable isotope Laboratory of Environmental Engineering Centre at the Queen's University of Belfast. The instrument used was a gas chromatograph that was coupled to an organic mass spectrometer and an isotope ratio mass spectrometer in continuos flow mode (CF-GC-MSD-IRMS). The GC column used was a CP Molseive 5A (25m x 0.32mm) and the helium flow rate was set to 3.0 mL/min (Hall et al., in press, b). The temperature of the injection port was 80 $^{\circ}$C and the column temperature was held isothermal at 30 $^{\circ}$C. The CO_2 concentrations were determined by calibrating peak areas on the IRMS with standards of known CO_2 concentration. The stable isotope results are expressed as a permil deviation from the Vienna Pee Dee Belemnite standard ($\delta^{13}C$ in ‰ VPDB) with a precision (σ) of ± 0.2 ‰.

RESULTS AND DISCUSSION
The isotopic composition of the cylinder CO_2 at the bottom of the column was -28.8 ‰, while the laboratory air at the top of the column was isotopically variable and ranged between values of –19.2 and –11.7 ‰. The maximum difference found between CO_2 at the lower end (P1) and the top of the column (P4) was -17.3 ‰ (Fig. 2, Table 1). However, this difference was only found at the start of the experiment and can be attributed to the isotopic difference between the CO_2 of the laboratory air and the cylinder CO_2. This is confirmed by direction of the isotopic trend which is from isotopically negative at the bottom of the column to isotopically more positive values at the top of the column (Fig. 2, t_0). Within the first hour (t_3) this trend reversed and gave a maximum isotopic difference of 5.2 ‰, slightly higher than the predicted minimal value of 4.4 ‰ (Cerling, 1991). After that the CO_2 concentrations in the column decreased again and the column began to homogenise isotopically (Fig. 2, t_8 and t_{11}).

LABORATORY AIR

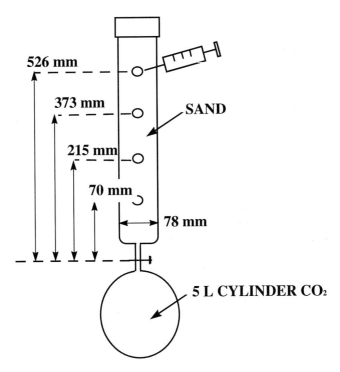

Figure 1 Experimental setup with a column filled with coarse sand (medium grainsize =3 mm).

The measured isotopic difference of 5.2 ‰ after one hour (at time t_3) clearly represents diffusional effects with the isotopically depleted $^{12}CO_2$ migrating faster and leaving the heavier $^{13}CO_2$ behind. As long as the CO_2 supply at the bottom of the column was sufficient the system was in a state of equilibrium. However, this equilibrium was maintained only for a short time, because supply of CO_2 at the bottom of the column was limited. This lead to a mixing of the two CO_2 endmembers at both ends of the column and resulted in a homogeneous $\delta^{13}C_{CO2}$ composition over its length. The theoretical value of 4.4 ‰ between bottom 'respired CO_2' and top 'soil CO_2' is a minimal estimate and stronger differences are possible due to higher flux rates in the soil column (Cerling, 1991). A higher than usual flux rate seems plausible for our experiment, because the concentration gradients were much stronger than in a natural soil system. The latter usually range between 1000 and 100000 ppmV (Clark and Fritz, 1997), while our experiment concentration differences between bottom and top CO_2 were at least twice as high (~200000 ppmV, Fig. 2, Table 1). Other discrepancies may arise when using soils with different grain sizes and moisture contents.

Time (min.)		t_0	t_3 54-65	t_8 150-169	t_{11} 214-227
Sampling port	Column height (mm)	CO_2 [ppmV]	CO_2 [ppmV]	CO_2 [ppmV]	CO_2 [ppmV]
P1	70	6291	238330	98753	69293
P2	215	1322	184300	83800	61283
P3	373	~500		52757	43747
P4	526	~500	19287	24352	21712
		$\delta^{13}C_{CO2}$ [‰]	$\delta^{13}C_{CO2}$ [‰]	$\delta^{13}C_{CO2}$ [‰]	$\delta^{13}C_{CO2}$ [‰]
P1	70	-29.0	-26.9	-27.0	-26.3
P2	215	-21.0	-28.2	-27.0	-26.3
P3	373	-13.8	-30.2	-27.4	-26.3
P4	526	-11.7	-32.1	-28.0	-26.4
	Difference $\delta^{13}C_{CO2}$ bottom-top	-17.3	5.2	1.0	0.1

Table 1 CO_2 concentrations and $\delta^{13}C_{CO2}$ isotopic compositions. The last line represents the isotopic difference between the CO_2 pool at the bottom (P1) and the top of the column (P4). At t_0 the CO_2 concentration was below detection limit at P3 and P4, but was assumed to be ~ 500 ppmV.

$$\delta^{13}C_{CO2} \ [‰ \ VPDB]$$

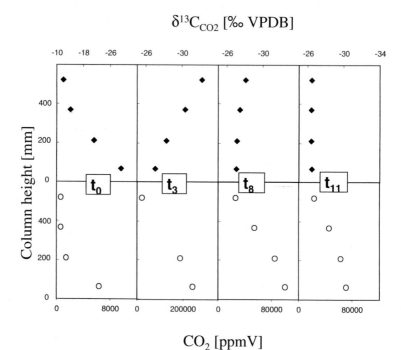

Figure 2 $\delta^{13}C_{CO2}$ isotopic compositions (top) and CO_2 concentrations (bottom) over the length of the column at sampling times t_0, t_3, t_8 and t_{11} at 0-13, 56-65, 150-169, and 214-227 minutes, respectively.

CONCLUSIONS

When using carbon stable isotopic compositions in soil systems as a monitoring tool for environmental applications, attention should be paid to understand secondary effects. The isotopic difference between upwards migrating and residual bottom CO_2 is an important effect that has to be accounted for. We were able to reconfirm the theoretically predicted trend that 'bottom' CO_2 (respired CO_2) is isotopically enriched in ^{13}C when compared to 'top' CO_2 (soil CO_2). Discrepancies between theoretical and measured diffusional isotopic differences are likely related to a higher CO_2 flux rate. Limitations of the column length and the porosity of the material used may further have contributed to this discrepancy. Further experiments with different porosity sands and a CO_2 producing biological cell with a defined CO_2 flux at the bottom of the column have to be carried out.

ACKNOWLEDGMENTS

This research was supported by EPSRC Grants GR/M26374 and GR/L85183, the Queen's University of Belfast, and the Department of Education (NI) and the QUESTOR Industrial Board. We thank David Clarke, David Lipscomb, Dr. Neil Ogle and Deidre Gibbons for their help during this work.

REFERENCES

Aggarwal, P.K.; Hinchee, R.E. *Environmental Science & Technology* **1991**, 25, 1178-1180.

Amundson R., Stern, L., Baisden, T. Wang, Y. *Geoderma* **1998**, 82, 83-114.

Baedecker, M.J.; Cozzarelli, I.M.; Siegel, D.I.; Bennett, P.C. *Geological Society of America Abstracts with Programs* **1988**, 20, 365.

Cerling T.E. *Geochimica et Cosmochimica Acta* **1991**, 55, 3403-3405.

Clark and Fritz *Environmental Isotopes in Hydrogeology*, Lewis, Boca Raton, Fl, **1997**.

Eganhouse, R.P.; Baedecker, M.J.; Cozzarelli, I.M. Symposium on Intrinsic Bioremediation of Ground Water, Denver, CO, August 30 - September 1, 1994, Environmental Protection Agency, Washington, DC, **1994**, p 111-120

Fritz, P., Mozeto, A.A., Reardon, E.J. *Chemical Geology (Isotope Geoscience Section)* **1985**, 58, 89-95.

Hall, J.A.; Kalin, R.M.; Larkin, M.J.; Allen, C.C.R.; Harper, D.B. **(in press, a)** *Organic Geochemistry*.

Hall, J.A.; Barth, J.A.C.; Kalin, R.M. **(in press, b)** *Rapid Communications in Mass Spectrometry*.

Hoefs, J. (Ed.) *Stable Isotope Geochemistry*, Springer-Verlag, Berlin, **1997**.

Lovley, D.R.; Lonergan, D.J. *Applied and Environmental Microbiology* **1990**, 56, 1858.

Landmeyer, J.E.; Vroblesky, D.A.; Chapelle, F.H. *Environmental Science & Technology* **1996**, 30, 1120-1128.

Suchomel, K.H.; Kreamer, D.K.; Long, A. *Environmental Science & Technology* **1990**, 24, 1824-1831.

Revesz, K.; Coplen, T.B.; Baedecker, M.J.; Glynn, P.D.; Hult, M.F. *Applied Biochemistry* **1995**, 10, 505-516.

The application of miniaturised electrical imaging in scaled centrifuge modelling of pollution plume migration

N. DEPOUNTIS[1], C. HARRIS[1], M.C.R. DAVIES[2]

[1] Cardiff University, [2] Dundee University

INTRODUCTION

The principal advantages of physical modelling using a geotechnical centrifuge are that correct stress conditions may be established in the model and that seepage of pore fluids in the prototype over a period of years can be simulated in the model in a matter of hours. As a result contaminant transport problems are increasingly being addressed through centrifuge-based physical modelling.

The aim of this study is to demonstrate the capabilities and limitations of a miniaturised electrical imaging technique (resistivity tomography) developed at Cardiff University to image contaminant plumes in scaled models of the vadose zone. A generic model of contaminant infiltration into partially-saturated sand was designed in order to explore the potential of centrifuge testing at elevated gravity of pollution plume evolution. Plume evolution was monitored using the miniaturised resistivity arrays.

The technique produces two-dimensional contoured images of the resistivity distribution before and during contaminant infiltration experiments. During the experiments dyed NaCl solution was released into the model and the change in resistivity associated with the contaminant plume was imaged as a function of time and g-level. Capillary pressure was monitored constantly during the experiments by matric potential probes (tensiometers) in order to investigate the effect of capillary forces on plume evolution. Tests at 1g (static conditions) and 10g are described.

This miniaturised technique represents a direct analogy to field-scale (prototype) electrical imaging surveys. Barker (1997) discusses the principles and applications of this technique in the area of environmental geophysics.

MINIATURISED ELECTRICAL IMAGING APPARATUS

The construction of the miniaturised electrical imaging apparatus followed that of the full-scale prototype, which has been widely used in field contaminant problems. The main elements of the apparatus are:

- the GEOPULSE earth resistance meter
- control of the GEOPULSE with a module (Imager-25) via a p.c.
- a multicore imager cable
- two 25-electrode arrays

The 25-electrode arrays are placed in the base and on the surface of the scaled model in order to allow two individual resistivity surveys, one from the surface downwards and the second from the base upwards. The electrode spacing is 3cm and each array is 72cm long. The electrodes consist of brass pins with a length of 0.8cm connected to individual cores of two screened 25-way cables. The screened cables are connected to a relay box via a connection box situated on the top of the centrifuge strongbox, in which the scaled model is assembled (Fig.1). The strongbox with the sand model and the resistivity equipment is placed in the centrifuge gondola and the relay box is operated from the control room with a 12V dc power supply unit. When the power supply unit is switched on the surface array is active, and when the power supply is switched off the base array becomes active.

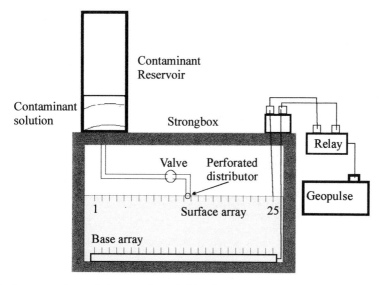

Figure 1. Apparatus for scaled contaminant infiltration tests

A multi-core cable connects the relay box and therefore the two electrode arrays with the Geopulse earth resistance meter. The 25 takeout cores of the cable are connected directly to a module (Imager-25) housed in the Geopulse. The Imager-25 control module enables automatic addressing of multi-electrode survey systems using any number of electrodes up to a maximum of 25 and the measurement of resistances using the GEOPULSE earth resistance meter. The resistance meter is placed inside the instrumentation panel of the centrifuge and can be operated in flight from the control room via a p.c. and the control software Image25.

The survey design is saved in a parameter file, which once created can be accessed by the Image25 whenever needed, and all the resistance measurements during the experimental flight can be stored in memory for later processing.

Data processing
The stored data can be used to plot a 2D image in the form of a pseudosection, which gives an approximate but distorted picture of the model subsurface resistivity. The electrode geometry can, however, distort the cross-section leading to errors in interpretation.

The correction of the distorted model of the subsurface is achieved with the RES2DINV program. RES2DINV is a computer program which will automatically determine a two-dimensional (2-D) resistivity model for the subsurface using the data obtained from electrical imaging surveys (Griffiths & Barker 1993). This programme eliminates electrode geometry effects so that the final processed images provide a good representation of the model subsurface resistivity. A forward modelling subroutine is used to calculate the apparent resistivity values, and a non-linear least-squares optimisation technique is used for the inversion routine (Loke & Barker 1995, 1996).

The program can automatically choose the optimum inversion parameters for a particular data set. However, the user can modify the parameters, which affect the inversion process. Though the algorithm used corrects the depth variation in resistivity and sharpens the image, resistivity contrasts may be inaccurate.

Since the physical model used during the tests is reasonably homogeneous (pluviated fine poorly graded sand) with a conductive anomaly created inside the model after the contaminant release, the unprocessed images are closely related spatially to the bodies giving rise to them. Thus, it might be prudent to choose as more representative the resistivity data obtained from the iteration that best suits the examined model. These data are saved in a xyz format and by using the Surfer contour package a contoured cross-section of the examined model can be plotted. The contoured section can be either in the form of absolute resistivity, or resistivity relative to the baseline pre-infiltration of contaminant values, as it can be seen in the later figures.

CONTAMINANT INFILTRATION EXPERIMENT AT MODEL SCALE
A generic model of contaminant infiltration into partially saturated sand was designed in order to explore the potential of testing pollution plume evolution at 1g (static) and 10g (Fig.1).

The apparatus used for this experiment is as follows:

- Rectangular strongbox filled with pluviated sand
- Aluminium liner inside the strongbox with perforations in the base and sides to enable uniform saturation of the pluviated sand
- Miniaturised electrical imaging apparatus
- Perspex cylinder (0.5m high) situated on the top of the strongbox, acting as the contaminant reservoir
- Solenoid valve to control the contaminant flow between the reservoir and the sand model
- Perforated distributor tube to allow uniform distribution of contaminant across the centre line of the model
- Two matric potential probes (tensiometers) placed inside a rectangular waterproof perspex box buried in the model in order to monitor the capillary pressure and therefore the degree of saturation during the experiment

The sand used for the tests was Congleton Sand which is poorly graded with $d_{10}=100\mu m$ and a coefficient of permeability $k=10^{-2}cm/sec$. The contaminant used was NaCl dissolved with tap water and accompanied with Eosin Y Disodium dye for more accurate visual observation of the contaminant plume evolution.

STATIC PROOF TEST

Experimental procedure
The sand was pluviated inside the strongbox. All the appropriate equipment was installed and the sand was wetted from the bottom of the liner until complete saturation was achieved. When the model saturation stabilised the valve controlling flow of contaminant was opened. The dyed NaCl solution flowed from the contaminant reservoir to the distributor placed between surface electrodes 13 & 14 allowing infiltration from a line source.

During this test 3.8 litres of dyed NaCl solution with a concentration of 0.18M were released gradually into the sand model on the lab floor at 1g. The contaminant solution was released from the line source, within a 2cm wide trench, across the model.

The test took place over three days, and the change in resistivity associated with contaminant plume evolution was imaged in 2-D perpendicular to the line source as a function of time (Figure 2). During the test capillary pressure changes were monitored closely with the tensiometers. As a result, the degree of saturation of the model could be assessed during and after the test.

Results and interpretation
The measured resistivity is always a function of the degree of saturation (moisture content) and conductivity of the pore fluid (concentration of contaminant). Thus, the anomaly shown in Figure 2 could be related with changes in both variables. However, a uniform saturation

was achieved prior to the contaminant release and, therefore, the anomaly shown is more related to the contaminant distribution inside the model.

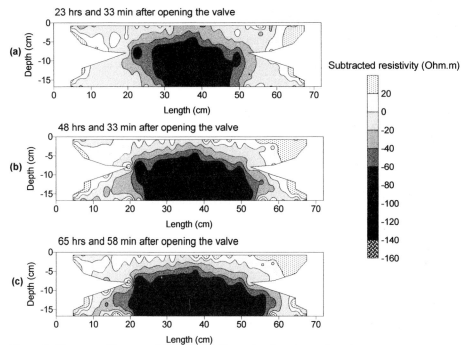

Figure 2. Tomographies associated with static contaminant experiment

The above models illustrate the apparent resistivity subtracted from the background recorded prior to the contaminant release. Therefore, the subtracted resistivity anomalies represent the passage of the contaminant plume.

Figure 3 shows the plume at the end of the experiment visible in two longitudinal sections of the model.

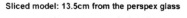

Figure 3. Geometry of the contaminant plume

The final images illustrating the changes in resistivity as a function of time shown in Figure 2 are similar with the observed plume geometry at the end of the test (Figure 3). This suggests that miniaturised electrical imaging can be used in scaled models and may be applied in centrifuge tests. From the tomographies shown in Figure 2 contaminant plume evolution appeared to be gravity-driven.

During the three-day experiment the degree of saturation of the sand model, estimated from the tensiometers, remained high. In the first model Figure 2(a), 23 hrs and 33 min after the contaminant release, the degree of saturation was 95-97%. In the third model Figure 2(c), 65 hrs and 58 min after the contaminant release, the degree of saturation was 85-87%. These values represent a capillary zone of fine sand where the liquid phase remains continuous.

The water retained initially in the vadose zone by capillary forces was displaced gradually by the denser contaminant during the experiment. At the end of the test some contaminant was retained in the smaller pores of the upper layers of the model by capillary forces. However, the main body of the plume moved towards the base of the model as a result of density-driven migration.

CENTRIFUGE TEST AT 10G
Based on the same principles and by following the experimental procedure described above, a scaled contaminant experiment was conducted at 10g. After model preparation and placement in the geotechnical centrifuge, the g level was increased to the selected value. Increasing g-level caused drainage of water from the sand and an increase of capillary pressure was recorded with the matric potential probes.

During this test 2.0 litres of dyed NaCl solution with a concentration of 0.1M were released into the sand model. The contaminant solution was released from a 2cm wide line source across the model, between surface electrodes 13 & 14. The contaminant was released only after the capillary pressure in the sand had stabilised.

The test took place over three hours, and the absolute resistivity values associated both with the drainage of the model and the contaminant plume evolution were imaged as a function of time (Figure 5). The tensiometers were placed at 10cm depth (LP/p1), and 2cm depth (LP/p2) respectively. Prior to the experiment, capillary pressure had been calibrated with different degrees of saturation for the same sand.

Results and interpretation
The capillary pressure of the model was monitored overnight prior to the centrifuge flight (Figure 4). The plotted curves represent almost complete saturation, with the degree of saturation before starting the experiment assessed as 98%.

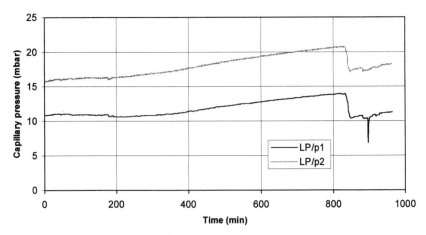

Figure 4. Capillary pressure curves prior to centrifuge test

After accelerating the centrifuge, water started to drain from the model until capillary equilibrium relevant to 10g conditions occurred. This was achieved after 1 hr and 30min, by observing the tensiometer response.

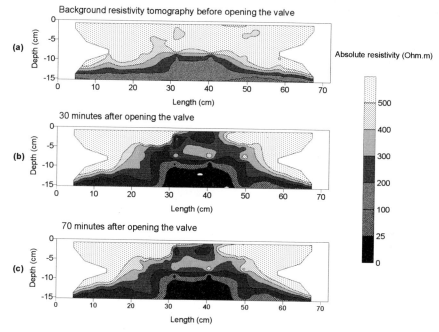

Figure 5. Tomographies associated with 10g contaminant experiment

The above models illustrate the absolute resistivity distribution during the 10g test. The scale used for imaging these models was different compared with the 1g proof test (Figure 2). An absolute resistivity scale was thought to be more reliable for imaging the conductive anomaly, since the degree of saturation varied towards the base of the sand model (no uniform saturation). This can clearly be seen in resistivity model (a), prior to the contaminant release, where resistivity decreased due to saturation increase.

In detail, model (a) illustrates the background absolute resistivity of the sand before opening the valve. Below 12cm the resistivity decreased dramatically, indicating an increasing saturation towards the base of the model. The zone with resistivity less than 100 Ohm.m corresponds with the continuous capillary zone, in which capillary water fills the soil pores. This occupied the lower 3.5cm of the sand model. The thickness of the continuous capillary zone decreases with increasing g levels (König et al. 1998, Depountis et al. 1999), and this is clearly shown in this experiment.

After 1hr and 45min the valve was opened and the contaminant was released across the model. The absolute resistivity associated with the contaminant plume evolution was recorded as a function of time (models b and c of Figure 5), using the miniaturised electrical imaging apparatus. Due to the lower degree of saturation in the sand model and the higher gravitational forces, the plume was driven into the model very rapidly compared with the 1g test.

Since the degree of saturation of the model as a whole did not change significantly after the contaminant release, the resistivity models (b) and (c) in Figure 5 indicate only the contaminant plume evolution. The lower resistivity distribution, which is below 25 Ohm.m, represents the higher contaminant concentration. The basic shape of the plume did not change significantly from 30 to 70 minutes after the opening of the valve, suggesting that at the end of 30 minutes the plume had already reached the base of the model. By comparing both models, we can only see downward migration of higher residual contaminant concentration zones.

The observed plume at the end of the 10g test is illustrated in Figure 6 with a black line and can be compared with the final resistivity model in Figure 5(c). As can be seen from the comparison of both figures the basic shape of the plume is similar, especially at the base of the model. The visual section shows dye retained in the upper part of the sand model, but the resistivity data suggests that little salt solution actually retained following passage of the plume.

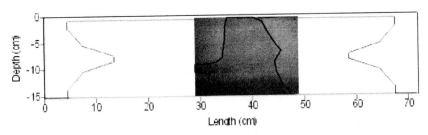

Figure 6. Geometry of plume at the end of the 10g experiment

The capillary pressure and, therefore, the degree of saturation, as mentioned above did not change significantly after the contaminant release (Figure 7). This suggests that the degree of saturation of the model remained in a range of 15-20% to the depth of 12cm, equivalent to a capillary pressure of 76 to 82 mbars (LP/p1). The break in both curves between 6000 and 7000 sec (Figure 7) indicates the end of the 10g test.

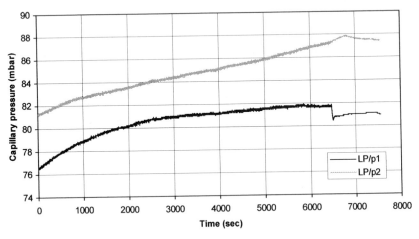

Figure 7. Capillary pressure curves during 10g test after opening the valve

The second tensiometer (LP/p2) indicated even higher capillary pressure (or lower saturation degree) due to its placement 2cm below the model surface. Here evaporation was significant even before the opening of the valve.

CONCLUSIONS
This study has investigated the effectiveness of a new miniaturised technique for imaging contaminant plumes in scaled models of the vadose zone. Comparison of resulting two-dimensional tomography with observed plume geometry at the end of a three-day static test showed the electrical imaging technique to be highly effective. The contaminant plume evolution during that test was gravity-driven, and its form was accurately represented in the resistivity models.

A second experiment was conducted at 10g with a lower volume of contaminant solution performed in a shorter time (3 hours). Although the contaminant concentration was only half that for the static 1g test, the plume migrated towards the base of the model much faster. This indicates that the plume migration is a function of degree of saturation and gravitational acceleration.

The results of the experiments show that contaminant evolution in the model of partially saturated sand is gravity-driven, with the plume migration and geometry therefore affected by the elevated gravity.

It is concluded that miniaturised electrical imaging can be a useful tool to monitor contaminant migration and can be applied during centrifuge tests associated with density-driven or other type of contaminant transport.

ACKNOWLEDGMENTS
This research was partially sponsored by NECER (Network of European Centrifuges for Environmental Research). This support is gratefully acknowledged. Mr. Depountis also acknowledges the continuous support of the Greek State Scholarships Foundation (SSF) which provided the funds for his training.

REFERENCES
Barker, R.D. 1997. Electrical imaging and its application in engineering investigations, *Modern Geophysics in Engineering Geology*, Geological Society Engineering Geology Special Publication, **No.12**, pp. 37-43.

Depountis, N., Davies, M.C.R., Burkhart, S., Harris, C., Rezzoug, A., König, D., Merrjfield, C., Craig, W.H. & Thorel, L. 1999. Scaled centrifuge modelling of capillary rise. *2nd BGS International Geoenvironmental Engineering Conference,* London 13-15 Sept. 1999 (in press).

Griffiths, D.H. & Barker, R.D. 1993. Two-dimensional resistivity imaging and modelling in areas of complex geology. *Journal of Applied Geophysics,* **29**, pp.211-226.

König, D., Rezzoug, A., Heibrock, G., Craig, W.H., Davies, M.C.R., Burkhart, S. 1998. Capillary phenomena in centrifuge testing. *Int. Conf. Centrifuge 98,* Tokyo 23-25 Sept. 1998, Spec. Lect. rep., 40-42.

Loke, M.H. & Barker, R.D. 1995. Least squares deconvolution of apparent resistivity pseudosections. *Geophysics,* **No.6**, pp.1682-1690.

Loke, M.H. & Barker, R.D. 1996. Rapid least-squares inversion of apparent resistivity pseudosections by a quasi-Newton method. *Geophysical Prospecting,* **44**, pp.131-152 .

Use of Resistivity Cone for Detecting Contaminated Soil Layers

Prof. M. FUKUE, Marine Science & Technology, Tokai University, Shimizu, Japan
T. MINATO, Graduate Student, Marine Science & Technology, Tokai University, Shimizu, Japan
M. MATSUMOTO, Koa Kaihatsu, Co., Tokyo, Japan
H. HORIBE, Koa Kaihatsu, Co.,Tokyo, Japan
N. TAYA, Koa Kaihatsu, Co., Tokyo, Japan

ABSTRACT
A cone-penetration technique has been developed to detect contaminated soil layers with electrolytes and NAPLs. In this study, the change in resistivity of soils was quantitatively examined by adding salt or oil to the soil samples. The results showed that the resistivity measurement was varied with an order of ppm for the electrolyte concentration of soil. It was found that the resistivity of sand increased with increasing of the concentration of oil. The effect of oil content was the stronger the lower water content of sand. In field, the actual cone penetration performances were made. The results showed that the resistivity cone can be used for detecting contaminated soil layer. Particularly, the instrument can be effectively used for examining the effect of remediation, by measuring the resistivity of ground before and after remediation.

INTRODUCTION
The chemical analysis for detecting contaminated soil layers may be essential, but it needs high cost and time. Therefore, at present requirements may be an easy, quick and low-cost detecting technique.

The resistivity of soils is primarily dependent on the resistivity of the pore water. Therefore, the change in the concentration of electrolytes or organic compounds influences the resistivity of soils. As pollutants are usually electrolytes or organic compounds, it is possible to distinguish between contaminated and non-contaminated soil layers from the variation of the resistivity. This means that contaminated plumes can be inspected by comparing the measured resistivity and the resistivity of non-contaminated soil with similar physical properties, such as particle size distribution, void ratio, water content, fabric, etc., except the type and concentration of solutes in the pore water. For such purpose, Campanella and Weemees (1990) and Campanella, Davies and Boyd (1993) developed a resistivity cone.

In this study, the resistivity cone developed is examined to detect contaminated zone in soil layer. In order to obtain the resistivity of soils, a relationship between electric current, voltage and resistivity was theoretically derived, so that no calibration was required (Fukue et al., 1998). If the theoretical relation is not obtained, the calibration using a material whose resistivity is known must be obtained.

INSTRUMENT
Cone
The resistivity cones used in this study are similar to those developed by Campanella and Weemees (1990). Four electrodes are located with a given interval, as shown in Figure 1. The

spacing r_M can be changed using different height of the plastic rings for electric insulation. The thickness of electrodes is 5 mm. The diameter of the cone rod, d is 42 mm. The maximum diameter of the cone part is 35.7 mm and the angle of the cone tip is 60 degree. The distance between the cone tip and the top electrode indicated by A in Figure 1, D is 250 mm.

A constant square current AC source, set at 1000 Hz, is applied to build the electric field in soil and the voltage yielded between the inner two electrodes located between the current and grounded electrodes (A and B), as shown in Figure 1, is measured.

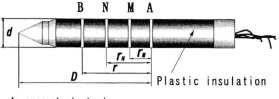

A: current electrode
B: grounded electrode

M and N : electrodes for potential measurement

Figure 1 Resistivity cone developed in this study

One-Dimensional Measurement

Since the cone penetration performance in the laboratory requires large amounts of soils and time for preparation. In this study, an one-dimensional test apparatus was used to examine effects of salt and oil on the resistivity of soils. The apparatus is shown in Figure.2.

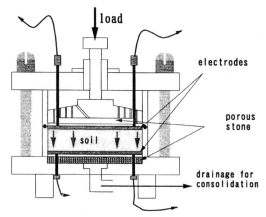

Figure 2   One-dimensional apparatus

RESISITIVITY

The electric field developed around the cone rod is complicated, because of the complicated boundary condition. For cone penetration, Fukue *et al* (1998) obtained the following theoretical relationship, based on the Ohm's law.

$$\rho = \pi^2 \Delta V / (C\,I) \tag{1}$$

where

$$C = 1/(d + r_M) - 1/(d + r - r_M) - 1/(d + r_N) + 1/(d + r - r_N) \qquad (2)$$

where ρ is the resistivity of the soil, ΔV is the potential difference to be measured and I is the current supplied. Eq.(1) was examined and was ensured by one dimensional experiments (Fukue et al., 1998).

For the one-dimensional electrical flow, the following Ohm's law is directly used;

$$\rho = (S / I)(\Delta V / L) \qquad (3)$$

where S is the sectional area and L is the height of specimen. The $\Delta V / L$ is the potential gradient.

RESULTS OF LABORATORY EXPERIMENTS AND DISCUSSION
Effects of Current and KCl Concentration for One-Dimensional flow

The resistivity was measured on two types of soils, i.e., quartz sand and Kibushi clay, using the one- dimensional apparatus. The current applied ranged from 0.01 to 200 mA. For pore water, various concentrations of potassium chloride (KCl) solution were used.

The potential difference was measured under the current applied. The resistivity was obtained using Eq.(3) and the measured potential difference. Figure 3 shows the relationships between resistivity and current in terms of KCl concentration of pore water. The result shows that the stable resistivity was obtained for current intensities between 0.01 and 5 mA. The unstable measurement for stronger current resulted from the limitation of amplifier capacity, which was dependent on both the magnitudes of potential difference and the intensity of current.

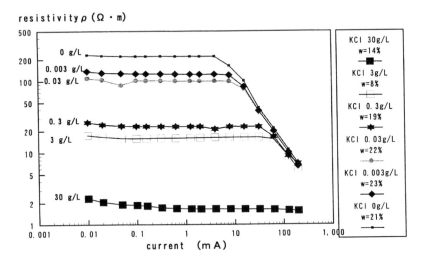

Figure 3 Relationships between resistivity and current, in terms of KCl concentration, for quartz sand

Figure 3 also shows that the KCl concentration influences on the resistivity of the sand. For the sand, the resistivity decreases from about 200 to 2 Ωm, by mixing with pore water with KCl concentration of 30 g/L.

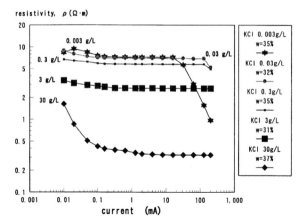

Figure 4 Relationships between resistivity and current of KCl concentration for Kibushi clay

A similar trend can be seen for Kibushi clay, as shown in Figure 4, but it should be noted that the resistivity of clay is much lower than that of sand.

Effects of NaCl Concentration for Cone Penetration

The model ground was prepared in a container, as shown in Figure 5. Basically, three soil layers were formed in the order of sand, clay and sand from the top. For each test case, a different concentration of sodium chloride was used as the pore water of the top sand layer, in order to examine the monitoring ability of the instrument.

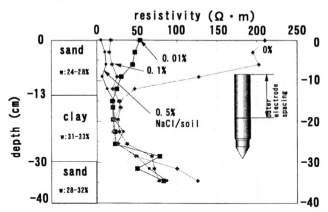

Figure 5 Resistivity change due to electrolyte concentration

The cone was penetrated into the model ground with a speed of 1.5 cm/s and potential difference ΔV between the two electrodes was measured every 3 seconds. The resistivity was then calculated using Eq.(1), where C was calculated using (2).

The experimental results are demonstrated in Figure 5. Figure 5 shows the resistivity profiles of the model ground. The resistivity of clean sand is approximately 200 Ωm. On the other hand, the resistivity of clay layer is 20 Ωm.   As the water content is similar for both the sand and clay, as indicated in Figure 5,   a large difference of the resistivity values for these soils may depend on the different structures, such as the pore size, particle size and the thickness of electric double layer formed on the surfaces of clay particles.

On the other hand, the adding of NaCl into pore water decreases the resistivity of the soil, as seen in Figure 5. It is noted that the NaCl concentration is expressed in (g/g x100 %; NaCl/soil). The increase of 0.01 % resulted in a decrease of the resistivity from 200 to 50 Ωm.

The resistivity measured does not necessarily show an abrupt change at the boundary between the two distinct soil layers, because the resistivity gives an average resistivity of two types of soil existing between the current and grounded electrodes. Theoretically, a smaller interval of electrodes may provide a better profile feature of soil strata. However, since cone penetration will disturb the soil around the cone and rod, a smaller interval will provide the resistivity of the disturbed soil. This is because of the disturbance of soil due to the surface friction of cone rod.

Effects of light oil
The effect of hydrocarbon fuels on the resistivity of soil was examined on sand mixed with light oil. Light oil was mixed with quartz sand with a water content of 5 and 20 %, respectively. First, a sample with 20 % oil content was prepared by mixing sand and oil. In order to obtain lower oil contents, a small amount of the 20 % oil-content soil was mixed with given amounts of sand with the same water content.

The sample was packed into the cell of the one-dimensional apparatus and the potential difference was measured under a current of 0.1 mA and 1000Hz.. The resistivity was calculated using Eq.(3).

Figure 6 shows that the resistivity increased with the increase in oil content. It is also seen that water content will strongly influences the effect of oil content. The higher the water content, the less effect of oil content will be. This is dependent on the structure variation relating to the soil-water system.

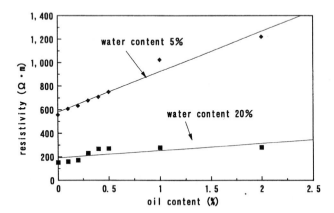

Figure 6 Relationship between resistivity and oil content in terms of water content

Since oil is LNAPL, the problem of oil-contaminated soils may concern with unsaturated soils. This may lead to efficient variation of resistivity with a degree of oil contamination, because of low water content of the soils.

RESULTS OF FIELD-EXPERIMENT AND DISCUSSION
Field experiment using the resistivity cone was performed at two sites, i.e., Fuji city and

Kawaguchi city. In the experiment in Fuji city, feasibility of the cone penetration was examined, while the possibility of soil contamination was investigated in Kawaguchi city.

Reliability of resistivity measurements –case in Fuji city

Four times of penetration were performed with a distance as shown in Figure. 7. Since the distance is only 80 cm, the soil properties are assumed to be very similar.

The penetration was made with a speed of 1cm/s. The spacing of electrodes, r_M was 5 cm. The current applied was 0.1 mA and 1000 Hz. The reading of measurement was made with a interval of 5 seconds.

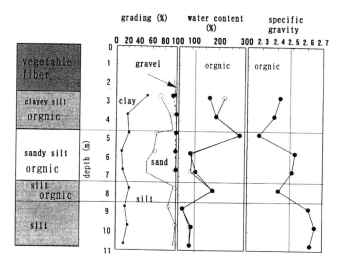

Figure 8   Soil profiles obtained by soil testing

The soil profile and resistivity are shown in Figures 8 and 9. Figure 8 shows that the soil layers contain a large amount of organic matter. The four profiles of resistivity shown in Figure 9 are very close each other, except the sandy soil layer. This is because the sand content of this sand layer varied in a short distance. It is noted that sand strata was observed to be inclined even in the boring core sample. Thus, it is seen that the values of resistivity obtained by the resistivity cone penetration performance are reliable.

The results also show that the resistivity reflects the types of soil. Generally, coarse-grained soils show a high resistivity.

Contamination of soil layer- case of Kawaguchi city

Cone penetration was performed near a place for leaving steel rods, steel implements, timber, etc. As rust patina and oily materials have permeated into the soil ground for a long time, the soils near the place may be partly contaminated with those substances.

The penetrations were performed near the place from one by one with a distance of 1 m. The cone was penetrated into ground with a speed of 1 cm/s and the potential difference was measured with every 5 seconds. The electrode spacing used was 3cm for Sites Nos. 1 and 2 and 5cm for Site No.3. The maximum penetration depth was about 6 m, where the cone penetration could hardly be achieved. Boring investigation was also made near the site location.

The experimental results obtained are shown in Figure 10. The soil profile observed from the soil samples was presented on the left-hand side of Figure 10. The resistivity profiles obtained from the cone penetration are demonstrated on the right-hand side.

Figure 10 indicates that the resistivity of clay or clayey soil at the top surface is a range between 12 and 25 Ωm, although the unsaturated top zone has a greater resistivity. The laboratory experiments using the one- dimensional apparatus showed that the resistivity values were 27 Ωm and 13 Ωm for the undisturbed and remolded clay samples obtained from the site, respectively. These values are comparable to the results obtained from the field experiment shown in Figure 10. It is seen that the resistivity values obtained from Nos. 1 and 2 are close to that for the remolded clay sample, while the resistivity from No. 3 accords with that of the undisturbed clay sample. Therefore, this was first considered to be the effect of electrode spacing. If so, the difference in resistivity has to be appeared through the whole profile. However, for the deep soil, the difference in resistivity for the different sites is not large. This may indicate that the upper clay layer for Sites No. 1 and 2 was contaminated with substances from rust, patina. and other materials.

Sand layer has usually greater resistivity than that of clay and silt, because of lower void ratio and lower water content. Figure 10 shows that the sand layer has a resistivity ranging from 40 to 60 Ωm. The sand with a lower resistivity at a depth around 1.5 m, may indicate a large amount of clay content or contaminants, where both the contents of clay and contaminants will decrease the resistivity.

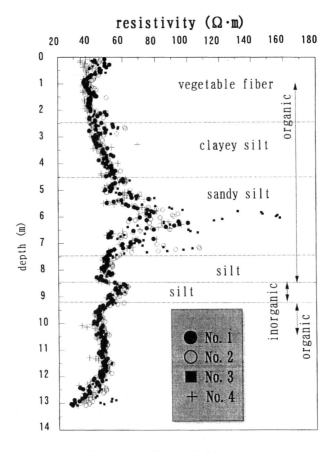

Figure 9 Resistivity of in-situ soils, measured with the resistivity cone

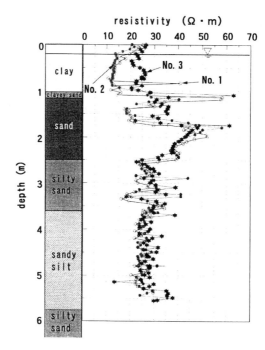

Figure 10 Resistivity of in-situ soil in Kawaguchi city

Below a depth of 2.5 m, silty sand and sandy silt mostly have a resistivity ranging from 23 to 30 Ωm, although the sand content influences the resistivity.

Thus, the resistivity depends primarily on type of soil, as shown in Figure 10, and the type and concentration of contaminants in the pore water. Therefore, the technique described in this study can be applied for inspecting and monitoring contaminated zones in soil profile. and can be more effectively used to examine remediation effects.

CONCLUSIONS
It was found from the laboratory and field experiments that the resistivity of soils is representative of soil type and degree of contamination. It is concluded that the resistivity cone developed can be used for detecting contaminated soil layer, based on the result that the concentration of electrolytes and hydrocarbon will change the resistivity from the inherent value.

ACKNOWLEDGEMENT
This study was partially executed with the financial support of the Ministry of Education, Japan, Grant-In-Aid for Scientific Research, No. 09650556.

REFERENCES
Campanella, R.G. and I. Weemees: 1990, Development and use of an electrical resistivity cone for groundwater contamination studies, Can. Geotech. J. 27, 557-567,.

Campanella, R.G., M. P. Davies and T. J. Boyd :1993, Characterizing contaminated soil and groundwater systems with in-situ testing, 1993 Joint CSCE-ASCE National Conference on Environmental Engineering, 1497-1504.

Fukue, M., Taya, N., Matsumoto, M. and Sakai, G :1998, Development and application of cone for measuring the resistivity of soil, Journal of Geotechnical Engineering, Japanese Society of Civil Engineers, No.596/3-43,   283-293, (Text in Japanese).

Use of an Interaction Energy Model to Predict Pb Removal from Illite

RAYMOND S. LI
Environmental Engineering Program, Univ. of Alberta, Alberta, Canada, T6G 2M8
RAYMOND N. YONG
Geoenvironmental Research Centre, University of Wales, Cardiff, U.K.
and LORETTA Y. LI
Dept. of Civil Engineering, Univ. of British Columbia, Vancouver, B.C. Canada, V6T 1Z4

ABSTRACT

The interactions between heavy metal contaminants and surface forces associated with reactive surfaces of soil particles are studied with a view to development of an analytical tool, which would permit prediction of heavy metal and soil sorption-desorption phenomena. The predictive model developed in this study considers the energies of interaction developed between a lead contaminant and an illitic soil. Since these energy calculations include both inner and outer sphere complexes, energy balance requirements can be realistically evaluated in respect to environmental mobility. For comparison between calculated values obtained from the predictive model and experimentally obtained measurements, laboratory batch equilibrium tests were used. These concerned themselves with sorption-desorption determination of illitic soil samples with the lead contaminant in the form of a lead nitrate solution. Force balance on lead ions indicates that lead ions in the so-called Stern layer, i.e. first two layers of water molecule, are un-removable. Good correlations between calculated and measured values are obtained.

KEYWORDS

interaction energy model, prediction, sorption-desorption, lead, clay mineralogy, illite

INTRODUCTION

There is increasing pressure to develop contaminated industrial sites for commercial or residential use. Decontamination of heavy metal contaminated sites and the ability to accurately predict the efficiency of the decontamination process are of major concerns to industry, consultant and governmental groups.

Most predictive models developed for heavy metal desorption in soil are based on either the advection-dispersion equation of contaminant transport for description of the movement of the contaminant (Shapiro and Probstein 1993), or coupled flow theory (Yeung and Delta 1995). Very few studies have considered the effect of interaction between heavy metal contaminant and surface forces associated with the reactive surfaces of clay soil particles. Only free ions in the bulk solution are removed, but not those metal ions held in the Stern layer or diffuse layer of the charged surface clays, unless the combined external applied forces (hydraulic and/or electrical) are large enough to overcome these surface forces of the clays.

Geoenvironmental engineering, Thomas Telford, London, 1999, 230–237

In this work, the interactions between heavy metal contaminants and surface forces associated with reactive surfaces of soil particles are studied with a view to development of an analytical tool, which would permit prediction of heavy metal and soil sorption-desorption phenomena. The predictive model developed in this study considers the energies of interaction developed between a lead contaminant and an illitic soil. Since these energy calculations include both inner and outer sphere complexes, energy balance requirements can be realistically evaluated in respect to environmental mobility. For comparison between calculated values obtained from the predictive model and experimentally obtained measurements, laboratory batch equilibrium tests were used. These concerned themselves with sorption-desorption determination of illitic soil samples with the lead contaminant in the form of a lead nitrate solution.

PREDICTIVE MODEL
The developed interaction energy model that describes the interaction mechanism between water molecules, ionic species and clay surfaces in terms of potential energy and concentration of ions is based on the DLVO (Derjaguin and Landau, Verwey and Overbeek) theory, i.e. long range energies. This model also considers the short-range energies when the distance between two particle surfaces is less than 3 nm (i.e. hydration energy). The distribution of adsorbed ions in any of the adsorptive layers can be calculated.

Determination of Ion Distribution
The adsorbed cations in this predictive model are distributed in two layers: Stern layer and diffuse layer. The Stern layer consists of partially hydrated ions which are in close proximity to the particle surface (the thickness is considered equal to the diameter of a water molecule). The diffuse layer consists of fully hydrated ions and is separated from the particle surface by at least one water molecule. With this assumption, the Stern layer is redefined to include only the inner Helmholtz Plane (IHP) while the outer Helmholtz Plane (OHP) is included in the diffuse layer. Possible configurations of ion distribution in this model that based on the structure of illite are shown in Fig.1. Many well-crystallized micas (illite minerals) have one-fourth of the silica replaced by aluminium (Grim 1968, p. 94). For simplicity, the replacement of silicon with aluminium is assumed in a regular pattern as shown in Fig.1b.

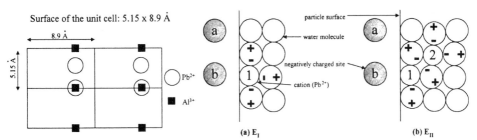

(a) Possible configuration of lead ion   (b) The configurations of cations at E_I and E_{II}
    adsorption in illite (View from the
    top or bottom of unit cells)

Fig. 1 Possible configuration of cations

The theoretical calculation of total surface charge density is expressed in terms of energies:

$$\sigma_t = \sigma_s + \sigma_d \qquad\qquad (1)$$

where σ_t = total surface charge density (C/m^2); σ_s = charge density of partially hydrated adsorbed cations in the Stern layer (C/m^2); σ_d = charge density of hydrated adsorbed cations in the diffuse layer (C/m^2).

From Boltzmann's distribution law, the concentration of cations in any layer of water is calculated as follows:

$$C_{j2} = C_{j0} \exp\left(\frac{-z_j e' \psi_2}{KT}\right) \tag{2}$$

or

$$C_{j2} = C_{j1} \exp\left[\frac{-(E_{II} - E_I)}{KT}\right] \tag{3}$$

where C_{j0} = bulk concentration of ions of type j (ions/m^3); C_{j1} = concentration of ions of type j in the first layer of water from particle surface; C_{j2} = concentration of ions of type j in the second layer of water from particle surface; z_j = valence number, either a positive or negative integer; e' = unit electron charge (1.602×10^{-19} C); ψ_2 = electric potential in the second layer of water from particle surface (OHP) (volt); $KT = 4.04 \times 10^{-21}$ J at 293° K; E_I, E_{II} = potential energies of a counter-ion at the first and second layer of water respectively (J).

For E_I - Mutual Energy for Cation in the First Layer
When a cation moves from the bulk solution into the first layer of water molecule, the potential energies of cations (E_I) in the IHP consist of several energies (Fig. 1b(a)). These include:
a) Electrostatic potential energy (E_{1s}) between the cation (1) and the negatively charged site (s) due to Coulomb forces in the IHP which may be written as:

$$E_{1s} = \frac{(z_1 e')(z_s e')}{4\pi \varepsilon' \varepsilon_0 r_{1s}} \qquad \text{unit:} \quad \frac{(C)^2}{m} \tag{4}$$

where z_1, z_s = valence of a cation and a negatively charged site respectively; r_{1s} = distance between the centre of cation (1) at the 1st layer and the negatively charged site. The subscript (s) represents site (a) or (b) as shown in Fig. 1b; ε' = dielectric constant of the first two layers of water from the particle surface and is equal to 5 (Ansoult et al. 1985); ε_0 = permittivity of free space (C^{2}J^{-1}m^{-1}).

b) Interaction energy ($E_{1d\theta}$) between the cation (1) and a dipole water molecule (d) at θ degree (Hiemenz and Rajagopalan 1997):

$$E_{1d\theta} = -\frac{(z_1 e')(\mu \cos \theta_1)}{4\pi \varepsilon' \varepsilon_0 r_{1d}^2} \tag{5}$$

where r_{1d} = distance between the centre of cation (1) and centre of the dipole; μ = dipole moment of water molecule; $\theta_1 = 90°$ (θ is the angle between the line of centres and axis of dipole, see Fig. 2).

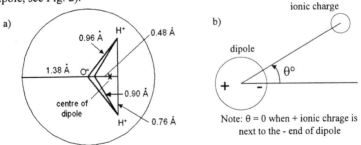

Fig. 2 a) Dipole centre of water molecule, and b) angle between line of centres and axis of dipole

c) Electrostatic potential energy (E_{ab}) between negatively charged site (a) and (b) in the lattice which may be calculated with Eq. (4).

d) Interaction energy ($E_{sd\theta}$) between dipole molecule (water) and negatively charged site (s) which can be calculated with Eq. (5) but with a different θ degree.

Because of its insignificantly small value and complexity, the interaction energy of the ion-dipole, site-dipole, and dipole-dipole are ignored when the ion, the charged site, or dipole is not in direct contact with the other dipole.

Therefore,

$$E_I = \Sigma E_{1s} + \Sigma E_{1d\theta 1} + E_{ab} \tag{6}$$

For E_{II} - Mutual Energy for Cation in the Second Layer

When another cation moves to the second layer of water molecule, the potential energies of cations (E_{II}) in the OHP are composed of the following energies:

a) Interaction energy (E_{2s}) due to Coulomb forces between cation (2) and negatively charged site (s) calculated with Eq. (4).

b) Interaction energy ($E_{2d\theta}$) between cation (2) and a dipole water molecule calculated by Eq. (5). Since the hydration number is equal to six, there are six water molecules around cation (2). Four of these water molecules with θ_1 equal to 0° and the other two with θ_2 equals 40°. The centre of the dipole is not at the centre of the water molecule. Hence, θ_2 is not equal to 45° as seen in Fig. 1b.

c) Interaction energy (E_{12}) between cations (1) and (2) (see Eq. 4).

d) Potential energy (E_{ab}) between negatively charged site (a) and (b).

e) Potential energy (E_{1s}) between cation (1) and negatively charged site (s).

f) Interaction energy ($E_{1d\theta}$) between cation (1) and a dipole molecule. Three of these water molecules have θ_1 equal to 90°, whilst other two have θ_2 equals 40°.

Therefore,

$$E_{II} = \Sigma E_{1d\theta 1} + \Sigma E_{1d\theta 2} + \Sigma E_{1s} + \Sigma E_{2d\theta 1} + \Sigma E_{2d\theta 2} + \Sigma E_{2s} + E_{12} + E_{ab} \tag{7}$$

Since $E_{2d\theta 1} = E_{1d\theta 1}$, $E_{2d\theta 2} = E_{1d\theta 2}$

$$(E_{II} - E_I) = (2E_{2d\theta 1} + 4E_{2d\theta 2} + \Sigma E_{2s} + E_{12}) \tag{8}$$

The charge density, σ_s, is represented by the equation

$$\sigma_s = \Sigma z_j e' \Delta x \, C_{j1} \tag{9}$$

where Δx = thickness of the water layer
Combining Eqs. (3) and (9) gives

$$\sigma_s = \Sigma z_j e' \Delta x C_{j2} \exp\left[\frac{(E_{II} - E_I)}{KT}\right] \tag{10}$$

Based on the Gouy-Chapman model, the charge density, σ_d, is written (Hiemenz and Rajagopalan 1997) as:

$$\sigma_s = L_1[\exp(Y_d) - \exp(-Y_d)] \tag{11}$$

where

$$L_1 = (2\varepsilon_0 \varepsilon KTC_0)^{1/2} \tag{12}$$

$C_0 = C_{j0}$ (assuming symmetrical electrolyte $z{:}z$)

$$Y_d = \frac{ze'\psi_d}{2KT} \tag{13}$$

$\psi_d = \psi_2$; ε = dielectric constant of the bulk solution (water) = 80.1 (Lide 1996).
This equation is simplified by considering a single binary electrolyte of valency z in the
solution (Kruyt 1952, p. 130). For simplicity in formulation, the valence of Pb^{2+} (z_1) ion is
used in Eq. (14). Since σ_s and σ_d must be positive values, the absolute values of σ_s in Eq. (10)
and σ_d in Eq. (11) are used. With this assumptions, Eqs. (1), (2), (10), and (11) may be
combined to give

$$\sigma_t = \left| L_2 \exp(-2 \, Y_d) \right| + \left| L_1 \left[\exp(Y_d) - \exp(-Y_d) \right] \right| \tag{14}$$

where

$$L_2 = z_1 e' \Delta x \, C_{j0} \exp\left[\frac{E_{II} - E_I}{KT} \right] \tag{15}$$

$C_{j0} = C_{10}$ = concentration of Pb^{2+} ions in bulk solution
Parameter Y_d in Eq. (14) is calculated by trial and error. The value of ψ_d can be obtained from
Eq. (13). Values of C_{12} and C_{11} are calculated by substituting the value of C_{10} into Eqs. (2)
and (3).

The variation in potential with distance from the OHP for the diffuse layer is expressed by the
Gouy-Chapman theory (Hiemenz and Rajagopalan, 1997):

$$\kappa x = \ln\left[\frac{[\exp(Y_p) + 1][\exp(Y_d) - 1]}{[\exp(Y_p) - 1][\exp(Y_d) + 1]} \right] \tag{16}$$

where:

$$Y_p = ze'\psi_p / 2KT \tag{17}$$

x = distance from the OHP; ψ_p = electric potential at point p (with a distance x from OHP).

$$\kappa = \left(\frac{(e')^2}{\varepsilon_0 \varepsilon KT} \Sigma_j z_j^2 C_{j0} \right)^{1/2} \tag{18}$$

For the accuracy of κ, the assumption of symmetrical $z{:}z$ is not applied in Eq.(18), i.e., z_+ for
Pb^{2+} is equal to 2 and $z.$ for NO_3^- is equal to -1.

By trial and error, parameter ψ_p in Eq.(17) can be obtained. The concentration of ions in layer
i at point p is calculated from Eq.(3).

Determination of Total Surface Charge for the Interaction Energy Model
Total surface charge (σ_t) is defined as ratio of cation exchange capacity (CEC) to its specific
surface area (S_a):

$$\sigma_t = CEC / S_a \tag{19}$$

The CEC is determined under a specific condition (pH7), resulting in only one CEC for each soil. Batch adsorption results in Fig. 3 indicate that adsorption capacity (q) of illite soil varies with equilibrium concentrations until the illitic soil reaches its maximum adsorption capacity. q becomes almost constant at this point. Lead ion retention in illitic soil is also a function of pH. The interaction energy model must incorporate with the adsorption isotherm to calculate the adsorption capacity at a known equilibrium concentration and pH value. The adsorption capacity (q) is used instead of the CEC value in Eq.14 to calculate the charge density in each water layer surrounding the particle surface. At different retention capacities, this model gives a different charge density.

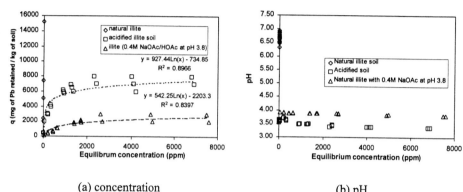

(a) concentration   (b) pH

Fig. 3 Batch equilibrium adsorption isotherm and pH of lead for Sealbond illitic soil

The log-function is adopted to fit the experimental data shown in Fig. 3. The relationship between adsorption capacity and equilibrium solution concentrations is given in Eq. 20.

$$q = a + b \exp\left(-0.5 \left[\frac{\ln\left(\frac{C_0}{c}\right)}{d} \right]^2 \right) \tag{20}$$

where q = adsorption concentration of Pb in soil or absorption capacity (g of Pb/g of soil); C_0 = equilibrium solution concentrations (ppm); a, b, c, and d are parameters to be determined.

q values in the batch test are based on the following equation:

$$q = (C_i - C_0) * V/M \tag{21}$$

where C_i = initial concentration of Pb in the solution (ppm); V = volume of solution in the batch equilibrium test; M = mass of soil in the batch equilibrium test.

Eqs. 20 and 21 may be combined as follows:

$$(C_i - C_0)\left(\frac{V}{M}\right) = a + b \exp\left(-0.5 \left[\frac{\ln\left(\frac{C_0}{c}\right)}{d} \right]^2 \right) \tag{22}$$

With a known initial adsorption concentration (q_i) of a soil sample, C_i is determined. The initial adsorption concentration (q_i) of a soil sample is obtained according to EPA3050 procedures. When C_i is available, C_o can be calculated from Eq. 22. It should be noted that each set of values for parameters (a, b, c, & d) of Eq. 22 are only applicable within a certain range of pH because soils with different pHs have different adsorption isotherms.

Use of Force Balance to Determine the Desorption of Lead Ions in Each Water Layer

To further expand the application of this model to predict desorption, the forces acting on each ions in the first layer can be calculated from the following equation:

$$E = Fr \qquad (23)$$

where E = potential energy, F = force, and r = distance between charges.

The mutual electric potential energy on cation 1 (as shown in Fig.1b(a)) includes E_{1a}, E_{1b}, and $E_{1d\theta}$. In order to remove the Pb^{2+} ions from the first layer of the charged particle surface (i.e. IHP), Pb^{2+} ions from the second layer (i.e. OHP) must be removed first. For every Pb^{2+} ion being removed from the second layer, there will be the same amount of + charge replaced to balance charges, e.g. in the batch test with acidic solution, the removed Pb^{2+} ion is replaced by $2H^+$ ions.

The interaction energy (E_{1H}) between Pb^{2+} ion in the first layer and H^+ ions in the second layer needs to be taken into consideration. Therefore, the total mutual electric energies on cation (1) is

$$E_{total} = E_{1a} + E_{1b} + \Sigma E_{1d\theta} + E_{1H1} + E_{1H2} \qquad (24)$$

Based on Eq. (23), the net force on each cation (Pb^{2+}) on the first layer can be calculated. Calculations for the cations in the second layer are similar to those in the first layer but follow the ion configuration shown in Fig. 1b(b).

As mentioned earlier, unless there is a net repulsive force between the particle surface and the ion or an external applied force that is larger than the attractive force, adsorbed cations in that particular layer cannot be removed. With a known net force for ions in each layer, it is possible to determine the potential for ion removal in a particular layer.

Based on the predictive model, the calculation indicates that ions in the first two layers of water molecule cannot be removed unless a large external force is applied.

RESULT AND DISCUSSIONS

Illitic soil (Sealbond illite from Canada Brick Company) with nitric acid solution (at pH 2) was used for batch equilibrium test. As shown in Table 1, there is good correlation between calculated and measured values. For example, batch equilibrium test data show that there is 47% removal of Pb^{2+} ions at pH 4.2 and 69% removal of Pb^{2+} ions at pH 3.2. Assuming a linear relationship between removal and pH, the removal of Pb^{2+} ions at pH 3.7 is 59%. The predicted Pb^{2+} ions removal value for illitic soil (with C_i = 500ppm or 5000 mg of Pb/kg of soil at pH 6.7) being acidified to pH 3.7 is 63%.

The percentage of error in these predictions is within ±10% for the concentration range between 5000 and 50000 mg of Pb/kg of illitic soil. Exceptions are at the lowest concentration at 2500 mg of Pb/kg of soil and the highest concentration at 80000 mg of Pb/kg of soil). The higher percentages of error are 50% and 25%, which might be due to experimental errors such as dilution of the solution and limitation of instrumentation used.

Table 1   Comparison of Pb^{2+} ion removal in illite soil determined by the model and batch equilibrium tests

C_i (ppm)	C_o (ppm)	% of ions in the Stern and the diffuse layer	% of free ions in bulk water	Ion concentration ratio of the Stern layer to the diffuse layer Stern : diffuse	% of removal (calculated by model, removal all except first 2 layers from particle surface)	% of removal (laboratory measurement)	% of error
250	63	74.8	25.2	30.8 : 69.2	66.2	44.0	50.8
500	183	63.4	36.6	39.8 : 60.2	63.1	58.6	7.7
750	335	55.3	44.7	44.5 : 55.5	64.0	61.3	4.5
1500	914	39.1	60.9	50.5 : 49.5	71.1	77.2	-7.8
2000	1356	32.2	67.8	52.3 : 47.7	75.4	82.9	-9.1
3000	2298	23.4	76.6	54.0 : 46.0	81.5	81.4	0.2
5000	4268	14.6	85.4	55.5 : 44.5	88.1	85.1	3.6
8000	7268	9.2	90.8	56.8 : 43.2	92.4	73.6	25.5

CONCLUSIONS

(1) The developed interaction energy model can predict the distribution of adsorbed Pb^{2+} ions within the Stern layer and diffuse layer of a charged particle. Based on the force balance, it can be used also to predict the removal of lead in the desorption process. The force balance calculation indicates that ions in the first two layer of water molecule from the particle surface cannot be removed unless an external force is applied. Theoretical calculations show that the predicted removal values are in good agreement with the results obtained from batch equilibrium tests (with dilute HNO_3).

(2) The proposed sorption-desorption model, incorporating the adsorption isotherm, was found to be a practical means to predict the amount of possible contaminant removal from the particle surface. In this model, the adsorption capacity should be used instead of the cation exchange capacity (CEC). In addition, the adsorption isotherm should be obtained in an environment similar to field remediation conditions.

REFERENCES

Ansoult, M., DeBacker, L.W., and Declereq, M. (1985). Relationship between Apparent Dielectric Constant and Water content in Porous Media. Soil Sci. Soc. Am. J. Vol 49, 47-50

Grim, R.E. (1968). Clay Mineralogy, 2nd Edition, McGraw-Hill, N.Y., 596 pp.

Hiemenz, P. C. and Rajagopalan (1997). Principles of Colloid and Surface Chemistry. 3rd Edition, Marcell Dekker, Inc., N.Y., 650 pp.

Kruyt, H.R. (1952). Colloid Science. Elsevier Press, New York. 130 pp.

Lide, D.L. (1996). CRC Handbook of Chemistry and Physics. 77th Edition, CRC Press.

Shapiro, A.P. and Probstein, R.F. (1993). Removal of Contaminants from Saturated Clay by Electroosmosis. Environ. Sci. Technol. Vol. 27, No.2, 283-291.

Yeung, A.T. and Dalta, S. (1995) Fundamental Formulation of Electro-kinetic Extraction of Contaminants from Soil. Canadian Geotechnical Journal, Vol. 32, No. 4, 569-583

The use of a laser induced fluorescent static cone to identify hydrocarbon contamination at a petrol station site

S.A. PARKHURST BEng MSc CEng MICE PE MASCE
Allott & Lomax, Reading, UK

SYNOPSIS

Laser Induced Fluorescent (LIF) static cone penetration tests are becoming established as a rapid and cost effective method of assessing the presence of petroleum hydrocarbon compounds spatially within a soil mass. The ROST™ LIF technique, operated by Fugro Ltd. of Hemel Hemstead, was used by Allott & Lomax as a rapid method of identifying petroleum hydrocarbon contamination at a petrol station and garage. This paper describes the use of the system as well as its benefits and disadvantages.

INTRODUCTION

Allott & Lomax were commissioned in 1996 to carry out a pre-acquisition audit of a site in North Walsham. At the time of the commission the site was a petrol station and vehicle repair garage. The site was bounded by residential and retail development and covered an area of approximately one hectare (Figure 1).

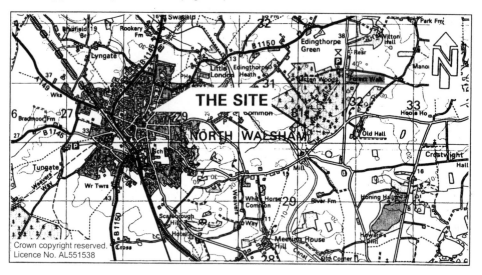

Figure 1. Site location plan

The determination of petroleum hydrocarbon contamination at a site can involve a time consuming multi-phased investigation to establish the spacial extent of the plume. A quicker and more flexible alternative is to use the Laser Induced Fluorescent (LIF) static cone penetration test to detect petroleum hydrocarbons in-situ.

This paper describes the use of a laser induced fluorescence static cone as part of a phased ground investigation at a petrol station site.

Geoenvironmental engineering, Thomas Telford, London, 1999, 238–243

The phase one investigation comprised five trial pits to 3.0m depth, one borehole to 20.5m depth, chemical testing, gas monitoring and ground water monitoring. The investigation was designed to provide a broad range of information with minimal disturbance to the commercial operation of the site. The phase two investigation comprised nine LIF probes to 12.5m depth and nine geo-probe boreholes to 2.5m depth (Figure 2).

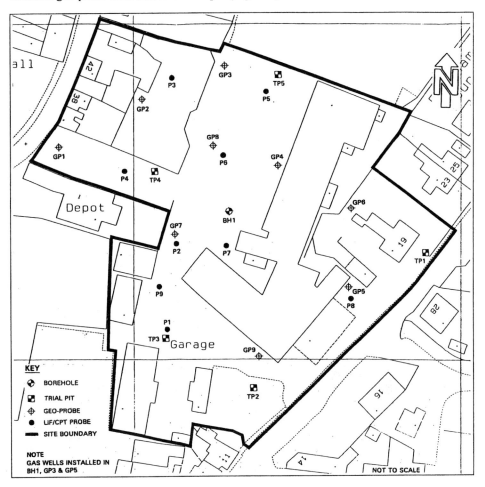

Figure 2. Site investigation layout and site plan

The use of the LIF cone was adopted because of the following main reasons:
- The main contaminant was hydrocarbon based.
- Speed of the test.
- Availability of real time results.
- The ground conditions suited its use.
- Geotechnical data was obtained from the standard static cone penetration test.
- Low water table.

The real time detection of petroleum hydrocarbon compounds provided flexibility to relocate and add LIF probes locations as the investigation proceeded.

INITIAL ASSESSMENT
An initial assessment of the contamination potential of the site was performed by completing a desk study and site walk over. The desk study utilised a variety of readily available sources to establish the past uses of the site. The walk over survey provided a visual confirmation of the desk study information.

The published geological map of the area indicated the site to be underlain by glacial deposits of the Corton Formation overlying Chalk. The Corton Formation comprised Corton Sand overlying Corton Till.

The hydrogeological map of the area indicated the piezometric water level within the chalk to be approximately 12 metres below the existing ground surface.

The initial assessment identified a number of high risk areas and likely site contaminants (Table 1).

Table 1. Risk areas

RISK AREA	SOURCE	POTENTIAL CONTAMINANTS
Petrol Station	Underground petroleum tanks	Hydrocarbons, additives Methane and carbon dioxide gases
Vehicle Repair	Paint shop Oil Storage Paint thinning Degreasing Spent fluid:antifreeze, brake fluid	Metals Hydrocarbons Chlorinated hydrocarbons Volatile organic compounds Ethylene glycol, polymerised glycols, ethers
Car Wash	Car washing	Detergents
Vehicle Breaking	Spent fluids, batteries	Acids and spent fluids as above
Original Garage	Building materials/asbestos roof	Asbestos

A preliminary risk assessment of the site was carried out to establish a targeting philosophy for the investigation (Table 2).

Table 2. Preliminary risk evaluation

SOURCE	PATHWAY	RECEPTOR	RISK SIGNIFICANCE
Petrol Station	Percolation	Ground Water	High
Vehicle Repair	Percolation Contact	Ground Water Site Users	Medium Medium
Car Wash	Percolation	Ground Water	Medium
Vehicle Breaking	Percolation Contact	Ground Water Site Users	Medium High
Original Garage	Contact Percolation	Site Users Ground Water	Low Medium

SCOPE OF THE INVESTIGATION
The aim of the investigation was to identify sources of contamination, determine their likely spacial extent and estimate the cost any remedial measures required. The scope of the investigation was influenced by the anticipated types of contamination, access restrictions on site and the time constraints imposed by the client.

A two stage targeted investigation was adopted primarily because of programme constraints. The phase one investigation was carried out to provide general information on the ground conditions and type of contaminants present.

The phase 1 site investigation indicated that the main source of contamination was hydrocarbon based. The phase two investigation was therefore carried out to establish in more detail the spacial extent of the hydrocarbon contamination present.

HYDROCARBON DETECTION USING THE LIF STATIC CONE

The LIF technique detects the fluorescence emitted by petroleum hydrocarbon compounds under exposure to an ultraviolet laser. Fibre optic cables fitted above the tip of a static cone are connected to an ultraviolet laser and detection system located in a standard static cone penetration truck (Figures 3 and 4).

Figure 3. Laser set up

Figure 4. Cone penetration rig

The LIF cone can be referenced to a dilute mixture of different petroleum hydrocarbon compounds or a specific hydrocarbon such as diesel or petrol. LIF cone referencing to a specific hydrocarbon will increase its sensitivity to that hydrocarbon.

The data provided by the LIF cone is presented in two forms:
- Fluorescence intensity
- Wave length time matrix

The fluorescence intensity is measured continuously as the cone penetration test is carried out. The data is displayed visually as fluorescence intensity verses depth and provides information relating to the depth over which the hydrocarbon is detected and its intensity relative to the reference mixture.

The wave length time matrix (WTM) displays fluorescence wave length, fluorescent decay and fluorescent intensity at any particular depth. The WTM signature is specific to the type of hydrocarbon compound present and is used for compound classification. For example a flat plot is indicative of a coal tar based hydrocarbon and a skewed plot indicative of a diesel based hydrocarbon.

The integrated LIF plot for CPT P2 (Figure 5) indicates the detection of hydrocarbons at four levels: coal tar contamination at 2.0 metres (point B) and diesel contamination at 3.3m and 6.5m (points C and D).

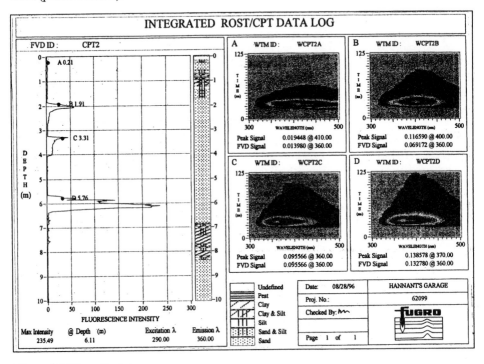

Figure 5. LIF result plot

The wave time matrix plot for CPT P8, carried out in the vicinity of the underground petrol tanks, indicated coal tar contamination at the surface (point A on Figure 5) but no contamination below this depth. This information was particularly useful in showing that petrol leakage from these underground tanks had not occurred.

The information obtained by the LIF cone was qualitative. It could not be used as an indication of the concentrations present. As such conventional laboratory testing of hydrocarbons of soil samples is required to provide quantification of the LIF results.

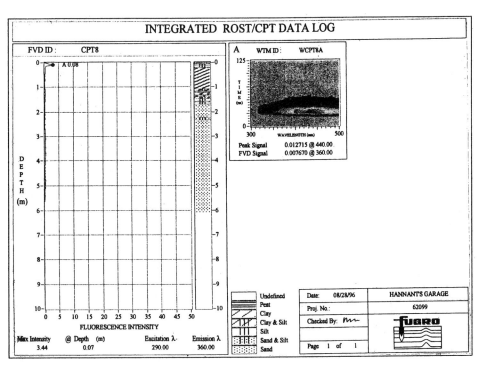

Figure 6. LIF result plot

CONCLUSIONS

The key benefits identified in using the system were:
- A real time indication of the petroleum hydrocarbon contamination present.
- The use of static cone penetration testing equipment was rapid.
- The system provided continuous data with depth.
- The wave length time response indicated the type of hydrocarbon compounds present.
- Static cone penetration testing was carried out at the same time as LIF cone testing.

The disadvantages of using the system were:
- Potential installation and reliability problems associated with using state of the art equipment in poor conditions.
- Secondary soil sampling and testing was required to quantify the hydrocarbon concentrations detected by the LIF cone.
- Only applicable to ground conditions which suited the static cone penetration test.

In summary the system provided a rapid and cost effective method of identifying the hydrocarbon contamination present beneath the site.

ACKNOWLEDGEMENTS

The author wishes to acknowledge the contribution that Fugro Limited have made in the preparation of this paper.

Transport, Persistence and Fate of Pollutants

Heavy metal retention capacity of natural clay liners of landfills.

RNDr. R. ADAMCOVÁ, Comenius University Bratislava, SK

INTRODUCTION

Due to the molecular diffusion, soluble pollutants are able to migrate even through quite "impermeable" bottom clay liners of landfills (waste disposal sites) if the clay is saturated with water. However, certain components of the clay liner can retard some pollutants either by chemical interaction, or by physical one, as described by many authors. Therefore it is necessary to test not only the permeability, but also the retention capacity when searching for the most suitable clay liner material. Usually, so-called sorption tests are used.

Batch tests are the most frequent type of sorption tests. Their main advantage is the speed, the main disadvantage is the strong dependence of the results on both, internal and external, test conditions. These use to differ from laboratory to laboratory, for it is impossible to compare results obtained by different authors. However, in the future a huge amount of data must be evaluated in order to elaborate a practical classification of retention properties. This would simplify the choice of natural clays for liners. Therefore, the method calls for unification. In this paper, the arrangement and the interpretation of batch tests will be explained how they were done in the cooperation between the Comenius University in Bratislava and the University of Agriculture in Vienna (Austria) and the Federal Technical University in Zurich (Switzerland) respectively.

THEORY

In the batch test, solution of known concentration of studied ions is added to the clay sample and mixed, still under monitoring of pH and temperature conditions. If the ions are retained in the clay, their concentration in the solution decreases. The test should be stopped when the equilibrium between ions retained ("sorbed") in the clay and of those in the solution was achieved. The difference between the initial solution concentration c_0 and the concentration of the equilibrium solution c_{eq} represents the concentration c of ions retained in the clay. The relation between c_{eq} (mol/l) and c (mol/kg) can be described by three different mathematical equations that represent three sorption models:

Henry model:
$$c = K_d.c_{eq}$$

Freundlich model:
$$c = K.c_{eq}^{N}$$

Langmuir model:
$$c = \frac{K_1. K_2 . c_{eq}}{1 + K_2 . c_{eq}}$$

where K_d is the distribution coefficient (l/kg), K (l/kg) and N (1) are the Freundlich coefficients, K_1 is the maximum sorption called sorption capacity (mol/kg) and K_2 is an equilibrium constant depending on the sorption energy.

Geoenvironmental engineering, Thomas Telford, London, 1999, 247–254

The experience showed that Henry model conforms to the one-layer sorption at very low concentrations of solutions. At higher concentrations, a multi-layer sorption occurs described by the Freundlich model. According to this model, sorption should be unlimited. However, at certain concentration the sorption capacity will be exhausted, that is expressed in the Langmuir model. So-called sorption isotherms are the graphical expression of the $c = f(c_{eq})$ relations. In an optimum case, it is possible to fit a trend line to the sorption isotherm points obtained in the batch test that conforms to one of the three mathematical functions mentioned above. Already the isotherm type gives some information to which extent is the sorption capacity of the clay exhausted. Isotherms enable also a qualitative comparison of retention properties of two and more clays. Furthermore, besides the qualitative interpretation, some quantitative retention parameters can be calculated from obtained constants, as well. The most important one is the sorption capacity from the Langmuir isotherm. From the Freundlich isotherm, the retardation factor R_f can be calculated (the knowledge about the effective porosity n_e of the clay is necessary for this calculation), however, the obtained values are much higher than retardation factors reported "in situ" or from diffusion tests (J.-F. Wagner, 1992).

BATCH TESTS

Many reports about sorption of diverse constituents on pure sorbents (clay minerals, zeolites, Fe-oxihydroxides) were published up to now. In most cases, tests were made with one constituent and under strict laboratory conditions (control of temperature and pH etc.). Their results, however, are not very useful when dealing with the retention capacity of natural polymineral clays as a whole concerning pollutants coming as a mixture in the landfil leachate. Studying heavy metals, different affinity of every metal to the clay, manifold competition reactions between the metals, as well as the changing pH due to reactions between the clay and the leachate (not only the sorption *s.s.*) determine the real retention. Therefore, standard sorption tests with only one metal under constant pH do not allow to compare which of selected clay types will be the best mineral liner of a landfill.

In the presented study, the retention properties of 12 clays were tested, concerning selected heavy metals. 0,5g of clay was mixed with 50 ml of solution (ratio 1:100) "over-head" with the speed of 40 rpm for 24 hours. This time should be enough to reach an equilibrium-like point. With the batch test arrangement, we tried to get more close to the natural processes in a clay liner of a landfill. Always mixtures of water solutions of lead, zinc and copper chlorides or nitrates were used. Also 1g NaCl/l was added in order to respect the effect of Na^+ ions (competing cations) and Cl^- anions (complexation) that are present in every leachate from municipal landfills in highest concentrations, as well as to prevent a rapid decrease of the ionic strength of the solution due to the sorption. For better simulation of the complex interaction between the leachate and the clay, real landfill leachate could be used instead of water, as well. The initial concentrations in our tests were 30mg/l, 100mg/l, 150mg/l, 0,002mol/l and 0,004mol/l from every metal (lead, copper and zink) in the solution testing first 8 clays and 0,0005mol/l, 0,001mol/l, 0,002mol/l and 0,004mol/l testing last 4 clays. Next time we would prefer same molar concentrations in order to offer equal amount of all studied cations, because this enables a better comparison of the clay affinity to various metals. In order to test the maximum retention capacity, we used solutions with pH between 5 and 5,25. Equilibrium solutions were decanted after centrifugation into PE bottles without filtration and the final pH was measured. Concentration of heavy metals were analyzed using AAS.

Sorption isotherms were constructed from the measured molar concentrations, first for every metal individually, then for all metals (Pb+Cu+Zn) together, so called "summary isotherms". Using the trend line function in the MS Excel program, a straight line (linear regression - Henry isotherm) or a parabola (power function - Freundlich isotherm) was fitted to the isotherm points. In some cases, sorption could be described by a broken Henry isotherm. Then a new graph $c_{eq}/c = f(c_{eq})$ was constructed and the Langmuir isotherm in the linear form of $c_{eq}/c = 1/K_1.K_2 + c_{eq}/K_1$ was searched for using linear regression. Quantitative retention parameters were determined only if regression coefficient was $R^2 \geq 0,75$.

CHARACTERISTICS OF STUDIED CLAYS

Clays of various genesis and age were included (tab.1). We tested mainly clays from Slovakia with some comparative foreign samples. Selected 12 clays should create a series tending to improve their retention properties referred to heavy metals. For outer limits, pure kaolin and natural bentonite of very high quality were taken for minimum and maximum retention. Other clays were natural fine-grained soils in geotechnical terms with different mineralogy, especially concerning the content of main clay minerals, carbonate and other sorbents like Fe-oxihydroxides and organic matter (tab.2).

Table 1. Geological characteristics of studied clays

Clay type	No.	Locality (country)	Genesis	Age
kaolin ("KGa-2, poorly crystallized")	1	Georgia (USA)	hydrolytic alteration of clastic sediments and industrial purification	Paleocene
kaolinite-dominated polymineral clay	2	Gregorova Vieska (SK)	fluviolimnic	Miocene
	3	Horn (A)	fluviomarine	Oligocene
	4	Breiteneich (A)	fluviomarine	Oligocene
montmorillonite - dominated polymineral clay with organic matter (pre-contaminated)	5	Banská Štiavnica (SK)	washed-up slope sediment cover of andesites	Holocene
polymineral clay with higher carbonate content	6	Ivanka p.Nitre (SK)	eluvial, loess	Pleistocene
montmorillonite-dominated polymineral clay with less carbonate	7	Modra (SK)	brackish	Miocene
montmorillonite-dominated carbonatic polymineral clay	8	Pozba (SK)	limnic to fluviolimnic	Pliocene
	9	Stupava (SK)	marine	Miocene
	10	Devínska Nová Ves (SK)	marine	Miocene
bentonitic clay	11	Zvolenská Slatina (SK)	re-deposited alteration products of andesite tuffs	Pleistocene up to Holocene
bentonite	12	Jelšový Potok (SK)	alteration of rhyolite and rhyolitetuffs	Miocene up to Pliocene

Table 2. Mineralogy of studied clays

No.	Content in the clay fraction <2µm (%)			Content in the complete sample (%)		
	montmorillonite	illite	kaolinite	carbonate	C_{org}	Fe-oxihydroxides
1	-	-	99-100	-	-	-
2	4	1	95	-	-	-
3	11	3	86	0-8	+	(+)
4	20	2	78	<1	(+)	+
5	80	5	9	<1	2,3	+
6	76	8	7	11	0,1	(+)
7	68-83[*]	[*]	5-13	5	<0,1	+
8	65	5	28	10	1,7	+
9	77-80[*]	[*]	5-6	20	0,3	(+)
10	75	12	7	20,5	1	+
11	75-90	-	11-13	<1	-	-
12	90-95	0-5	0-5	-	0,2	+

[*] swelling clay minerals together as "montmorillonite" (illite and/or M/I-mixed layers included);
- not present or negligible amount; (+) probably present; + present, amount not determined.

Table 3. Some physical properties of studied clays

No.	Clay fraction <2µm (%)	Plasticity index $I_P=w_L - w_P$ (%)	Geo-technical class	Filtration coefficient k_f (m.s^{-1})	Cation exchange capacity (CEC) mmol/100g
1	70-75	N	N	N	3
2	52	38	CH	N	16
3	30	34	CS	$(1,03 \div 1,27).10^{-11}$ triax. test undisturbed sample	5-9
4	42,5	25	CI	N	36
5	35	20	CI	9.10^{-9} triax. test undisturbed sample	16
6	25	16,5	CI - CL	$3.10^{-9} \div 1.10^{-8}$ triax. test undisturbed sample	25
7	57	36	CH	7.10^{-9} consolid. test undisturbed s.	32
8	35	29	CH	$2,44.10^{-10}$ triax. test compacted sample	14
9	51	43	CH	$10^{-11} \div 10^{-12}$ triax. test compacted sample	19
10	48	44	CV	$(8,12 \div 9,8).10^{-11}$ triax. test compacted sample	20
11	47	37	CH -CV, ev. MH	$2,06.10^{-11}$ triax. test compacted sample	41
12	70	74	CE	N	80

N -not determined

In the batch tests, untreated (i.e. natural) complete clay samples were used. However, a special sample is there as well - from the locality called Banská Štiavnica. This sample was highly pre-contaminated having been situated in the subsoil of a pond with flotation residual from ore processing for ca 30 years. We were interested if the clay with already 11 gPb/kg (!) 280 mgZn/kg, 50 mgCu/kg and other heavy metals sorbed from the tailings leachate is still able to retain more ions.

ontent of the clay fraction <2μm, as well as some physical properties and the geotechnical
lassification (Slovak Technical Standard 73 1001) are summarized in tab.3.

ESULTS AND DISCUSSION
orption isotherms of all clays are compared in fig.1 and 2. They show clearly that the
ffinity to a certain metal changes from clay to clay. But also the affinity of certain clay
hanges from metal to metal. Due to competitive reactions, increasing "sorption" of the
referred metal decreases usually the "sorption" of other metals, however, the "summary
orption" is increasing till the retention capacity is exhausted.

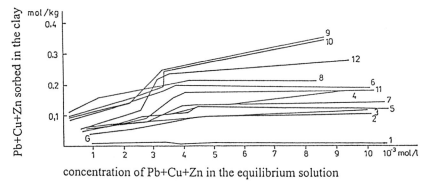

concentration of Pb+Cu+Zn in the equilibrium solution

Figure 1. "Summary sorption isotherms" (numbers indicate the clay type according to tab.1)

ead, the most preferred metal from our triplet, is retained best of all in carbonate-rich clays.
esides the buffering effect of carbonates, precipitation of lead carbonate was proved using
RD. Further retention of lead depends on the type and amount of clay minerals, the content
f montmorillonite seems to be the second controlling factor. The retention processes are by
ar not as clear when dealing with copper and zinc. The presence of less noble metals in the
olution fosters the precipitation of copper that is slightly better retained than zinc in most
lays. Zinc is also known as one of the most mobile metals up to high pH values. On the other
and, it is interesting that zinc is retained well in kaolinite-dominated polymineral clays (but
ot in pure kaolin), even better than copper. The zinc isotherms of these clays are still
creasing while those of copper are already decreasing, as well as the zinc isotherms of other
lays. It is difficult to explain, no correlation to the clay fraction content, CEC, amount of
ontmorillonite or carbonate was found, nor that to the pH value of the equilibrium solution.

resented differences in the heavy metal retention prove that it is insufficient to evaluate the
tention properties according to one particular metal. The "summary sorption", as defined
bove, seems to be more useful. Using "summary sorption isotherms", a qualitative
valuation of the retention capacity could be done. Because in most cases the retention
apacity was not exhausted yet, the contents of metals sorbed in all clays were compared at
e same equilibrium concentration of the solution (0,008mol/l). Descending order according
 the summary retention follows: montmorillonite- dominated carbonatic polymineral clay
amples 9 and 10> bentonite s.12 > montmorillonite- dominated carbonatic polymineral clay
8 > polymineral clay with higher carbonate content s.6 > bentonitic clay s.11 > kaolinite-
ominated polymineral clay s.4 > montmorillonite- dominated polymineral clay with less
arbonate s.7 > montmorillonite- dominated polymineral clay with organic matter pre-
ontaminated s.5 > kaolinite-dominated polymineral clay s.2 and 3 > kaolin s.1.

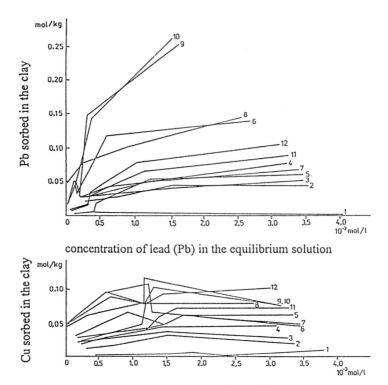

concentration of lead (Pb) in the equilibrium solution

concentration of copper (Cu) in the equilibrium solution

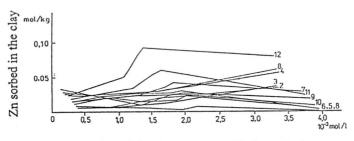

concentration of zinc (Zn) in the equilibrium solution

Figure 2. Sorption isotherms of studied metals (numbers according to tab.1)

As expected, retention corresponds with the mineral composition of the clay, it increases wit
increasing content of carbonates and montmorillonite, carbonates are dominant. Because of
very low content, the importance of other sorbents is not as much obvious. However, they
contribute to the cation exchange capacity significantly. A sequence analysis of the highl
pre-contaminated sample No.5 showed that up to 19% of Zn was originally bound to Fe
oxihydroxides, 20% of Zn and 29% of Pb to easily oxidable phases including organic matter
but only 9% of Zn and 12% of Pb due to ion exchange (R.Adamcová et al., 1997). Test
showed, that this clay is still able to retain further metals. Due to the real heavy metal conten
in this sample, the sorption isotherm should be positioned much higher than was calculate

from the equilibrium concentration of the solution. The surprisingly high retention capacity of the kaolinite-dominated polymineral clay No.4 is also due to the unexpected high CEC.

The qualitative comparison at certain equilibrium concentration is only of orientation value, therefore a quantitative interpretation was attempted (fig.3, tab.4).

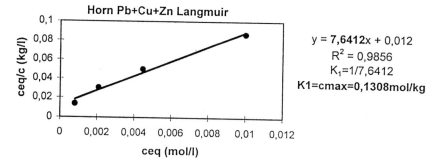

Figure 3. Example of line-fitting procedure and calculation of sorption capacity

Table 4. Quantitative retention parameters

No.	Type and parameters of the sorption isotherm							
	Henry - broken			Freundlich			Langmuir	
	K_{d1} (l/kg)	K_{d2} (l/kg)	R^2	K (l/kg)	N	R^2	$K_1 = c_{max}$ (mol/kg)	R^2
1	X	X	X	X	X	X	0,011	0,99
2	X	X	X	0,6462	0,3922	0,89	0,120	0,965
3	X	X	X	0,4059	0,2785	0,98	0,131	0,99
4	X	X	X	1,1696	0,416	0,94	0,238	0,895
5	X	X	X	0,447	0261	0,87	X	X
6	X	X	X	0,540	0,204	0,87	0,190	0,99
7	X	X	X	0,917	0,384	0,87	0,165	0,96
8	X	X	X	0,834	0,2341	0,93	0,292	0,94
9	X	X	X	1,581	0,366	0,76	X	X
10	X	X	X	1,491	0,338	0,92	X	X
11	42,13	0,55	0,88	2,224	0,506	0,84	0,224	0,90
12	63,48	9,43	0,86	6,350	0,628	0,87	0,414	0,80

X - insufficient regression

Comparing the calculated sorption capacity c_{max}, the retention changed only very little: nontmorillonite- dominated carbonatic polymineral clay samples 9 and 10 > bentonite s.12 > nontmorillonite- dominated carbonatic polymineral clay s.8 > bentonitic clay s.11 $\cong$ kaolinite-dominated polymineral clay s.4 > polymineral clay with higher carbonate content s.6 > nontmorillonite- dominated polymineral clay with less carbonate s.7 > (montmorillonite-dominated polymineral clay with organic matter pre-contaminated s.5) > kaolinite-dominated polymineral clay s.2 and 3 > kaolin s.1.

Both, the qualitative as well as the quantitative comparison show, that the estimate of the retention properties according to the clay fraction content, plasticity and geotechnical class is very inaccurate. The most important factors of the retention capacity are the carbonate content (ev. the buffering capacity of the clay) and the cation exchange capacity due to all sorbents *s.s.* in the clay (not only clay minerals). We plotted our clays in a graph where carbonate content and CEC were two independent variables (fig.4). Comparing the calculated retention capacity K_l (tab.4), we tried to border very approximately the fields of low, intermediate, high and very high retention capacity. Much more data are necessary for the calculation of the real direction of the bordering lines, as well as for the definition of the limits for every retention class, e.g. what is „high retention capacity". However, this plot represents one of the first attempts to classify clays generally according to their retention properties. After revision, it could be well applied in practice.

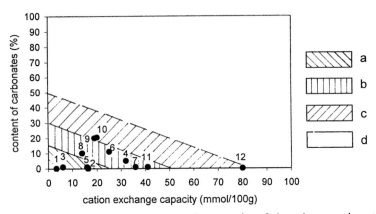

Figure 4. Qualitative estimate plot of the retention capacity of clays due to carbonate content and cation exchange capacity (numbers indicate the tested clay type according to tab.1). Retention capacity: a) low, b) medium, c) high, d) very high

CONCLUSION

Except of pure kaolin, all tests could be interpreted also as Freundlich isotherms, the retention capacity was probably not exhausted yet. But, the parameters of the Freundlich isotherms were not used as comparative criteria because no clear trend was observed. Neither one of the isotherms could be interpreted as a simple Henry isotherm! Therefore, the use of constant distribution coefficients K_d in the migration calculations is mostly incorrect when dealing with highly contaminated solutions! The experiments also showed that testing only one metal, no verdict about the retention of other metals can be stated. Therefore the "summary sorption" of several metals from a mixture solution is a better criterion for the evaluation of the retention properties of clays in contact with landfill leachates. The sorption capacity calculated from the Langmuir isotherms can be used for the classification of clays according to their retention properties. However, first of all, standardization of batch method, more test results, as well as certain convention about the class limits of retention capacity are necessary.

REFERENCES

Adamcová R. et al., 1997: Sludge deposits around the mining town of Banská Štiavnica – environmental and geotechnical problems. In: GREEN 2. Contaminated and derelict land. Thomas Telford Publishing, London, 150-157
Slovak Technical Standard 73 1001 Subsoil under shallow foundations (in Slovak)
Wagner J.-F., 1992: Verlagerung und Festlegung von Schwermetallen in tonigen Deponiebasisabdichtungen. Ein Vergleich von Labor- und Geländestudien. Schriftenr. Angew. Geol., Karlsruhe, Nr.22

pH Influence on Sorption Characteristics of Heavy Metal in the Vadose Zone

M. Elzahabi
Department of Civil Engineering and Applied Mechanics, McGill University, Que., Canada.

ABSTRACT

Sorption is an important process in the modelling and prediction of the movement of heavy metals in unsaturated clay barriers. This experimental study investigates the effect of pH changes in the acidic range on the adsorption characteristics of heavy metals such as: Lead, Copper and Zinc in an unsaturated soil. A series of one dimensional coupled solute and moisture leaching column tests, using different heavy metal solutions, were conducted on an unsaturated illitic soil at varying pH values. Variations of volumetric water content with distance were measured for different time durations, and concentrations of heavy metals in the liquid and solid phases were analysed. The migration and retention profiles of contaminants along the soil column were determined for each individual layer in the soil along with the partitioning coefficients.

Results from column leaching tests showed that the adsorption characteristics of heavy metals are highly dependent on the soil pH, the presence of carbonates, and the initial conditions of the soil. The presence of carbonates in the natural illitic soil increased the retention of heavy metals at high pH and also enhanced the buffering capacity of the soil. At high soil pH and carbonate content, heavy metals are retained in the soil if the buffering capacity is high enough to resist the acidic leachate input, and sorption processes will prevail in the carbonate phase. As the soil pH decreases, the dissolution of carbonates increases and cation exchange capacity becomes the more dominant process in heavy metal retention. In acidic soil, cation exchange capacity of the soil decreases slightly with a lowering the soil pH. This however does not detract from the ability of the soil to retain heavy metals under unsaturated conditions.

INTRODUCTION

Heavy metals (H.M.) found in sludge and landfill leachates are highly toxic to humans, animal and aquatic life. The most common heavy metals found in leachates are Lead (Pb), Copper (Cu), Zinc (Zn), Cadmium (Cd), Chromium (Cr) and Nickel (Ni). The concentrations of these heavy metals vary from 0-100 ppm in municipal solid waste leachate to 100-10,000 ppm in sewage sludge, mining wastes and various industrial wastes (Yong & Diperno, 1991). The solubility of H.M. is highly pH dependent and increases with decreasing soil pH. The same is true for adsorption and retention of heavy metals in mineral soils.

Movement of chemicals in the ground is primarily a liquid phase process involving the movement of water and dissolved solutes. Chemicals that are not sorbed will exist primarily in the dissolved

Geoenvironmental engineering, Thomas Telford, London, 1999, 255–263

phase and their movement in the ground water will be controlled by the relative amount of water, and soil processes that govern the fate of the contaminants. Soil is an excellent adsorbent for both organic and inorganic chemical compounds. The major adsorbing surfaces in soil are clay particles and organic matter. The interactions between contaminants and soil fractions can be expected to influence the physical and physico-chemical behaviour of the material.

Adsorption data are needed to determine the transport properties of the soil with respect to the contaminants under consideration. Test data permits one to calculate the partition coefficient K_d required in the contaminant transport equations. Most transport models use K_d obtained from 'linear' adsorption isotherms – i.e. as a constant parameter (Freeze and Cherry, 1979; Rowe, 1988). Simplification of (K_d) as a constant and linear function may lead to an improper evaluation of the adsorption/desorption phenomena. Davidson et al. (1976) concluded that at high concentrations, the assumption of a linear isotherm can lead to serious errors in predicting contaminant migration. 'Although the convenience of the approach is beyond dispute, its validity as a means of developing reliable predictions of the behaviour of inorganic contaminants in actual ground water systems is questionable in many situations,' (Cherry et al., 1984).

The partition coefficient (K_d) reflects the degree of retardation by reversible ion exchange, and may also include the effects of solute adsorption. Batch equilibrium tests used for determination of adsorption isotherms utilize small portions of soil and representative contaminants. The problems arising therefrom relate to the question of whether or not a small quantity of totally disturbed soil are appropriate to simulate the field conditions or situations, and the variability of K_d from one soil to another. The batch technique does not necessarily reflect actual leachate soil interaction. The ratio of solution to soil and the time required to attain equilibrium does not always give a good estimate of the migration and adsorption of heavy metals through the clay barrier, (Darban, 1997). Adsorption characteristics obtained from leaching column tests provide a better means for evaluation of soil sorption performance. The results obtained include the effects of contaminant precipitation, and effective exposed surface area of the soil system.

This study on unsaturated soil-contaminant interaction investigates the effect of pH changes in the acidic range on the adsorption of heavy metals such as: Lead, Copper and Zinc. A series of one dimensional coupled solute and moisture leaching column tests, using different heavy metal solutions, were conducted on an unsaturated illitic soil at varying pH values. Variations of volumetric water content with distance were measured for different time durations, and concentrations of heavy metals in the liquid and solid phases were analysed. The migration and retention profiles of heavy metals along the soil column were determined for each individual layer in the soil along with the partition coefficients (K_d).

METHODOLOGY

One dimensional coupled solute and moisture flow tests, were conducted on an unsaturated illitic soil at varying pH values. Samples were tested in horizontal leaching column tests designed to simulate slow flow of leachate through unsaturated clay (Elzahabi and Yong, 1997). The soil was then statically compacted into the leaching column cell to its maximum dry density (1.88 Mg/m³) and optimum moisture content (14.16%). Each cell was leached with two different permeants: 2500mg/l and 5000mg/l (pH 2.8) of lead, copper and zinc nitrate solutions at constant atmospheric pressure and under a negligible hydraulic gradient. Tests were conducted at room temperature and for two time durations. For each time duration, the test sample was removed and sectioned horizontally into a number of layers for chemical analysis. Variations of volumetric water content with distance were measured for different time durations. Metal partitioning analysis of heavy metals in the pore fluid (soluble ions) and the solid phases (extractable ions) along the soil column were determined for each individual layer in the soil along with the partition coefficients (K_d). The adsorption characteristics profiles along the soil column are presented in Figures 1 and 2.

Metal Partitioning Analysis
The metal partitioning analyses were conducted according to the Environmental Protection Service procedures EPS and ASTM standards. Two different soil suspension methods were used to determine the concentration of heavy metals in the soil: (a) the pore fluid removal or the batch shaker test was used to determine the soluble cations concentration in the liquid phase, and (b) the acid digestion method was used to measure the metals concentration in the solid phase. For both methods, analysis of the fluid for heavy metals ions (Zn^{+2}, Pb^{+2}, Cu^{+2}) associated with the sample solution was performed using the Perkin Elmer Model 3110 atomic adsorption spectrophotometer (AAS). The cations concentrations were expressed in mg/100g.

Soluble Phase
Soluble cations were determined using the pore fluid removal method or the batch shaker test. Duplicate suspensions with a ratio of 1:10 soil weight to solution volume of distilled water were prepared as recommended by the United States Environmental Protection Agency (USEPA, 1987) to estimate the attenuation of chemicals from batch tests. For each batch test, 30 ml of distilled water was added to 3 g of soil and mechanically shaken in the end-over-end shaker for 24 hours. After agitation, the sample was then centrifuged in plastic Nalgene centrifuge tubes at 10,000 rpm for 20 min and the clear supernatant was collected for cations (Na $^+$, K $^+$, Ca $^{+2}$, Mg^{+2}) and heavy metals (Zn $^{+2}$, Pb $^{+2}$, Cu $^{+2}$) analysis by means of atomic adsorption spectrophotometry (AAS). The concentration of a particular species found in the supernatant or in the pores of each slice was calculated using the following formula:

$$C_{mg/100g} = \frac{DF \ x \ Cs_{mg/l} \ x \ V_{ml} \ x \ 100_g}{1000_{ml} \ x \ WDS_g} \qquad (1)$$

Where C is the soluble species concentration found in the supernatant and is expressed in mg per100g of the soil, DF is the dilution factor, Cs (ppm) or (mg/l) is a particular species concentration in the pore fluid and obtained from AAS reading, V is the volume of distilled water used during the test (where 30 ml was used, 1000 ml is to convert from 1l to ml and WDS is the weight of dry soil in g (where 3 g was used).

Solid Phase
Heavy metals and existing cations in the solid phase were measured using the pore fluid extraction or the acid digestion method (3050) of the United States Environmental Protection Agency (USEPA, 1986). This method is an acid digestion method used to dissolve all the sample metal bonds and to prepare soil samples for analysis by atomic adsorption spectrophotometry (AAS) in order to determine the background level of the trace metals in soil. Hot diluted nitric acid (HNO_3) was used to dissolve the inorganic-metal bonds and to ensure complete oxidation. The hydrogen peroxide (H_2O_2) was added to the soil, and was warmed until the effervescence was minimal or the general sample appearance was unchanged – to ensure complete destruction of the organic-metal bonds. After cooling, the sample was diluted up to 100 ml and then allowed to settle over night. The sample was then centrifuged at 5,000 rpm for 10 min to clear the supernatant from particulates that may clog the nebulizer. The concentration of a particular

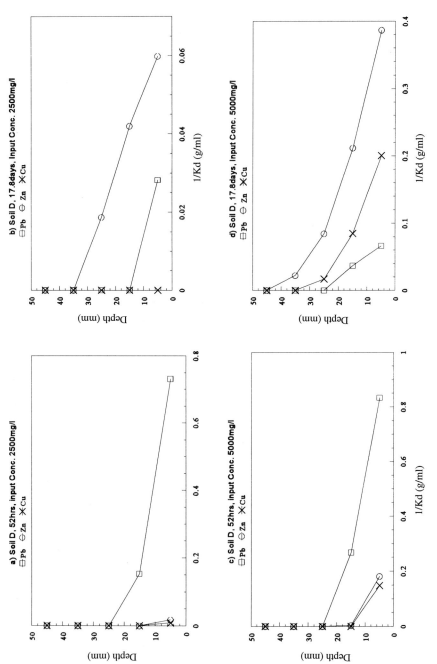

Figure 1 Kd Profiles of Pb, Zn and Cu for illitic Soil (pH 3.5)

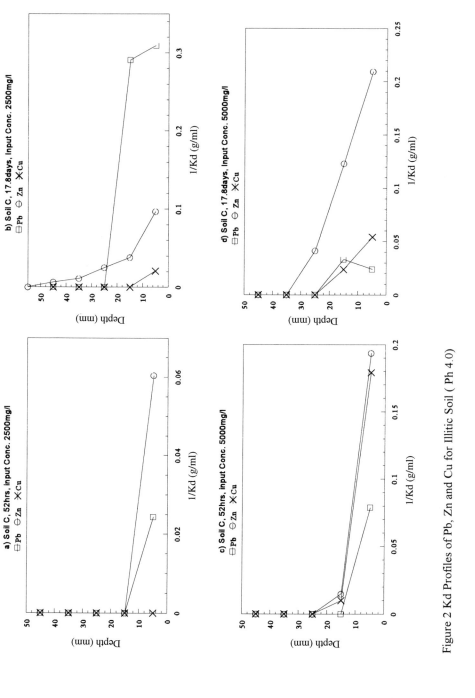

Figure 2 Kd Profiles of Pb, Zn and Cu for Illitic Soil (Ph 4.0)

species adsorbed by the clay and found in the supernatant for each slice was obtained using the following formula:

$$Ct_{mg/100g} = \frac{DF \; x \; Cs_{mg/l} \; x \; V_{ml} \; x \; 100_g}{1000_{ml} \; x \; WDS_g} \tag{2}$$

Where Ct is the total adsorbed concentration of metals by the clay which was found in the supernatant and is expressed in mg per100 g of the soil, DF is the dilution factor, Cs (ppm) or (mg/l) is a particular species concentration in the supernatant and obtained from AAS reading, V is the volume of distilled water used during the test (100 ml was used), 1000 ml is used to convert from 1 liter to 1 ml and WDS is the weight of dry soil expressed in g (where 1 g or 2 g was used). Thus,

$$S \; = \; Ct \; - \; C \tag{3}$$

Where (S) is the total adsorbed or retained concentration of metals in the solid phase and can be obtained by subtracting the soluble species concentration C from the total adsorbed concentration (Ct) of metals by the clay. The units of S, Ct and C are expressed in mg of constituent adsorbed per100 g of dry soil.

Adsorption Characteristics

K_d is generally identified as the distribution coefficient or the partition coefficient, and is used to describe contaminant partitioning between the liquid and solids only if the reactions that cause the partitioning are fast and reversible, and only if the isotherm is linear, (Yong et al., 1992).

$$K_d \; = \; \frac{S}{C'} \tag{4}$$

Where, S represents the ratio between the mass of the contaminant adsorbed onto the solid phase divided by the mass of the soil to obtain a measure of the relative mass of the constituent adsorbed on the solid phase. The units of the numerator are expressed in mg of adsorbed constituent / 100g of soil. The denominator C' represents the concentration of the contaminant in solution and its unit is expressed in mass of constituent / volume of moisture or g/ml. The distribution coefficient (K_d) unit is expressed in ml/g. According to Yong et al., (1992) and Mohsen (1993), the mass of constituent in moisture / volume can be computed from:

$$C' \; = \; \frac{C}{n \; S \; V} \; = \; \frac{C}{\theta \; V} \; = \; C \; \frac{\rho_d}{\theta} \tag{5}$$

C is the concentration of the contaminant in solution and is expressed in (mass/mass) or (mg/100g), ρ_d is the bulk density of the soil (g/ml), θ is the volumetric water content (fraction), n is the porosity (fraction), S (fraction) is the water saturation and V (ml) is the total volume of soil.

The volumetric water content (θ) is defined as volume of water per volume of moist soil:

$$\theta = \frac{V_w}{(V_s + V_v)} = \frac{\dot{V}_w}{V} \qquad (6)$$

Where V_s is the solid volume and V_v is the void volume.

The relation between the water content (w) and the volumetric water content (θ) is:

$$\theta = \frac{w}{100} \rho_d \frac{1}{\rho_w} \qquad (7)$$

where ρ_w is the water density and ρ_d is the bulk density or the dry density, and is defined as the weight of soil divided by the total volume of the soil: (Yong & Warkentin, 1975).

$$\gamma_d = \frac{W_s}{V_s + V_V} \qquad (8)$$

RESULTS AND DISCUSSION

Heavy metal in illite is mostly retained in the carbonate or hydroxide phase at soil pH values above 4. Below pH 4, retention of heavy metals is mostly by exchange mechanisms, (Yong et al., 1992). In order to study the movement of heavy metals in acidic soil, the buffering capacity of the illitic soil was reduced by using nitric acid. Lowering the pH value of the soil increases the solubility of heavy metals in the soil water by redissolving the heavy metal precipitates (e.g., carbonate and hydroxides) into the soil water. The greater availability of the ions will enhance mobilization and the rate of their loss by leaching, (Yong & Phadungchewit, 1993; Yong et al., 1992). The retention of heavy metals increases when the pH of the soil solution exceeds the value required for precipitation, (Farrah and Pickering, 1976a, 1976b, 1977 and 1979). Since the focus of this study is on the retention of heavy metals by exchangeable mechanisms, and since sorption of lead onto illite soil at high pH values is by precipitation mechanisms (Macdonald & Yong, 1997, Yong et al., 1993) the soil pH in this study was reduced to values below pH 4 to avoid H.M. precipitation.

The test results show moisture and solute distribution along the soil column from the source point due to the internal suction gradients -- as expected. The volumetric water content along the soil column shows higher values near the source point. This decreases as one proceeds along the column. For longer time periods, the increase in volumetric water content is seen to be highly dependent on the soil pH and the fluid input concentration. Other factors which control the adsorption characteristics of heavy metals are carbonate presence, degree of saturation, and influent H.M. concentrations. Variations of the adsorption characteristics with time and distance for illitic soil at pH 3.5 and pH 4.0 are provided in Figures 1 and 2 respectively. As shown, the partitioning coefficient K_d increases with increasing soil pH and distance from H.M. source, and decreases with increasing the time duration of wetting and the degree of saturation. Generally, at initial time of wetting (52 hrs) and high input concentration of (5000 mg/l), values of K_d at shallow depths (1 to 2 cm) in soil D (pH3.5) where the soil is fully saturated, tended to be in the range of 1.2 - 3.73 mg/l (Pb) < 5.5 - 269 ml/g (Zn) < 6.74 - infinite (Cu) ml/g as shown in Figure 1c, to 1.3 - 6.5 ml/g (Pb) < 55.9 - infinite ml/g (Zn) < 117 - infinite ml/g (Cu) for low input concentration (2500mg/l), Figure 1a. Whereas for soil C (pH 4.0), values of 12.68 - infinite ml/g (Pb) > 5.58-101 ml/g (Cu) > 5.1622-67.73 ml/g (Zn) were associated with high concentration (5000mg/l) (Figure 2c) while values of infinite - infinite (Cu) > 41.03 - infinite ml/g (Pb) > 16.54 - infinite ml/g (Zn) were associated with low concentration (2500mg/l) (Figure 2a). Since H.M. are highly soluble in acidic saturated conditions, one expects to have a total desorption from the soil to the soluble phase at low pH. However, results showed that at initial time, and due to 5000 mg/l of input concentration, 84% of lead was retained and only 16% migrated from the first section of soil column D to the soluble phase. While for soil C, 94% of lead was retained and only 6% was released to the soluble phase within the first section. The release of lead at initial time might be due to the amount of water entrapped in the first section of soil C and D resulting in an increase of the volumetric water content to near saturation (v.w.c. increased from 26.69% to 41.75% for the first section of soil D, while for soil C, v.w.c increased from 26.69% to 35.81%), (Elzahabi & Yong, 1997).

At shallow depths (above 4 cm) and as the time of wetting increases (17.8days), the partitioning coefficient K_d decreases with increasing the degree of saturation . As shown in Figure 1d, for the first two sections of soil D and for high input concentration, values of K_d ranged from 15.06 - 27.19 ml/g (Pb) > 4.98 - 11.78 ml/g (Cu) > 4.7 - 11 ml/g (Zn), to infinite - infinite (Cu) > 35.5 - infinite (Pb) > 16 - 23 ml/g (Zn) for low input concentration (Figure 1b). While for soil C, K_d values ranged from 41.64 - 30.11 ml/g (Pb) > 18.46 - 42 ml/g (Cu) > 4.77 - 8.11ml/g (Zn) when soil is leached with high H.M. concentration (Figure 2d), to 48.35- infinite ml/g (Cu) >10.32 - 26 ml/g (Zn) > 3.23 - 3.43 ml/g (Pb) Figure 2b. However, as the distance and time of wetting from H.M. source increased, (Figure 1 a,b,c,d and Figure 2 a,b,c,b,d), and within the unsaturated phase (below 4cm depth), the amount of H.M. released to the soluble phase decreases sharply indicating that H.M. migrates only a very short distance from the input source during the specified migration time, resulting in infinite K_d values. This might be related to the decrease in the degree of saturation along the soil column and may also be explained by the cation exchange capacity of the soil, which was still able to almost completely retain the introduced H.M. under unsaturated conditions, as demonstrated in the present study.

The presence of carbonates in the illite soil increases the retention of heavy metals at high pH and also enhances the buffering capacity of the soil. The higher the carbonate content of the soil, the greater is the retention of heavy metals by the carbonate phase (Yanful et al., 1988). Therefore, because of its high carbonate content and high cation exchange capacity, heavy metals were accumulated in the first section of soil with pH 6.9 resulting in a calculated partition coefficient value of "infinity".

Concluding Remarks
This research study provides the experimental information necessary for numerical analyses on transport of H.M. in the vadose zone. Particular attention to the effect of degree of saturation, the presence of carbonate, soil pH and heavy metals concentrations are given.

REFERENCES
Cherry, J.A., Gillham, R.W., and Barker J.F. (1984). University of Waterloo, Canada. "Contaminants in Groundwater: Chemical Process". Studies In Geophysics, Groundwater Contamination. National Academy Press. Washington, D.C., pp. 46-64.
Darban, A. K. (1997). "Multi-component Transport of Heavy Metals in Clay Barriers". McGill University, Montreal. 307pp.
Davidson, J.M., Ou L.T., and Rao, P.S.C. (1976). "Behavior of High Pesticide Concentrations

in Soil Water Systems". Proceedings of the Hazardous Waste Research Symposium, Residual Management by Land Disposal, EPA-600/9-76-015, U.S.Environment Protection Agency, Washington D.C., pp. 206-212.

Elzahabi M. and Yong R.N., (1997), " Vadose Zone Transport of Heavy Metals", Proceedings of the first Geoenvironmental Eng. conference organized by the British Geotech.Society and the Cardiff School of Engineering, University of Wales, Sept. 1997, pp. 173-180.

Farrah, H., and Pickering, W.F., 1976a, " The Sorption of Copper Species by Clays". I.Kaolinite. Aust. J. Chem., Vol. 29, pp. 1167-1176.

Farrrah, H., and Pickering, W.F., 1976b. "The Sorption of Copper Species by Clays". II. Illite and Montmorillonite. Aust. J. Chem., vol.29, pp. 1177-1184.

Farrah, H., and Pickering, W.F., 1977. "The Sorption of Lead and Cadium Species by Clay Minerals". Aust.J.Chem., Vol.30, pp.1417-1422.

Farrah, H., and Pickering, W.F., 1979. "pH Effects in the Adsorption of Heavy Metals Ion by Clays", Chem. Geol., Vol. 25, pp.317-326.

Rowe,R.Kerry, 1988. "Eleventh Canadian Geotechnical Colloquim:Contaminant Migration Through Groundwater- The role of Modelling in the Design of Barriers". Canadian Geotechnical, J., Vol. 25,pp.779-798.

Freeze, R.A., and Cherry, J.A. (1979). "Groundwater". Prentice-Hall, N.J., 604 pp.

MacDonald E. and Yong R.N., (1997), "On The Retention of Lead by Illitic Soil Fractions", Proceedings of the first Geoenvironmental Eng. conference organized by the British Geotechnical Society and the Cardiff School of Engineering, University of Wales, Sept. 1997, pp. 116-127.

Mohsen, Mohamed F.N. (1993). "Adsorbed Concentration vs. Lab-Reported Concentration in Soil". Technicale Notes in Ground Water. Vol.32, No.3, pp. 499-500.

USEPA (1986). Method 3050, Acid Digestion of Sediments, Sludges and Soils. Test Methods for Evaluating Solid Wastes, SW846, 3rd ed. U.S. Government Printing Offices, Washington, D.C.

USEPA (1987). Batch Type Adsorption Procedures for Estimating Soil Attenuation of Chemicals. Office of Solid Waste and Emergency Response, Washington, D.C, EPA/530-SW-87-006

Yanful, E.K., Quigley, R.M., and Nesbitt, H.W., 1988. Heavy Metal Migration at Lland Fill Site, Ontario, Canada-II:Metal partitioning and Geotechnical Implications Applied Geochemistry, Vol. 3, pp. 623-629

Yong, R.N., and Diperno, N., 1991. Sources and Characteristics of Waste with Specific Reference to Canada. McGill University. Geotechnical Research Centre. Geo-Engineering series 91-1.

Yong, R.N., Galvez-Cloutier, R., and Phadungchewit, Y., 1993." Selective Sequential Extraction Analysis of Heavy Metal Retention in Soil". Can. Geo. J., vol 30:834-847.

Yong, R.N., Mohamed, A. M. O. and Warkentin, B. P., 1992. "Principles of Contaminant Transport in Soils". Elsevier, Amsterdam.

Yong, R.N., and Phadungchewit, Y., 1993. "PH Influence on Selectivity and Retention of Heavy Metals in Some Clay Soils". Canadian Geotechnical J., Vol. 30, pp.821-830.

Yong, R. N. and Warkentin, B. P., 1975. " Soil Properties and Behaviour", Elsevier, New York.

Scaled Centrifuge Modelling of Capillary Rise

N. DEPOUNTIS[1], M.C.R. DAVIES[2], S. BURKHART[1], C. HARRIS[1], A. REZZOUG[3], D. KÖNIG[3], C. MERRIFIELD[4], W.H. CRAIG[4], L. THOREL[5]

[1] Cardiff University, [2] Dundee University, [3] Bochum University, [4] Manchester University, [5] LCPC Nantes

INTRODUCTION

Contaminant plume evolution in the unsaturated zone depends strongly on capillary forces existing at the water-air interface. For this reason, the basic understanding of capillary phenomena is essential in any attempt to model contaminant behaviour in this zone.

This paper reports the experimental results from tests investigating capillary rise at different g-levels. The experimental programme is part of the NECER project (Network of European Centrifuges for Environmental Geotechnic Research). Tests were performed at the geotechnical centrifuge laboratories of Cardiff, Bochum, Manchester, and LCPC Nantes. The aim was to determine the scaling laws of capillary rise under static and dynamic conditions.

In all laboratories capillary rise was investigated by performing tests from, dry to wet, in which columns of dry sand soils were inundated with water from their base. Capillary rise above the phreatic surface of the sand model was distinguished in a continuous phase (almost completely saturated) and a discontinuous phase (partially saturated). Matric potential probes (tensiometers) monitored the capillary rise of the continuous phase and the rise of the visible wet discontinuous front was monitored by video cameras. Pore pressure transducers (ppts) were used to record the change of the water level inside the sand model during the tests.

APPLICATION OF THEORETICAL LAWS IN A COLUMN MODEL

The basic model scheme used for the wetting experiments is illustrated in Figure 1. In this kind of experiments water is permitted to flow into the dry sand column until the phreatic surface is established inside the sand at the same level with the head of water inside the measurement tube. The water moves under the action of buoyancy forces towards the sand model. These forces are balanced by viscous forces, which act as the frictional force. After water equilibrium the capillary forces start to act and capillary rise takes place.

Capillary Rise

Capillary rise in porous media is quite similar to water rise in a capillary tube, assuming that the pore size distribution of the examined media is characterised by a uniform pore size. This is the case for uniform soils and glass beads. Consider the ideal situation of one pore size sand placed in the column of Figure 1, which is connected to a water tube, both subjected to a uniform N-g acceleration field.

As soon as capillary rise takes place a continuous liquid phase (saturated fringe) starts to develop, depending strongly on the acceleration field. When the saturated fringe has being fully completed, the pressure at the base of both columns is equal to:

$$p_0 - \frac{2\gamma}{r_0} + \rho_w h_2 Ng = p_0 + \rho_w h_1 Ng \tag{1}$$

The equation above leads to the following formula, which estimates the height of the continuous phase for a perfect wetting fluid at Ng acceleration field.

$$h_{CNm} = \frac{2\gamma}{r_0\rho_w Ng} = \frac{h_{CNp}}{N} \qquad (2)$$

where the subscripts m and p are referred to the model and prototype respectively.

The soils used for the experiments were uniform sands so that the scaling law from above can be applied to the model. After the establishment of the continuous phase it is anticipated that a discontinuous phase (partially saturated fringe) will develop following the same law.

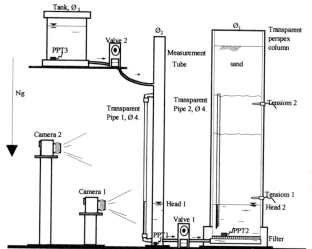

Figure 1. Common basic scheme of column model apparatus (König et al. 1998)

Rate of Capillary Rise

Several authors (Landau et al. 1967, Bikerman 1970) have investigated the flow of a viscous liquid in a cylindrical tube, with the liquid flowing along the tube under the action of pressure difference Δp maintained at the ends of the tube. A uniform porous medium is similar to a very narrow cylindrical tube, so that the liquid movement can be assumed to be laminar. Thus, the flowing liquid has a velocity of:

$$u = \frac{r_0^2 \Delta p}{8 h_t \mu_w} \qquad (3)$$

When capillary forces cause a pressure difference at the ends of the narrow tube, then according to Poiseuille (1841) a perfect wetting fluid will rise with a rate of:

$$u = \frac{r_0^2 \rho_w g (h_{CN} - h_t)}{8 \mu_w h_t} \qquad (4)$$

Equation (4) can be applied to a uniform porous medium and, therefore, to the sand model. If we substitute $u = dh_t/dt$ in the above formula and integrate we obtain:

$$t = \frac{8\mu_w}{r_0^2 \rho_w g}(h_{CN} \ln \frac{h_{CN}}{h_{CN} - h_t} - h_t) \qquad (5)$$

It is apparent from equations (4) and (5) that in the case of a centrifuge model capillary velocity and time can be scaled as N and $1/N^2$ respectively.

DESCRIPTION AND OBJECTIVES OF THE WETTING TESTS

The objectives of the wetting experiments is to verify the influence of g-level on capillary rise and define the scaling laws for modelling capillary phenomena in a geotechnical centrifuge. Tests from dry to wet conditions were performed at Cardiff and Manchester Universities (UK), Bochum University (Germany), and LCPC in Nantes (France). All laboratories used column models, except LCPC, which used a rectangular model.

Test Design and Measurement Techniques

The basic set-up of the apparatus used is illustrated in Figure 1, and Table 1 are summarises the test specifications for each laboratory. Column test apparatus consist of a transparent perspex column with an internal diameter $\emptyset_1$, which is filled with pluviated sand. The sand model receives water through a system of filters either directly from a measurement tube (diameter $\emptyset_2$) or indirectly from a tank (Bochum). Alternatively a rectangular box instead of the perspex column may be used (LCPC).

Table 1. Laboratory specifications for dry to wet tests

	Bochum	Cardiff	Manchester	LCPC
Sand d_{10}/d_{50} (mm)	Bochum sand 0.12/0.18	Congleton sand 0.1/0.15	HPF5 sand 0.05/0.2	Fontaineblau sand 0.15/0.23
Sand Column $\emptyset_1$/height (cm)	Perspex Column 20/100	Perspex Column 20/100	Perspex Column 4/50	Rectangular container
Measurement tube $\emptyset_2$ (cm)	4.5	18	Over flow system (water tank)	Marriotte bottle device
Video cameras	Camera1 & 2	Camera 2	Camera 1	CCD camera 1
Tank $\emptyset_3$ (cm)	19	-	-	-
Valve	Valve 1 & 2	Valve 1	Valve 1	Valve 1
ppts & tensiometers	ppt1 & ppt3	ppt1 & ppt2 tensiometer 1 & 2	-	ppt2
Ng-level applied location	phreatic surface in the sand	phreatic surface in the sand	phreatic surface in the sand	base of the model

The instrumentation used for recording the experimental data were:
- Pore pressure transducers (ppts) were used in all laboratories but Manchester, in order to measure the hydraulic head in the column, the tube, and the water tank (Bochum).
- Matric potential probes (tensiometers) were used at Cardiff to measure the suction above the phreatic surface in the sand.
- Video cameras were used by all participants for optical measurements of the rising of the visible wet front inside the sand column.

Test Procedure

After model preparation and placement in the geotechnical centrifuge, the g level is increased to the selected value; valve No.1 is then opened and water flows from the measurement tube towards the sand column (or the rectangular container). The water level inside the model increases as the water level decreases inside the tube until equilibrium of both levels is achieved. At this point capillary forces start to act at the water-air interface and capillary rise occurs inside the sand model. The test procedure varied slightly in each laboratory, and variations are described below.

At Bochum the tests were started with an initial head of water in the measurement tube (Head 1 in Fig.1). Then at the desired g-level valve No.1 was opened and water started to flow as it has already been described in the test procedure. A constant water level was maintained inside the tube by opening valve No.2 to allow water to enter the tube from the water tank each time it was needed by observing video camera No.1. As a result the phreatic surface of the model remained constant during the tests. The water transfer from the tank to the tube was monitored by the ppts and the wet front rising was recorded by camera No.2. Tests were conducted at different g-levels, with a new sample being used for each test.

At Cardiff the tests were performed by transferring dyed deaired water from the tube to the sand column; by opening valve No.1 at the desired g-level. A single sample was used for tests conducted at different g-levels. Each test was conducted after the centrifuge had been accelerated to the highest g-level for the series of experiments. When the required g-level had been reached the valve was opened to permit water to pass from the tube into the column. On reaching equilibrium the valve was closed and the centrifuge acceleration reduced to the next test level prior to the valve being opened and the procedure repeated. The ppts and the tensiometers monitored changes in the phreatic surface and capillary pressure respectively. The continuous liquid phase of the fringe was detected by the tensiometers, whereas the top of the fringe (i.e. extend of the discontinuous phase) detected by camera No.2.

At Manchester instead of a measurement tube an overflow tank with a constant head of water was used to maintain a constant head of water inside the sand model after water equilibrium. This has the advantage that the location of the phreatic surface was established in the model before the onset of capillary rise. The overflow tank was continuously supplied with water through the hydraulic slip rings of the centrifuge. The constant phreatic surface inside the model was monitored by observing the head of water in pipe 2 (air vent tube). As at Bochum a new sample was used for each test and the capillary rising was observed by camera No.1.

At LCPC with the rectangular box containing pluviated sand was connected to a system based on a Mariotte bottle in order to maintain a constant phreatic surface throughout the tests. Comparison between the pore pressures at the base of the model obtained from ppt2 and the observed water table recorded by a CCD camera (camera No.1), shows a good response of the Mariotte bottle. At LCPC several tests were performed under different g-levels.

BOCHUM TEST RESULTS

Presented here are the main results from the capillary rise observed by the video cameras for the Bochum and Congleton Sands. The "capillary wet front" is defined as the visible contrast between dry and wet sand, and "capillary rise" is defined as the height measured from the water table to the average capillary wet front (top of the capillary fringe).

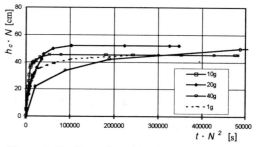

Figure 2. Capillary rise of Bochum Normsand

Figures 2 & 3 illustrate the prototype capillary rise ($h_{Cm}.N$) of the fringe versus the prototype capillary rise time ($h_{Cm}.N^2$) at each g-level. Each curve represents a centrifuge test with a different sample each time. The curves are drawn from the constant phreatic surface of the model. As it can be seen from Fig. 3 & 4 it seems that capillary rise can be scaled in a model by a factor of 1/N and time can be scaled by a factor of $1/N^2$. Concerning the time scaling law it seems that all curves get closer more consistent in Bochum Sand than in Congleton Sand.

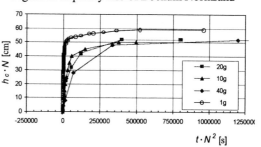

Figure 3. Capillary rise of Congleton Sand

The first 4 readings of the 40g curve in Fig.3 were made 2 minutes after the beginning of the test. In this case the superposition of the water flow combined with a dynamic effect may lead to a deviation of the measurements shown. Fig.4 groups the optical measurements from all tests and shows that the prototype values of h_C for Bochum Sand present a small slope with increasing g's, whereas in Congleton Sand the curve is almost horizontal.

The prototype values of h_C are fitted constantly for Congleton Sand compared with the values for Bochum Normsand, which deviate slightly from the scaling law. This may be a result of difficulties in the measurement of small capillary rise for such soils at high g's. Another reasons may be the effect of the gravity on the shape of the menisci and on capillary pressure (Schubert 1982), and the possible effect of stress on capillary rise.

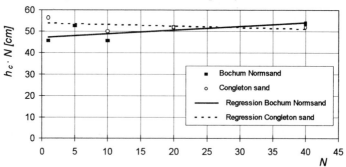

Figure 4. Linearity of $h_C.N$ versus Ng-level

CARDIFF TEST RESULTS

At Cardiff two zones of the capillary fringe were distinguished, one continuous where the sand is almost saturated, and one discontinuous where the sand is partially saturated. The pore size distribution of Congleton sand varied slightly, but still was centred on a characteristic pore size $d_0=124\mu m$, which was defined from the tensiometer readings. As a result, the continuous liquid phase developed first by wetting all the pores with the minimum characteristic d_0. Then the discontinuous liquid phase developed by drawing further water from the reservoir and wetting the smaller pores above the continuous capillary surface.

Figure 5 illustrate the prototype capillary rise of the continuous phase (h_{CNp}) versus N^2.time for a continuous test (from 11.15 to 3.1g). The phreatic surface in the sand was decreasing due to capillary forces throughout the test. The curves for each g-level were drawn from an initial time, which was corresponding, to the re-opening of the valve. Therefore, due to the nature of the experiments (continuous tests), no real conclusion can be drawn regarding the rate of growth of the capillary fringe. However, it is evident that the time for completion of the continuous phase increased as the g-level decreased. As for the discontinuous phase it seems that it needed more time for complete development, especially at lower g's (i.e. this phase is still rising at 3.1g as it can be seen in Figure 5).

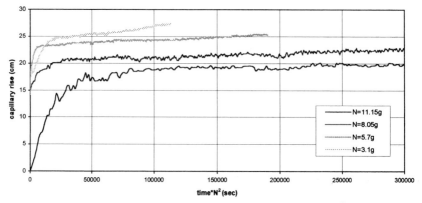

Figure 5. Prototype continuous capillary rise of Congleton Sand versus N^2.time

Both phases seem to be scaled by a factor of 1/N in a centrifuge model, and this can be seen clearly in Fig.6, where the prototype capillary rise (continuous phase:h_{CNp} monitored by ppts and tensiometers; discontinuous phase:h_{Cp} monitored by the video camera) tend to be linear. However, if we consider that the expected rise of the liquid continuous phase for this sand is 23.7cm (as defined from the tensiometer readings during the tests) we can see that only at the lower g levels this becomes more evident. Similar to the observations of Bochum, this might be due to the increasing stress effect at higher acceleration fields.

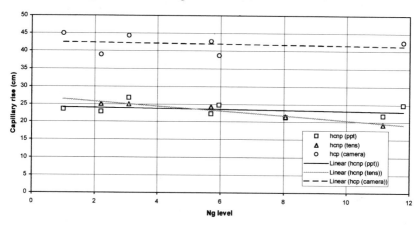

Figure 6. Prototype capillary rise of continuous and discontinuous phase

MANCHESTER TEST RESULTS

The movement of the wetting front was observed by video cameras over periods up to 8 hours for PFA and HPF5 Sand. The capillary rise in PFA, which is finer and less permeable than HPF5, was approximately 6m at static conditions over a period of many months. However, the coarser HPF5 sand had a rise of approximately 3m at static conditions.

Tests in both materials over a range of acceleration fields up to 60g confirm the same scaling laws for capillary rise and time as before. In Fig.7 the prototype capillary rise is expressed as N times the observed model capillary rise versus N^2.time. Figure 7 shows very good correlation between the tests at different g-levels. However, at acceleration levels greater than 50g anomalies were observed (König et al. 1998). Therefore at lower g's the experiments confirm the scaling laws.

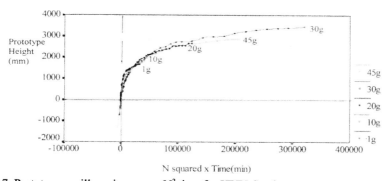

N squared x Time(min)

Figure 7. Prototype capillary rise versus N^2.time for HPF5 Sand

LCPC TEST RESULTS

At LCPC capillary rise was measured by using a CCD video camera, and interpreted later with image processing to obtain grater accuracy. The test configuration assumed that what is observed along the perspex window is the same as inside of the model. This is not totally true, so that cross analysis has to be done in the future with tensiometer measurements.

Nevertheless, rough results are shown in Figure 8, where the dry part (top) and the wet part (bottom) of the sample are clearly visible with the grey contrast. The front itself is not very sharp, and is influenced by the stratigraphy of the thin layers of pluviated sand. A small pipe located outside show the water table, maintained constant with the Mariotte bottle device.

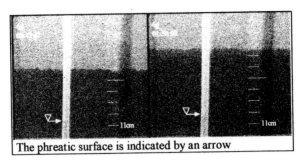

The phreatic surface is indicated by an arrow

Figure 8. Capillary rise in Fontainebleau sand under 6g (LCPC)

The basic theory concerning capillary rise is given by a Laplace Law (equation 2), where capillary rise is calculated for a narrow tube, assuming a perfect wetting fluid (Adamson et al. 1997). Reporting the results of the theoretical behaviour and the centrifuge results on the same graph (Figure 9), it can be shown that the Laplace Law is acceptable as a first approximation for these few results. The correlation with the sand granulometry is not obvious, but the equivalent tube radius appear to be far less than $d_{10}/2$. From a theoretical point of view, if sand could be modelled as an equivalent capillary tube, the expected results would be on a vertical line on the graph. The scattering observed on the results might be due to natural variation of soil properties between different pluviations, including a possible ageing.

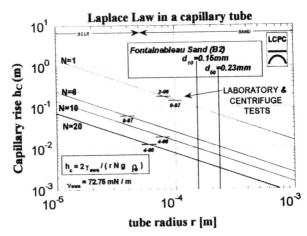

Figure 9. Theoretical capillary rise following the Laplace Law, and experimental results

CONCLUSIONS

This study has investigated the scaling laws of capillary rise in fine poorly graded sands. Two zones within the capillary fringe have been distinguished, one continuous where the sand is almost saturated, and one discontinuous where the sand is partially saturated. Both phases seem to be scaled in a centrifuge model by a factor $1/N$, following the theoretical analysis. A slight deviation from the scaling law has been observed either due to a stress effect on capillary pressure or due to different model conditions and properties which existed in each test. These external parameters are expected in geotechnics, so we still can accept the $1/N$ as the apparent scaling law for capillary rise.

From the theoretical point of view in a centrifuge model capillary velocity and time can be scaled as N and $1/N^2$ respectively. However, the scaling law for the rate of growth of the capillary fringe was difficult to assess during the tests. Generally, the time for completion of the continuous liquid phase increases as the gravitational acceleration decreases. It seems from the experimental results that capillary time tends to be scaled as $1/N^2$, but it needs more investigation to verify this scaling relationship by performing more tests with a new sample for each test.

As for the discontinuous liquid phase, this takes more time to develop. This follows the theoretical rule of inverse proportionality between time and pore size, considering that discontinuous phase develops only in the smaller pore sizes. This aspect also requires further investigation.

NOTATION

d_0	characteristic pore size
d_{10}	effective grain size
g	gravitational acceleration
h_1	piezometric water level in measurement tube
h_2	piezometric water level in sand column
h_C	capillary rise
h_{CN}	capillary rise of continuous phase
h_t	height of liquid lifted by capillary forces at a random time
p_0	atmospheric pressure
r_0	characteristic pore radius
t	capillary time
u	capillary velocity
N	scale factor (g-level)
γ	surface tension
ρ_w	density of water
μ_w	viscosity of water
m, p	subscripts referred to model and prototype respectively

REFERENCES

Adamson, A.W. & Gast, A.P. 1997. *Physical Chemistry of Surfaces.* Wiley. 784p.

Bikerman, J.J. 1970. *Physical Surfaces.* Academic Press. 476p.

König, D., Rezzoug, A., Heibrock, G., Craig, W.H., Davies, M.C.R., Burkhart, S. 1998. Capillary phenomena in centrifuge testing. *Int. Conf. Centrifuge 98,* Tokyo 23-25 Sept. 1998, Spec. Lect. rep., 40-42.

Landau, L.D., Akhiezer, A.I. & Lifshitz, E.M. 1967. *General Physics. Mechanics and Molecular Physics.* Pergamon Press Ltd. 372p.

Schubert, H. 1982. *Kapilarotät in porösen Feststoffsystemen.* Springer Verlanger. Berlin, Heidelberg; New York.

Finite element analysis of transport and fate of pollutants in complex hydrogeological systems

I KAZDA[1] and I VANÍČEK[2]

[1] REAT – Research and Educational Academy, Prague, Czech Republic
[2] Czech Technical University, Faculty of Civil Engineering, Prague, Czech Republic

INTRODUCTION

Groundwater resources, as a rule, represent a high-quality source of water supply for the population, industries and agriculture. In industrial countries, including the Czech Republic, however, these resources are often negatively affected or even deteriorated by anthropogeneous activity. Groundwater is, in numerous cases, contaminated by toxic organic or non-organic substances of various origin. That is why groundwater ecology has presently become the centre of considerable attention (Gibert, Danielopol, Stanford 1994), and a number of research works have been devoted to the interaction between the solutes and the percolated rock medium during their transport in groundwater (Appelo, Postma 1994).

The prevention of undesirable impacts of human activities or elimination of their consequences tends to be technically difficult and demands high costs, as these are often hydrogeological systems with a complex geological structure. The pollution mostly affects shallow aquifers into which contaminants may easily penetrate. Some cases of contamination, however, also include aquifers lying deep below the ground surface.

Whenever an aquifer is polluted in such a way which leads to a present-day or a future disaster (i. e. the users of the aquifer water are under a direct threat or under a threat resulting from remote solute transport), suitable remediation has to be designed and carried out. The applied remediation methods may be divided into technical and biological ones. Among the technical methods (Charbeneau, Bedient, Loehr 1992), there are namely pumping and subsequent treatment of contaminated groundwater, vapour extraction, air sparging, the use of treatment walls and immobilisation. The application of the above mentioned methods, to a certain extent, depends on the local situation and, consequently, technical remediation methods are continually complemented and improved. Biological remediation methods are usually friendlier to the environment, but their development and implementation are rather demanding (Baker, Henson 1994). Not only do they imply the selection of a suitable species of microorganisms, but they also require suitable living conditions to be secured for them throughout the whole remediation time. That is why the implementation of biodegrading methods is usually limited to smaller areas of shallow aquifers.

BASIC CONSIDERATIONS

As we may presume from the references above, remediation methods of contaminated aquifers are technically well developed, and their practical implementation has already brought considerable experience. It is much more difficult, however, to select a suitable remediation method for each particular locality, which has to provide a reliable achievement of the necessary results, being economically the most feasible as well. Experience gained at other localities may be applied, even though hydrogeological conditions, the extent of contamination and the nature of contaminants usually vary to such a degree that the final solution cannot be accepted without a detailed analysis. Remediation measures for polluted hydrogeological systems have to be designed with utmost care. The solution accepted for implementation has to satisfy not only technical aspects, but also economical ones. In assessing and considering potential alternatives, required by design optimisation, numerical modelling based on the finite element method has proved extremely effective (Anderson, Woessner 1992, Kazda 1990, 1997). During more than forty years of its existence, the finite element method has gone through a long development to be currently used not only in technical, but also in natural sciences.

We have to underline, however, that although the problems of groundwater flow have been very well elaborated by using the finite element techniques, and much progress has been achieved in modelling the solute transport as well, nevertheless, by far not all possibilities of the finite element method with reference to modelling remediation methods have been used. The problem to be solved consists namely in creating special finite elements which would account for modelling individual remediation methods (above all in three-dimensional domains), respecting the regional character of shallow aquifers which are affected by contamination most frequently.

A successful use of the finite element method for the modelling of remediation of an environmental problem depends on:
- variational formulation of a given class of boundary value problems or initial value problems,
- applied type of finite elements,
- applied meshing of the solved domain into finite elements.
It is only the successful fulfilment of all three requirements that constitutes the sine qua non condition of any finite element analysis.

The present-day development of finite elements is characterised by focusing on new and new applications, while relatively little attention is paid to this method's apparatus itself. The main attention, therefore, is focused on the first mentioned requirement, while the remaining two are relatively neglected.

This fact is of great importance in numerical modelling. In assessing environmental problems related to the abstraction and protection of groundwater resources, three-dimensional problems have to be solved still more often, including the problems of a regional character. This requires high-level professional skills on the part of the problem solver, which may in no way be guaranteed by purchasing commercial software only. For, as it is, numerical modelling of the above given problems requires namely individual approach to each problem and unconditional satisfaction of all three basic demands in order to assure a sufficient degree of precision of the given numerical model and, therefore, its usability for decision-making.

The main challenge consists namely in the development and verification of special techniques of the finite element method, suitable for the solution of three-dimensional problems of groundwater flow and solute transport. In particular, new techniques must be developed allowing for adequate modelling of those parts of the area, which are affected by physical singularities or which substitute the impact of the domain boundary parts which, from the point of view of a conceptual model, lie at infinity. Derivation of three-dimensional finite elements with special interpolation functions, as well as transition elements facilitating local variations of the detail of the mesh of three-dimensional elements by means of interpolation polynomial modification will be also of great importance.

Numerical modelling of practical problems dealing with solute transport necessarily implies a suitable method. The method that proved right consists in starting with a simpler model which serves for the verification of the basic conception of the given problem and for the potential development of requirements for further hydrogeological prospecting of the given locality. Only then is it appropriate to prepare a complex model whose preparation will require more time. By observing the procedure described in the following paragraph, a great deal of time may be saved during the simulation of solute transport in particular. The following part of the paper includes an example showing how the representation of suitably selected water particle flow paths may be used for the development of a preliminary design of a hydraulic barrier, while the final part is devoted to the description of a simple a posteriori indicator of numerical precision of the solution and its significance.

SIMULATION OF SOLUTE TRANSPORT IN REGIONAL DOMAINS

The simulation of solute transport in a given domain requires, in the first place, finding a vector field of seepage velocities, which, in their turn, serve for the solution of contaminants transport. As the decisive share in the transport is mostly that of advection, a dependable definition of groundwater flow patterns is of utmost importance. In case of steady groundwater flow no more than seepage velocities are necessary to be established before the solution of the solute transport time pattern is continued.

In case of a regional character of a given domain it is of advantage to separate both parts of solution, applying a different element mesh for each of them. Groundwater flow modelling usually requires an extensive domain to be considered in order to be able to set correct boundary conditions at its boundary. This domain may be generally subdivided into larger elements than those required by the solute transport, whose simulation depends on keeping to the limiting values of Peclet and Courant number. At the same time, however, the transport accounts for a smaller domain to be considered than that necessary for solving groundwater flow. This holds true, e.g. for the simulation of time-limited accidents of landfills and underground repositories or for aquifers with a low concentration of contaminants.

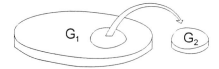

Figure 1 Two stages of solute transport solution: G_1 – domain for solving groundwater flow, G_2 – domain for solving solute transport

The suggested procedure, therefore, consists of two stages. In the first stage a sufficiently large domain G_1 is considered (Fig. 1), subdivided into elements in such a way to allow for solving a boundary value problem of groundwater flow with due precision. In the second stage of computation a smaller domain G_2 within the domain G_1 is selected so as to meet the needs for the solute transport simulation. A detailed subdivision of this domain into elements follows according to the requirements for the numerical stability of the initial value problem of transport simulation. Boundary conditions for groundwater flow at the boundary of the domain G_2 are derived from the solution carried out in the domain G_1 during the first stage.

The given procedure is of advantage namely when the domain G_1 models a homogeneous shallow aquifer. In such case horizontal flow with a free surface may be considered in G_1, so that the applied mesh is created by planar elements, the numerical model being of a quasi-three-dimensional character. The transport in the domain G_2, however, is usually of a three-dimensional character and, therefore, three-dimensional elements are used for modelling within this domain.

The given procedure not only saves the computer time, necessary for the transport simulation, but it also allows to assess in advance whether the flow pattern has been displayed with sufficient precision. Using a two-stage solution is also beneficial in the situation when, for example, advection and hydrodynamic dispersion characteristics are not known with sufficient precision and the effect of their probable dispersion has to be assessed in a qualitative way.

DESIGN OF A HYDRAULIC BARRIER

While designing a hydraulic barrier to provide temporary protection of a water resource from contaminated groundwater, it is necessary to establish the most suitable position of the barrier, the number of bore holes it will consist of, the bore hole depth, the pumping rate, as well as other parameters. In order to find the optimum solution, a large number of suitably selected alternatives have to be solved within this type of design. As each alternative of the solute transport simulation requires a great number of time steps, the design sets high demands not only on the computer time necessary, but also on the evaluation of efficiency of individual alternatives.

The design of a hydraulic barrier may be substantially sped up and simplified by evaluating the effect of individual alternative solutions on groundwater flow only. For this purpose it is advisable to analyse the effect of the barrier on seepage flow in a graphical way by using water particle flow paths and piezometric head contours (Havlíková, Kazda 1997). Here the design of a hydraulic barrier implemented for the protection of a water reservoir serving as a source of drinking water from groundwater contaminated by leachate from a landfill may be given as an example. An extensive analysis of this problem is further limited to the results of two alternative solutions, characteristic for the barrier design.

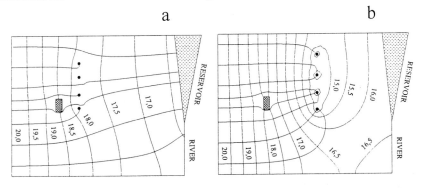

Figure 2 Hydraulic barrier design: a – unsatisfactory alternative, b – acceptable alternative

In Fig. 2a the landfill is marked by a hatched rectangle and the hydraulic barrier, formed by four bore holes, is located at a distance of 550 m from the landfill. The effect of the barrier on the seepage flow is displayed by means of flow paths and piezometric head contours. Fig. 2a makes it clear that some flow paths pass between the bore holes of the barrier, and so this alternative cannot be considered as convenient. Locating the barrier at a distance of 800 m from the landfill with a simultaneous increase in the pumping rate, however, proved more convenient. As displayed in Fig. 2b, the seepage flow in this alternative has been dramatically changed by the hydraulic barrier: the groundwater flow is directed away from the reservoir towards the barrier, limiting thus the contaminant transport by advection.

All solved alternatives are assessed in a similar way. The simulation of contaminant transport will be subsequently carried out only for one or more resulting alternatives affecting the groundwater flow in the most beneficial way.

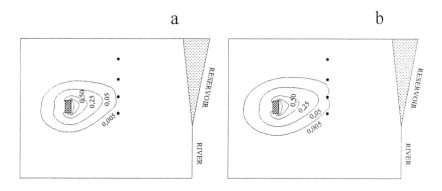

a   b

Figure 3 Solute transport time pattern: *a* – elapsed time of 12 months, *b* – elapsed time of 18 months

In Fig. 3a related to the second alternative of the hydraulic barrier, the isolines showing relative contaminant concentration represent the extent of aquifer pollution one year after the accident started, while Fig. 3b displays the same case another half a year later. It was presumed that the landfill was not subjected to any remediation for the whole time. Fig. 3 makes it evident that the hydraulic barrier fulfils its function quite well; at the same time, the effect of contaminant diffusion, which was of considerable degree in this case, can be seen.

When the solute transport shows a dominating effect of advection, flow paths may also be used for the representation of the contamination time stepping. This method of graphical analysis fits namely non-homogeneous three-dimensional domains (Kazda 1997).

RESIDUAL ERROR INDICATOR AND ITS APPLICATION
In solving groundwater flow and solute transport it is important to assess the degree of numerical precision of the applied model. This problem is presently becoming a centre of great attention (Křížek, Neittaanmäki, Stenberg 1998). One of the suitable methods of numerical results precision assessment is the usage of a posteriori residual error indicators.

The finite element method applies residua as a natural choice for estimating numerical errors. Let us consider a linear differential equation

$$Lu = f \tag{1}$$

where L is a linear differential operator, u is the function sought in the domain G and fulfilling the given boundary conditions at its border and f is the given function. Let us denote the solution of this boundary problem, obtained by using the finite element method, u_h. The solution error e is given by the difference

$$e = u - u_h \tag{2}$$

and it holds true that

$$Le = r \tag{3}$$

The residuum r may be easily computed from the equation

$$r = f - Lu_h \tag{4}$$

By using the finite element method, the residuum r_i for each element mesh node may be easily found and the values r_i ($i = 1, 2, \ldots, n_u$, where n_u is the mesh node number) may be used as an error indicator. The implementation of this indicator's computation into a standard code for the finite element method does not present any difficulty. Moreover, residua r_i also have their physical meaning; in solving groundwater flow, for example, r_i represents the discharge, necessary for a precise fulfilment of a continuity equation in a sub-domain, formed by elements with a common given node.

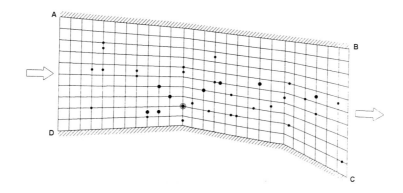

Figure 4 Maximum values of residual error indicator in the nodes for single precision solution

The residual error indicator has proved effective in assessing numerical precision of practical problems and it has proved to be highly sensitive. The solution of a steady groundwater flow with a free surface in a domain, displayed in Fig. 4, may serve as an example. At first, the problem was solved with single precision and the assessment of precision applied residua, modified according to the formula

$$r_i^* = \frac{c}{k}|r_i| \qquad i = 1, 2, \ldots, n_u \tag{5}$$

where c is a suitably chosen constant and k is hydraulic conductivity. Solid circles in Fig. 4 represent the nodes whose residua fulfil the condition

$$r_{max}^* \geq r_i^* > 0.8 \, r_{max}^* \qquad \text{(8 values of } r_i^*) \tag{6}$$

and small circles represent the nodes with residua in the interval

$$0.8 \, r_{max}^* \geq r_i^* > 0.6 \, r_{max}^* \qquad \text{(26 values of } r_i^*) \tag{7}$$

The node in which r^{*}_{max} is found is marked by a double circle.

The computation was repeated with double precision. As expected, the maximum residuum value fell down by four orders, but the value distribution of r^{*}_{i} went through significant changes as well. Within the interval (6) there were only four values and within the interval (7)

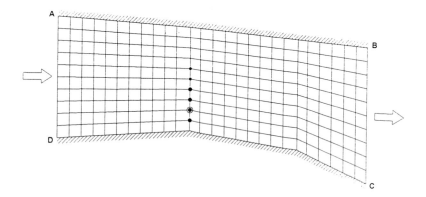

Figure 5 Maximum values of residual error indicator in the nodes for double precision solution

there were no more than two (see Fig. 5). In Fig. 6 the distribution of r^{*}_{i} is compared in a histogram. The values r^{*}_{i} are distributed among ten intervals and the constant c in equation (5) is by four orders higher when computing with double precision in order to make the comparison practical. The histogram shape makes it evident that the distribution of r^{*}_{i} within the solved domain is probably of exponential or Weibull type.

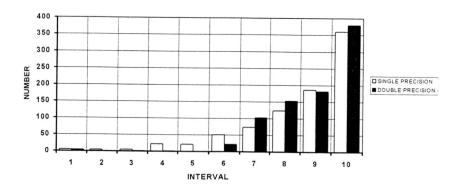

Figure 6 Distribution histogram of residual error indicator values for solutions with single and double precision

CONCLUSIONS

The finite element method allows solving the solute transport in groundwater even in such three-dimensional domains that have a regional character and a complex geological structure. The practical application of this method, however, requires sufficient theoretical knowledge and experience in numerical modelling.

As contaminant transport is greatly affected by groundwater flow patterns, a careful analysis of groundwater flow in a given domain has to be made first of all. In designing remediation it is advisable to assess the impact of designed measures on seepage flow and carry out the solute transport simulation only for the most suitable alternatives.

A detailed precision assessment of the applied numerical model always has to be performed. For this purpose an a posteriori residual error indicator has proved effective, which is sufficiently sensitive and whose value may be easily established in all nodes of the applied element mesh.

ACKNOWLEDGEMENTS

This paper was prepared as part of the research project *Numerical modelling of environmental problems in general multi-parameter rock systems* and it was supported from grant No. 103/99/0751 by the Grant Agency of the Czech Republic. The support is gratefully acknowledged.

REFERENCES

Anderson, M, P. – Woessner, W. W.: Applied Groundwater Modeling – Simulation of Flow and Advective Transport. Academic Press, San Diego 1992.

Appelo, C. A. J. – Postma, D.: Geochemistry, Groundwater and Pollution. A. A. Balkema, Rotterdam 1994.

Baker, K. H. – Henson, S. D.: Bioremediation. McGraw–Hill, New York 1994.

Charbeneau, R. L. – Bedient, P. B. – Loehr, R. C.: Groundwater Remediation. Water Quality Management Library, Vol. 8, Technomic Publishing, Basel 1992.

Gibert, J. – Danielopol, D. L. – Stanford, J. A.: Groundwater Ecology. Academic Press, San Diego 1994.

Havlíková, S. – Kazda, I.: Quasi-three-dimensional Numerical Model of Protection of a Water Reservoir from Contaminated Groundwater (In Czech). Stavební obzor (Civil Engineering Survey), 1997, 6, No. 5, pp. 143 – 146.

Kazda, I.: Finite Element Techniques in Groundwater Flow Studies. Elsevier, Amsterdam 1990.

Kazda, I.: Groundwater Hydraulics in Environmental and Engineering Applications (In Czech). Academia – Publishing House of the Czech Academy of Sciences, Prague 1997.

Křížek, M. – Neittaanmäki, P. – Stenberg, R. (eds): Finite Element Methods – Superconvergence, Post-processing and a Posteriori Estimate. Lecture Notes in Pure and Applied Mathematics, vol. 196, Marcel Dekker, New York 1998.

Diffusion and sorption experiments with DKS permeameter

CLAUDIO FERNANDO MAHLER, COPPE (Graduate Courses of Engineering)/UFRJ
RAQUEL QUADROS VELLOSO, EE (Engineering School)/UFRJ

The analysis of contaminant transport through clay liner is a relevant aspect in the design of industrial, urban and mining waste disposal systems, since these areas must be designed and operated to prevent contaminating substances from reaching underground water systems in unacceptable concentrations. The design requires an estimate of the potential contaminant transport rate. However, before any attempt at quantification can be made, values for transport mechanism control parameters must be established.

Clayey materials are frequently used as contaminant barriers. In these materials, which have low hydraulic conductivity, the main contaminant transport mechanism is molecular diffusion. Parameters controlling transport for these conditions are the diffusion coefficient and sorption parameters. These parameters depend on soil constituents and characteristics as well as on the chemical constitution of the contaminant. The great complexity of the factors involved makes it necessary to determine the parameters of each type of soil.

This paper presents an equipment called DKS Permeameter (Diffusion, Konvektion, Sorption), for the study of soil-contaminant transport mechanisms, developed at the Institute for Soil Mechanics of the Ruhr-University Bochum and some results obtained from its use in COPPE/Federal University of Brazil (UFRJ), Brazil. This equipment allows the determination of the effective diffusion coefficient and the sorption parameter in conditions that better reflect field conditions.

The soil under study is a mixture of sodium-bentonite with low hydraulic conductivity ($k=10^{-9}$ cm/s) with adequate liner characteristics. The result led us to observe the relevance of determining the parameters of sorption for structured soils, since sorption determined from batch test results with pulverized soil represents maximum soil capacity. Projects based on this parameter would overestimate the attenuation capacity of the liner.

KEY WORDS: DIFFUSION, SORPTION, LINER, CONTAMINANT

1. INTRODUCTION

In a project industrial, domestic and mining waste disposal systems, a contaminant transport analysis through the liner is very important, because in these areas it is expected that no contaminants reach the ground water system with unacceptable concentrations. However, before any quantification, it is necessary to establish values for the parameters that control the mechanism transport.

The DKS Permeameter (Diffusion, Konvektion, Sorption) is an equipment developed at the Institute for Soil Mechanics of the Ruhr-University Bochum for the study of soil-contaminant transport mechanisms in liners. With the permeameter it is possible to determine the hydraulic conductivity before and after the contamination, the effective coefficient of diffusion and the parameter of sorption.

The materials used as liners have normally a very low hydraulic conductivity and a high percentage of clay. In this case the main transport processes are the molecular diffusion and sorption (Goodwall and Quigley, 1977; Desaulniers et al , 1981; Shackelford, 1991, Jessberger and Onnich, 1994, Rowe et al, 1995).

2. THEORETICAL PRINCIPLES OF THE DKS PERMEAMETER

As observed by Jessberger and Onnich (1994) the theoretical principles to describe the mass transport phenomena have been presented by soil physicists and hydro geologists. As observed by these authors Van Genuchten (1974) proposed fundamental transport equation that indicates the concentration alteration of substances as a function of time:

$$D.\delta^2 c/\delta z^2 = V. \; \delta c/\delta z + \delta c/\delta t + \delta n/(a\delta t) \qquad (1)$$

where

c is the concentration of adsorbate in the fluid stream (moles per unit volume of solution);
n is the amount of adsorbate on the adsorbent (moles per unit volume of volume of packed bed);
V is the velocity of fluid through interstices of the bed;
z is the distance variable along the bed;
D is the diffusion coefficient of the adsorbate in solution in the bed; and
a is the fractional void volume in the bed.

Figure 1 shows a scheme of the DKS Permeameter.

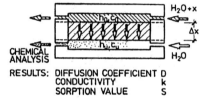

RESULTS: DIFFUSION COEFFICIENT D
CONDUCTIVITY k
SORPTION VALUE S

Figure 1. Scheme of the DKS Permeameter (apud Jessberger and Onnich, 1995)

A soil sample is moulded in the centre of the permeameter in a space of 80x80x20 mm. A solution with constant concentration percolates in the upper channel percolates and distilled water percolates in the lower channel. This gives a constant concentration gradient in the sample and permits that the diffusion happens according to Jessberger and Onnich (1993). As proposed by them all parts of the permeameter coming in contact with the experimental solutions are made of resistant materials that do not interact with the used solutions. A scheme of the complete experimental situation is presented in Figure 2.

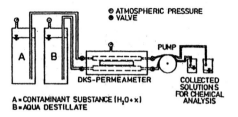

Figure 2. Scheme of the Complete Experimental System (apud Jessberger and Onnich, 1993)

In the test two five-litre tanks, one containing a contaminant solution that fills the upper channel and the other with distilled water that flows through the lower channel (Figure 1 and 2). The tanks are situated on one side of the system where they feed it and on the other side two peristaltic pumps are inserted to control the velocity of the flow. The concentration of the fluids on both sides of the system is taken from time to time so that the chemical concentration of the fluids is controlled.

The results of the chemical analyses are plotted in graphs that relates the emission to time.

Figure 3 shows the phases resulting from the interaction between the effects of diffusion and sorption as proposed by Jessberger and Onnich (1994):

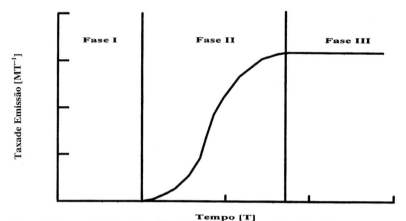

Figure 3. Phases of the test (modified from Jessberger and Onnich, 1994)

Phase 1 – there is no emission of contaminant through the soil sample, which means that all ions are retained by the sample;

Phase II – non-stationary phase. The capacity of sorption of the sample has not yet been exhausted, but there is no increase in the quantity of contaminant that flows through the sample;

Phase III – Stationary Phase – The total capacity of sorption has been reached and the contaminant flows through the soil without any retention.

The diffusion coefficient is obtained when the test gets the Phase III (stationary). In this phase

$-\Delta m_o / \Delta t = \Delta m_1 / \Delta t$ = constant   (2)

where $\Delta m_1 / \Delta t$ is the emission rate (MT^{-1}). That means that the decrease in mass of a chemical is equal to the increase in chemical mass in the lower tank. (Shackelford, 1991)

The flow of mass of a chemical in a saturated soil under combined effects of hydraulic gradient and concentration for the one dimensional transport is given by Shackelford (1991) in the following equation:

$F = nvC - nD_h \delta C / \delta x$ (3)

Where
     F is the flow of mass $(ML^{-2}T^{-1})$
     n is the soil porosity
     v is the velocity of flow of solvent (normally water) $(LT^{-1}+)$
     C is a concentration of chemical in the liquid phase in the soil (ML^{-3})

x is the direction of the transport (L); and
D_n is the coefficient of hydrodynamic dispersion (L^2T^{-1}).

Two processes are involved in the hydrodynamic dispersion: mechanical dispersion that occurs during the advective flow and molecular diffusion caused by the thermo-kinetic energy of the particle of the solute (Freeze and Cherry, 1979). The coefficient of hydrodynamic dispersion can be represented by:

$$D_h = D_m + D_e \qquad (4)$$

Where D_m is the coefficient of mechanical dispersion and D_e is the effective diffusion coefficient.

Mechanical dispersion occurs during advection and is caused by the movement of the fluids through the pores and by the heterogeneity of the medium, creating variations in the velocity of the flow.
The molecular diffusion is a phenomena of transport of mass resulting from the variation of chemical concentration of the soluble in the liquid phase. The process of diffusion only finishes when there is no more concentration gradient (Freeze and Cherry, 1979).
A molecular diffusion may occur when there is no more movement of fluid. However if there is advective flow, the molecular diffusion is the simultaneous process to the mechanics dispersion that causes a moisture of the ionic and molecular components. For problems involving liners of low permeability it seems that the mechanical dispersion is not important ($D_h=D_e$) (Rowe, 1987). In the case that there is only concentration gradient, the flow velocity is zero and D_m is also zero and the mass flow can be given by the 1st Fick Law:

$$F_D = -nD_e\delta C/\delta x \qquad (5)$$

where
F_D is the flow mass   due to molecular diffusion $(ML^{-2}T^{-1})$.

For the stationary condition equation (5) is applied to obtain the effective diffusion coefficient, D_e. As F_D is the mass flow, the effective diffusion coefficient can be given by the following equation:

$$D_e =- \delta x \Delta m/(n\delta CA\Delta t) \qquad (6)$$

Where $\Delta m/\Delta t$ is the emission rate (MT^{-1}), A is the cross-section area of the sample (L^2).

Assuming that the concentration gradient through the sample is constant (Figure 4), as proposed by Shackelford (1991) we have:

$$D_e =- \Delta x \Delta m/(n\Delta CA\Delta t) \qquad (7)$$

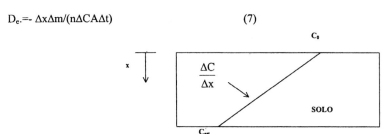

Figure 4. Concentration Profile in the soil sample

When the test reaches stationary condition (Phase 3 of Figure 3) it is possible to obtain the effective diffusion coefficient using equation (7).

To reactivate chemicals the coefficient of effective diffusion is initially obtained from the emission factor of the stationary phase. Figure 5 shows the plotted line using this coefficient and the results of the tests. The difference between the two lines indicated the mass adsorbed by the soil S_{pk}.

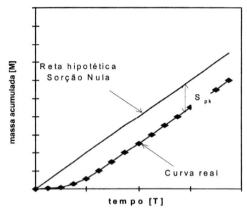

Figure 5. Determination of the mass of the studied element retained in the soil, S_{pk}

3. TEST RESULTS

The material used in the tests consisted of moisture of 80 % of a natural pure kaolim and 20 % of commercial sodium bentonite. An attempt was made with this moisture to reproduce the characteristics of the foundation of the Gramacho Landfill, which receives the domestic waste from Rio de Janeiro. Tables 1 present the characterisation of the samples used in two tests. The pollutant used was the potassium chlorine solution (KCl). In the E1 test it 3000 mg/L and in test E2 2000 mg/l was used.

Table 1: Characterisation of the samples

	Test 1	Test 2
LL [%]	139	139
LP [%]	34,5	34,5
G [-]	2,637	2,637
w [%]	224	212
e [-]	6	5,5
n [-]	0,86	0,85
S [%]	98,5	100
γ_t [g/cm^3]	1,213	1,265
Changeable ions [meq/100g]		
P		3,0
Al		-
Ca		2,8
Mg		2,1
K		0,11
Na		9,88
H+Al		-
CTC		14,8
pH [-]		9,8

The samples were prepared by adding distilled water to the moisture since the water content was about 210 % corresponding to 1,5 LL. After this procedure, the moisture was left in a humid chamber acquire homogeneity.

The test begins after moulding the soil in the DKS permeameter. The test is stopped when the stationary phase is reached. Figure 6 shows results of the emission factor during the tests, Table 2 present the values obtained to determine of the effective diffusion coefficient.

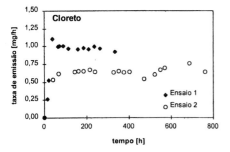

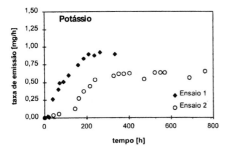

Figure 6. Graphs of the emission factor

Table 2: Determination of D_e.

	Chlorine		Potassium	
	Test 1	Test 2	Test 1	Test 2
C_0 [mg/L]	1559	1040	1525	980
C_{est}* [mg/L]	284	225	273	216
ΔC [mg/L]	1275	815	1252	764
$\Delta m/\Delta t$* [mg/h]	0,96	0,65	0,90	0,61
D_e [m²/a]	0,025	0,027	0,024	0,027
D_e x 10^{16} [m²/s]	7,9	8,5	7,6	8,5

* middle values obtained in the stationary phase.

Using the effective diffusion coefficient computed before, it is possible to obtain the graphs of accumulated potassium retained in the soil (Figure 7).

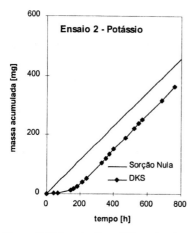

Figure 7. Determination of the potassium mass retained in the soil

Ritter (1998) did batch tests on diffusion and equilibrium for the same soil and the same solution. Even for the chloride ions, as for the potassium, the effective diffusion coefficient that best adapted to the test data was $D_e = 0,03$ m^2/ a.
Figure 8 shows the results of the sorption parameters of the test in the DKS apparatus and in the portion test.

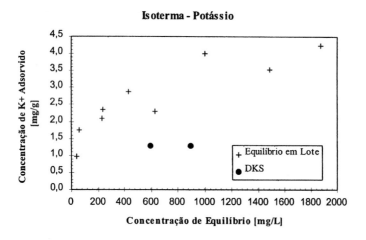

Figure 8. Isotherm of adsorption for potassium – DKS and Portion Test

From the isotherm of adsorption it is possible to conclude that for the same level of equilibrium concentration (between 500 and 1000 mg/l of K$^+$), a concentration of K$^+$ adsorbed in the portion equilibrium test is much greater than the K$^+$ concentration adsorbed in the DKS test. This difference can be explained considering the reason that in the portion equilibrium test the soil is pulverised, This means that all surfaces of the particles are disposable for the adsorption. In the DKS test the soil is structured with less surface of disposable particles. Tests done at the Ruhr University showed similar results (Jessberger et al, 1995).

3. CONCLUSIONS

The analysis of the results of the DKS test is simple and the determination of the parameters is direct.

The values of the diffusion coefficients encountered are consistent. It seems to be a wide difference between the sorption parameters of each type of test. The reason may be the structured condition of soil in the DKS test. Voice et al (1983) concluded that the relation of solid and water influences the value of the partition coefficient obtained in equilibrium batch tests. From this reason these coefficients can not be directly applied to actual systems.

Finally, it could be written that further studies with the DKS, with lower concentration levels, should be done to obtain more information about the influence of the soil structure in the absorption process.

AKNOWLEDGMENTS

The authors would like to thank the Brazilian Research Council for its continuous aid and the colleagues Maria Claudia Barboza and Elisabeth Ritter for their kind support. The authors would also like to give their heartfelt thanks to Ruhr University, Bochum, and Prof. H.-L. Jessberger and Dr. Gunnar Heibrock for lwnding the DKS Permeameter.

APPENDIX I. REFERENCES

Desaulniers, D.D., Cherry, J. A. and Fritz, P. (1981). Origin, age and movement of pore water in argillaceous quaternary deposits at four sites in Southwestern Ontario. Journal of Hydrology, 50, p. 231-257.

Freeze, R. A. and Cherry, J.A (1979). Groundwater, Prentice Hall.

Goodall, D.C. and Quigley, R.M. (1977). Pollutant migration from two sanitary landfill sites near Sarnia, Ontario, Canadian Geotechnical Journal, V 14, no. 2, p. 223-236.

Jessberger, H.-L. and Onnich, K. (1993). Calculations of pollutant emissions through mineral liners based on laboratory tests. 10th Clay Conference, July'93, Adelaide, Australia.

Jessberger, H.-L. and Onnich, K. (1994). Determination of pollutant transport parameters by laboratory testing. XIII ICSMFE, New Delhi, India, p. 1547-1552.

Jessberger, H.-L., Onnich, L., Finsterwalder, K. and Beyer, S. (1995). Versuche und Berechnungen zum Schadstofftransport durch mineralische Abdichtungen und daraus resultierende Materialentwicklungen (in German)

Ritter, E. (1998). Effect of the salinity of the soil in the diffusion and sorption of inorganic ions. Ph.D. Dissertation, COPPE/UFRJ-Brazil (in Portuguese).

Rowe, R.K., Quigley, R.M. and Booker, J.R. (1995). Clayey Barrier Systems for Waste Disposal Facilities, E & FN Spon, London, UK, 390 p.

Shackelford, C.D. (1991). Laboratory diffusion testing for waste disposal – a review. Journal of Contaminant Hydrology, 7, p. 177-217.

Voice, T.C., Rice, C.P. and Weber, W.J. (1983). Effect of solid concentration on the sorptive partitioning of hydrophobic pollutants in aquatic systems. Journal of Environmental Science and Technology, 17, p. 513-518.

Van Genuchten, M.T. (1974). Mass Transfer Studies in Sorbing Porous Media, Diss. New Mexico State University, 1974.

Chemical aspects of the strengthening of contaminated material using lime

J. D. MCKINLEY[1], H. R. THOMAS[1], K. P. WILLIAMS[1] and J. M. REID[2]
[1]Geoenvironmental Research Centre, Cardiff School of Engineering, Cardiff University, Cardiff, U. K.
[2]Transport Research Laboratory, Crowthorne, U. K.

ABSTRACT
Strengthening of contaminated materials using inorganic cementitious agents is becoming more widely used in the UK. The method has particular advantages for bulk fill operations such as highway earthworks. Research has been done into the chemical characteristics of leachates and leached solid samples from a study into the long term durability of a lime strengthened silt / pfa mixture amended with sewage sludge. This involved determinations of chemical composition and mineralogy, and geochemical modelling using MINTEQA2. None of the heavy metals tested is present in the leachate at a concentration likely to pose a significant environmental threat, although some were present to a higher degree than expected based on the inorganic chemistry. This is ascribed to complexation of the heavy metals with dissolved organic matter.

INTRODUCTION
The Transport Research Laboratory (TRL) is currently undertaking a large research project for the Highways Agency on processing contaminated land in highways schemes. This includes work on lime stabilisation of contaminated material, and the work to March 1997 is summarised in Reid and Brookes (1997), where the development at TRL of a dynamic leaching test as a method of examining the long term integrity of stabilised materials is reported. The contaminated material is a mixture of lightly contaminated river silt and pulverised fuel ash, from the A13 highway scheme at Rainham Marshes, to which heavily contaminated sewage sludge was added and the resulting mixture stabilised with lime. Cardiff University's Geoenvironmental Research Centre has undertaken fundamental research on behalf of TRL into the processes taking place inside lime stabilised contaminated soils, with particular regard to the chemical aspects and their effect on contaminant mobility. The investigation included: a review of the literature relating to chemical aspects of stabilisation of contaminated soils and waste sewage sludges with lime; chemical composition and mineralogical analysis of leachates, stabilised solid and unstabilised solid samples supplied by TRL; and geochemical assessment modelling of the leachates using MINTEQA2. The purpose of this paper is to summarise the major findings of that investigation, and to highlight the importance influence of mineralogy and organic content on contaminant mobility in lime stabilised contaminated materials.

BACKGROUND
Soil stabilisation using lime or cement has long been used to improve the handling and mechanical characteristics of soils for civil engineering purposes (Sherwood, 1993). Of more recent interest is the stabilisation of contaminated soils and sewage sludges for use in bulk fill operations for highway earthworks (Reid and Brookes, 1997). There is a strong suggestion

that in materials with a low organic content heavy metal mobility is reduced due to physical micro-entrapment and not to hydroxide formation (Albino et al., 1996). Stabilisation has both a physical aspect involving changes to the mechanical properties of the material, and a chemical aspect involving changes to the form and mobility of the contaminants present. The term 'lime' is used in the literature to refer to calcium oxide, calcium hydroxide and calcium carbonate. Here, the term should be understood to mean either calcium oxide or calcium hydroxide.

Calcium oxide has a strong exothermic reaction with water, which has an additional drying effect on the soil, but otherwise the reactions are similar to those occurring when calcium hydroxide is used (Sherwood, 1993). The principal reaction is one between the lime and the clay minerals of the soil, or other fine pozzolanic components, to form a tough, water insoluble gel which cements the soil particles. Cementing between particles due to surface coating by this gel is a major mechanism in lime stabilisation (Ingles, 1970), the other being reinforcement due to the formation of solid reaction products in the soil pore spaces. Stabilisation can occur only if the reaction products are insoluble, and the pH must rise to about 12·4, so that silicate and aluminate dissolve from the clay minerals (Bell, 1996). The development of the reaction products therefore depends on the lime's ability to attack the minerals in the clay fraction. In their overview article outlining the possibilities for lime stabilisation of heavy metal contaminated sludges, Boardman and MacLean (1996) also stated that care must be taken over the sulphate, chloride and organic content of the sludges, since these can interfere with the cementitious reactions.

Poon and Boost (1996) performed Toxicity Characteristics Leaching Procedure tests on sewage sludge stabilised with pfa and calcium oxide, and found that more copper and less zinc leached from the stabilised sludge than from the unstabilised sludge, although stabilisation generally reduced the metal leaching potential. The ability of soluble organic anions to mobilise certain metals in alkaline soils must be considered when interpreting data on mobility. McBride and Blasiak (1979) studied the solubility of copper and zinc as a function of pH in the range 4–8·5, for an acidic mineral soil. They found that zinc solubility increased above pH 7·2, probably due to the dissolution of zinc-organic complexes. In addition, the solubility of copper, which is known to have a high affinity for organic matter, increased in solubility above pH 6·5. Tyler and McBride (1982) eluted mineral and organic soils with calcium chloride and found that the mobility order was Cu<<Zn<Ni<Cd. The mobility of all these metals was reduced when the mineral soil was limed with calcium hydroxide if the final pH was less than 6·5, which is much lower than the pH of lime stabilised soils. Hsiau and Lo (1997) have recently published an extensive study into metal leachability from calcium hydroxide stabilised sewage sludge. They measured copper, lead, chromium and zinc leachability, and found that the more the heavy metals were fractionated to the organically bound form in the unlimed sludge, the more unstable they were after lime stabilisation. Liming increased the solubility of copper and reduced that of zinc, an effect ascribed to irreversible dissolution of organics at high pH and the higher affinity of some metals, such as copper, for organics.

The Transport Research Laboratory has been examining the use of calcium oxide stabilisation for the improvement of bulk fill, for a variety of applications. In the United Kingdom it is necessary to demonstrate that pollution of controlled waters will not occur either in the short or long term from the use of stabilised, contaminated material. TRL wished to address the form in which the various contaminants are present in both the stabilised and unstabilised materials, and the nature and extent of development of cementitious compounds in the stabilised materials. The aim was to get beyond the 'black box' approach by examining what is actually happening during lime stabilisation of these contaminated materials and the effect on contaminant mobility. To this end TRL set up two further leaching tests with distilled water as the leachant. The leachate was collected and stored separately for each sample. The two leached stabilised material samples, and leachate samples from each leaching column, were delivered to Cardiff University in February 1998 for specialised chemical and mineralogical testing.

Figure 1. Characterisation of leachate samples from first leaching test

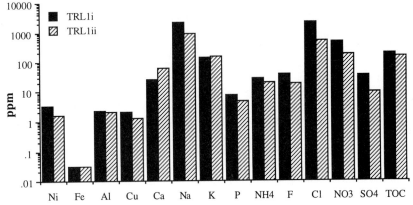

CHEMICAL ANALYSIS

After filtration through a 0·2μm pore size filter the leachate samples were analysed by Inductively Coupled Plasma Atomic Emission Spectroscopy (ICP) on a Perkin Elmer 400 Plasma Spectrophotometer and by ion chromatography (IC). Total organic carbon (TOC) was determined using a Shimadzu Total Organic Carbon Analyser. The stabilised solid samples were size fractionated by wet sieving at 20μm and 45μm using the minimum of water, and then specific gravity separated using 1 g of each size fraction in bromoform, whose specific gravity is 2·9. X-ray diffraction (XRD) analysis of the bulk and significant fractioned solids was performed to determine mineral content.

Figure 1 shows the results of the chemical analysis of the two leachates from the first leaching test, where samples TRL1i and TRL1ii are the first and second leachate sample. Figure 2 shows the corresponding results for the second leaching test. Each leachate sample corresponds to approximately one leachate kilogram per litre of stabilised material.

Table 1. Characterisation of leachate samples

	Sample number			
Determinand	TRL1i	TRL1ii	TRL2i	TRL2ii
pH	12·17	12·59	12·65	12·77
Eh (V)	+0·037	+0·009	-0·002	-0·017

The leachate samples were highly alkaline, with the second sample from each test being higher in pH than the first. The leachate samples were observed to have a pale yellow tinge, consistent with light contamination by dissolved organic matter (Hem, 1970), as shown by the highly significant TOC values. Table 1 shows the pH and redox potential measured for each leachate sample. The redox values are near zero, and given the high pH indicate a very low dissolved oxygen content (Langmuir, 1997, p.408-410).

The principal chemical components in the leachate samples are sodium, chloride, nitrate, potassium and calcium, with sodium and chloride being dominant. Sulphate, fluoride, ammonium and phosphorus are also significant. The calcium concentrations in the TRL1 leachates are much lower than in TRL2i, but otherwise the principal chemistry of the TRL1 samples is similar to that of the TRL2 samples. Comparison between TRL1i and TRL1ii, and between TRL2i and TRL2ii, shows a dramatic reduction in the concentrations of sodium, chloride, sulphate and nitrate, and a smaller reduction in the organic carbon, fluoride, ammonium and phosphorus concentrations. In contrast, the potassium concentration is relatively steady and the calcium concentration increases with continued leaching. For the minor components, principally the heavy metals, only nickel, copper and aluminium are present above the 1ppm level, being generally in the range 1 to 3.5ppm, with copper and nickel showing a marked reduction in concentration as leaching progresses. This indicates an early mobilisation of the sodium and chloride, longer term leaching of the calcium and a steady reduction in the concentrations of organic carbon, copper and nickel over time. In addition, arsenic, selenium, chromium, zinc, cadmium, lead, tin and boron were detected in trace amounts close to or below the detection limits of the ICP-AES apparatus used, while manganese and magnesium were not detected.

Figure 2. Characterisation of leachate samples from second leaching test

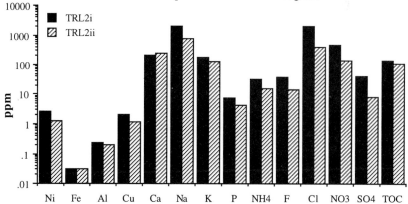

Table 2 shows the results of particle size and XRD analysis of the stabilised solid material. Only about 33% passed through the 20µm filter, and large aggregations of particles were observed on the 45µm filter. This indicates the coarsening of the materials which results from the cementing action of the lime stabilisation. The proportion of sinks was very low, and was negligible for the -20µm size fraction. The sink fractions should be dominated by the heavy metal oxide constituents, since soil organic material, quartz, calcite and clay minerals have specific gravities less than 2.9. The cementing products of the lime stabilisation will be similar to those in cement, whose specific gravity lie between 2.1 and 2.5 (Soroka and Relis, 1976), and are not expected in the sinks. It was impossible to satisfactorily identify all of the peaks on the XRD traces, but the principal crystalline phases are believed to have been identified. The mineralogy of the bulk stabilised solid samples is dominated by quartz and calcite, neither of which generally has a strong affinity for heavy metals, with small amounts of muscovite and possibly kaolinite being evident. There is relatively more calcite and muscovite in the finest fractions, and the sink fractions are dominated by donathite, $(Fe,Mg)(Cr,Fe)_2O_4$, and hematite,

Fe_2O_3. The mineralogy of the stabilised material was found to reflect the mineralogy of the unstabilised silt/pfa mixture, which is the main parent material.

Table 2. Major mineral types in the stabilised solids by XRD

Fraction	% of bulk by weight	Major mineral types identified
Bulk	100	quartz, calcite, muscovite, kaolinite
$-20\mu m$	33	quartz, calcite, muscovite
$-45+20\mu m$	7	quartz, calcite
$+45\mu m$	60	quartz, calcite
$-45+20\mu m$ sinks	0·2	donathite, hematite
$+45\mu m$ sinks	0·5	donathite, hematite

Neither of the main XRD peaks for calcium silicate hydrate of the tobermorite type, at 3·03 and 1·83Å (Copeland et al., 1960), where observed, although these may be masked by peaks for calcite and quartz respectively. Neither of the calcium sulfoaluminate ettringite peaks at 9·73 and 5·61Å were found, although Copeland et al. (1960) point out that these are often weak in cement pastes, nor the peaks listed by Copeland et al. (1960) as characteristic of calcium aluminate hydrates. This suggests that the amount of crystalline cementing material formed is low in proportion to the mineral phases already present.

SPECIATION MODELLING

Geochemical assessment modelling of leachate samples was undertaken using version 3·11 of MINTEQA2 (Allison et al. 1991), the US EPA's speciation code. MINTEQA2 is an equilibrium speciation model, and the results therefore indicate the thermodynamically most stable species, not the kinetically favoured. The determinands from the chemical analysis with concentrations greater than 1 ppm were entered into MINTEQA2 as components, and the equilibrium speciation at fixed pH calculated. The organic carbon content, and therefore the possibility of heavy metal complexation with dissolved organic matter, was not modelled. The results should provide insight into the major chemical aspects of these aqueous samples, and their probable behaviour should the dissolved organic matter break down. It is recognised however that the speciation modelling will not allow an assessment of the degree of metal complexation with the organic matter, identified by McBride and Blasiak (1979) and by Tyler and McBride (1982) as a significant mechanism in heavy metal leachability from alkaline stabilised materials. To assess the sensitivity of the results to pH, speciation was calculated at the measured pH minus 0·5, the measured pH, and the measured pH plus 0·5.

Table 3. Speciation results for leachate sample TRL2i

	pH= 12·15		pH= 12·65		pH= 13·15	
Ion	Aqueous %	Main solid	Aqueous %	Main solid	Aqueous %	Main solid
Ni	0·1	$Ni(OH)_2$	0·8	$Ni(OH)_2$	2·6	$Ni(OH)_2$
Cu	2·7	Tenorite	5·3	Tenorite	26·1	Tenorite
Ca	99·6	Calcite	99·7	Calcite	40·9	Portlandite

Table 3 shows some of the results of the MINTEQA2 modelling of leachate sample TRL2i. For this inorganic system, at the measured pH, the copper and the nickel present are predicted to be almost completely precipitated, the copper as tenorite, which is copper oxide, CuO, and the nickel as nickel hydroxide, $Ni(OH)_2$. The speciation modelling also indicates that sodium, potassium and ammonium are completely soluble over the range modelled, and that for pH less than about 12 the aluminium present will precipitate as diaspore, which is aluminium oxide hydrate, $Al_2O_3.H_2O$, but be completely dissolved above that value. This recalls the observation that a pH above about 12·4 is necessary in lime stabilisation if the silicate and aluminate phases of the clay minerals are to dissolve (Bell, 1996). Approximate modelling of the stabilised solid material shows that at the leachate pH zinc, lead, nickel, iron, manganese, magnesium and copper should be completely precipitated.

MINTEQA2 was also used to model the effect of equilibrating leachate sample TRL2i with atmospheric carbon dioxide and reducing the pH to 9. This produced a complete dissolution of the nickel as an aqueous carbonate, precipitation of nearly all of the calcium as calcite and practically all the aluminium as diaspore, but little change in the aqueous copper content. Repeating these simulations at a pH of 8 results in complete dissolution of the calcite but little change in the distribution of the other metals.

MINTEQA2 predicts that in the absence of dissolved organic matter the copper and nickel in the leachate should be precipitated. However, their presence in the leachate samples indicates that they are in a mobile form, which must be due to causes not considered in the speciation modelling. At the leachate pH zinc, lead, nickel, iron, manganese, magnesium and copper are predicted to precipitate completely, and of these only nickel and copper are found in significant quantities in the leachates from TRL's testing. The other heavy metals considered in the model are known to complex less strongly with organic carbon, were predicted to be precipitated, and were not found in the leachates.

DISCUSSION
Lead, arsenic, selenium, mercury and iron concentrations in the leachates are very low, and so these metals are not being leached from the stabilised contaminated material in significant quantities. Sodium chloride leaches out rapidly, as TRL observed in their first study. The principal substances of concern in the leachates would appear to be the copper, the nickel, and the organic content and therefore the BOD. However, the concentrations in the leachates are not high, and are reducing as leaching progresses. The aluminium, copper and nickel concentrations are higher in the leachates from the stabilised materials than from similar tests on unstabilised materials, indicating increased mobility as a result of lime stabilisation. Calcium is leaching out of the stabilised material, and the results indicate that it will be present in the leachates at a concentration of about 400ppm for a considerable time. Speciation modelling indicates that the copper and nickel should be precipitated at the high pH of the leachates, and yet both are found in significant concentrations in the samples collected.

The results of previous studies suggest that if metals are strongly complexed with organic matter, as may occur in sewage sludge, they remain so when the material is limed, but the organic matter itself dissolves at the resulting high pH. Hence, the metal moves into solution as an metal-organic complex. Copper, cadmium and nickel in particular appear to be solubilised by this mechanism. There is, therefore, a potential hazard from soluble forms of heavy metals in highly alkaline environments where the organic content dissolves. McBride (1994) listed a typical affinity sequence of divalent metal ions for soil organic matter at pH 5, the affinity order appearing to correlate with the Pauling electronegativity, a factor related to an ion's ability to form covalent bonds with organic bases. Part of this sequence is shown in Table 4, which also lists the ratio of the percentage of each constituent mobilised in an NRA leaching test on the stabilised material to that on the unstabilised material, from data supplied by TRL. The degree to which the contaminants become more mobile follow the same order as McBride's sequence. However, it must be remembered that this sequence is for a soil organic environment under acidic conditions.

Table 4. McBride's affinity order and TRL leaching ratios

Affinity sequence	Cu	>	Ni	>	Pb	>	Zn	>	Mn	>	Mg
Electronegativity	2·0		1·91		1·87		1·65		1·55		1·31
NRA leaching ratio	7·4		3·6		1·1		0·49		0·17		0·004

The leachates from the stabilised solid samples contain significant concentrations of dissolved organic matter, probably originating from the contaminated sewage sludge, which provides an excellent substrate for microbial growth. The cofactors phosphorus and sulphate are also present, and the heavy metals are not present at high enough concentrations to be toxic, so microbial growth is expected. Bacterial growth in the leachate samples was observed under laboratory storage conditions, and would be expected to also occur in the field. In the field the bacteria will digest the organic component, both dissolved and solid. Such biodegradation processes often produce acid, which will accelerate the reduction in pH of the leachate, affecting the speciation of the system, and could start to attack the cementing phases in the stabilised material. During biodegradation of the organic material heavy metals which are complexed with it may be released into solution, or mineralised, or metabolised by the bacteria.

The organic content of the leachates is much higher than was found in TRL's previous study and significant bacteria growth was observed in the present leachates. The implications of this could be quite significant. For example, long term facilitation of release of sorbed metals through bacterial activity may occur. Microbial activity in the leachates was not anticipated at the outset, but is well supported by the high organic carbon content and the reduced concentrations of some ions which form part of the nutrient package for the microbes. The microbial behaviour is important because it is necessary to know whether the microbes contribute significantly to the degradation of the organic matter, causing the release of any heavy metals complexed with the soil's organic matter.

CONCLUSIONS

Stabilisation of contaminated materials using inorganic cementitious agents is becoming more widely used in the UK. In materials with a low organic content, it is generally believed that heavy metal mobility is reduced due to micro-encapsulation, and not to hydroxide formation. However, if metals are strongly complexed with organic matter they may become mobilised as metal organic complexes at high pH as a result of dissolution of the organic matter. The results of the chemical determinations support the conclusions drawn by TRL in their previous study that the contaminant concentrations in the leachates are relatively low, such that the leachates are unlikely to pose a serious environmental threat. In general, the concentrations found were comparable to those from TRL's previous study, except that the organic content of the leachates is much higher and significant bacteria growth was observed in the present leachates. The long term implications of this could be quite significant, because the microbes may contribute significantly to the degradation of the organic matter, causing the release of any heavy metals complexed with the soil's organic matter. The soil mineralogy is primarily silica, calcite and hematite, none of which should sorb significant quantities of heavy metals.

Copper and nickel are present in the leachates to a higher degree than expected based on the inorganic chemistry. The available evidence points towards heavy metal complexation with organic material, which becomes dissolved at the high pH of lime stabilisation. Long term, mechanical stability of the material will probably not be compromised by microbial activity but the environmental impact of microbial degradation has not yet been addressed. However, it must be emphasised that none of the heavy metals tested is present at a concentration high enough to pose a significant threat to controlled waters and the concentrations of those elements which could cause concern should decrease with time.

ACKNOWLEDGEMENTS

This paper is based on research carried out for the Highways Agency. Any views expressed in this paper are those of the authors and not necessarily those of the Highways Agency.

REFERENCES

Albino, V., Cioffi, R., de Vito, B., and Santoro, L. (1996). Evaluation of solid waste stabilization processes by means of leaching tests. Env. Tech., Vol 17, 309–315.

Allison, J.D., Brown, D.S., and Novo-Gradac, K.J. (1991) MINTEQA2/PRODEFA2, a geochemical assessment model for environmental systems. Tech. Rept. EPA/600/3-91/021, United States Environmental Protection Agency.

Bell, F.G. (1996). Lime stabilization of clay minerals and soils. Eng. Geol., Vol 42, 223–237.

Boardman, D.I. and MacLean, J.A. (1996). Lime treatment of metal contaminated sludges. In *Lime stabilisation*, Rogers, C.D.F., Glendinning, S., and Dixon, N., 115–126.

Copeland, L.E., Kantro, D.L., and Verbeck, G. (1960). Chemistry of hydration of Portland cement. Proc. Fourth Int. Symp. on the Chemistry of Cement, 429–468.

Hem, J.D. (1970) Study and interpretation of the chemical characteristics of natural water. Tech. Rept. 1473, Water Supply Paper, US Geological Survey.

Hsiau, P.C. and Lo, S.L. (1997). Effects of lime treatment on fractionation and extractabilities of heavy metals in sewage sludge. J. Env. Sci. and Health Part A, Vol 32, 2521–2536.

Ingles, O.G. (1970). Mechanisms of clay stabilization with inorganic acids and alkalis. Australian J. Soil Res., Vol 8, 581–595.

Langmuir, D. (1997). Aqueous environmental geochemistry. Prentice Hall.

McBride, M.B. and Blasiak, J.J. (1979). Zinc and copper solubility as a function of pH in an acid soil. Soil Sci. Soc. of Am. J., Vol 43, 866–870.

McBride, M.B. (1994). Environmental chemistry of soils. Oxford University Press.

Poon, C.S. and Boost, M. (1996). The stabilization of sewage sludge by pulverized fuel ash and related materials. Env. Int., Vol 22, 705–710.

Reid, J.M. and Brookes, A.H. (1997). Stabilisation of contaminated material using lime. Geoenvironmental Engineering Conference 97, Yong, R.N. and Thomas, H.R., 409–414.

Sherwood, P.T. (1993). Soil stabilization with cement and lime. HSMO.

Soroka, I. and Relis, M. (1976). On the density of Portland cement early hydration products. In *Hydraulic cement pastes: their structure and properties*, 87–90.

Tyler, L.D. and McBride, M.B. (1982). Mobility and extractability of cadmium, copper, nickel and zinc in organic and mineral soil columns. Soil Sci., Vol 134, No 3, 198–205.

Examining competitive sorption of copper and nickel in London Clay with centrifuge leaching tests

J. D. MCKINLEY and V. ANTONIADIS
Geoenvironmental Research Centre, Cardiff School of Engineering, Cardiff University, Cardiff, U. K.

ABSTRACT

Sorption of contaminants on to soil constituents often has a significant effect on contaminant transport. The sorption equilibrium distribution coefficient, K_d, is often used in solute transport models when predicting the movement of a hazardous material in the ground, to calculate a retardation factor, R_d, for the contaminant. Centrifuge leaching tests were conducted to directly study the transport of Cu and of Ni, both singly and in combination, through compacted London Clay. The study used a falling head permeameter-style leaching apparatus in a laboratory centrifuge, simulating the migration of these heavy metals from landfill leachate through this low permeability clay. R_d was estimated from the breakthrough curves. Competitive sorption was found to have an important effect on the transport, causing the mobility of Ni in the London Clay to be significantly higher when present with Cu than when present on its own. The mobility of Cu was also higher when present with Ni, but the difference was less significant.

INTRODUCTION

Predicting the fate and transport of contaminants in natural soils and compacted clay barriers is a complex problem that requires an understanding and quantification of several processes such as advection, diffusion, dispersion, sorption, complexation, precipitation and filtration. The pattern of contamination in contaminated land and wastes is generally complex, with mixtures of different organic and inorganic materials often encountered. Heavy metals form a group of contaminants common in waste leachates (Sharma and Lewis, 1994) and many of them constitute a significant health hazard because of their toxicity. One important process affecting heavy metal mobility in soil is competitive sorption, in which the soil solids preferentially sorb one solute instead of another, which is therefore more mobile than if it alone were present.

The Geoenvironmental Research Centre is currently undertaking an experimental investigation quantifying the effect of competitive sorption on the mobility of selected heavy metals in London Clay. The investigation uses a laboratory centrifuge to conduct large hydraulic gradient leaching tests on compacted clay, with leachate sampling to determine breakthrough curves. The purpose of this paper is to present initial leaching test results for Cu and Ni, both singly and in combination, demonstrating the impact of competitive sorption on heavy metal mobility in this clay.

BACKGROUND

Sorption of metals is a competitive process between ions in solution and those sorbed on to the soil surface through either electrostatic attraction or surface complexation. As a result, pore water chemistry strongly affects ion retention on soil solids (Echeverría et al., 1998), and therefore mobility, since strongly retained metals ions are removed from the pore water and become less mobile than weakly retained ions. Griffin and Au (1977) found that the amount of lead adsorbed by a calcium montmorillonite was dependent on the ratio of the equilibrium concentrations of the two exchanging cations and not upon the actual concentrations in solution, a result which is consistent with Boyd et al.'s (1947) expression for a competitive Langmuir adsorption equation. Kuo and Baker (1980) reported that sorption of Cu, Cd and Zn by three acid surface soils increased with increasing pH at low pH, Cu being preferentially sorbed over Zn and Cd, such that the presence of Cu interfered strongly with the sorption of the other two heavy metals.

Contaminant mobility is often estimated from sorption characteristics quantified by batch equilibrium experiments on soil suspensions. If adsorption of contaminants by the solid phases in soil is to be modelled, the applicability of terms which describe adsorption from soil suspension studies needs to be examined. Leaching column tests are more representative of natural conditions, but are technically more challenging to run and, for low hydraulic conductivity soils, very time consuming. Comparison between batch equilibrium and leaching column experiments gives mixed results even in low sorption capacity soils (Grolimund et al., 1995; Yong et al., 1992).

Small, laboratory centrifuges and large, geotechnical centrifuges have both become widely used to study the migration patterns of contaminants in soil, for a range of practical and scientific applications (Mitchell, 1998). In particular, centrifuges permit the systematic examination of the relative importance of different mechanisms in contaminant transport under controlled and reproducible testing conditions. Centrifuge leaching tests have shown great promise for the experimental study of reactive contaminant transport processes in fine-grained soils (Celorie et al., 1989; McKinley et al., 1998), because they greatly increase the fluid flow rate through compacted soils using relatively simple apparatus. As Mitchell (1998) points out, a centrifuge leaching test sample is subjected to an effective stress regime similar to that in the field but scaled in length and so has a lower risk of boundary leakage than if the sample were simply subjected to a large hydraulic gradient.

MATERIALS AND METHODS
London Clay was chosen for use in this study. London Clay is a stiff to very stiff, fissured clay with a hydraulic conductivity between 10^{-10} and 10^{-12} m/s (Chandler et al., 1990; Dewhurst et al., 1998). Blocks of the soil were recovered from tunnelling work below the Thames in Westminster, central London. X-ray diffraction analysis indicates that the principal minerals in the recovered samples are quartz, kaolinite, montmorillonite and muscovite. Of these, the montmorillonite will have the biggest influence on the sorption of heavy metals from solution, because it has a high surface area and a high cation exchange capacity. The blocks of clay were then broken up, allowed to dry in air, and crushed to a fine powder.

Three leaching tests on compacted London Clay have been conducted, one using a solution of 500 mg/L Cu as $Cu(NO_3)_2$, one using a solution of 500 mg/L Ni as $Ni(NO_3)_2$, and one using a solution of 500 mg/L Cu and 500 mg/L Ni as their nitrates. Figure 1 shows the centrifuge leaching cell used for this study. The cell has an upper section with a porous base, which retains the soil, and a lower section within which the leachate collects. In use, the cell resembles a falling head permeameter. The centrifuge equipment permits four such leaching cells to be used at once. Dried clay was slurried with deionised water, left overnight and then placed in the leaching cell. The four cells were centrifuged at 5285 gravities until one pore volume was collected, producing a layer of compacted clay on the porous filter. This took approximately two hour, typically. The supernatant fluid was then replaced with leachant solutions and leaching tests were performed at 5285 gravities and 21°C. Periodically, fresh leachant was added to the top of the cell to replace fluid passing through the clay, the collected leachate being removed at the same time. This leachate was acidified with 5% nitric acid for storage, and the heavy metal content determined by ICP-AES. Duplicate tests were conducted to examine variability.

Figure 1. Centrifuge leaching cell

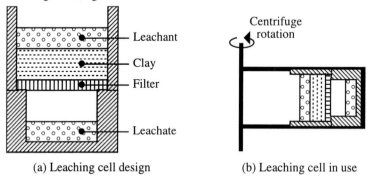

(a) Leaching cell design   (b) Leaching cell in use

Figure 2. Breakthrough curves for Cu, influent 500 mg/L Cu

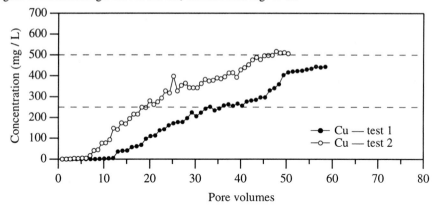

RESULTS AND DISCUSSION

Figures 2 and 3 show the breakthrough curves for the leaching of 500 mg/L Cu and of 500 mg/L Ni through London Clay, respectively. Although both duplicates were mixed from the same sample, prepared in the same apparatus, and tested at the same time under the same conditions, there is still a considerable difference in the variation in effluent concentration with the number of pore volumes. Moreover, each duplicate was tested for the same length of time, so it is clear that the first clay specimen for Cu had a slightly higher hydraulic conductivity than did the second, while the first clay specimen for Ni had a hydraulic conductivity about twice that of the second. Time constraints meant that neither test could be run for significantly longer than the time taken to reach the leachant concentration, marked on the figures by the upper dashed line.

Figure 3. Breakthrough curves for Ni, influent 500 mg/L Ni

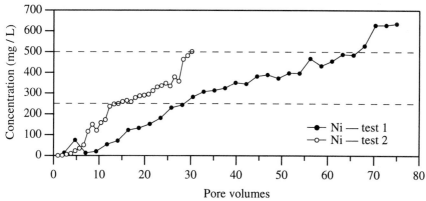

The number of pore volumes required for the leachate concentration to rise to half of the leachant concentration is a measure of the mobility of the contaminant, being approximately equal to the retardation factor (Shackelford and Redmond, 1995). For these tests, this level corresponds to 250 mg/L, which is shown on the graphs by the lower dashed line. For Cu, 20 pore volumes are required in one duplicate and 35 in the other, while for Ni 15 and 28 pore volumes are required. This suggests that Cu sorbs slightly more strongly to the clay than Ni does.

Figure 4. Breakthrough curves for Cu, influent 500 mg/L Cu + 500 mg/L Ni

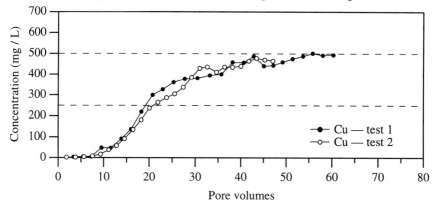

Figures 4 and 5 show the breakthrough curves for Cu and Ni, respectively, when both are present in the leachant at 500 mg/L. There is now excellent agreement between each of the two duplicate tests, with the leachate concentration rising to steady values similar to those in the leachant. Approximately 20 pore volumes are required for the Cu concentration to rise to 250 mg/L, which is equal to the lower of the two values found for Cu on its own. In contrast, approximately 7 pore volumes are required for the Ni concentration to rise to 250 mg/L, compared to 15 and 28 for Ni on its own. This indicates that in the presence of

Ni the mobility of Cu is increased only slightly, if at all, while in the presence of Cu the mobility of Ni is increased by a factor of 2–4. Cu appears to be sorbing preferentially on to the soil, which is consistent with the observation from the single contaminant leaching tests that Cu appears to bond more strongly with the clay than Ni does. Competitive sorption effects are therefore having a significant effect on the migration rate of the less preferred heavy metal, behaviour similar to that noted by Alloway (1997) for surface soils.

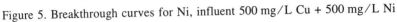

Figure 5. Breakthrough curves for Ni, influent 500 mg/L Cu + 500 mg/L Ni

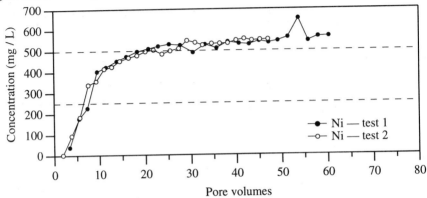

The work reported here is part of an on-going research project to examine Cu, Ni and Zn mobility in London Clay and the effect of competitive sorption. The results here are from the development stages of this project, which may account for the variation between the duplicate tests for each single element. Any conclusions drawn from the results are therefore preliminary, and additional analysis of this data is needed. Tests at lower centrifuge speeds are also planned, providing data on the dependence of mobility on leachant flow rate. The apparent distribution coefficients calculated from the breakthrough will be extrapolated to zero flow rate, and the estimated distribution coefficients compared with those measured in conventional batch sorption tests.

CONCLUSIONS

The rate of migration of Cu and Ni, individually and in combination, through compacted London Clay has been successfully measured using a permeameter style leaching cell in a laboratory centrifuge. The centrifuge technique provides the opportunity to simulate confining stresses similar to those in a natural soil deposit. For the first tests conducted as part of this on-going research project, in which the heavy metals were present in the leachant on their own as nitrates, 20–35 pore volumes of flow were needed for the Cu concentration in the leachate to rise to half of the leachant value, while 15–28 pore volumes were needed for the Ni concentration to rise to this level. In the second test, where both heavy metals were present in the leachant, the number of pore volumes required to reach half of the leachant value was 20 for Cu and 7 for Ni. This indicates that the presence of the preferentially sorbed Cu ion has interfered with the sorption of the Ni ion on to the soil solids and resulted in a two to fourfold increase in the migration rate of Ni.

ACKNOWLEDGEMENTS

This work is sponsored by the Engineering and Physical Sciences Research Council, grant number GR/M27067. This support is gratefully acknowledged.

REFERENCES
Alloway, B.J. (1997). The mobilisation of trace elements in soils. Contaminated soils — 3rd international conference on the biochemistry of trace elements, Prost, R., pp 101–113.

Boyd, G.E., Schubert, J., and Adamson, A.W. (1947). The exchange adsorption of ions from aqueous solutions by organic zeolites: 1. ion exchange equilibria. J. Am. Chem. Soc., Vol 69, No 11, pp 2818–2829.

Celorie, J.A., Vinson, T.S., Woods, S.L., and Istok, J.D. (1989). Modelling solute transport by centrifugation. ASCE J. Env. Engng., Vol 115, No 3, pp 513–527.

Chandler, R.J., Leroueil, S., and Trenter, N.A. (1990). Measurements of the permeability of London Clay using a self-boring pressuremeter. Géotechnique, Vol 40, No 1, pp 113–124.

Dewhurst, D.N., Aplin, A.C., Sarda, J.P., and Yang, Y. (1998). Compaction-driven evolution of porosity and permeability in natural mudstones — an experimental study. J. Geophysical Res. — Solid Earth, Vol 103, No B1, pp 651–661.

Echeverría, J.C., Morera, M.T., Mazkiarán, C., and Garrido, J.J. (1998). Competitive sorption of heavy metals by soils: isotherms and fractional factorial experiments. Environmental Pollution, Vol 101, No 2, pp 275–284.

Griffin, R.A. and Au, A.K. (1977). Lead adsorption by montmorillonite using a competitive Langmuir equation. Soil Sci. Soc. Am. J., Vol 41, No 5, pp 880–882.

Grolimund, D., Borkovec, M., Federer, P., and Sticher, H. (1995). Measurement of sorption isotherms with flow-through reactors. Env. Sci. Tech., Vol 29, No 9, pp 2317–2321.

Kuo, S. and Baker, A.S. (1980). Sorption of copper, zinc and cadmium by some acid soils. Soil Sci. Soc. Am. J., Vol 44, pp 969–974.

McKinley, J.D., Price, B.A., Lynch, R.J., and Schofield, A.N. (1998). Centrifuge modelling of the transport of a pulse of two contaminants through a clay layer. Géotechnique, Vol 48, No 3, pp 421–425.

Mitchell, R.J. (1998). Centrifugation in geoenvironmental practice and design. Can. Geotech. J., Vol 35, No 4, pp 630–640.

Shackelford, C.D. and Redmond, P.L. (1995). Solute breakthrough curves for processed kaolin at low flow rates. ASCE J. Geotech. Engng, Vol 121, No 1, pp 17–32.

Sharma, H.D. and Lewis, S.P. (1994). Waste containment systems, waste stabilization, and landfills. New York: John Wiley and Sons Inc..

Yong, R.N., Mohamed, A.M.O., and Warkentin, B.P. (1992). Principles of contaminant transport in soils. Amsterdam: Elsevier.

Multi-component reactive solutes in partially saturated soil

H. R. THOMAS (1), H. MISSOUM (1), K. BENDANI (2)

(1)Geoenvironmental Research Centre, Cardiff School of Engineering, Cardiff University, Newport Road, Cardiff CF2 1XH.
(2)University of Mostaganem, Dept. of Civil Engineering, BP :227, ALGERIA

ABSTRACT

The objective of this work is to describe a model incorporating multicomponent equilibrium chemistry into multiphase transport in partially saturated soil. This process includes the effect of temperature and chemical osmosis in the transport phenomena. The algebraic equations for the multicomponent reactive solute set for chemical equilibrium will contain the phase exchange necessary to describe significant chemical processes affecting transport. These chemical reactions, such as surface or aqueous complexation, ion exchange, mineral precipitation and dissolution and acid/base or redox are implemented in the model. The model obtained is described by two sets of equations ie: partial differential equations for multiphase flow and the algebraic equations for equilibrium chemistry. These will be solved by an iterative procedure.

INTRODUCTION

As contaminants are usually chemically active, it is necessary to include chemical reactions in the study of contaminant transport in soils. The interaction between chemical species and transport, within a geothermal system may be very important in describing the state of the system.

Simulation models that couple the geochemical processes to the physical transport phenomena under realistic boundary conditions can be effective for remediation techniques. Modelling multicomponent reactive transport requires the solution of equations that simulate the complex relationship between solute transport and the various chemical reactions, such as acid/base reactions, reduction and oxidation, complexation, ion exchange, precipitation and dissolution of minerals. The chemical equations depend on local conditions at each point of the system. All these processes of solute transport must satisfy the principle of conservation of mass and energy, and the principle of conservation of charge. The simulation of multiple reacting chemical substances are governed by the equilibrium equations of thermodynamics.
The two related but distinct problems, multiphase transport versus multicomponent reactive solute behaviour, are strongly coupled.

The objective of this work is to develop and to describe a model incorporating multicomponent equilibrium chemistry into multiphase transport in partly saturated soil, including, the effect of temperature, chemical osmosis and air pressure. A two step iterative procedure is adopted to solve the two sets of equations. The fully coupled non-linear differential equations of transport are discretised via a Galerkin weighted residual approach in the space domain and an implicit integrating scheme in the time domain. The non-linear algebraic chemical equations are solved by a Newton-Raphson iterative method. A solution for all variables is obtained after convergence of the computations between the two sets of equations.

Geoenvironmental engineering, Thomas Telford, London, 1999, 302–309

MODEL FORMULATION

The equations presented here, to describe nonisothermal multiphase flow together with the transport of a reacting chemical species system in partly saturated soils, are limited to the following assumptions :

a)   The soil is considered as a multiphase porous media composed of solid, liquid, and gas phases . The latter phase is a binary mixture of water vapour and dry air.
b)   Flow in the liquid phase depends on the gradients of matric potential, temperature and salt concentration
c)   The reacting chemicals may be aqueous or solid phases.
d)   Darcy's law is used to govern the flow of dry air and water, whereas Philip and de Vries type vapour flow is used to govern the flow of vapour.
e)   All phases are in local equilibrium.
f)   The soil matrix is rigid ie: no deformation is taken into account.

$$V_w^l = -K_w \left(\nabla \left(\frac{u_m}{\gamma_w} + \nabla z \right) \right) \tag{1}$$

GOVERNING EQUATIONS FOR TRANSPORT

The motion of pore water in partially saturated soils depends on soil type, the water content and the solute in the pore water. The pore water velocity due to the hydraulic gradient can be expressed by Darcy's law as:

where K_w is the hydraulic conductivity, u_m is the matric suction, γ_w is the weight of water and z is the elevation.

Osmosis under chemical solute concentration gradients may be important and it will drive water to flow in the direction of increasing solute concentration. This part of the contribution to water flow can be written as (Mitchell, 1976):

$$V_w^c = -K_w^c (\nabla u_o) \tag{2}$$

where K_w^c is the permeability with respect to chemical solute concentration. u_o refers to the osmotic potential.

The osmotic potential can be expressed by Van't Hoff law as (Yong et al., 1975):

$$u_o = -RT\sum C_i f_i \tag{3}$$

where T is the temperature, R is the specific gas constant, C_i is the concentration of solute i and f_i is the coefficient of osmotic efficiency.

Philip and de Vries' (1957) approach concerning vapour flow is adopted, so the velocity of water vapour can be written as :

$$V_v = D_v \nabla \rho_v \qquad (4)$$

where D_v is the coefficient of water vapour diffusion and ρ_v is the water vapour density.

By applying the mass conservation principle to the moisture phase, the following governing differential equation is obtained:

$$\frac{\partial(\rho_1 \theta_1)}{\partial t} + \frac{\partial(\rho_v \theta_g)}{\partial t} + \nabla.(\rho_1 V_w^l) + \nabla.(\rho_1 V_w^c) + \nabla.(\rho_1 V_v) + \nabla.(\rho_v V_g) = 0 \qquad (5)$$

θ_l and θ_g are respectively the volumetric water and gas contents, ρ is the density. The subscripts *l* and *v* refer to liquid and vapour respectively. t is the time.

The velocity of air flow can be described by Darcy's law as :

$$V_a = -K_a (\nabla u_a) \qquad (6)$$

where K_a is the pore air permeability of the soil and u_a is the pore air potential.

The equation of mass conservation of pore dry air can be written as:

$$\frac{\partial(\rho_{da}(\theta_a + H\theta_1))}{\partial t} + \nabla.\left[\rho_{da}\left(V_a + H(V_w^l + V_w^c)\right)\right] = 0 \qquad (7)$$

ρ_{da} is the density of dry air and H is the volumetric coefficient of solubility of air in water.

The principle of energy conservation gives :

$$\frac{\partial \phi}{\partial t} + \nabla.Q = 0 \qquad (8)$$

where Q is the total heat flux and ϕ is the heat content of the soil, defined respectively as:

$$Q = (C_{pl}\rho_1(V_w^l + V_w^c) + C_{pv}\rho_1 V_v + C_{pv}\rho_v V_g + C_{pa}\rho_a V_g)(T - T_r) - \lambda_T \nabla T + (\rho_1 V_v + \rho_v V_g)L \qquad (9)$$

$$\phi = H_c(T - T_r) + \rho_v \theta_g L \qquad (10)$$

H_c is the specific heat capacity of the soil, T is the reference temperature , C_{pl}, C_{pv} and C_{pa} are the specific heat capacity of soil water, soil vapour and dry air respectively, λ_T is the coefficient of thermal conductivity of the soil and L is the latent heat of vaporization of the soil water.

The partial differential equation governing advective-dispersive chemical transport under water flow conditions in partially saturated soil using the mass conservation principle can be expressed as :

$$\frac{\partial}{\partial t}\left[\theta_i T_i + (1-n)\rho_s (S_i + P_i)\right] = \nabla.\left(DnS_i \nabla C_i - (V_w^l + V_w^c)C_i\right) + \Gamma \tag{11}$$

$$i = 1,2,\ldots\ldots\ldots, N_a$$

where T_i is the total analytical concentration, C_i is the total dissolved concentration of the aqueous component i, S_i is the total sorbed concentration of the aqueous component, P_i is the total precipitated concentration of the aqueous component i, ρ_s is the density of the soil grain, Γ is the production or removal term and N_a is the number of the aqueous component.

The coefficient of hydrodynamic dispersion D accounts for the various transport-controlled processes which include dispersion and diffusion transport of contaminants and can be defined as (Bear and Verruijt 1987) :

$$D = D_d + D_m \tag{12}$$

where D_d and D_m are respectively the mechanical dispersion and molecular diffusion coefficients.

CHEMICAL REACTIONS

The processes of surface complexation and ion exchange are usually accounted for by linear or nonlinear Freundlich isotherms, where all reactions between solid and liquid phases are simulated by the distribution coefficient. Other processes such as precipitation/dissolution of minerals are accounted by a source or sink term.

In this section, a general formulation of various chemical reactions such as acid/base and redox reactions, surface and aqueous complexation, ion exchange, and precipitation/dissolution of minerals in equilibrium is presented.

Redox and acid/base reactions
The oxidation and reduction potential can be specified by the activities of electroactive species through the Nernst equation as:

$$E_h = E_0 + \frac{RT}{nF}\ln\left(\frac{\chi_{red}}{\chi_{ox}}\right) \tag{13}$$

where E_h is the solution potential relative to the standard hydrogen electrode, E_h is the standard potential for the reaction, n is the number of electron transferred, F is the Faraday constant, and χ_{red} and χ_{ox} are the activities of the reduced and oxidized species respectively.

Chemical reactions (acid/base reactions) involving hydrogen and hydroxide can be represented by the concentration of H^+ or OH^- as a primary variable in the soil solution.

Surface or aqueous complexation
Complexation reaction can be written as :

$$A + B \longrightarrow AB$$

where A and B are the reactants and AB is the product.
The law of mass action can be expressed by :

$$K_{AB}^a = \frac{[A][B]}{[AB]}$$ (14)

where K_{AB}^a is the equilibrium constant. The square brackets represent the solution activity 0.

Ion exchange
Ion exchange reaction can be written as:

$$n_B A^{nA} + n_A B(ad) \leftrightarrow n_B A(ad) + n_A B^{nB}$$

where nA and nB represent the ionic charges. Consequently the stoichiometric coefficients of the exchanging species, and A(ad) and B(ad) are the adsorbed masses of components A and B respectively. The mass action equation for the above expression is ;

$$K_{AB} = \left(\frac{A(ad)}{[A]^{nA}} \right)^{nB} \left(\frac{[B]^{nB}}{B(ad)} \right)^{nA}$$ (15)

where the square brackets represent solution activity. K_{AB} is the selectivity coefficient of species A with respect to species B.

Mineral precipitation and dissolution
The general form of precipitation/dissolution reaction of a solid is :

$$A_a B_b (s) \leftrightarrow aA(aq) + bB(aq)$$

where (aq) refers to the aqueous phase and (s) to the solid phase. This reaction is described by a solubility product of the form:

$$K^p = \frac{[A]^a [B]^b}{[A_a B_b]}$$ (16)

The square brackets again refer to solution activity. For pure solid, the solution activity equals unity. The solubility product is an equilibrium constant specific to a solid.

CHEMICAL EQUILIBRIUM EQUATIONS

The total dissolved concentration, total sorbed concentration, total precipitated concentration, and total analytical concentration are obtained respectively from the mass balance equations of chemicals as follows:

$$C_k = \sum_{j=1}^{N_a} \alpha_{kj}^c c_j \qquad k = 1,2,\ldots\ldots,I \tag{17}$$

$$P_k = \sum_{j=1}^{N_p} \alpha_{kj}^p p_j \qquad k = 1,2,\ldots\ldots,I \tag{18}$$

$$S_k = \sum_{j=1}^{N_s} \alpha_{kj}^s s_j \qquad k = 1,2,\ldots\ldots,I \tag{19}$$

$$T_k = \sum_{j=1}^{N_a} \alpha_{kj}^a c_j + \sum_{j=1}^{N_s} \alpha_{kj}^x s_j + \sum_{j=1}^{N_p} \alpha_{kj}^p p_j \qquad k = 1,2,\ldots\ldots,I \tag{20}$$

where c_j is the concentration of species j in the aqueous phase, s_j is the concentration of sorbed species j in the solid phase and p_j is the precipitated species in the aqueous phase.

α_{kj}^c, α_{kj}^s and α_{kj}^p are respectively the stoichiometric coefficient of the kth aqueous component in the jth aqueous species, jth sorbed species, and jth precipitated species. N_a, N_p and N_s represent the number of aqueous, precipitated and sorbed species respectively.

The concentration of dissolved species j can be obtained from the complexation reaction equations as:

$$c_j = \left(\frac{K_j^a \prod_{k=1}^{I} \gamma_{ak}^{\alpha_{kj}^a}}{\gamma_{aj}} \right) \prod_{k=1}^{I} c_k^{\alpha_{kj}^a} \qquad j = I+1, I+2, \ldots\ldots, N_a \tag{21}$$

where K_j^a is the equilibrium constant of jth dissolved species and γ_{aj} is the activity coefficient of jth dissolved species.

The concentration of jth sorbed species can be obtained from the ion exchange or sorption and charge balance equations as:

$$s_j = K_{jJ}^{1/Z_j} \left(\frac{\gamma_{sJ} Z_J s_J}{N_{eq} \gamma_{aJ} c_J} \right)^{Z_j/Z_J} \left(\frac{N_{eq} \gamma_a c_j}{\gamma_{sj} Z_j} \right) \qquad j = 1,2,\ldots\ldots,N_s \tag{22}$$

$$\sum_{j=1}^{N_s} Z_j s_j = N_{eq} \tag{23}$$

where K_{jJ} is the selectivity coefficient of the *jth* species with respect to the *Jth* species, or the effective equilibrium constant of *jth* sorbed species. Z_j is the valence of the *jth* species, γ_{sj} is the activity coefficient of the *jth* sorbed species and N_{eq} is the number of equivalents of surface sites available for ion exchange reaction.

The solubility product can be obtained from the precipitation/dissolution equations as :

$$(\prod_{k=1}^{I} \gamma_{ak}^{a_{kj}^{p}})\prod_{k=1}^{I} c_{k}^{a_{kj}^{p}} = K_{j}^{p} \quad i = 1,2,\ldots\ldots\ldots,N_{p} \tag{24}$$

where K_j^p is the the solubility product of the *jth* precipitated species.

MODEL PARAMETERS
The extended version of the Debye-Huckel equation for the determination of the activity coefficient of any chemical species is expressed as (Truesdell and Jones, 1974).

$$\ln \gamma = -\frac{Az^2\sqrt{I}}{1+Ba\sqrt{I}} + bI \tag{25}$$

A and B are constants depending only on the dielectric constant, density and temperature, z is the ionic charge, a and b are two adjustable parameters and I is the ionic strength. The above expressions can be used for moderately saline solution.

The ionic strength can be defined as:

$$I = 0.5\sum_{i=1}^{M} z_i^2 m_i \tag{26}$$

where M is the number of species in the solution mixture and m_i is the molality.

The thermodynamics equilibrium constants depend on the temperature and pore air pressure (Simunek and Suarez, 1994). When the air pressure is atmospheric, the thermodynamic equilibrium constant can be written as:

$$\log K = a_1 + \frac{a_2}{T} + a_3 T + a_4 \log T + a_5 T^2 \tag{27}$$

where a_1, a_2, a_3, a_4 *and* a_5 are empirical constants

NUMERICAL IMPLEMENTATION
All formulations require the solution of a set of nonlinear equations. Two families of approaches for numerical solution are proposed in the literature : the Sequential Iteration approach (SIA) also called the two-step approach (Simunek and Suarez, 1994) and (Yeh et al., 1996), and the Direct Substitution Approach (DSA) (Yeh and Tripathi, 1989). The first method is adopted here; a solution of the transport and chemical equations is achieved in an iterative way. Concentrations of sorbed and precipitated species in the transport equations are treated as a source or sink term, which are updated after each iteration by means of the chemical equations.

A Galerkin finite element method associated with a finite difference recurrence relationship is used to obtain simultaneous solutions to the non-linear differential equations for the transport equations. The Newton-Raphson method is then used to solve the non-linear algebraic equations for chemical equations.

CONCLUSIONS

The work presented describes a model for the analysis of the coupled transport of heat, moisture, air and multicomponent reactive solute in partly saturated soil, taking into account the effect of osmotic potential. The solution of the coupled multiphase chemical transport problem based on a chemical system in equilibrium, the chemical reactions and transport processes is described. A sequential iteration approach, to solve sets of nonlinear partial differential equations and nonlinear algebraic equations is adopted. Such a model can be extended to deal with mixed chemical kinetics and equilibrium chemistry.

REFERENCES

Bear J. and Verruijt A. 1987. Modeling groundwater flow and pollution, Dordrecht, D. Reidel Publishing Company.

Mitchell, J.K. 1976. Fundamentals of soil behaviour, New York : John Wiley & sons, Inc.

Philip, J.R. and de Vries D.A. 1957. Moisture movement in porous materials under temperature gradients. Trans. Am. Geophys. Un., 38, 222-232

Simunek, G., and Suarez, D.L. 1994. Two dimensional transport model for variably saturated porous media with major ion chemistry. Water Resour. Res., 30(4), 1115-1133

Truesdell, A.H. and Jones, B.F. 1974. Wateq, a computer program calculating chemical equilibria of natural waters. J. Res. Un.Geol. Sur., 2(2), 233-248.

Yeh,G.T. and Tripathi V.S. 1989. A critical evaluation of recent developments in hydrochemical transport models of reactive multi-chemical components. Water Resour. Res., 25(1), 93-108

Yeh, G.T., Salvage, K. and Choi, W. 1996. Reactive chemical transport controlled by both equilibrium and kinetic reactions, Computations Methods in Water Resources, vol. 11, edited by A.A. Aldama et al., pp. 585-592, Comput. Mech., Billerica, Mass.

Yong, R.N. and Warkentin, B.P. 1975. Soil properties and behaviour, New York, American Elsevier Publishing Company.

Transport of Lead in Partially Saturated soil

H. R. THOMAS, R. N. YONG AND A. A. HASHM
Geoenvironmental Research Centre, Cardiff School of Engineering,
Cardiff University, Newport Road, Cardiff, CF2 1XH, Wales

ABSTRACT
This paper presents an investigation into the transport of Lead in a profile of unsaturated soil with the aim of providing an understanding of aspects of both the transport process and the fate of the contaminant. A series of leaching experiments were conducted under different moisture contents, lead concentrations and soil pH. Other experiments were carried out to determine the main physical and chemical characteristics of the soil. The moisture content, the lead concentration and the soil pH were found to have an effect on the hydraulic conductivity of the soil. Lowering the pH of the soil was found to reduce the retention capacity of the material.

INTRODUCTION
The need for studies of contaminant transport in soil arises from concerns regarding environmental protection. The potential danger of contamination of the soil and groundwater system, might in some cases have serious consequences for humans, plants and animals. Much effort has been directed toward a better understanding of contaminant migration and the factors affecting the process (Mangold and Tsang, 1991).

Several factors and processes combine to control the advance or growth of a contaminant plume. These include, in addition to the hydrology of the system the multitude of complex interactions and reactions which occur between the contaminants and the soil-water system. Studies on various sets of interactions between selected contaminants and the soil may shed light on the inter-relationships between the two kinds of participants (Yong et al., 1992).

An investigation of contaminant transport mechanisms in partially saturated soils is necessary to achieve an improved understanding of pollutant migration in the vadose zone. This can differ significantly from fully saturated behaviour. This is partly because the forces of contaminant-soil interaction are more dominant transport control processes, because of the smaller amount of water present within the system. Also current field and theoretical investigations have shown that spatial variability in the physical parameters is significant for field scale flow and solute movement through the unsaturated zone. However much less effort has been directed toward unsaturated soils.

The presence of lands that are contaminated with heavy metals as a result of mineral resources exploitation, metal mining and smelting processes is well established. Heavy metals are also commonly found in several kinds of wastes including sludge and landfill

Geoenvironmental engineering, Thomas Telford, London, 1999, 310–317

(Municipal and industrial waste) leachates. Heavy metals are highly toxic to humans, animals, and aquatic life (Yong et al, 1992). They suggested that the concentration of heavy metals may range from 0-100 ppm in municipal solid wastes to 100-10,000 ppm in sewage sludge, mining wastes, and various industrial wastes originating from the electroplating, pulp and paper, and chemical industries. The retention of heavy metals is governed by a complex collection of processes that depend on the properties of both the soil and the leaching solution. These can vary significantly with changes in the properties.

MacDonald and Yong (1997) reported that removal of heavy metals from soil pore water occurs by mechanisms such as adsorption, complexation , precipitation as Oxides, hydroxides, carbonates, and filtration. However sediments and soils are neither permanent nor irreversible sinks for contaminants and may become sources of contamination to surface water and ground water when changes in the local environment occur. Farrah and Pickering (1978) identified increased soil acidity as one of the possible mechanisms that may mobilise heavy metals in soils. At high pH levels, the aqueous metal cations hydrolyse. This hydrolysis results in precipitation of metal hydroxides onto soils, which is difficult to distinguish from removal of metals from solution by adsorption (Yong et al., 1992).

This study is directed towards an investigation of Lead transport in a profile of partially saturated soil, with the aim of providing an understanding of aspects of contaminant transport and fate. This is achieved by conducting a series of leaching experiments designed to investigate experimentally the process of transport in terms of migration and retention of Lead. Tests at different degrees of saturation, concentrations, and soil pH were performed under isothermal conditions. The leaching tests were conducted in horizontal leaching cells under zero pressure head, so gravitational effects were minimal. Flow occurred by suction effects alone.

Experimental work has also been carried out to determine the physical and chemical properties of the soil.

MATERIALS AND METHODS

The soil used in this research work was obtained by mixing a group of different clayey soils with sand. The group of clayey soils used were originally collected from landfill sites in South Wales. Their general properties, attenuation characteristics and use as landfill liners were assessed and reported by Yong et al., (1997). The sand used is a well graded washed sand. Sand was added to improve the permeability of the soil and hence increase the flow rate. The presence of the clay ensured that the soil had certain chemical retention capacity.

In order to study the effect of the soil pH on the transport process, as a single parameter without the interference of other factors, the same soil had to be used with different pH values. To achieve this, it was decided to wash the original soil with 10% Nitric acid several times over a long period of time to ensure complete soil pH equilibrium. Lower pH values of 4.5 and 3.5 were chosen to investigate the suggestion, that at pH values ranging between 1 and 4, the ability of a soil to adsorb and retain contaminants deteriorates since dissolution effects are dominant (Yong et al., 1992). The soil was air dried and ground to pass through a 2-mm sieve.

Soil	Original soil	T.S 1[*]	T.S 2[*]
Specific Gravity	2.54	2.64	2.65
Soil pH	7.8	4.5	3.5
Soil EC (μ S)	1102	269	492
Liquid Limit%	40.5	36.6	36.5
Plastic Limit%	21.7	18.8	19.6
Gravel%	0.0	0.0	0.0
Sand%	63.2	65.6	65.2
Silt%	21.9	25.2	23.8
Clay%	14.9	9.2	11.0
Compaction density Mg/m^3	1.69	1.76	1.76
Organic Contents by weight%	2.30	2.34	2.41
Carbonate contents by weight%	4.0	1.0	0.75
Specific Surface Area m^2/g	67.04	90.74	88.13
Cation Exchange Capacity meq/100gm	18.99	3.64	2.53

Table 1. Chemical and physical properties of the soil.
T.S 1,2 Original soil treated with acid for modification of pH.

A series of physical and chemical analyses were carried out to determine the properties of the soil, as summarised in table 1. Measurement of specific gravity, water content , atterberg limits, optimum moisture content/maximum dry density relationship, particle size distribution, soil pH, carbonate content and organic content were performed according to BS1377:1990. The cation exchange capacity was determined using the ammonium acetate exchange method (Bache, 1976). The specific surface area was measured using the ethylene glycolmonoethyl ether (EGME) adsorption method (Carter et al, 1965). The total heavy metal concentration was determined using the XRF method with calibration by the acid digestion method (Williams, 1997). The concentration of heavy metals in solution was determined using Induced Couple Plasma Emission Spectroscopy.

TEST PROCEDURE
Soil leaching tests study both adsorption and diffusion of contaminants through the soil profile. In leaching tests, the leachate is passed through the soil sample as in permeability tests. Because of the soil structure, pore geometry and pore continuity, only a fraction of the total surface area of the soil particles comes in direct contact with the permeating leachate(Yong et al., 1992).

A total of nine horizontal leaching cells were designed to simulate flow of leachates through unsaturated soil. Details of the testing schedule are presented in Table 2.

Experiment No.	Soil pH	Initial degree of Saturation	Leachate	Concentration ppm
1		0.2		
2	7.8	0.4	Water	
3		0.6		
4		0.2		
5	7.8	0.4	Lead	500
6		0.6		
7	7.8	0.2	Lead	1000
8	4.5	0.2	Lead	500
9	3.5	0.2	Lead	500

Table 2 Testing Schedule

Before testing, the dry soil was mixed with different volumes of distilled water to achieve three different degrees of saturation of 0.2, 0.4 and 0.6. The material was then placed in environmentally controlled conditions for at least 24 hours to prevent moisture losses and to allow for uniform moisture distribution. The soil was then compacted directly into the leaching cells to the desired density. Each cell was then leached with the required solution (pH of the lead solutions =1.5), at constant atmospheric pressure and under zero head. Tests were conducted at room temperature until the cell reached full saturation. Readings at regular intervals were taken daily for the volume of the solution transported in the soil. The system was checked regularly for trapped bubbles that had been released from the soil voids when filled with water. On completion of the test, the soil was removed and sectioned horizontally into 15 slices. Each slice was then tested for the total and the soluble concentrations of Lead.

RESULTS AND DISCUSSION
Table 1 present the general physical and chemical properties of the soil before and after the washing process. It can be seen that the process used to change the pH value of the soil inevitably caused some changes in other properties. The results show reduced carbonate content in the washed soil samples. This is to be expected, as washing with acid obviously dissolves carbonates. The ammonium acetate exchange test shows that the CEC of the original soil sample is in the range of the CEC values of the clay minerals present in the soil samples. However, it can be seen that the value of this parameter is lower in soil samples with lower pH. This might be explained by the acid treatment which might cause protonation of the functional groups on the surfaces and the edges of the mineral particles and on the hydrous oxides, hence reducing the net negative charge. This process will reduce the exchange capacity of the soil.

Variations of the flow rate are plotted in Figure 1 in terms of volume of water passed through the soil as a function of time for various initial degrees of saturation. Water moves along the soil columns due to the internal suction gradient from a location of high energy to low energy. Results show that the flow rate decreases with time, as expected. Furthermore, it also varies with the soil's initial degree of saturation. It can be seen that in the early stage of the experiment the flow rate in the sample having an initial degree of saturation of 0.6 is higher than in that with an initial degree of saturation of 0.4. Similarly both have higher flow rates than in the soil with an initial degree of saturation of 0.2, at the early stages of the leaching processes. This pattern changes at later stages of the leaching process where the flow rate become higher in the samples of lower degree of saturation. These changeable patterns of flow can be explained by the fact that according to Darcy's Law there are two factors which can determine the flow rates. These are the suction gradient of the soil and the unsaturated hydraulic conductivity. Changes in the flow pattern are dictated by the change in the contribution of each of these factors at different degrees of saturation.

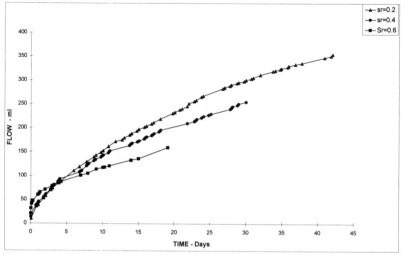

Fig. 1. Flow rate for different initial degrees of saturation (soil pH: 7.8, leachate: water)

Figure 2 illustrate the results of leaching the soil with two different concentrations of lead and with water, holding all other parameters of the soil constant. The figure shows that the slowest flow rate occurs when the was soil leached with water, the lower concentration of 500 ppm yielded a faster flow rate, with the fastest achieved with the higher concentration of 1000 ppm of lead. This can be explained by the effect of the chemical in the solution on the effective porosity of the soil. This behaviour was noticed by Hird et al., (1997) through numerous studies into the behaviour of clay soils. They reported that the interaction between the component clay particles and the pore fluid has a significant influence on overall permeability.

It has been suggested (Hird et al., 1997) that a change in pore fluid chemistry may lead to the osmotically induced shrinkage of the diffuse double layers and hence of the aggregates of clay particles. This would result in the opening of macropores, leading to an increase in permeability. However in these experiments this effect would only apply to the first two slices of the soil where most of the lead is sorbed. Therefore other effects needs to be considered. An analysis of the soil slices for other cations, after the leaching process, showed an increase in the concentrations of Magnesium and Calcium in comparison with their initial concentrations in the soil. This increase might be explained by the exchange of Lead with these cations resulting in their release to the bulk solutions. This in turn obviously increases their concentration. Changes in pore fluid chemistry, arising in this way might again cause osmotically induced shrinkage of the diffuse double layer, leading to increased permeability.

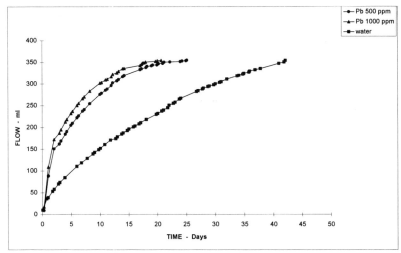

Fig. 2. Flow rate at different concentrations (soil pH: 7.8)

Figure 3 demonstrate the effect of leaching a solution of 500 ppm Lead in soil samples that have different pH when all other characteristics of the leaching process are kept constant. The figure shows that the flow rate increases with decreases of the soil pH. These results are compatible with those presented by Yong et al (1992), who reported that inorganic acids may solubilise some constituents that comprise the clay soil structure. Acid has been reported to solubilise aluminium, iron, alkali metals and alkaline earth (Grim, 1953). Since clay minerals contain Alumina in large quantities, they are susceptible to partial dissolution by acids.

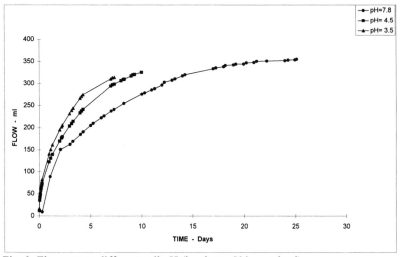

Fig. 3. Flow rate at different soil pH (leachate: 500 ppm lead)

The retention profile of lead within the soil columns as a function of soil pH is illustrated in figure 4. Despite the high percentage of sand in the soil, it can be seen that the soil had a good retention capacity.

Since heavy metals are highly soluble in acidic conditions, very little or no retention is expected at low soil pH (Yong et al., 19992). The pH of the soil was found to have an effect on the retention capacity of the soil. A decrease in pH values results in a reduction in the retention capacity. However, the results showed that the retention capacity of low pH soil was still relatively high. For a 500 ppm input concentration , nearly 100% of the Lead was retained in the first section of the soil, when the pH was 7.8. When the soil pH was 4.5 , 92% of the Lead was retained in the first section. 8% migrated and was retained in the second section. At a pH of 3.5 the first section of the soil retained 80%, with the remaining 20% migrating to be retained in the second section.

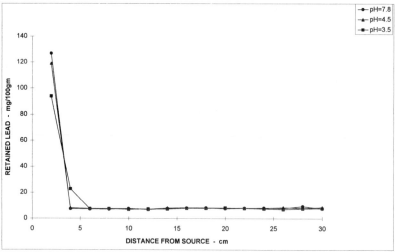

Fig 4. Amount of Lead retained as a function of the soil pH (leachate: 500ppm lead)

CONCLUSIONS
Sorption and migration behaviour of Lead was examined under different degrees of saturation, concentrations and soil pH. The initial degree of saturation was found to have an effect on the flow rate. The presence of Lead, in increased concentration in soil was found to increase the flow rate in the soil sample. Leaching with lead was also found to have an effect on the redistribution of other chemicals, for example, Calcium and Magnesium. Reducing the soil pH was also found to increase the flow rate in the samples. On examination of the retention capacities of the soil, it was found that despite the high percentage of sand in the soil, the material had a good retention capacity. The pH of the soil was found to have an effect on the retention capacity causing a reduction with lower pH values. However the results showed that the retention capacity remained high.

ACKNOWLEDGEMENT
This research was partly funded by an EPSRC Total Technology Award to the third author. The financial support received is gratefully acknowledged.

REFERENCES

Bache, B. W. (1976). The measurement of cation exchange capacity of soils. J. Sci. Food Agric. 27:273-280.

Carter D., Heilman T. and J. Gonzalez. (1965). Ethylene Glycol Monoethyl Ether for determining surface area of silicate minerals. Soil Sci. J. March, 356-361.

Farrah, H., and W. F. Pickering. (1978). Extraction of heavy metal ions sorbed on clays. Water Air Soil Pollut. 9:491-498

Grim, R. E. (1953). Clay mineralogy. McGraw Hill. Inc. New York, 296p.

Hird, C. C., Smith, C. C. and V. Prakash (1997). The influence of physicochemical stresses on permeability and volume change behaviour of bentonite. Proc. 1st Int. BGS Conf. on Geoenvironmental Engineering. organised by the Geoenvironmental Research Centre, Cardiff School of Engineering, Cardiff University. Cardiff (Ed. R. N. Yong and H. R. Thomas), 331-336, Thomas Telford

Mangold, D. C. and C. F. Tsang. (1991). A summary of subsurface hydrological and hydrochemical models. Review of Geophysics, 29(1), 51-79.

Yong, R. N., Mohammed A. M. O. and B. P. Warkentin. (1992). Principles of contaminant transport in soils. Elsevier Pub. Amsterdam.

Yong, R. N., Tan, B. K., Bentley, S. P., Thomas, H. R., Williams, K. P., Pooly, F. D. and W Zuhairi. (1997). Attenuation characteristics of natural clay materials in South Wales and their use as landfill liners. Proc. 1st Int. BGS Conf. on Geoenvironmental Engineering. organised by the Geoenvironmental Research Centre, Cardiff School of Engineering, Cardiff University. Cardiff (Ed. R. N. Yong and H. R. Thomas), 331-336, Thomas Telford

Macdonald, E. M. and R. N. Yong. (1997).On the retention of lead by illitic soil fractions. Proc. 1st Int. BGS Conf. on Geoenvironmental Engineering. organised by the Geoenvironmental Research Centre, Cardiff School of Engineering, Cardiff University. Cardiff (Ed. R. N. Yong and H. R. Thomas), 331-336, Thomas Telford

Williams, K. P. -Private communication- Testing protocol applied in the Environmental Engineering Laboratory.

The influence of capillarity on multiphase flow within porous media. – A new model for interpreting fluid levels in groundwater monitoring wells in dynamic aquifers.

DR.-ING. MATTHIAS VOGLER, Ingenieursozietät Prof. Dr.-Ing. Katzenbach und Dipl.-Ing. Quick,
PROF. DR.-ING. ULVI ARSLAN and PROF. DR.-ING. ROLF KATZENBACH, Darmstadt University of Technology, Institute of Geotechnics, Germany

INTRODUCTION

The ground and groundwater are contaminated by different contaminants and groups of contaminants. Due to the large number of spills with light non aqueous phase liquids (LNAPL) the knowledge of the behaviour of these spills in the environment is very important. LNAPL's consist of different hydrocarbons called oil in the following as oil which are immiscible with the water phase and soil air phase. The multiphase flow of LNAPL's is strongly influenced by the porous media's capillary properties.

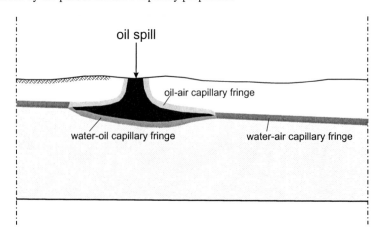

Figure 1. Oil spill

CAPILLARITY IN MICROSCOPIC SCALE

The capillary rise is defined by the equilibrium of the capillary pressure and the gravity pressure of the fluid column in a capillary. The capillary rise is reversely proportional to the diameter of the capillary. The phase saturation of the different phases is defined by the ratio of the phase volume to the pore volume in the control volume. The wettability is rising from water to oil and to air. Therefore water is always the wetting phase and soil air the non-wetting phase. Oil is the non-wetting phase to water and the wetting phase to soil air. In porous media the water phase is wetting the surfaces of the media. The soil air phase is in the middle

Geoenvironmental engineering, Thomas Telford, London, 1999, 318–325

of the larger pores and the oil phase is between the water and the soil air phase. The capillary pressure-saturation-relation is hysteretic due to the irregular pore geometry and due to the effect of the trapped non-wetting phase. When a capillary tube with varying diameter, a so-called Jamin-tube, is wetted by a rising water table the capillary rise is limited by the larger diameter of the Jamin-tube (fig. 2). In a drainage process with a falling water table the menisci of the water remain higher at the position of a narrow diameter (Taylor, 1948). The entrapment of the non-wetting phase can be shown by mean of the double capillary model (fig. 3). The left capillary tube has a small diameter, the right capillary tube has a larger diameter and the capillary tube at the top has a medium diameter (Moore and Slobod, 1956). Due to the falling of the wetting phase table from the fully saturated state (fig. 3a) the wetting phase will fall down according to the diameter of the capillary tubes. The right tube will drain more than the left one (fig. 3b). If the wetting phase table is rising from this state the wetting phase will flow from the left tube into the tube at the top. This leads to the entrapment of the non-wetting phase in the capillary tube at the right side.

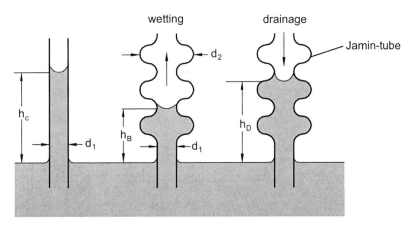

Figure 2. Capillary hysteresis due to irregular pore geometry (Jamin-tube)

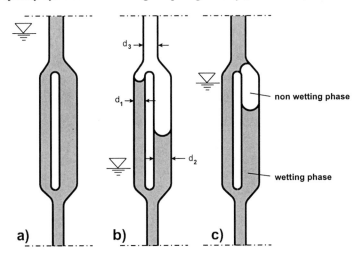

Figure 3. Entrapment of non wetting phase (double capillary model)

CAPILLARITY IN MACROSCOPIC SCALE

In a macroscopic scale it is necessary to include the observations from the microscopic scale in the macroscopic model. The capillary properties in a macroscopic scale are described by the capillary pressure-saturation relations and the relative permeability-saturation relations of the 3 2-phase systems oil-air, oil-water, water-air and the relative permeability-saturation relation in the 3-phase system water-oil-air. The capillary pressure-saturation relations and the relative permeability-saturation relation are hysteretic relations (Lenhard et al.,1991). The hysteresis is caused by the two effects observed at a mircoscopic scale, the irregular pore geometry and the entrapment of the non-wetting phase. The following characteristic curves of the capillary pressure-saturation relations (fig. 4) and the relative permeability-saturation curves for the wetting (fig. 5) and the non-wetting phase (fig. 6) can be distinguished:

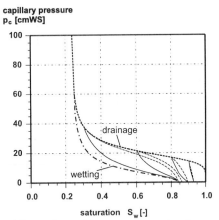

- PDC   Primary Drainage Curve

- MDC   Main Drainage Curve

- MWC   Main Wetting Curve

- SDC   Scanning Drainage Curve

- SWC   Scanning Wetting Curve

Figure 4.   capillary pressure-saturation relation

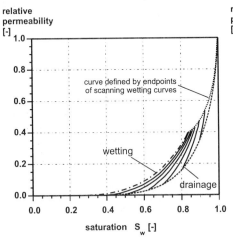

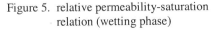

Figure 5.   relative permeability-saturation relation (wetting phase)

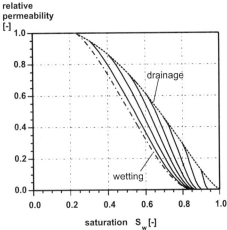

Figure 6.   relative permeability-saturation relation (non wetting phase)

This characteristic curves are transferable to any 2-phase system. The amount of phase entrapment is dependent on the maximum saturation of the non-wetting phase prior to the entrapment (Land, 1968).

ESTIMATION OF FREE OIL VOLUME FROM FLUID LEVELS IN MONITORING WELLS

Due to the absence of capillary effects in monitoring wells it is impossible to evaluate the distribution of oil in the subsoil surrounding a monitoring well directly from the fluid levels in the monitoring well. Van Dam (1967) and de Pastrovich et al. (1979) reported that the thickness of the oil in the monitoring well was 4 times the thickness of the oil within the subsoil. Schiegg (1979) found out in laboratory investigations that there was a transition zone with a changing oil saturation over the depth within the subsoil. First theoretical approaches to determine the oil volume within the formation under consideration of the capillary pressure-saturation relations are given by Lenhard and Parker (1990). The oil saturation is calculated from the capillary pressure-saturation relation of the water-oil and the oil-air system each related to the different interfaces (fig.7).

Luckner (1995) modified the approach of Lenhard and Parker. He criticized that their approach calculates a negative oil saturation above the intersection point of the oil-air and the water-oil capillary pressure-saturation curves (fig. 7). The approach of Luckner is to scale the oil-air capillary pressure-saturation curve to a modified porous medium which consists of the solid particles and the water phase. This approach calculates a larger oil volume especially with a small oil thickness in the monitoring well. With this approach the zone with oil saturation reaches up to the ground surface. If this distribution of oil in the underground would be realistic it would be possible to detect the size of the oil body in the underground by taking samples from the ground surface. Real oil spills and the investigations of Schiegg (1979)show that this is not realistic. This contradiction is due to the fact that in Luckner's approach there are only interfaces between water and oil and between oil and air. Vogler (1999) showed that there exist pores with interfaces of water-oil, water-air and oil-air within the subsoil. The parallel existence of these different interfaces lead to an approach in which the oil volume is calculated from the oil-water and oil-air capillary pressure-saturation relations and the phase distribution above the intersection point is calculated from the water-air capillary pressure-saturation relation (fig. 8). With this approach it is possible to describe the

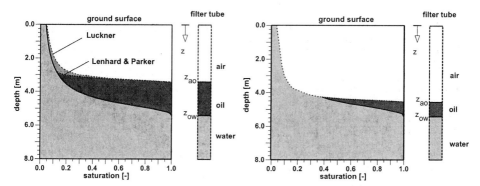

Figure 7. phase saturations over depth   Figure 8. phase saturations over depth
           Lenhard and Parker / Luckner   new approach

phase distribution continuously over the depth. The main advantage of this new approach is that the hysteretic capillary pressure-saturation relations can be included to take into account saturation history e.g. the effects of irregular pore size distribution and phase entrapment.

LABORATORY INVESTIGATION

With a laboratory experiment in a multiphase column the influence of a fluctuating ground-water table was studied. The column had an inner diameter of 15 cm and a height of 80 cm. A 2 '' filter tube was attached to the column to model a groundwater monitoring well (fig. 9). The column was filled with a medium grained sand with grain size 0.2 mm – 1 mm. The water level was controlled by an inlet at the base of the column. First the column was filled from the base slowly with water until a 1 cm thick water layer remained at the top of the sand. Then the water level was lowered to h=30 cm. An oil volume of 334 cm³ was infiltrated from the top of the sand into the column after 4 days. After the infiltration and distribution of the oil within the model the oil formed a 24.2 cm thick oil layer after 12 days. Thereafter a water volume of 220 cm³ was drained out of the column. The thickness of the oil layer in the filter tube remained nearly constant with 24.1 cm after 14 days. The infiltration of 320 cm³ water after 14 days lead to a reduction of the oil layer thickness to 7.1 cm after 21 days. After draining 680 cm³ of water out of the column the thickness of the oil layer in the filter tube increased with 27 cm after 32 days almost to the initial thickness.

To simulate the laboratory experiment the new approach including hysteresis has been implemented in the resevoir simulator ECLIPSE 100, which is based on the finite difference method. The mesh for the numerical simulation consists of two columns of elements (fig. 10).

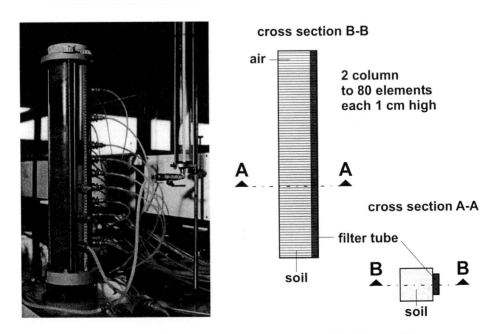

Figure 9. Experimental setup   Figure 10. Simulation model

The element column on the left side is representing the soil with the hysteretic model to simulate the capillary effects in the multiphase column and the element column on the right side is representing the filter tube without capillary effects. The parameter of the simulation are given in table 1. The boundaries are impervious for all phases with the exception of the edge at the top which is permeable for the air phase. The infiltration and drainage of water was controlled with a water-well in the lowest soil element. The infiltration of the oil phase was controlled with an oil well at the top of the soil.

Table 1. Parameter of the numerical simulations

parameter	unit	laboratory experiment	case study
oil density ρ_o	[g/cm³]	0,845	0,802
oil viscosity η_o	[Pa s]	$5,07 \cdot 10^{-3}$	$2,0 \cdot 10^{-3}$
transformation parameter $^d n$	[-]	4,3	2,25
transformation parameter $^i n$	[-]	2,7	2,25
scale parameter $^i \alpha_{wo}$	[1/cm]	0,81	0,938
scale parameter $^i \alpha_{oa}$	[1/cm]	0,178	0,390
scale parameter $^d \alpha_{wo}$	[1/cm]	0,174	0,469
scale parameter $^d \alpha_{oa}$	[1/cm]	0,063	0,195
void ratio n	[-]	0,35	0,35
permeability k_w	[m/s]	$2,05 \cdot 10^{-4}$	$1,0 \cdot 10^{-3}$
residual water saturation $S_{r,w}$	[-]	0,1	0,1
max. water saturation drainage $^d S_{max,w}$	[-]	0,95	1,0
max. water saturation wetting $^i S_{max,w}$	[-]	0,85	0,855
residual oil saturation in oil-air system $S_{r,o}$	[-]	0,05	0,051

After 14 days the simulated phase distribution shows a residual oil saturation in the oil-air system caused by the infiltration of the oil from the top and a layer of mobile oil with changing saturation at the top of the water saturated zone (fig. 11). The calculated thickness of the oil layer in the filter tube is 21.2 cm. Due to the missing capillarity the oil is in a different depth in the filter tube than in the soil column. According to the hysteretic approach it represents the mobile oil volume in the soil column. The simulated phase distribution after 21 days shows a residual oil saturation in the depth of approx. 40 cm to 53 cm. This oil saturation is fixed and therefore immobile. The thickness of the oil layer has reduced to 12 cm during the

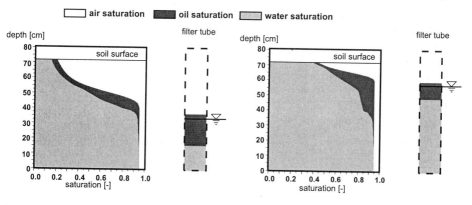

Figure 11. Phase saturations, t = 14 d

Figure 12. Phase saturations, t = 21 d

calculation. A second calculation without accounting for the hysteresis leads to a nearly unchanged oil thickness of 25 cm during the hole simulation run. The simulations show that the changes in the oil layer thickness in the filter tube and the distribution of the oil in the soil can only be verified if the hysteresis is taken into consideration.

CASE STUDY

To monitor an oil spill at an airport site several goundwater monitoring wells where drilled. A typical time series of fluid levels in a monitoring well is shown in Fig 13. The measured oil layer thickness in the monitoring wells is shown in Figure 14. The maximum of the oil thickness was approx. 50 cm. Due to the increase in groundwater level after approx. 400, 900 and 1400 days the thickness of the oil in the monitoring well decreases. The decrease in groundwater level in the timesteps between causes an increase in the measured oil thickness. After approx. 1700 days the thickness of the oil in the monitoring wells has reduced to nearly zero. A short decrease in groundwater level causes the remobilisation of a thin oil layer after approx. 1900 days. With the following increase in groundwater level the oil layer in the monitoring wells disappears.

The mesh for the numerical simulation also consisted of two element columns. The left column represents the soil body and the right column a monitoring well in the middle of the oil spill. The parameters are given in table 1. Starting with an initial oil thickness of d = 45 cm resp. d = 50 cm the observed changes in groundwater level were given as boundary conditions on the lower edge of the mesh. The calculated thickness of the oil layer compared to the observed thickness of the oil layer is shown in Figure 14. There is a good agreement between the calculation and the observations. Only the model which incorporates the hysteresis due to irregular pore size and the hysteresis due to phase entrapment is able to explain the marked changes in the thickness of the oil layer.

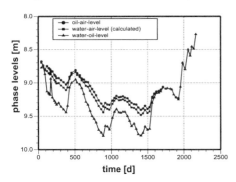

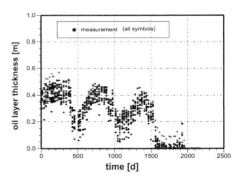

Figure 13. Time series of fluid levels in monitoring wells

Figure 14. Observed and calculated thickness of oil layer in the monitoring wells

SUMMARY AND CONCLUSION

A new approach is presented to calculate the volume of oil in the underground of an oil spill from fluid levels in monitoring wells. The approach includes the effects of hysteresis due to irregular pore geometry and to phase entrapment. It is possible to explain the drastic changes in the oil thickness in a monitoring well due to the decrease and increase in the groundwater table. A correct evaluation of the oil volume infiltrated into the underground from an oil spill and the effective control of remediation works can only be done by using the newly developed approach under consideration of the saturation history.

REFERENCES

de Pastrovich, T.L., Baradat, Y., Barthel, R., Chiarelli, A., Fussell, D.R. (1979):
  Protection of groundwater from oil pollution. CONCAWE, Report 3/79, Den Haag, Netherlands.
Land, C.S. (1968):
  Calculation of Imbibition relative Permeability for Two- and Three-Phase Flow from Rock Properties. Trans. AIME, Vol. 243, 149-156
Lenhard, R.J., Parker, J.C. (1990):
  Estimation of Free Hydrocarbon Volume from Fluid Levels in Monitoring Wells. Ground Water, Vol. 28, Heft 1, 57-67
Lenhard, R.J., Parker, J.C., Kaluarachchi, J.J. (1991):
  Comparing Simulated and Experimental Hysteretic Two-Phase Transient Fluid Flow Phenomena. Water Resources Research, Vol. 27, Heft 8, 2113-2124
Luckner, L. (1995):
  Erfassung von Ölkontaminationen im Untergrund durch Flüssigkeitsspiegelmessungen in Pegeln. Wasserwirtschaft, Jg. 85, Heft 6, 304-308
Moore, T.F., Slobod, R.L. (1956):
  The Effect of Viscosity and Capillarity on the Displacement of Oil by Water. Producers Monthly, Vol. 20, 20-30
Schiegg, H.O. (1979):
  Verdrängungs-Simulation dreier nicht mischbarer Fluide in poröser Matrix. Mitteilungen der Versuchsanstalt für Wasserbau, Nr. 40, Hydrologie und Glaziologie, ETH Zürich.
Taylor, D.W. (1948):
  Fundamentals of Soil Mechanics. Wiley, New York.
van Dam, J. (1967):
  The Migration of Hydrocarbons in a Water-Bearing Stratum. The joint Problems of the Oil and Water Industries. Inst. Petrol., London, 55-96
Vogler, M. (1999):
  Einfluß der Kapillarität auf die Mehrphasenströmung bei der Sanierung von Mineralölschadensfällen im Boden. Mitteilungen des Institutes und der Versuchsanstalt für Geotechnik der Technischen Universität Darmstadt, Nr. 45

Influence of humic acid on the electromobility of copper in contaminated soil

J. WU
Leeds Environment Centre, University of Leeds, Leeds, UK
L. J. WEST
School of Earth Sciences, University of Leeds, Leeds, UK
D. I. STEWART
School of Civil Engineering, University of Leeds, Leeds, UK

ABSTRACT
An investigation of the influence of humic acids on copper mobility in a model soil is reported. Humic acids were extracted from peat with alkali and separated by acid precipitation. Electrokinetic column experiments were then carried out on consolidated sand/kaolin specimens spiked with humic acids and/or copper (II) nitrate. The transport behaviour of copper and humate species in the electric field is interpreted in terms of humate-copper-clay interactions at a range of pH values. The results show that cupric ions form a positively charged, soluble complex with humic acids.

INTRODUCTION
The mobility of toxic metals is strongly influenced by organic matter (OM) present in soils, river sediments, and aquifers. Formation of soluble complexes with OM enhances the mobility of many metal ions, for example iron, mercury, copper, nickel, vanadium and lead, and organic complexes are the dominant dissolved forms of such metals in many natural waters. For example, Sauve et al (1997) report that almost all soluble copper in contaminated soils from Montreal, New York and Denmark was complexed to OM. However, metals may also accumulate in industrially contaminated soils and river sediments due to complexation with solid phase OM (e.g. see Wu et al., 1998).

Much soil OM is present as humic substances, which can be solids, coatings on soil particles or be dissolved in the pore fluid. This study focuses on the sub-group called humic acids, which have a large influence metal mobility. Humic acids (HA) are defined as the organic component which is extractable with alkali but which precipitates on acidification to pH2 (another sub-group, fulvic acids are defined as that part of the alkaline extracted OM that does not precipitate in acid). Humic and fulvic acids are probably produced by microbial degradation of plant matter, and represent a range of chemical compounds whose detailed chemical structure is unknown. Humic acids consist of several aromatic rings with carbon chains, typically with a molecular weight of about 50,000, and with carboxyl and hydroxyl as the main functional groups (Bloom, 1981). Their precipitation as pH is reduced below 2 is probably the result of protonation of dissociated carboxyl ($-COO^-$) groups and formation of inter-molecular linkages, which results in increased hydrophobicity (Ritchie and Posner, 1982). Fulvic acids differ from humic acids in having lower molecular mass and a higher proportion of hydrophilic (e.g. carboxyl) functional groups to hydrophobic groups (e.g. alkyl chains), making them soluble over the entire pH range.

Geoenvironmental engineering, Thomas Telford, London, 1999, 326–333

According to Bloom (1981), carboxyl functional groups are dominant over hydroxyl groups as receptors for metal binding. Most metal cations form 'outer sphere' complexes with these functional groups, which means that they remain fully hydrated and that there is no direct co-ordination with the ligand. However, cations such as Cu^{2+} and Pb^{2+} form 'inner sphere' complexes which means that dissociated carboxyl groups (-COO⁻) replace some of the hydration water molecules, and as a result of this their humate complexes are very stable.

The influence of HA on toxic metal mobility depends on humate solubility and on electrical charge (because charge affects sorption onto clay minerals and iron oxides). However, the electrical charge on metal-HA complexes, its pH dependence, and humate sorption behaviour remains poorly constrained. This paper reports a study of the mobility and electrical charge of humate complexes of cupric ions (Cu^{2+}) in an artificial kaolinite soil, using electrokinetic column experiments. Here, a voltage is applied across the soil using electrodes, and the induced flow of electrical current and water are monitored (e.g. see Hamed et al, 1991; West and Stewart, 1995). The electrically induced pore fluid flow provides data on the surface electrical properties of the soil particles (zeta potential), whereas the post testing distributions of chemical species (HA, metals, etc.) provides data on their mobility and electrical charge.

MATERIALS AND METHODS

HA was extracted from Irish Moss Peat (supplied by Erin Horticulture Ltd, Ireland) following the procedures of Yong & Mourato (1988). The peat was extracted with 1 N HCl at temperature of 65°C and supernatant was discarded. The residue was then extracted with 0.5N NaOH. The resulting supernatant was adjusted to pH2 using 4 N HCl, and the HA precipitate was removed by centrifugation. Samples for chromatographic elemental analysis were air-dried whereas those used for electrokinetic tests were stored at 4°C. The air-dried extracts contained 55.3%C, 3.6%N, 3.3%H, and 0.7%S, which is similar to the composition of the International Humic Substances Society (IHSS) peat HA standards.

Electrokinetic column experiments were carried out on a 4:1 w/w mixture of Speswhite kaolin and the 0.1-2mm fraction of Sherburn yellow building sand (see Table 1), contaminated with HA and/or copper (II) nitrate. Kaolin was selected because it has been used in previous electrokinetic studies (West and Stewart, 1995; West et al, *in press*). Cupric ions were selected because these are particularly strongly bound to OM. The relatively high Cu loading (~300mg/kg) used is representative of industrially contaminated soil/sediment. Where both cupric ions and HA were added, the HA was added first, to simulate the situation where a natural humic soil is contaminated with copper.

The electrokinetic apparatus consists of a 200mm long, 90mm internal diameter acrylic tube with stainless steel voltage monitoring electrodes (see Figure 1). Slurries prepared at 100% moisture content were de-aired under vacuum and consolidated vertically (18.4kPa for 48 hours) into this apparatus using a 100mm long extension tube (not shown in Figure 1). After consolidation, extension tube was removed and the specimens were trimmed to fit the apparatus (i.e. to 200mm length). Soil from the extension tube was retained for analysis.

The column apparatus was then set up horizontally with graphite power supply electrodes at the ends as shown in Figure 1. The anode and cathode reservoirs were filled with distilled water, and a voltage of 7.5 volts was applied across the power electrodes for 15 days. During testing, electrical current, weight of the anode reservoir, reservoir pH and the voltages at the monitoring electrodes were monitored daily. After fifteen days of voltage application, the specimens were extruded and sliced into ten equal portions.

Samples from the electrokinetic column apparatus and of the consolidated soil from the extension tube were homogenised and their pH was measured directly using a glass electrode. A portion of soil from each slice was oven dried (to 105°C) for gravimetric moisture content determination, copper determination by X-ray-fluorescence (Rh Tube), and HA determination by loss-on-ignition (LOI) at 550°C (for specimen KS5, which had the lower HA level, LOI was conducted on an alkali extract rather than on the dried soil). The effect of loss of bound water from the kaolinite lattice was removed by subtraction of blank specimen data.

Pore fluid was extracted by centrifugation from the portions of the specimens that were not oven dried (samples from each pair of adjacent slices were combined to produce sufficient soil). Pore fluids were filtered at 0.45μm and their pH and conductivity were determined electrometrically. The samples were then acidified with a few drops of 5% nitric acid, and pore fluid copper level was measured using a Varian Spectra (AA-10) Flame Atomic Absorption Spectrometer (the detection limit of this technique is about 1mg/L for copper).

RESULTS
The pH of the soil was about 4.7 in the specimens contaminated with Cu but no HA (KS3 and 8), which is similar to that of uncontaminated Spewhite kaolinite (West et al, 1995). Where HA is present the pH was lower (around 4 for soil with 1000mg/kg HA and 2 for that with 10,000mg/kg HA). As HA is defined as the humic fraction that precipitates at pH2, it is likely that some HA in the latter specimens was in the solid phase.

Table 1. Properties of materials

Speswhite Kaolin			Sherburn yellow sand		
Composition [a] Kaolinite		94%	Composition [c] Silica		90 %
Mica		4%	Ferric oxide		0.64 %
Montmorillonite		1%	Alumina		4.56 %
Feldspar and quartz		1%	Magnesia		0.45 %
Surface area (m^2g^{-1})		14.0	Lime		0.33 %
Percentage of particles($<2\mu m$)[b]		80%	Percentage of particles		
Plastic limit [b]		38%	>2mm		1 %
Liquid limit[b]		80%	<100 μm		10 %
CEC (meq/100g at pH7)		4-5	Specific Gravity [c]		2.67
Exhangeable ions	$Na^+, Ca^{2+}, Mg^{2+}, K^+$				

[a] McGuffog, ECC International Ltd   [b] Al-Tabbaa (1987)   [c] Tom Langton and Son Ltd

Table 2. Initial conditions in electrokinetic column experiments

Test	Humic acid added mg/kg	Cu added mg/kg	Total humic acid mg/kg	Total Cu mg/kg	Pore fluid Cu mg/L	pH soil	pH pore fluid	Cond. pore fluid mS/cm	Init. m.c. %	Ke** $x10^{-5}$ cm^2/Vs
KS3	0	338	0	296	<1	4.65	6.44	1.99	66	1.0
KS4	10,000	338	8590	248	140	2.05	2.17	5.20	70	0
KS5	1,000	338	960	303	55	4.35	4.26	3.28	64	2.0
KS6	10,000	0	8620	26	*	2.10	*	*	69	0
KS8	0	338	0	304	<1	4.70	6.35	2.06	65	2.8

* pore fluid was not extracted from specimen KS6   ** average for first week of testing

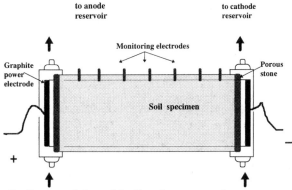

Figure 1. Schematic diagram of electrokinetic column apparatus

The Cu levels in the pore fluid of the specimens which contained no HA were below the detection limit, which indicates that most of the Cu was sorbed to the clay. In contrast, significant Cu was present in the pore fluid of the specimens containing both HA and Cu (about 30% of the total Cu in the specimen with the higher HA level, KS4, and 15% of the total in the specimen with the lower HA level, KS5, see Table 2).

Specimens containing HA or copper separately (KS3, 6, 8) showed similar electrical current levels, rising to about 8mA in the first week and then falling towards 2mA (Figure 2). This is similar behaviour to uncontaminated kaolin (West et al., *in press*). Specimens containing both copper and HA (KS4 and 5) showed higher current levels (maximum 50mA for the specimen containing 10,000ppm HA; KS4, and 20mA for that containing 1000ppm HA; KS5).

The electroosmotic flow behaviour (Figure 3) of the specimens containing up to 1000kg/kg of HA (KS3, 5, 8) was similar, with 0.25 to 0.5 pore volumes per week flowing from the anode to the cathode. The electroosmotic permeability values (1 to 2 x10^{-5} cm^2/Vs, see Table 2) are similar to those for uncontaminated kaolinite (West et al, *in press*). This indicates that clay zeta potentials are predominantly negative, and similar to those seen in uncontaminated kaolinite (about -30 mV, West and Stewart, 1995). In contrast, there was hardly any electroosmotic flow in specimens with 10,000mg/kg of HA (KS4 and KS6), regardless of whether copper was present. This suggests that their zeta potential were close to zero. This is probably as a result of the low pH produced by these levels of HA (the zeta potential of kaolinite is usually negative when the pH is above about 3, but zero at lower pH, West and Stewart, 1995). Electroosmotic flow did not result in significant moisture content change.

During testing, the pH of the catholyte quickly climbed to values between 10 and 12, which were maintained through the test. These values reflect the production of hydroxyl ions at the cathode by electrolysis and their subsequent reaction with atmospheric carbon dioxide (West et al, *in press*). The anolyte pH quickly fell to about 2 or slightly below, reflecting the production of H$^+$ ions by electrolysis at the anode (Hamed et al., 1991). All the specimens showed reduced pH near the anode and increased pH near the cathode (Figure 4). However, the pH profiles differ according to the amount of HA present. In the specimens containing no HA, the anode third of the specimens had a soil pH of about 2, which rose sharply to about pH4 in the middle and more gradually to about pH6 near the cathode (Figure 4a). The pore

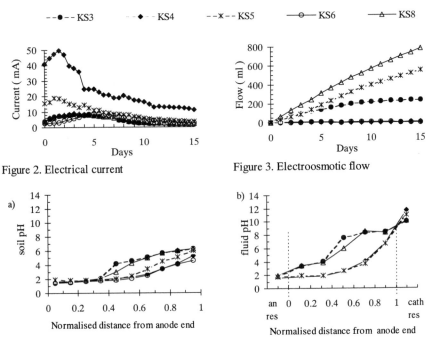

Figure 2. Electrical current

Figure 3. Electroosmotic flow

Figure 4. a) soil pH b) fluid pH

fluid pH showed a similar pattern but was between 1 and 2 units higher than that of the soil (Figure 4b). The pH behaviour is typical of electrokinetic column experiments and is caused primarily by the electromigration of H^+ and OH^-/HCO_3^- from the anode and cathode reservoirs respectively. The sharp change in pH in the middle of the specimens marked the point where these species meet and neutralise one another.

In the specimens containing HA, the soil pH in the anode half is below 2 whereas in the cathode half it rose smoothly to between pH 4.5 and 6 (the pH of the pore fluid was similar to the soil pH near the anode, but higher than the soil pH near the cathode). These graphs indicate that (i) entry of H^+ ions from the anode did not have very much effect on the pH of the specimens with 10,000mg/kg HA because their pH was already low and (ii) the HA buffered the pH near the cathode, by neutralising OH^-/HCO_3^- ions from the cathode reservoir.

The final normalised copper levels (i.e. total soil Cu / initial soil Cu) are shown in Figure 5a. In the specimens containing copper but no HA (KS3 and KS8), copper had become depleted near the anode and accumulated near the middle of the specimens (there was no change in the copper level near the cathode). The total normalised copper profiles for the specimens containing copper and HA (KS4 and 5) show a different pattern, with a broader zone of copper depletion (anode two thirds of the specimens) and copper accumulation occurring immediately adjacent to the cathode. In all the specimens, copper accumulates where the pore fluid pH rises above about 5 (see Figure 4b). Mass balance indicates that about 90% of the Cu remained within the specimens (the remainder probably accumulated in the cathode reservoir where the blue colour of basic copper carbonate, azurite, was observed).

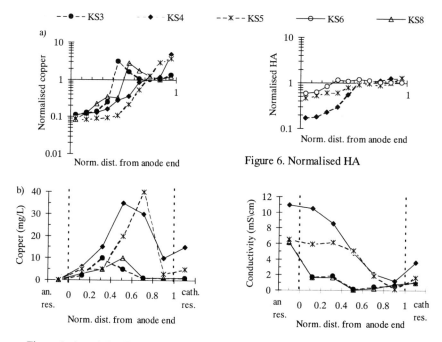

Figure 5. a) total Cu b) pore fluid Cu   Figure 7. Pore fluid conductivity

The final pore fluid copper levels are shown in Figure 5b. In specimens with no HA (KS3 and KS8), pore fluid Cu levels were elevated from their initial values in the anode half, due to desorption caused by acidification of the soil. They peaked in the middle of the specimens and were low near the cathode. Specimens with HA (KS4 and 5) had much higher initial pore fluid copper (see Table 2). After testing these levels were well below their initial values, although they still exceeded those in the non-HA tests. Copper was detected in the pore fluid even near the cathode and in the cathode reservoir, where conditions where neutral/alkaline.

In all four specimens that contained copper, the highest pore fluid Cu level occurred immediately to the left (anode side) of the peak in total Cu. This is consistent with transport of positively charged copper species towards the cathode and their subsequent precipitation.

Normalised HA levels (i.e. final HA / initial HA) are shown in Figure 6. The specimens containing both HA and copper (KS4 and 5) showed depletion of HA in their anode halves and slight accumulation near the cathode, with the specimen containing the highest level of HA (KS4) showing the larger change. Mass balance indicates that about one quarter of the HA in each specimen had been lost. In contrast, the specimen containing 10,000ppm HA and no copper (KS6) showed only slight depletion of HA near the anode and slight accumulation in the middle, and no HA had been lost from the specimen. These data indicate that the presence of copper in KS4 and KS5 resulted in enhanced transport of HA.

The voltage profiles measured using the monitoring electrodes along the specimens are shown in Figures 8a-e and the final pore fluid conductivities in Figure 7. The regions of copper precipitation had relatively resistive pore fluid and show up as regions of steep

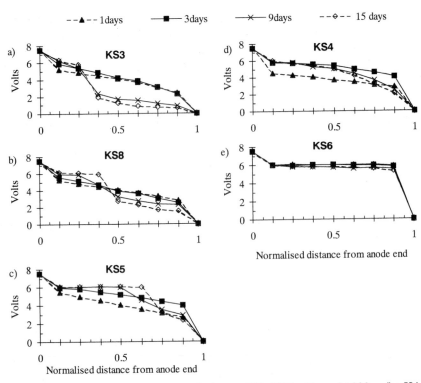

Figure 8. Voltage profiles (KS3 & 8 - Cu but no HA, KS5 - Cu and 1000mg/kg HA, KS4 - Cu and 10,000mg/kg HA, KS6 – no Cu and 10,000mg/kg HA)

gradient in the voltage data. This indicates a reduction in the pore fluid ionic strength and possibly an increase soil tortuosity due to precipitation.

DISCUSSION

Where no humic acid was present, copper was sorbed because the kaolinite component of the sand-clay mixtures had sufficient exchange capacity (see Table 1), and copper is relatively strongly sorbed compared to the natural ions present on the clay. During electrokinesis, H^+ generated by anode electrolysis entered the specimen from the anode, reduced the pH and caused copper desorption, probably producing uncomplexed Cu^{2+} ions (the dominant form of dissolved Cu(II) in air saturated water at pH2 is Cu^{2+}, Dirkse, 1986).

These Cu^{2+} ions were then transported towards the middle of the specimens by electromigration (direct movement of charged species in an electric field) and electroosmosic advection with the pore fluid. They were precipitated at the pH jump in the middle of the specimens, probably as a copper oxide (Cu^{2+} precipitates as CuO from air saturated water at about pH5 where present at the aqueous concentrations seen here, Dirkse, op.cit.).

The presence of copper enhanced the transport of HA towards the cathode (Figure 6), which indicates that HA and Cu^{2+} ions formed a soluble Cu-humate complex. Enhanced aqueous copper levels indicate that the presence of HA resulted in copper desorption (which may result partly from reduced pH, but also from complex formation). Together, HA and copper were transported towards to cathode even in the absence of electroosmotic flow (test KS4), which indicates that the Cu-humate complex was positively charged. An explanation for this behaviour may be that although HA is undissociated at pH2, and therefore uncharged, the Cu^{2+} co-ordinates with carbonyl oxygens to produce a positive complex.

In the specimens containing Cu and HA, both accumulate near the cathode where the pore fluid pH rises above 5 (Figure 5a). This indicates that some Cu enters the solid phase here (i.e. it is sorbed or precipitated). However, aqueous Cu levels remain above those in the specimens containing no HA (KS3 and 8), and significant Cu was measured in the catholyte. This indicates the presence of a (possibly different) aqueous Cu-humate complex in neutral to alkaline conditions.

CONCLUSIONS

A series of electrokinetic column experiments on sand-kaolinite mixtures contaminated with copper and humic acids (HA) showed that:

- in the absence of HA, copper was desorbed by the entry of H^+ produced by anode electrolysis, transported towards the cathode by electromigration and electroosmotic advection, and precipitated where the pH rose in the middle of the specimens.

- the presence of high levels (10,000mg/kg) of HA prevented electroosmotic flow, probably by reducing the zeta potential of kaolinite towards zero.

- the presence of HA at levels of both 1,000 and 10,000 mg/kg enhanced the mobility of Cu(II) ions and their electrokinetic transport. The enhanced transport of both HA and Cu(II) ions towards the cathode, in the absence of any pore fluid flow, shows that they form a positively charged, soluble Cu-humate complex in acidic conditions (pH2-4).

REFERENCES

Al-Tabbaa, A. (1987) Permeability and stress strain response of Speswhite kaolinite. Ph.D thesis . University of Cambridge, Cambridge, UK.

Bloom, P.R. (1981) Metal-organic matter interactions in soil. *In* Chemistry in the Soil Environment (eds. M. Stelley and D.M. Kral), 129-150. Am. Soc. Agronomy and Soil Soc. of Am.

Dirkse, T.P. (1986). Copper and hydroxide. *In* Solubility Data Series - Copper, Silver, Gold and Zinc, Cadmium, Mercury, Oxide and Hydroxides (ed. A.S. Kertes), 10-82. Pergamon Press, Oxford.

Hamed, J., Acar, Y.B. and Gale, J. G. (1991) Pb (II) removal from kaolinite by electrokinetics. ASCE Journal of Geotechnical Engineering, 117 (2), 241-271.

Rashid, M.A. (1985) Geochemistry of marine humic substances. Springer-Verlag, New York.

Ritchie, G.S.P. and Posner, A.M. (1982) The effect of pH and metal binding on the transport properties of humic acids. Journal of Soil Science 33, 233-247.

Sauve S., McBride M. B., Norvell, W. and Hendershort, W. H. (1997) Copper solubility and speciation of in situ contaminated soils: effects of copper level, pH and organic matter. Water, Air & Soil Pol., 100, 132-149.

Weber, J. H. (1988) Binding and Transport of metals by humic materials. *In* Humic Substances and their role in the environment (eds. F.H. Frimmel and R.F. Christman), 165-178. John Wiley & Sons.

West, L. J. and Stewart, D. I. (1995) Electrokinetic decontamination: the effect of zeta potential. Geoenvironment 2000. American Society of Civil Engineers Special Publication. 46, 1535-1549.

West, L. J. Stewart, D. I., Binley, A.M. and Shaw, B. (in press). Resistivity imaging of soil during electrokinetic transport. Engineering Geology. Elsevier.

Wu, J., West, L. J. and Mannings, S. (1998) Sequential extraction on metal contaminated sediments for risk assessment. Proceedings of The Fifth International Conference on Reuse of Contaminated Land and Landfills (ed. M.C. Forde), 93-96. Engineering Technics Press.

Yong, R.Y. and Mourato, D (1988) Extraction and characterization of organics from two Champlain Sea surface soils. Canadian Geotechnical Journal 25, 559-607.

A Case Study of Groundwater Pollution due to Copper Mining (Sohar, Sultanate of Oman)

R S SHARMA & T S AL-BUSAIDI
University of Bradford, UK

ABSTRACT

In 1982, the Oman Mining Company (OMC) commenced copper mining and smelting operations in the area of Sohar (Wadi Suq), Sultanate of Oman. Seawater was used for mining operations until 1993. During this period, 11 million tonnes of tailings had been deposited behind an unlined tailings dam. This has resulted in major groundwater pollution problem.

This paper presents results from investigation of groundwater pollution. Data pertaining to various parameters such as pH and TDS were obtained from the OMC. Most of the data were for a period of 12 years since 1984 and were gathered from an extensive network of field monitoring wells covering a distance of 14 km downstream of the tailings dam. The trends of groundwater pollution are presented with explanation. The issue of the existing remedial measures (cut-off walls for containment of the pollution source) is briefly addressed.

INTRODUCTION

Groundwater pollution, from sources such as waste disposal sites, is a world wide problem. Mining of mineral ores and disposal of resulting waste pose significant risk to the groundwater (Sharma, 1994). Sultanate of Oman is one of the countries, which face a serious industrial groundwater pollution. In 1982, the Oman Mining Company (OMC) commenced copper Mining and Smelting operations in the area of Wadi Suq, west of Sohar. Mining, and consequently the processing and accumulation of waste tailings, continued until 1994.

During this time 11 million tonnes of sulphide-rich tailings, and 5 million cubic metres of seawater have been disposed of behind the unlined tailings dam (Al-Busaidi, 1998). This has caused two serious environmental problems involving the downstream water quality: salinity (the most immediate problem) and the potential for mobility of toxic metals. It is emphasised that heavy metals are not a problem in Wadi Suq at present, but could develop into a serious problem if the pH control is relinquished. This paper addresses briefly the origin, development, current status and remediation of groundwater pollution.

THE AREA OF SOHAR
Location

The mine processing plant and tailings dam lie immediately to the north of the Sohar-Buraimi road in the foothills of the Oman Mountains. Currently mined areas lie about 3.5 km to the south (Lasail) and about 4 km to the north (Aarja and Bayda) of the plant. Wadi Suq is a 71 square kilometre catchment, lying immediately north of the lowest reaches of Wadi Jizzi in the northern Batinah plain.

Geology

The geological formations of the area have undergone substantial tectonic activity involving much low angle faulting and thrusting which makes the geology of the area complex. The geological column is shown in Table 1.

Age	Stratigraphy	Description
Quaternary	Modern alluvium	Wadi channel deposits: loose and poorly sorted
	Terrace deposits	Poorly sorted, well-rounded, calcite-cemented mainly coarse sand, gravels and conglomerates
Tertiary	Limestones and mudstones	not present in the catchments within the vicinity of the mine
Mesozoic	Hawasina Melange	Shales and lime-stones mainly in the lower part of the catchments
	Ophiolitic Complex	Volcanic extrusive rocks containing ores covering the large parts of the upper reaches of the wadis
	Hawasina Group	Limestones and chert mostly in Upper Wadi Jizzi

Table 1: Schematic Geological Succession of Oman Mountains

The mine area and the upper wadi catchment, which run east from the mine area, lie mainly on the Ophiolitic Complex volcanics. Parts of the catchments are also on the Hawasina Melange and the Hawasina Group sediments, though at least half of the area of Wadi Suq, for example, is covered by alluvium. The cemented alluvium lies above the bed level of Wadi Suq downstream of the tailings dam, forming low plateau area between and adjacent to the wadi courses (Gass, 1980).

The wadi and terrace alluvium are the main aquifers. Although the intrinsic permeability of the bedrock is negligible, groundwater movement through the fractures in the body of the bedrock is known to occur. The permeability of the bedrock beneath the wadi alluvium is probably enhanced by weathering so there may be ample opportunity for groundwater movement.

MINING OPERATIONS AND SOURCES OF POLLUTION

Copper ore occurs as chalcopyrite, a copper/iron pyrites ore, in which copper occurs at about 1% by weight. Copper ore was first worked from the Lasail area, starting in 1983, since then working has extended to the north to the Aarja and Bayda area to exploit additional reserves. The extraction of the ore produces mine drainage water which is typically acidic and sulphate rich and which has, in the past, found its way into the wadi systems. Most of the pumped mine drainage water is now used as a transport medium for grinding and milling avoiding the need to use the seawater supply (OMC report, 1996). The mining process involves excavation and transport of the ore to the process plant for grinding and milling. The ore material is passed through air flotation treatment to separate the copper sulphide from the iron sulphide and barren rock. The copper sulphide concentrate is further processed in preparation for smelting and refining. The waste rock and iron sulphide residues are pumped as 'tailings' to storage behind a specially constructed tailings dam. The sequence of the processes is set out in Fig.1.

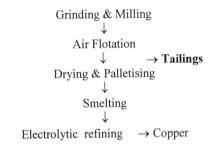

Fig. 1: Sequence of Mining Operations in Sohar Copper Mines

HISTORY OF THE SITE AND ENVIRONMENTAL PROBLEMS

Table 2 shows a brief history of the site and environmental problems (OMCO reports, 1996).

Mine Activity	Year
• Construction of the Lasil Mine and Mill	1980
• Commissioning of the Tailings Dam	1982
• Commissioning of Seawater Pipeline	1983
• Detection of Groundwater Contamination	1983
• Installation of Cut-off Trench No. 1	1983
• First Large Scale Importation of Seawater.	1984
• Cessation of Discharge of Lasil Mine Water to Surface Sediments in Wadi Lasil and Awainah	1986
• Installation of the Grout Curtain and Cut-off Trench No. 2	1992
• Cessation of Seawater Importation	1993
• Mining Terminated, and Concentrator De-commissioned	1994
• Smelter and Refinery Continue Operation Indefinitely	1994

Table 2: Summary of the Activities Related to Groundwater Pollution

FIELD DATA

There are three mines located in the Wadi Suq catchment area and over 30 monitoring sites are used to monitor the water quality in that areas. The monitoring points listed in Table 3 have been selected for the analysis presented in this paper, mainly to illustrate the problem.

Area	Selected Points
Wadi Suq	MW(1), MW(2), MW(3), MW(4) and MW (KM14)

Table 3: The Selected Monitoring Wells

Monitoring wells MW(1) to MW(KM14) are at increasing distance from the tailings dam with MW(KM14) located farthest at 14 km downstream of the tailings dam. It should be noted that monitoring of the wells has been done through the history of the problem starting from 1983 until now, the results mentioned here are for quarterly basis for each alternate year, i.e.1984, 86, 88,…96. The reason for this selection is to simplify the analysis of the available data.

The values of concentration of metals reported in the OMC report (1996) are given below in Table 4.

Substance	Concentration (mg/l)
Pb^{+2}	0.3
Cr^{+2}	<0.02
Zn^{+2}	94
Ni^{+2}	0.04 – 1.6

Table 4: Metal Concentrations Observed in the Mining Area

The available rainfall data for the area of Wadi Suq is presented in Fig 2. The highest rainfall was during Jan 1988, but the rainfall in Jan 1996 was also relatively high and reported to be widespread in most parts of Sultanate of Oman (Al-Busaidi, 1998).

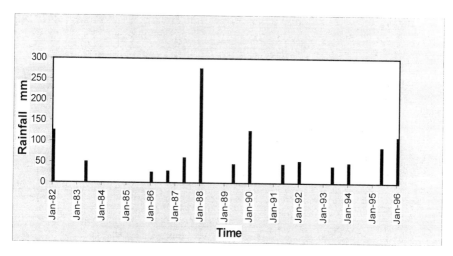

Fig.2: Rainfall in the Upper Wadi Suq Catchment

RESULTS AND DISCUSSION
Variation of pH
For all the monitoring wells, the variation of pH with time is given in Fig.3. Inspection of Fig.3 shows that the pH was observed to be generally between 6.5 – 8.5. However, there is a clear reduction in January 1996 for most of the monitoring wells except for (KM14). The lowest point was observed in MW(3) where pH dropped to 4.5. It seems that this drop in pH

was only in the vicinity of the tailings dam as the monitoring wells MW(1) to MW(4) are nearby the dam whereas the MW(KM14) is 14 km downstream of the dam.

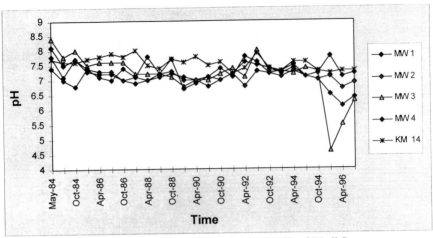

Fig.3: pH Variation in Monitoring Wells of Wadi Suq

The reason behind this reduction in pH could be due to the following:
- The heavy rain which occurred in the winter of 1996, was exceptional in many areas of Oman and this might have accelerated the chemical reaction in the tailings (oxidation of pyrite) according to the following reaction:

$$2\ Fe\ S_2 + 2\ H_2O + 7\ O_2 \rightarrow 2\ Fe\ SO_4 + 2\ H_2\ SO_4$$

Pyrite   Water   Oxygen   Sulphuric Acid

And so oxidation of pyrite produces sulphuric acid, which causes a drop in pH. The tailings dam was unlined and seepage through the dam was observed, so the possibility of accelerated oxidation of pyrite was high.
- Another cause of pH drop could be an inadvertent malfunction of the pH adjustment, which was achieved by adding lime to the tailings. Neutralisation of the pH occurred as shown in the following reaction.

$$H_2\ SO_4 \quad + \quad Ca\ (OH) \quad \rightarrow \quad Ca\ SO_4.\ 2\ H_2O$$

Sulphuric Acid   Lime   Gypsum

Generally, the oxidation of pyrite should be controlled in the tailings, as it is obviously the cause of pH drop. If the drop continues, then the heavy metals can become mobile resulting in pollution of the groundwater.

Sodium Concentrations
Concentrations of sodium ions have been used to monitor the salinity of groundwater for the monitoring wells of Wadi Suq. Fig.4 shows that the concentration of Na^+ ions are generally high in all monitoring wells. A common trend of increasing concentrations of Na^+ ions was observed in all monitoring points starting from 1984, this was apparently due to the large use of seawater.

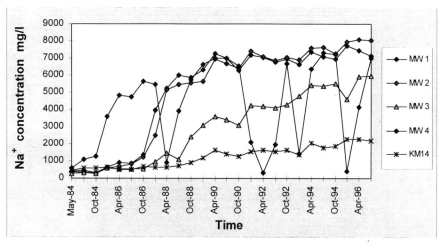

Fig. 4: Sodium Concentration in Monitoring Wells of Wadi Suq

By comparing the trends of all monitoring wells, it is obvious that the closer the well location to the tailings dam the higher the values of Na^+, an exception was for MW (4), where the pattern varies in irregular manner. Inspection of Fig.4 shows that the monitoring well (KM14) has the lowest concentration of Na^+ ions compared to the other monitoring wells. This lower concentration of Na^+ observed in the MW(KM14) appears to be resulting from natural attenuation as the monitoring well KM14 is farthest from the tailings dam.

Total Dissolved Solids

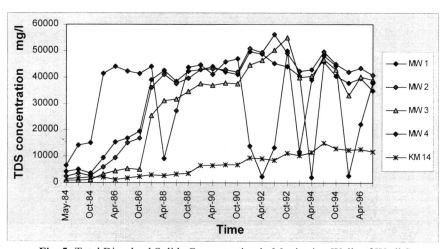

Fig. 5: Total Dissolved Solids Concentration in Monitoring Wells of Wadi Suq

The total dissolved solids (TDS) test has been used to monitor the total salinity of the groundwater. From Fig. 5, it is clear that the values of TDS varied over a wide range between 1000 - 55000 mg/l which indicates a highly saline groundwater. Comparison of Figs 4 and 5 indicate that the patters observed for Na^+ and TDS are consistent.

The large use of seawater since 1984 increased the salinity of the groundwater, which evident from the rapid increase in TDS from Oct.1984 for all monitoring wells. Furthermore the slight decrease in TDS (after seawater use was stopped in 1993) confirms that use of seawater in processing of the copper ore caused groundwater pollution in terms of salinity. The reason for the sudden reduction in the TDS of MW (4) in 1988, 1992 and 1996 was not immediately clear since other monitoring wells do not follow the same pattern of reduction in these periods. For MW (KM14), although the salinity is less than other monitoring points, it increased from about 1000 mg/l in 1984 to 11000 mg/l in 1996.

REMEDIAL MEASURES

The cut-off trench was constructed in 1992, just below MW (3). Results for TDS for the monitoring well at 14 km are plotted in Fig.6. Inspection of Fig.6 suggests that the cut-off trench was not effective in preventing the movement of salinity downstream Wadi Suq since there is no evidence of reduction in the values of TDS or Na^+ after 1992.

Following calculations further illustrate the problem of salinity:
Generally, salinity (TDS) of seawater is 35000 mg/l, (Bouwer, 1979) then
In mid (1992) TDS of (KM14) $\cong$ 7500 mg/l $\cong$ 21% of seawater salinity
In mid (1996) TDS of (KM14) $\cong$ 10000 mg/l $\cong$ 31% of seawater salinity.

From the above calculations it is clear that the salinity of MW (KM14) has increased significantly within four years despite the construction of the cut-off structure.

This apparent increase in salinity despite the construction of cut-off structure could be due to the seepage occurring through the cut-off structure or the water just passed it. However, there could be other reasons for the ineffectiveness of the cut-off structure such as the location of the structure and/or the delayed construction of the cut-off trench.

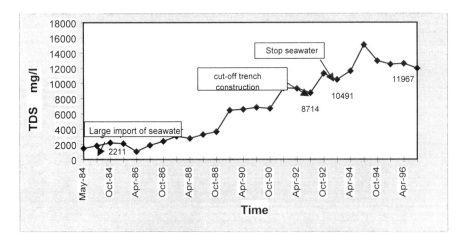

Fig. 6: Salinity Development in Monitoring Well (KM 14)

Despite the discontinuation of importing seawater (1993), the concentration of Na^+ ions did not decrease (see Fig.4). An explanation for this might be that a large amount of salt load is still in the aquifer and water movement is towards downstream of Wadi Suq. The reported leakage of seawater pipeline could be another factor for increased concentration of Na^+.

The construction of cut-off structure downstream of the dam is apparently less successful than was originally hoped, and has failed to halt the development of a saline plume, which currently extends continuously some 14 km downstream.

CONCLUSIONS
The analysis of the field data has shown that the groundwater pollution in the area of Sohar is due to copper mining and storage of tailings. Salinity is the main form of groundwater pollution, but there can be potential for heavy metal pollution depending upon the control of pH. The saline plume has extended at least up to 14 km downstream of the tailings dam. The movement of saline plume downstream of Wadi Suq indicates that there is continuity in the aquifer between upper and lower regions of the Wadi. The remedial measures in terms of cut-off trench and stoppage of usage of seawater in the processing of copper ore appear to be less successful than expected.

ACKNOWLEDGEMENTS
The Ministry of Water Resources, Sultanate of Oman and Oman Mining Company generously provided various data. Their help for the field visit of the second author is also acknowledged.

REFERENCES
Al-Busaidi T.S (1998). A case study of groundwater pollution due to copper mines in the Sultanate of Oman. MSc thesis, University of Bradford.
Bouwer, H. (1979). Groundwater hydrogeology. McGraw-Hill.
Gass I.G. (1980). Oman Geological Ophiolite Project. Open University, Milton Keynes, UK.
Oman Mining Company (1996). Environmental reports (for the period of 1983-1996).
Sharma R.S.(1994). Some aspects of liquefaction of tailings dams. MSc thesis, Imperial College of Science, Technology & Medicine, London (University of London).

Mass transport and consolidation of bentonite by the method of unified molecular dynamics and homogenization analysis

Y. ICHIKAWA, Associate Professor, Department of Geotechnical and Environmental Engineering, Nagoya University, Nagoya 464-8603, Japan

K. KAWAMURA, Professor, Department of Earth and Planetary Science, Tokyo Institute of Technology, Meguro-ku, Tokyo 152-8551, Japan

M. NAKANO, Professor, Department of System and Information Engineering of Bioproduction, Kobe University, Kobe 657-8501, Japan

N. SAITO, Nuclear Power R&D Center, Tokyo Electric Power Company, Yokohama 230-8510, Japan

INTRODUCTION

Bentonite is a key component for preventing transport of hazardous radioactive species in high-level radioactive waste (HLRW) management. In addition recently it has been used as an engineered barrier for disposal and containment of industrial and municipal wastes as geosynthetic clay liner. When designing the disposal facilities we commonly apply macro-phenomenological models to predict water flow and pollutant transport. However the existing models are not sufficiently effective, because these do not always reflect the atomic-based true physical and chemical behavior, which is essentially important for the transport phenomena in bentonite. Furthermore in the HLRW management it is required that the whole disposal system works more than a hundred thousand years. We cannot provide such the long-term experimental data, though the phenomenological model works for interpolation-based prediction, so the accurate data is essential. In this sense a new scheme is required for analyzing the true behavior of clay.

Bentonite is a typical micro-inhomogeneous material. That is, it consists of nanometer size of clay minerals (mainly sodium montmorillonite), micrometer size of macro-grains such as quartz and feldspar particles, pore water, and air in its microscopic level. A montmorillonite mineral is of lamellar shape with size of approximately $100 \times 100 \times 1\text{nm}$, and a group consists of several montmorillonite lamellae and interlamellar water.

There are two essential issues for analyzing the behavior of such micro-inhomogeneous materials: One is how to determine the characteristics of constituent components of the micro-continuum which are directly affected by their molecular movement, and another is how to relate the microscopic characteristics to the macroscopic behavior. For solving the first problem, we apply the Molecular Dynamics method (MD; Allen & Tildesley 1987, Kawamura 1990), then we employ the Homogenization Analysis (HA; Sanchez-Palencia 1980) for estimating the micro- to macro-behavior. This procedure is called the *unified MD/HA method* (Kawamura et al. 1997, Ichikawa et al. 1999). The predicted hydraulic conductivity is quite compatible with experiment-based data of Pusch (1994). In this paper we discuss the seepage problem in pure- and salt-water, the diffusion characteristics of Na^+ and Cl^- ions, and the consolidation behavior of bentonite.

Geoenvironmental engineering, Thomas Telford, London, 1999, 342–349

MOLECULAR BEHAVIOUR OF MONTMORILLONITE HYDRATE IN PURE AND SALT WATER

The structure and physical properties of clay minerals are hardly known by means of experimental methods because of their poor crystallinity. We have applied molecular simulation methods for specifying the true physical and chemical properties of montmorillonite hydrate (Kawamura *et al.* 1997, Ichikawa *et al.* 1999).

The molecular simulation methods belong to a field of computational physics and chemistry. There are two major tools in this field, that is, the metropolis Monte Carlo method (MC) and the MD. We find a few contribution for the molecular simulation of clay minerals. The most of works were performed by MC (Delville & Sokolowski 1993; Delville 1995; Skipper, Chang & Sposito 1995; Skipper, Sposito & Chang 1995; Boek *et al.* 1995-1, 1995-2; Karaborni *et al.* 1996). In all of these studies, rigid clay layer structures and rigid water molecules were employed. Teppen *et al.* (1997) gave an (NPT)-MD calculation for gibbsite, kaolinite, pyrophyllite and beidellite using a flexible molecular model. Here N denotes the number of molecules, P the pressure, and T the temperature. Note that the molecular formula of Na-montmorillonite hydrate with n- interlamellar water is given by $Na_{1/3}Al_2[Si_{11/3}Al_{1/3}]O_{10}(OH)_2 \cdot nH_2O$. We call this nH_2O system, and if $n = 0$, it is called the "dry montmorillonite".

In MD the motion of every molecule is given by the Newton's equation, and the force is calculated by differentiating an inter-atomic potential function. The key issue is to determine the interatomic or intermolecular interactions quantitatively. We use a new empirical interatomic potential model. That is, the potential function for all atom-atom pairs (i.e., the 2-body term) is composed of the Coulomb, short-range repulsion, van der Waals and Morse terms, and a 3-body term is added to the H-O-H interaction because of its sp^3 hybrid orbital. Detail are found in Kawamura (1992) and Kumagai, Kawamura & Yokokawa (1994).

Swelling Property of Montmorillonite Hydrate

We can now calculate a wide variety of physical properties by using the MD results. In the calculation of swelling property of montmorillonite hydrate we employ an (NPT)-ensemble MD scheme under condition of 300K temperature and 0.1MPa pressure. The Verlet algorithm is used for time-integration ($\Delta t = 0.4\,\text{fs}$), and by the Ewald method the electrostatic energy and force in long range interaction are calculated.

We plot the calculated swelling property of Na-montmorillonite in Figure 1 comparing with experimental data (Fu *et al.* 1990) of Na^+-Wyoming montmorillonite with the formula $Na_{0.75}[Si_{7.75}Al_{0.25}](Al_{3.5}Mg_{0.5})O_{20}(OH)_4 \cdot nH_2O$. Both are remarkably corresponding.

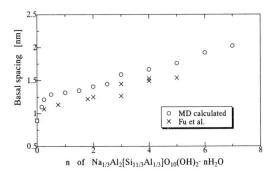

Figure 1. Swelling property (= basal spacing) for our model compared with experimental data for the Wyoming montmorillonite given by Fu *et al.* (1990).

Diffusivity and Viscosity of Pure- and Salt-Water

The MD models are of one clay-mineral layer with 3,000 water molecules for the pure water case, and with 2,970 water molecules and 30 NaCl molecules for the saltwater case, which is equivalent to the seawater. The size of a basic cell is ca. (3.1, 2.7, 15.9)[nm]. An (NVE)-ensemble MD (V the volume and E the internal energy) is carried out ($\Delta t = 0.4$ fs) after equilibrating the system with 50,000 steps of (NPT)-ensemble MD calculation. A snap shot for the clay-water system with saltwater is shown in Figure 2.

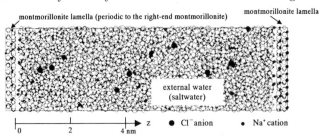

Figure 2. Snap shots for montmorillonite hydrate with salt-water.

We divide the clay-water system into 50 slices with 0.186nm thickness in z-direction, then we can calculate the mean square displacement (m.s.d.) and the diffusivity (slope of the m.s.d.) for each molecule in the slice. In Figure 3 the m.s.d. in the 16th slice accounted from the left-end side is shown for the saltwater case.

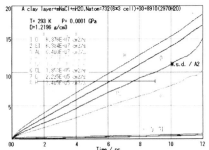

Figure 3. Mean square displacement (m.s.d.) in a slice for salt-water.

By applying the Stokes-Einstein relationship with its diffusing spare $\delta = 0.152$[nm], which is obtained by our MD calculation for pure water (without clay mineral), the viscosity of water at each sliced region is determined. Figure 4(a) shows the diffusion coefficient and viscosity for pure water in each slice, and Figure 4(b) does for saltwater.

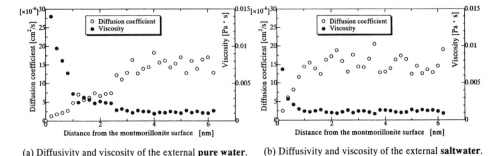

(a) Diffusivity and viscosity of the external **pure water**.   (b) Diffusivity and viscosity of the external **saltwater**.

Figure 4. Distribution of diffusivity and viscosity in pure- and salt-water.

We find the structurally ordered water layer in contact with the clay-surface, which is called the "ice sheet". Thickness of the sheet is ca. 0.5nm for the pure water case, and is equivalent to two layers of water molecules. In the diffusion layer of 3 to 4nm thickness, the diffusion coefficient is rapidly changed with distance from the clay surface. The viscosity is also changed in this region. We call such a water property the *iceberg effect*. For the saltwater case the ice sheet becomes thinner as shown in Figure 4(b).

SEEPAGE PROBLEM BY HOMOGENIZATION ANALYSIS (HA)

HA is a new type of the perturbation theory developed for a micro-inhomogeneous material with periodic microstructure (Figure 5). We here apply HA to the seepage problem in bentonite with distributed water viscosity in vicinity of montmorillonite. For this problem we start with the Navier-Stokes equation, and obtain a macroscopic seepage flow equation including the effect of spatial distribution of viscosity.

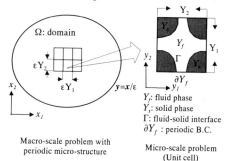

Figure 5. Macro- and micro-scale problems in HA.

HA Formulation of Seepage Problem with Distributed Viscosity

We think a flow problem in porous media whose domain is microscopically periodic (Figure 5). Here the local coordinate system y is related to the global coordinates x by $y = x/\varepsilon$. Let us think the following incompressible Navier-Stokes flow field:

$$-\frac{\partial P^\varepsilon}{\partial x_i} + \eta \frac{\partial^2 V_i^\varepsilon}{\partial x_k \partial x_k} + F_i = 0 \ \text{ in } \Omega_{\varepsilon f}, \quad \frac{\partial V_i^\varepsilon}{\partial x_i} = 0 \text{ in } \Omega_{\varepsilon f}, \quad V_i^\varepsilon = 0 \text{ on } \partial\Omega_{\varepsilon f} \quad (1)$$

where V_i^ε is the velocity vector with the shearing viscosity η, P^ε the pressure, F_i the body force vector, and $\Omega_{\varepsilon f}$ the water flow region in the global coordinate system ($\partial\Omega_{\varepsilon f}$ its boundary). We introduce an asymptotic expansion

$$V_i^\varepsilon(x) = \varepsilon^2 V_i^0(x, y) + \varepsilon^3 V_i^1(x, y) + \cdots, \quad P^\varepsilon(x) = P^0(x, y) + \varepsilon P^1(x, y) + \cdots, \quad (2)$$

where $V_i^\alpha(x, y)$ and $P^\alpha(x, y)$ ($\alpha = 0, 1, \ldots$) are Y-periodic functions such as $V_i^\alpha(x, y) = V_i^\alpha(x, y+Y)$, $P^\alpha(x, y) = P^\alpha(x, y+Y)$ with the size of a unit cell Y. And new variables $v_i^k(y)$ and $p^k(y)$ ($k = 1, 2, 3$), called the characteristic functions, are defined by

$$V_i^0 = \left(F_k(x) - \frac{\partial P^0(x)}{\partial x_k}\right) v_i^k(y), \quad P^1 = \left(F_k(x) - \frac{\partial P^0(x)}{\partial x_k}\right) p^k(y). \quad (3)$$

Then we have the following *micro-scale equations* [MiSE] of only y:

$$-\frac{\partial p^k}{\partial y_i} + \eta \frac{\partial^2 v_i^k}{\partial y_j \partial y_j} + \delta_{ik} = 0 \ \text{ in } Y_f., \quad \frac{\partial v_i^k}{\partial y_i} = 0 \text{ in } Y_f, \quad v_i^k = 0 \text{ on } \Gamma. \quad (4)$$

Now we introduce a mass averaging operation for Eqn$(3)_1$, and get the Darcy's law:

$$\tilde{V}_i^0 = K_{ji}\left(F_j - \frac{\partial P^0}{\partial x_j}\right), \qquad K_{ji} = \tilde{v}_i^j = \frac{1}{|\boldsymbol{Y}|}\int_{Y_f} v_i^j\, dy \qquad (5)$$

where $\tilde{V}_i^0$ is the averaged mass velocity in the unit cell ($|\boldsymbol{Y}|$: volume of the unit cell).

Averaging of ε^{-2}-term of the mass conservation equation yields the following *macro-scale equation* [MaSE], called the HA-seepage equation:

$$\frac{\partial \tilde{V}_i^0}{\partial x_i} = 0 \quad \text{or} \quad \frac{\partial}{\partial x_i}\left\{K_{ji}\left(F_j - \frac{\partial P^0}{\partial x_j}\right)\right\} = 0 \quad \text{in } \Omega. \qquad (6)$$

The first order approximations of pressure P^ε and velocity V_i^ε are given by

$$V_i^\varepsilon(\boldsymbol{x}) \simeq \varepsilon^2 V_i^0(\boldsymbol{x}, \boldsymbol{y}), \qquad P^\varepsilon(\boldsymbol{x}) \simeq P^0(\boldsymbol{x}). \qquad (7)$$

In geotechnical engineering we usually use the following empirical Darcy's law

$$\tilde{V}_i' = -K_{ij}'\frac{\partial H}{\partial x_j}; \qquad H = \frac{P}{\rho g} + \zeta \qquad (8)$$

where $\tilde{V}_i'$ is the average velocity, H the total head and ζ the elevation head. Compared this with Eqns (5)-(7), we know the correspondence

$$\tilde{V}_i' = \tilde{V}_i^\varepsilon \simeq \varepsilon^2 \tilde{V}_i^0, \qquad (9)$$

so we have the following interpretation between the HA-*permeability* K_{ij} and the conventional one (called the C-*permeability*) K_{ij}':

$$K_{ij}' = \varepsilon^2 \rho g K_{ij}. \qquad (10)$$

where ρ the mass density of water which is assumed to be constant because of incompressibility, and g the gravitational acceleration.

Numerical Results and Discussion
As a finite element model of MiSE for the montmorillonite hydrate with pure- and salt-water, we employ the unit cell as shown in Figure 6. Here the viscosity at a Gaussian point of FE is specified by using the data shown in Figure 4(b) and 4(c). The calculated C-permeability transformed from the HA-permeability by Eqn(10) is given in Figure 7.

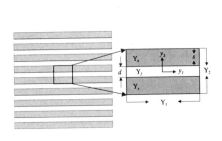

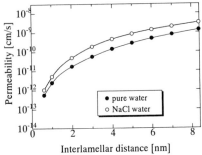

Figure 6. Unit cell for plane flow.   Figure 7. Permeability for pure- and salt-water.

CONSOLIDATION UNDER CHANGE OF PERMEABILITY

We calculate the long-term behavior of water-saturated bentonite on the basis of elasto-plastic consolidation theory under the change of permeability, which is obtained by the preceding MD/HA method for seepage problem. We assume the followings:
1) Bentonite consists of pure montmorillonite, and the involved water is only of interlayer type. This is supported by experiments for the compacted bentonite (Pusch 1994).
2) Groups of montmorillonite lamellae are located in random direction, so the water flow in bentonite is isotropic.
3) The permeability changes depending on the volumetric strain ε_v. It can be assumed that there is no volume change of the solid part (i.e., montmorillonite minerals), so for the montmorillonite lamellae we have

$$\varepsilon_v = \frac{d - d'}{2s + d},\tag{11}$$

where d is the interlayer distance before deformation, d' the distance after deformation and s the thickness of a montmorillonite lamella.

We use the Cam clay model (see Wood 1990) as an elastoplastic model whose yield function can be written as

$$f = \frac{\lambda - \kappa}{1 + e_0}\left\{\ln\left(\frac{p'}{p'_0}\right) + \frac{\eta}{M}\right\} - \varepsilon_v^p = 0\tag{12}$$

where p' is the mean stress, $\eta = q/p'$ the stress ratio (q the deviator stress) and ε_v^p the volumetric plastic strain. Material parameters are shown in Table 1 for Kunigel VI (a bentonite clay produced in Japan) with its dry density 1.8g/cm^3.

Table 1. Cam clay parameters for compacted bentonite (PNC 1997).

Slope of normal compression λ in $v : \ln p'$ plane	9.12×10^{-2}
Slope of unloading-reloading κ in $v : \ln p'$ plane	4.78×10^{-2}
Shape factor M for ellipse/slope of critical state line	0.58
Initial void ratio e_0	0.53
Reference size of yield locus p'_0	4.7(MPa)

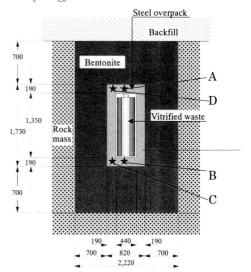

Figure 8. An engineered barrier system for HLRW.

By using a model of the engineered barrier system (EBS) for HLRW shown in Figure 8, we calculate the long-time behavior of EBS including the surrounding rock mass of granite. The FE calculation is performed under plane strain condition, and during the deformation the permeability is changed as followed the value shown in Figure 7 with its initial value $K_0' = 4.81 \times 10^{-13}$cm/s. Time dependent subsidence is given in Figure 9 at each point A, B, C and D. The distribution of permeability at $t = 1,000$days and $t = 30,000$years is found in Figure 10.

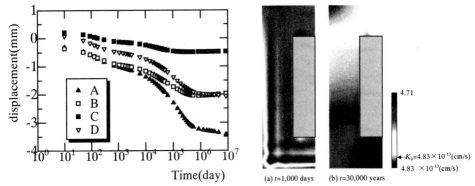

Figure 9. Calculated subsidence.   Figure 10. Permeability distribution.

CONCLUSIONS

For analyzing the seepage problem in bentonite clay in pure- and salt-water we developed a unified MD/HA procedure. The method provides the integrated interpretation of micro-inhomogeneous material behavior from the molecular level to the micro/macro-continuum level. In the unified MD/HA method we applied MD for determining properties of each constituent component, then HA is used for relating the microscopic characteristics to the macroscopic behavior. That is, in this seepage problem we calculate the profile of water viscosity near clay surface by MD, and we derive the Darcy's law and macroscopic seepage equation by HA in relation to the conventional seepage problem.

We next calculate consolidation behavior of bentonite for a model of EBS in HLRW management. We introduce the permeability calculated by the unified MD/HA method, which is changed corresponding to the volumetric strai.

Our results can be summarized as follows: 1) The *iceberg effect* is quantitatively calculated by MD, that is, water molecules are constrained at the surface of clay mineral like ice, and in the vicinity of the surface the water viscosity is rapidly changed. The profile of the salt-water case is different from the pure-water one. 2) The *close-distance effect* of neighboring clay minerals is obtained by HA, that is, the water flow in the interlamellar space is extremely restricted because the distance of a montmorillonite mineral to adjacent ones is very narrow. Note that in highly compacted bentonite it is understood that the most of water is of the interlamellar type. This fact is shown by the numerical solution of HA. Because of the coupled phenomenon of these two effects, we can conclude that the water flow in the bentonite clay is crucially prevented. The permeability for the salt-water case is rather higher than one for the pure-water case. 3) We can calculate the long-time deformation behavior of bentonite by using the MD/HA seepage model and the Cam clay type of consolidation model.

It is important to understand that by this unified MD/HA method we can determine the true velocity field of water in the microscopic point of view, so it is easy to combine this result to the mass transportation problem in bentonite, and on the very long-time behavior of bentonite we need to consider chemical change of bentonite.

The authors are grateful that Tokyo Electric Power Company supported a part of this research.

REFERENCES

Allen, M.P., & Tildesley, D.J. (1987); *Computer Simulation of Liquids*, Oxford Sci. Pub.

Boek, E.S., Coveney, P.V., & Skipper, N.T. (1995-1); "Molecular modeling of clay hydration: A study of hysteresis loops in the swelling curves of sodium montmorillonite", *Langmuir*, 11, 4629-4631.

Boek, E.S., Coveney, P.V., & Skipper, N.T. (1995-2); "Monte Carlo molecular modeling studies of hydrated Li-, Na-, and K-smectite: Understanding the role of potassium as a clay swelling inhibitor", *J. Am. Chem. Soc.*, 117, 12608-12617.

Delville, A., & Sokolowski, A. (1993); "Adsorption of vapor at a solid interface: A molecular model of clay wetting", *J. Phys. Chem.*, 97, 6261-6271.

Delville, A. (1995); "Monte Carlo simulations of surface hydration: An application to clay wetting", *J. Phys. Chem.*, 99, 2033-2037.

Fu, M.H., Zhang, Z.Z., & Low, P.F. (1990); "Changes in the properties of a montmorillonite-water system during the adsorption and desorption of water hysteresis", *Clays and Clay Minerals*, 38, 485-492.

Ichikawa, Y., Kawamura, K., Nakano, M., Kitayama, K., & Kawamura, H. (1999): "Unified molecular dynamics and homogenization analysis for bentonite behavior; Current results and the future possibility", *Engineering Geology*, to be appeared.

Karaborni, S., Smit, B., Heidug, W., Urai, J., & van Oort, E.(1996); "The swelling of clays: Molecular simulations of the hydration of montmorillonite", *Science*, 271, 1102-1104.

Kawamura, K. (1990); *Molecular Dynamics Simulation Using Personal Computer*, Kaibundo (in Japanese).

Kawamura, K. (1992); "Interatomic potential models for molecular dynamics simulations of multi-component oxides", in *Molecular Dynamics Simulations* (ed. F. Yonezawa), Springer-Verlag, 88-97.

Kawamura, K., Ichikawa, Y., Nakano, M., Kitayama, K., & Kawamura, H., (1997); "New approach for predicting the long term behavior of bentonite: The unified method of molecular simulation and homogenization analysis", *Sci. Basis for Nuclear Waste Management XXI*, Material Research Soc., 359-366.

Kumagai, N., Kawamura, K., & Yokokawa, T. (1994); "An interatomic potential model for H_2O: Applications to water and ice polymorphs", *Mol. Simul.*, 12(3-6), 177-186.

PNC (1997); *Consolidation Characteristics of Buffer Material*, PNC TN8410 97-015 (in Japanese).

Pusch, R. (1994); *Waste Disposal in Rock*, Elsevier.

Sanchez-Palencia, E. (1980); *Non-Homogeneous Media and Vibration Theory*, Springer-Verlag.

Skipper, N.T., Chang, F.-R.,C., & Sposito, G. (1995); "Monte Carlo simulation of interlayer molecular structure in swelling clay minerals. 1. Methodology", *Clays and Clay Minerals*, 43(3), 285-293.

Skipper, N.T., Sposito, G., & Chang, F.-R.C. (1995); "Monte Carlo simulation of interlayer molecular structure in swelling clay minerals: 2. Monolayer hydrates", *Clays and Clay Minerals*, 43(3), 294-303.

Teppen, B.J., Rasmussen, K., Bertsch, P.M., Miller, D.M., & Schaefer, L. (1997); "Molecular modeling of clay minerals. 1. Gibbsite, kaolinite, pyrophyllite, and beidellite", *J. Phys. Chem. B* 101, 1579-1587.

Wood, D.M. (1990); *Soil Behaviour and Critical State Soil Mechanics*, Cambridge Univ. Pr.

Containment and Remediation

In groundwater treatment of polluted groundwater

P J Barker, Bachy Soletanche Ltd, Godalming, UK, and J J Kachrillo, Soletanche Bachy, Nanterre, France.

Summary: The combination of conventional vertical containment barriers and drain panels containing treatment material allows the concept of active containment to be proposed as an original and economic solution for insitu treatment of industrial pollution of groundwater. Associated with instrumentation of the site, the pollution is isolated from the environment and treated at low cost.

INTRODUCTION

The remediation of polluted sites, and the treatment of that pollution, constitutes a complex problem, and every case is special. Many solutions are proposed to clean or to isolate pollution, such as bioremediation, soil washing or flushing, electrochemical, vacuum extraction, solidification/stabilisation, containment etc.

The choice of remediation technique(s) at a particular site takes into account not only the chemical but also the geological and hydrogeological conditions. In many cases it may be necessary to preserve the groundwater flow.

In order to avoid spreading of industrial or accidental pollution, watertight containment barriers are often installed. They are often combined with a pumping or extraction system to enable the ground water levels to be controlled and/or to allow treatment at surface before return to the groundwater outside the contained/contaminated area. This commonly applied concept has a number of disadvantages, primarily the need for pumping and exsitu treatment and monitoring on the site for what may be a considerable period, until such time as the pollution has reached acceptable levels.

Bachy Soletanche have developed several innovative techniques in this field and it is the combination of two of these that permits insitu treatment of pollution at sites where maintenance of the groundwater regime is critical. Those techniques are "panel drain" and "ecosol," which together with conventional in ground barriers allow the implementation of active containment schemes.

PANEL DRAIN

This technique originally comprised a series of drainage panels, separated by watertight bentonite/cement slurry panels. A pipe, at the base of the watertight panels, provides connection between the panels, whilst a valve system, operated from the surface, can be installed to regulate the connection. This technique has been used in France to solve a number of problems posed by the construction of deep drainage trenches on sloping ground, and to intercept and control destabilising groundwater flow, especially where traditional techniques would cause additional instability during construction.

Geoenvironmental engineering, Thomas Telford, London, 1999, 353–360

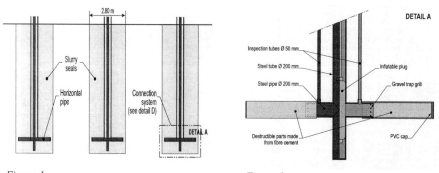

Figure 1   Figure 2

The construction method used involves constructing a panel under bentonite cement slurry, and before the slurry has set, a horizontal pipe (200 to 400mm diameter), linked to a vertical standpipe, is installed at the base of the panel. The pipe has, at both ends, sacrificial parts sealed with caps, and a temporary inflatable plug prevents any flow occurring through the pipe (Figs 1&2). When the slurry has set, the intermediate panels are excavated using bio-degradable drilling mud as temporary support. The set slurry immediately adjacent to the panels is excavated at the same time as the sacrificial pipe ends, thus providing the potential for drain continuity (by removal of the inflatable plugs) once the installation is complete (Figs 3&4).

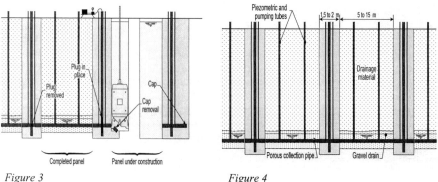

Figure 3   Figure 4

The patented process has been awarded the innovation prize of the "Fédération des Travaux Publics" in 1997 and has been used to stabilise large landslips near Lyon and Marseilles.

ECOSOL
A number of physical and chemical processes can be used to trap and remove pollutants from groundwater. Examples are precipitation, adsorption and ionic exchange.

Precipitation
Chemical precipitation is mainly used to eliminate dissolved heavy metals such as iron, nickel, copper, lead, trivalent and hexavalent chromium, etc. These cations are often precipitated in the form of metallic hydroxides in chemical reactions controlled by the pH. In cement based materials, lime solubility dictates a pH of 12-12.5. However, this reserve of

alkali in cements, although large, is not infinite, and alkaline buffers may be incorporated in trapping materials to extend their precipitation ability with time.   In some cases, specific reagents are employed to react with pollutants and to precipitate into insoluble mineral forms.

Adsorption
Adsorption is a physical mechanism based on the properties of some porous materials to fix molecules on their surface. Specific surface governs this mechanism.

Attractive forces have different origins:
* physical bonds by pores of similar sizes than the molecules to catch,
* Van der Waals forces, electrical trapping is applicable to polarised molecules,
* surface affinity : hydrophobic organic molecules have affinities for sites on activated carbon.

Adsorption on activated carbon is a widely used method for removing organic pollutants dissolved in waste water. Adsorption capacity values vary depending on the compounds to be removed. For instance adsorption capacity of chlorophenol is six times higher than that of butylacetate in same conditions.   Some clays are able to adsorb cations such as cadmium, strontium, mercury, nickel, zinc ...

Ion Exchange
This process is employed for removing dissolved anions in polluted groundwater. Removal method can be percolation through specific resins or clays. Elimination of cyanides by such a mechanism is well understood.

Ecosol is the name given to the general concept of utilising the above retention mechanisms in active containment schemes, and its effectiveness has been demonstrated a number of times.   For example, a conventional active barrier was constructed in 1994 in northern France near Lille.   In that case material utilised for the construction of the A22, many years earlier, was found to contain hexavalent chrome which was leaching away from the highway embankment and towards adjacent business premises.   An active barrier was designed to have a trapping potential of 15kg of chromium 6 per linear metre of trench.   Monitoring has shown that chromium concentrations remain high upstream of the barrier, but are negligible downstream.

BARRIER TREATMENT CHANNELS
A further refinement of the concept is to contain the pollution using low permeability ($< 10^{-9}$ m/sec) vertical barriers, and to place panel drains at intervals around the containment, across the line of the barrier, to provide treatment channels to allow controlled passage and treatment of contaminated groundwater.   A typical arrangement is shown (Fig 5). In this instance the site would not be capped, as is usual for conventional containment, but percolation of rainfall into the site would provide the driving head through the barrier treatment channels. A particular example, at Auby, near Lens in France, is described below.

Introduction
The site is an old cokeworks, and during many years of production it had become polluted with PAH, BTEX and various aromatics. The owner of the site, Charbonnages de France, had begun to dismantle the works, and during preparatory works had mobilised the contamination at the site and the groundwater downstream had become polluted.

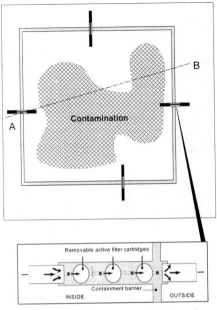

The site is situated near the Canal de la Haute Deule, and is in hydraulic contact with the canal by means of a small stream, which passes down gradient of the site. Concentrations of PAH have been found in the canal sediments in the past.

In the short term, clay underlying the site isolates the shallow polluted groundwater from the chalk aquifer below. However the thickness of the clay is as little as 2.5m, and so the potential of migration of pollution into the chalk in the longer term is a real possibility.

It was therefore decided to put into place an active barrier, under the protection of which all the remaining demolition and subsequent decontamination works could take place without prejudicing the quality of the groundwater downstream.

Figure 5

Site Description

Investigations have revealed fill overlying clayey sands to a depth of between 4.5 and 7.5m, all containing pollution. This is underlain by a clay layer between 2.5 and 10m thick, within which are lenses of clayey sands. Below the clay lies a white flinty chalk.

On the hydrological map for the area two distinct water tables are shown. A study of hydraulic gradients and flows confirms that they are separated by the relatively impermeable clay layer at depths of between 4.5 and 7.5m.

Contamination

Investigations have revealed significant contamination, with levels locally higher than Dutch intervention thresholds in PAH, BTEX and phenols. There are also locally high levels of heavy metals such as cadmium, zinc and lead. The concentrations recorded are consistent with the history of the site and the odours from the ground.

From analysis of groundwater samples from the site, it is evident that levels of PAH, BTEX and phenols increased significantly after removal of storage tanks from the site. The heavy metals were not evident and appeared to be relatively immobile.

Although not presenting the same dangers to public health as the PAH and BTEX, sulphates and ammonium were also found to be present at concentrations aggressive towards bentonite/cement slurries. The average concentration of sulphates and ammonium in groundwater were found to be 2000 mg/l and 650 mg/l respectively. Thus, in relation to the sulphates the groundwater is identified as aggressive (level A3 of the standard P 18-011) whilst in relation to the ammonium the groundwater is identified as very aggressive (level A4 of the standard P 18-011).

The Solution

Having regard to the above, the targets for the works were set as follows:

- to isolate the site from its surroundings such that the groundwater beyond the immediate confines of the site will no longer be polluted, specifically the stream on the north boundary of the site. This will be particularly the case during further groundwork at the site to remove foundations etc. This is to be achieved whilst causing minimum disruption to the existing groundwater regime;
- to fix pollutants contained in the groundwater, notably PAH, phenols and BTEX;
- to make the system durable in relation to the aggressive nature of the untreated groundwater;
- maintenance and monitoring of the works to be as simple and as cost effective as possible.

The conventional means of achieving the above would be to construct a vertical containment barrier around the entire site and to pump and treat the contaminated groundwater at a treatment plant on site before disposal. However the barrier length required would be in excess of 1200m, and as described earlier, the active pump and treat system would be relatively expensive to set up and run.

Taking into account the groundwater movements and hydraulic gradients at the site it was decided to construct a partial containment along the northern site boundary parallel to the stream, with short returns at each end. The containment would be by conventional bentonite/cement slurry, single-phase construction. Three barrier treatment channels were constructed within the barrier length, using the drain panel process to afford maintenance of groundwater flow and treatment of groundwater exiting the site. The general layout of the site is shown (Fig 6).

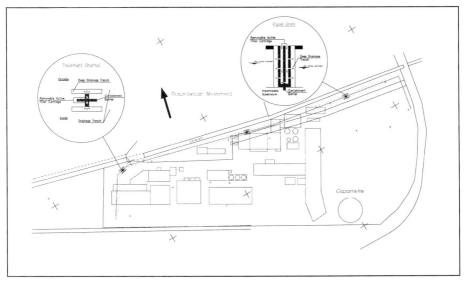

Figure 6

Aggressive Nature of Groundwater

As previously noted, the ammonium and sulphates in the groundwater at the site have the potential to adversely affect the bentonite/cement slurry forming the vertical barrier, as follows:

$$2NH_4NO_3 + Ca(OH)_2 \rightarrow Ca(NO_3) + 2NH_4OH$$

or

$$(NH_4)_2SO_4 + Ca(OH)_2 \rightarrow CaSO_4.2H_2O + 2NH_4OH$$

The lime contained in the mix is therefore dissolved. Laboratory immersion tests have demonstrated a change in colour of a traditional mix from blue/green to brown and loss of any measurable strength in a period of 50 days. In addition the ammonia formed passes partly to the gaseous state which can result in an increase in permeability.

The mix design for the bentonite cement slurry therefore included a high proportion of PFA, in addition to CLK cement (containing OPC and GGBFS), bentonite and specific additives to aid mixing and resist the effects of the ammonium.

Barrier Construction

The scope of the wall was as follows:

Length	-	436 m
Thickness	-	600 mm
Depth	-	6 to 8.5m
Elevational area	-	3322 m^2

The wall was continuously excavated under bentonite cement slurry by long reach backhoe using well-established techniques. The required 1m penetration into the clay layer was achieved by recording the depth at which clay was first brought to the surface, and excavating for a further metre.

Barrier Treatment Channel Construction

Three treatment channels were constructed, spaced at approximately 100m intervals. Their locations are shown in Fig 6. Each treatment channel is made of an "H", whose central part is the panel drain itself and the branches are drainage channels which lead into the central portion. The panel drain is made up of two cylinders of 400mm diameter connected so as to allow them to work in series, in parallel or individually. This allows the flow through the treatment channel to be adjusted if necessary (for example, in periods of high rainfall). The pipework is made of stainless steel.

After completion and initial set of the barrier wall, cross panels of bentonite cement slurry were constructed at each gate location and, before the slurry had started to set, the pre-fabricated panel drain with sealed inlet and outlet pipes, was lowered into position in the slurry. Once the channel slurry had set, the sacrificial pipe ends were removed affording flow paths through the panel drain. Fig 7 shows a panel drain being placed.

The branches, each 50m long, to each side and end of the panel drain, were 5m deep and backfilled with single size stone. They were constructed by drain trenching equipment.

Each cylinder within the panel drain has a removable cartridge which contains the treatment material.

Figure 7

Control and Instrumentation

It is anticipated that after the first year, each cartridge will need to be replaced annually. Regular visits and associated costs for analysis to check the effectiveness of treatment and the need to replace the treatment cartridges would be a significant cost to the client, particularly during the first year. This has been minimised by using automated techniques wherever possible. To achieve this, Soldata, part of the Soletanche Bachy group of companies, has installed instrumentation which automatically monitors the PAH in the groundwater at each treatment channel using fluorescence.

In addition groundwater levels upstream and downstream are monitored. The data acquisition centre is in a cabin on site, and the results can be monitored remotely using a modem. Each treatment channel is monitored for four hours in each 24 hours.

Results

After 2 months the results can be briefly summarised as follows:

Figure 8

- The PAH level has been close to zero downstream of the treatment channels since completion of the installation.
- Upstream, after an initial period of very high levels due to ground disturbance during construction, the PAH level has decreased to the pre-construction level (approx. 100 ppb).
- With two filters working in parallel, the difference in head between upstream and downstream of the treatment channel is no more than 150mm.

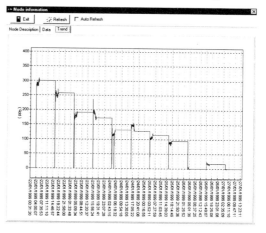

The monitoring is carried out at 6 locations, upstream and downstream of each treatment channel, by pumping to a central unit at the site, where analysis and reporting is automatically carried out. Each location is monitored for one two hour period in every twelve hours.

Figures 8,9 & 10 show results from the remote monitoring at the site, and all relate to the same barrier treatment channel. Figure 8 shows the results on the upstream (or dirty) side of the gate over an initial period of seven days, and indicates steadily reducing PAH contamination over that period, as noted above. The lengths of horizontal traces indicate the periods when other points are being monitored.

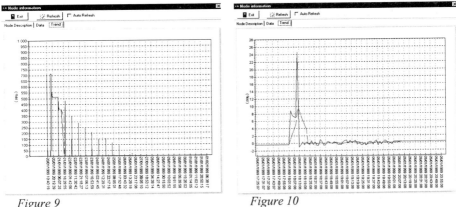

Figure 9   *Figure 10*

Figure 9 shows the results from the clean side over a period of 12 days. The peaks are due to the changeover from monitoring the dirty side, as it takes a short while for the pipes to be flushed and for the clean side groundwater to reach the central analysis unit. After this initial period, the PAH level is shown as zero. This is further demonstrated by on Fig 10, which is a zoom on one, two hour monitoring period. Pumping commences at 18.01 hours on 26 January and stops at 20.01 hours. Between 18.01 and 18.13, the high values indicate "contamination" from the dirty side of the channel, whilst from 18.13 to 20.01 values are shown as oscillating close to zero.

Whilst PAH is not the only contamination at the site, it has been agreed with the client, and with the French administration, that only the PAH need be monitored regularly, as that is the major pollution at the site. The active material in the treatment channels will also adsorb other contaminants at the site. They are checked for by the client at lesser intervals by "traditional methods".

As monitoring proceeds, the cartridges containing the active treatment media can be changed as each become saturated. This will be carried out by a single person using a specially designed hoist sited adjacent to each treatment channel.

Acknowledgement
The help and assistance of Charbonnages de France is gratefully appreciated.

References
P18-011. Normalisation française. Classification des environnements agressifs.
A DENIAU. De la paroi drainante au procédé panneau-drain. Travaux No.725 Novembre 1996 pp 42-47.
G EVERS. Pracicals solutions for the treatment of groundwater. NATO/CCMS Pilot Phase Study Phase III, EPA 542-R-98-003.
M RAT et al. Etude et traitement d'une pollution au chrome due à l'utilisation de déchets dans des remblais autoroutiers. Symposium International "Exemples majeurs et récents en Géotechnique de l'Environnement". Février 1996. Paris.

Electrokinetic treatment of cadmium spiked clays. Experiments.

Pascal MARCEAU, Paul BROQUET, Laboratoire de Géologie Structurale et Appliquée, Université de Franche-Comté, Besançon, France, and Pascal BATICLE, EDF, Direction des Etudes et Recherches, Service Applications de l'Electricité et Environnement, Moret-sur-Loing, France.

INTRODUCTION

Recently, attention has turned to electrokinetic techniques that make it possible to move and extract contaminants from fine-grained soil under an electric field (Figure 1). The application of a constant electric current has several effects : (1) electrolysis of water, plating reactions and gas formation occur at the electrodes. H^+ and O_2 are produced at the anode, and OH^- and H_2 at the cathode (2) the electric potential difference leads to electroosmosis , the pore water flow is toward the cathode, since most soils have a negative surface charge (3) the electric field initiates electromigration of species available in the pore fluid and of those introduced at the electrolytes. These phenomena change the chemical pore fluid composition and induce sorption reactions in the soil. Bench scale studies showed that inorganic species and heavy metals such as arsenic, cadmium, chrome, copper, iron, lead, mercury, nickel and zinc can be efficiently extracted from polluted or spiked soils (Pamucku and Wittle, 1992; Acar and al., 1995; Cox and al., 1996; Reed and al., 1995; Li and al.,1996; Acar and Alshawabkeh, 1996, Marceau and al., 1999). Extraction rates of over 90 % are reported. But species seem to precipitate with OH^- near the catholyte. Organic species such as benzene, phenol and acetic acid can also be removed (de Marsily and al., 1992; Bruell and al., 1992; Acar and al., 1992). The objectives of these pilot-scale studies are (1) to investigate the feasibility and efficiency of extracting cadmium (Cd) from an artificial soil under an electrical field at dimensions more representative than bench-scale studies (2) to avoid Cd hydroxide formation near the catholyte in order to extract all the Cd from the soil (3) to investigate influence of heterogeneities.

EXPERIMENTAL PROCEDURE

Soil composition

Two different materials were used in the tests, the KAOLIN P300, an industrial material, and the SILT77, a natural soil. Their granulometries are similar with about 50 % silt (63 to 2 μm) and 50 % clay (< 2 μm). Their mineralogies are very close, with about 40 % quartz and 55 % phyllosilicates. The main difference is the phyllosilicate composition with 95 % kaolinite for KAOLIN P300, and 75 % kaolinite plus 25 % chlorite for SILT77. Their maximum cation-exchange capacity (CEC) at pH greater than 6.5 for Cd is 2.5 to 2.6 meq/100 g (1meq Cd = 56.2 mg) for KAOLIN P300 and 2.75 meq/100 g for SILT77. Below pH 2 and for a concentration under 1 meq/l, CEC is 0.2 meq/100 g for KAOLIN P300 whereas for SILT77 it is 0.8 meq/100 g. That difference is probably due to presence of chlorite in the SILT77, which has higher CEC than kaolinite.

Geoenvironmental engineering, Thomas Telford, London, 1999, 361–367

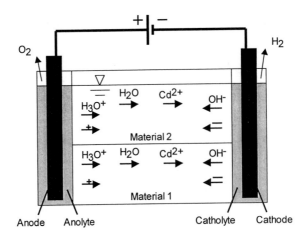

Figure 1. Electrokinetic treatment schematic view. Case of two different layers.

Equipment
The pilot-scale experiment takes place in a container open on the upper side. Its dimensions are 270 cm in length, 120 cm in height and a cathode- anode spacing of about 100 cm. The inside of the container is covered with a thin layer of impermeable and non-conductive fiberglass. Carbon electrodes are used at the anode, and steel electrodes at the cathode. A power supply that provides 0-150 V DC and 0-16 A current is used. Vertical and horizontal drains are used to pump in situ pore fluid for analyses. pH of the catholyte is monitored and controlled by the addition of sulfuric acid.
During the test , a constant electric current of 8.25 A is applied, i.e., a resulting current density of 0.3 mA/cm^2. Sulfuric acid added in the catholyte maintains the pH near 3 and prevents hydroxide formation. Concentrations of Cd^{2+}, sulfates (SO$_4$$^{2-}$), pH and electrical potential between cathode and anode are monitored. After the test, a core sampling is realized at depths of 25 cm and 75 cm to measure the final Cd concentration.

Testing program
Two pilot-scale tests are conducted. The first one is conducted using KAOLIN P300. The soil is put in the container in ten successive identical layers. Each layer is composed of a mixture of deionized water, Cd-nitrate in solution and clayey material. After consolidation, the height of the soil is 102 cm. Before the start of the experiment, the container contains 3.25 t of clayey material, 2.87 kg of Cd, for a water content of 49 % and an average Cd concentration of 882 mg/kg of dried matter (mg/kg DM).
The second pilot-scale test is conducted using KAOLIN P300 and SILT77, in order to investigate the influence of a simple heterogeneity, i.e. two superposed layers of different materials. The soil is put twice in the container in five successive identical layers. The five upper layers are composed of KAOLIN P300, the five lower ones are composed of SILT77. Each layer is composed of a mixture of deionized water, Cd-nitrate in solution and material. After consolidation, the height of the soil is 103 cm. Before the start of the experiment, the container contains 1.775 t of SILT77 and 1.662 t of KAOLIN P300, 3.16 kg of Cd, and an average Cd concentration of 919 mg/kg DM.

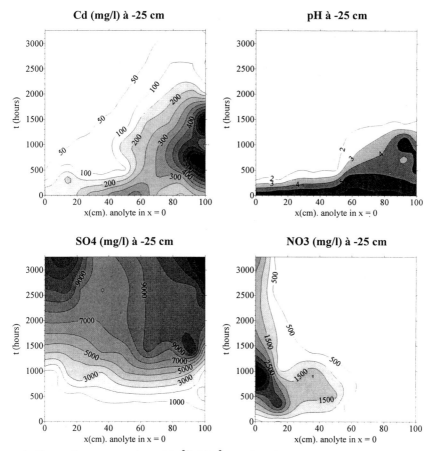

Figure 2. Change in concentration of Cd^{2+}, $SO4^{2-}$, $NO3^-$ and pH at −25 cm during the monolayer experience. At -75 cm, the results are analogous.

RESULTS AND COMMENTS

Monolayer pilot-scale test

The electro-osmose due to the process induced a consolidation phenomena. At the end of the experiment, a 5 cm average settling of the soil surface was measured.

Due to electrolysis of water, the pH of the anolyte decreased from 6 to 2 after 50 h and the pH of pore fluid from 6-7 at the beginning to 2 after 1, 500 h (Figure 2). These protons allowed the desorption of Cd^{2+} ions, which were adsorbed on the clay surface. This acid front moving toward the cathode coupled with electromigration flushed Cd^{2+} to the catholyte, where it plated onto the steel electrodes as Cd^0 (metallic form). Final Cd^{2+} concentrations in the pore fluid are below 50 mg/l over 90 % of the cathode-anode spacing.

Nitrates were also flushed to the anolyte where they concentrated. The initial concentration comes from the Cd-nitrate used to spike the clayey material. Sulfates, progressively introduced in the catholyte with sulfuric acid, moved toward the anode. Final concentrations vary from 6,000 to 13,000 mg/l.

After the 3, 259 h of electrokinetic process, the electrical potential decreased from 44.5 to 9 V, due to the increasing ionic strength of the pore fluid. The energy expenditure of the test was 430 kWh, i.e., 132 kWh per ton of dried material or 159 kWh/m^3.

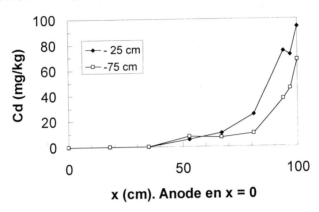

Figure 3. Monolayer test. Cd final concentrations in the clayey material at 25 and 75 cm depth.

The final Cd concentrations and pH compared with CEC predicted a residual Cd concentration of less than 137 mg/kg DM. The results given by the soil samples were better (Figure 3). The residual Cd concentration was below 10 mg/kg DM on half of the soil near the anolyte and increased to 93 mg/kg DM at a depth of 25 cm and 68 mg/kg DM at a depth of 75 cm. These values correspond to an average Cd concentration of 12 mg/kg DM and a Cd removal of 98.5 %.

Bilayer pilot-scale test
The electro-osmose due to the process induced a consolidation phenomena. At the end of the experiment, a 5 cm average settling of the soil surface was measured.

Due to electrolysis of water, the pH of the anolyte decreased from 6 to 2 after 50 h. This acid front moving toward the cathode coupled with electromigration flushed Cd^{2+} to the catholyte, where it plated onto the steel electrodes as Cd^0 (metallic form) (Figure 4). But the acid front evolution was not the same in the two materials. Whereas in KAOLIN P300 the pH of pore fluid was lower than 2 after 1, 500 h, in SILT77 the acid profil was stabilized after 1, 500 h, only a part of that material was acidified. Final Cd^{2+} concentrations in the pore fluid were below 50 mg/l on 90 % of the cathode-anode spacing in KAOLIN P300, but in SILT77 they are over 50 mg/l in the half of the cathode-anode spacing near the catholyte.

Sulfates, progressively introduced in the catholyte with sulfuric acid, moved toward the anode. As with protons, after 1, 500 h a part of SILT77 was ignored by the movement of the sulfate ions, whereas sulfates went through the KAOLIN P300.

After the 3, 260 h of electrokinetic process, the electrical potential had decreased from 40 to 10.3 V, due to the increasing ionic strength of the pore fluid. The energy expenditure of the test was 478 kWh, i.e., 139 kWh per ton of dried material or 172 kWh/m^3.

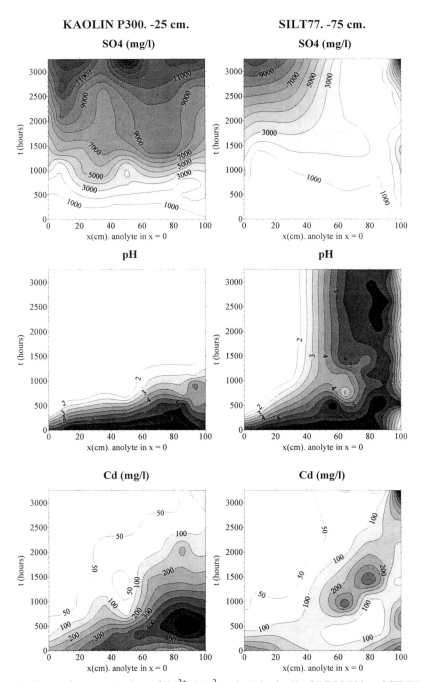

Figure 4. Change in concentration of Cd^{2+}, SO_4^{2-} and pH in the KAOLIN P300 and SILT77 layers during the bilayer experience.

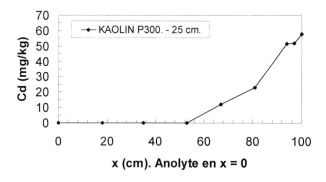

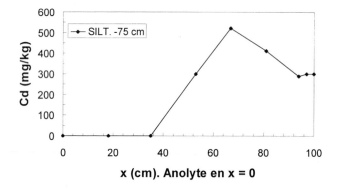

Figure 5. Bilayer test. Cd final concentration in the KAOLIN P300 and SILT77.

The final Cd concentrations and pH compared with CEC predicted a low residual Cd concentration in KAOLIN P300 and in the acidified part of SILT77. The residual Cd concentration was below 5 mg/kg DM on half of KAOLIN P300 near the anolyte and increased to 58 mg/kg of dried matter (DM) at a depth of 25 cm (Figure 5). In SILT77 at depth of 75 cm, the residual Cd concentration was below 5 mg/kg DM in the first 35 centimeters, increased to 520 mg/kg DM at 65 cm and decreased to 300 mg/kg near the catholyte. For the entire pilot, these values correspond to an average Cd concentration of 109 mg/kg DM and a Cd removal of 88 %.

CONCLUSION
The feasibility and efficiency of transporting Cd with electrokinetic treatment were investigated in two pilot-scale test. The first one was a 3.25 t clayey medium specimen spiked with Cd-nitrate solution at a concentration of 882 mg/kg DM. No scale effect was noticed. Conditioning the catholyte by adding sulfuric acid avoided hydroxide formation and gave high removal rates of Cd, here 98.5 %. The preliminary study of CEC compared with the contaminant concentration measured in situ and pH made it possible to follow the electrokinetic process.
The second test was a 3.44 t bilayer medium specimen spiked with Cd-nitrate solution at a concentration of 919 mg/kg DM An average removal rate of Cd of 88 % was reached. Nevertheless, it appeared that the two layers did not evolve in the same way during the

experiment. The SILT77 layer conductivity increasing less fast than the KAOLIN P300 layer one, the electric current was progressively focused in that one and in the part of the SILT77 layer sufficiently conductive. Consequently, a part of the SILT77 layer near the catholyte was isolated of the process. In that area, there were not enough protons to permit the metallic ions extraction, so residual Cd concentration was high.

That pilot-scale study showed that a simple heterogeneity perturbed the electrokinetic process. It seems interesting to examine closely the different kinds of heterogeneities, both for the geometry and the materials composition, and their implications. There are a lot of heterogeneities in soil, such as stratification, mineralogic composition variation, or permeability variation. These major or diffuse heterogeneities play an important role in a electrokinetic treatment because they control the current density repartition, for example the water table. As a result, some parts of the soil can be ignored by the electroremediation and remain polluted. For these reasons, it seems more realistic in most cases to excavate and to homogenize the polluted soil and to electroremediate it in tanks instead of using an in situ treatment.

Cadmium has been removed, but the soils are polluted by very low pH and sulfates. To solve that problem, the electrokinetic process can be carried on, but without pH control. Firstly, no new chemical species are introduced in the electrolytes and sulfates are progressively flushed to the anolyte where they concentrate. Secondly, the electrolytes are connected each other, the pH of the electrolytes becomes neutral. Like sulfates and nitrates, excess H^+ are progressively flushed to the catholyte, and a neutral pH is reached for the soil.

REFERENCES

Acar Y. B., Alshawabkeh A. N., 1996. – Electrokinetic remediation, I : Pilot-Scale Test with Lead-spiked kaolinite, J. Geotech. Eng., 122(3), 173 – 185.

Acar Y. B., Gale R. J., Alshawabkeh A. N., Marks R. E., Puppala S., Bricka M., Parker R., 1995. – Electrokinetic remediation: basics and technology status, J. Hazardous Mat., 42(2), 117 –137.

Acar Y. B., Li H, Gale R. J., 1992. – Phenol removal from kaolinite by electroosmosis, J. Geotech. Eng., ASCE, 118(11), 1837 – 1852.

Bruell C. J., Segall B. A., Walsh M. T., 1992. – Electroosmosis removal of gasoline hydrocarbons and TCE from clay, J. Envir. Eng., 118(5), 84 – 100.

Cox C. D., Shoesmith A., Ghosh M. M., 1996. – Electrokinetic remediation of Mercury-Contamined Soils using Iodine/Iodide lixiviant, Envir. Sci. Technol., 30(6), 1933 – 1938.

Marceau P., Broquet P., Baticle P., 1999 – Dépollution par méthode électrocinétique d'un matériau argileux dopé au cadmium. Essai pilote, C. R. Acad. Sci. Paris, 328, 37 – 43.

de Marsily G., Lancelot F., Londiche H., 1992. – Récupération assistée d'hydrocarbures dans des milieux poreux par application d'un champ électrique, Revue générale de l'électricité, R.G.E., 48 rue de la Procession, Paris, 3, 20 – 23.

Li Z., Yu J.-W., Neretnieks I., 1996. – A new approach to electrokinetic remediation of soils polluted by heavy metals, J. Contaminant Hydrology, 22, 241 – 256.

Pamukcu S., Wittle J. K., 1992. – Electrokinetic removal of selected heavy metals from soils, Environ. Progress, 11(3), 241 – 250.

Reed B. A., Berg M. T., Thompson J. C., Hatfield J. H., 1995. – Chemical conditionning of electrode reservoirs during electrokinetic soil flushing of Pb-contamined silt loam, J. Envir. Eng., 121(11), 805 – 815.

Delineation and bioremediation of chlorinated hydrocarbons for field sites in the Netherlands

Dr MARK DYER, University of Durham, School of Engineering; MARCUS VAN ZUTPHEN, TNO Netherlands Institute of Applied Geoscience; JAN GERRITSE, TNO Netherlands Institute of Environmental Technology

ABSTRACT

In the early 1990's the Dutch established a national R&D programme called NOBIS to create a market for in situ remediation technologies to treat contaminated land. The acronym stands for *Nederlands Onderzoeksprogramma Biotechnologische In-situ Sanersing*. The majority of sites under investigation in the NOBIS programme are contaminated with organic pollutants. The contaminants range from hydrocarbon fuels such as benzene and toluene that float on the water table to the denser chlorinated hydrocarbons. The first project described in the paper concerns the development of 'Dynamic Monitoring' as a novel technique for interpreting field data for an improved delineation of pollution from CHC's at a site. The second project investigates the choice and delivery of carbon substrates for the insitu bioremediation of 1,2-dichloroethane (1,2-DCA) by anaerobic respiration.

INTRODUCTION

Of the many types of pollutants that can pose a risk from contaminated land, organic pollutants are often the most troublesome. The chemicals can be environmentally significant at low aqueous concentrations (e.g. drinking water limits of $10\mu g/l$ for trichloroethylene) and extremely difficult to remove. A particularly problematic group of contaminants are dense non-aqueous phase liquids (DNAPLs), which include chlorinated hydrocarbons (commonly used as degreasers and cleaners or bulk chemicals for production of plastics).

Chlorinated hydrocarbons (CHC's) are characterised by relatively low viscosity, high density and low aqueous solubility. As a result, spillages of CHC's typically lead to deep-seated pollution, where the immiscible liquid penetrates the aquifer as a finger or ganglia that pools at obstructions (Pankow and Cherry 1996). The obstructions are often due to changes in permeability, for example at the boundary of clay or mudstone strata. Subsequently, the contaminant pollutes an aquifer by a slow rate of dissolution from residuals sorbed to the surface of soil particles or more significantly from the surface of pools of pure product (Schwille 1988).

GROUNDWATER MODELLING AT TILBURG

This scenario was encountered at a housing estate in the City of Tilburg. Redevelopment of the site uncovered an historic spillage of trichloroethylene from a former textile factory. The spillage had lead to deep-seated pollution of the underlying aquifer to a depth of 12 metres. A later ground investigation using 37 CPT soundings characterised the underlying aquifer system as principally gravely sands from the Nuenen Group, with two layers of loam at 8 and 10m metre depth.

In 1989, an initial remediation strategy involved excavation and replacement of the top 3 metres above the water table with granular fill; followed by installation of vapour and groundwater extraction wells to a depth of 12 metres. During this early stage of site remediation, no clear distinction could be made between pools and plumes of TCE, even though concentrations of up to 50,000 μg/l indicated the presence of pure product. Between 1990 and 1996, the pump and treat system extracted approximately 4300 kg of TCE, with 577 kg of TCE alone being extracted in 1996 as shown in Table 1

To investigate why the pump and treat operation had stagnated and to optimise a future remediation strategy, a consortium was formed under the NOBIS programme in 1997 to analyse seven years of groundwater monitoring data from wells already installed at the site. A 3-D groundwater model was developed for the site using the programme MODFLOW (McDonald & Harbaugh 1988, Pollock 1989). The model was used to predict changes in the direction of the groundwater flow caused by large variations in pumping rates from five extraction wells. The results enabled measurements of TCE from groundwater samples to be linked to areas up stream of monitoring wells. The approach has been termed 'Dynamic Monitoring'.

Year	P1	P2	MD1	MD2	MD3	Vapour Extract	Excava -tion	Total Mass
1990	75	1200				238	157	1670
1991	42	200				105		347
1992	6	192	5	356	51	18		975
1993			28	175	222	14		439
1994			82	201	230	9		522
1995			33	137	194	1		365
1996			11	270	289	7		577

Table 1 Mass of TCE (kg) extracted from excavation works, groundwater pumps (P1, P2, MD1, MD2, MD3) and vapour extraction wells btween 1990 and 1996.

A total of ten events were modelled involving a) significant fluctuations in pumping rates for individual deep wells, b) temporary shut down of individual deep wells and c) the change over from two deep-wells (18 metres deep) to three mini-wells (10 metres deep). Pronounced variations in groundwater pumping between these ten events caused significant changes in the direction of groundwater flow, at times leading to a reversal of flow. As a result it was possible to accurately locate high concentrations of TCE up stream of individual wells as shown in Figure 1, and to delineate the site into areas of high levels of pollution synonymous with the pooling of TCE. The results are being used to plan future remediation using steam injection with localised groundwater extraction.

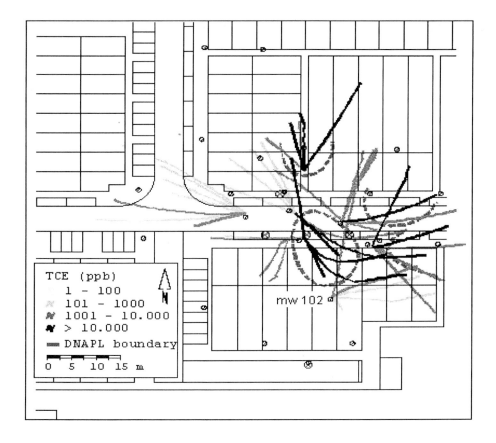

Figure 1 Plan view of the Tilburg housing estate showing variations in groundwater flow related to changes in measured concentrations of TCE at individual monitoring wells.

ANAEROBIC BIOREMEDIATION OF 1,2-DICHLOROETHANE

Similar to the housing estate in Tilburg, the second NOBIS site investigated was polluted with a CHC. On this occasion the contaminant was 1,2-dichloroethane and the location was an industrial site in the Botlek area of Rotterdam. Again, pump and treat was used to temporarily control the spread of pollution off site. However at this location, the site owners were interested in developing a long-term remediation strategy using in-situ bioremediation. In particular, the NOBIS research consortium investigated the development of a biologically active zone to transform 1,2-DCA by reductive dechlorination as illustrated by the left-hand pathway in figure 2. (Bosma et al 1998, Gerritse et al 1999, Dyer et al 1999).

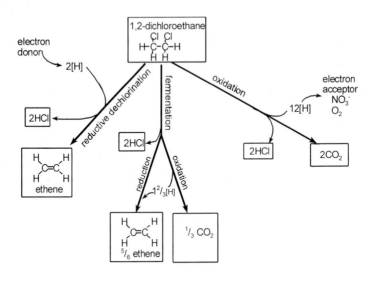

Figure 2 Transformation pathways for bioremediation of 1,2-Dichloroethane (after Gerritse et al 1999)

As part of the investigation, preliminary laboratory tests were carried out to examine the performance of different carbon substrates for the in-situ bioremediation of a chlorinated solvent 1,2-dichloroetane by reductive dehalogenation (Dyer et al 1999). The biochemical process involved the supply of electrons and protons from the metabolism of a carbon substrate for substitution of chloride ions on 1,2-DCA. The preliminary study comprised a series of anaerobic soil column tests (Bosma et al 1997). The soil columns were prepared using soil and groundwater samples from boreholes at the site. Groundwater was flushed through the columns under anaerobic conditions. A comparison was made between the transformation of 1,2-DCA without a carbon substrate and in the presence of sugars (molasses) and alcohol (methanol) respectively. In addition, different modes of delivery were investigated. In the case of molasses, the material was injected into the column as a plug to simulate grout injection in the field, where as methanol was delivered as a constant flow dissolved in the influent.

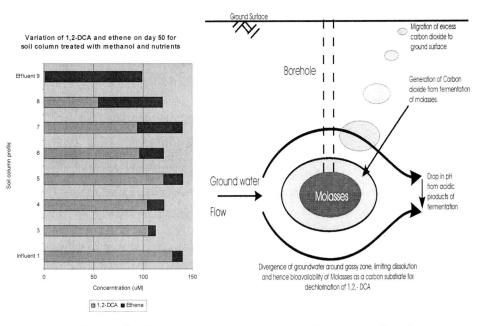

Figure 3 Preliminary soil column test results for the transformation of 1,2-DCA to ethene by dihaloelimination

Figure 4 Potential adverse effects from the fermentation of molasses as a carbon substrate for in-situ bioremediation of chlorinated solvents

Both carbon substrates led to transformation of the pollutant 1,2-DCA to principally ethene as shown in Figure 3. However, striking differences were observed between the rate and efficiency of degradation, as well as secondary effects from the fermentation of the molasses. In particular, fermentation led to a drop in pH and an excessive production of carbon dioxide, which temporarily blocked the flow of groundwater. It would be difficult to extrapolate the results to a full-scale remediation project. Nevertheless, the results highlight the potential for gas generation from fermentation to isolate molasses from the surrounding groundwater and so limit its bioavailability as a carbon substrate as illustrated in Figure 4. Furthermore, the metabolism of molasses into shorter chain organic acids and alcohols caused a temporary fall in pH, which could inhibit microbial activity.

CONCLUSION

The site investigation and remediation of land polluted with chlorinated solvents is fraught with problems. The NOBIS project carried out at Tilburg served to underline these difficulties. Firstly it is extremely awkward if not impossible to locate the pure product at depth. The relatively high density, low viscosity and low aqueous solubility leads to deep seated and localised pooling. As a result it proved impossible to sample the pure product at Tilburg. The geological setting of the site would have typically lead to pooling of the solvent in thickness of a few centimetres according to the limited data provided by Schwille from sand tank tests. As a result, the engineer is left with having to sample the dissolved product which in turn could be less than one percent of the maximum aqueous solubility because of the shallow dissolution profiles above solvent pools (i.e. less than 13000 $\mu g/l$).

Dynamic Modelling of Groundwater provided a valuable tool for interpreting the large volume of monitoring data by capturing significant changes in either the measured concentration of TCE or changes in pumping operations. Using ten events, it was possible to construct contours of locations yielding concentrations of TCE between 100$\mu g/l$ and 10,000 $\mu g/l$. Although this approach did not precisely located the pure product, it provided a greatly improved delineation of the site for targeting either more detailed groundwater sampling for locating individual pools of pure product or for future remediation using localised extraction methods or in-situ bioremediation.

With regard to the devlopment of a bio-active zone at the site in Rotterdam, soil column tests graphically showed that molasses is a complex carbon substrate for in-situ bioremediation. Unlike methanol, it was firstly fermented into to short chain organic acids and alcohols for subsequent metabolism by respirative bacteria. The fermentation process generated excessive amounts of carbon dioxide, which periodically blocked the flow of water and the supply of electron donors for the treatment of 1,2-DCA. These side effects show that a complex sugar(s) ach as molasses requires at least one extra metabolic cycle before being used as a primary ubstrate for the bioremediation of a chlorinated solvent.

would be difficult to extrapolate the laboratory results to a full-scale remediation project. owever the findings do raise a number of questions about the build up of gas pressures around a ug of molasses in the ground, its distribution and seepage through the soil and groundwater as lustrated in Figure 4. All of these issues affect the dissolution and hence bioavailability of olasses for the degradation of chlorinated hydrocarbons.

EFERENCES

Bosma TNP, Aalst-Van Leeuwen MA, Gerritse J and Van Heiningen E. Intrinsic dechlorination of 1,2-dichloroethane at an industrial site. 1998 Proc 1st Int Conf Remediation of chlorinated and recalcitrant compounds, Monterey, California.

Chapelle FH. (1993) Groundwater microbiology & geochemistry. John Wiley & Sons

Dyer MR, Gerritse J, Van Heiningen E and Bosma T. 1999. In-situ bioremediation of 1,2-dichloroethane under anaerobic conditions. ICE Proc Geotechncial Engineering (submitted for publication)

Gerritse J, Borger A, Van Heiningen E, Rijnaarts HM, Bosma TNP, Taat J, Van Winden B, Dijk J And De Bont Jam. (1999) Assessment and monitoring of 1,2-dichloroethane dechlorination. Battelle Bioremediation Conference, San Diego.

McDonald, M. G. & Harbaugh, A. W. (1988) A Modular Three-Dimensional Finite Difference Groundwater Flow Model. U.S. Geological Survey.

Pankow, J. F. & Cherry, J. A. (1996) Dense Chlorinated Solvents and Other DNAPLs in Groundwater. Waterloo Press, Portland Oregon.

Pollock, W. (1989) Documentation of Computer Programs to Compute and Display Pathlines Using Results from the U.S. Geological Survey Modular Three-Dimensional Finite Difference Groundwater Flow Model. U.S. Geological Survey.

Schwille, F. (1988) Dense Chlorinated Solvents in Porous and Fractured Media: Model Experiments (Translation by J. F. Pankow), Lewis Publishers Inc, Chelsea Michigan.

Remediation of Contaminated Ground using Soil Mix Technology: From Research to Commercialisation

DR C. W. EVANS, May Gurney (Technical Services) Ltd, Norwich, UK, and

DR A. AL-TABBAA, Engineering Department, University of Cambridge, UK

ABSTRACT

The first part of the paper presents an overview of a research and development project carried out in 1994-95 which led to the commercialisation of an in-situ treatment methodology of contaminated ground using soil mix technology. The second part of the paper presents a case study in which the process was used commercially, for the first time in the UK in 1997, to control organic pollutants in contaminated soils and groundwater on an environmentally sensitive site. The three stages of the research project, namely treatability study, site trial and assessment of the treated contaminated ground are presented and the three different treatment/containment scenarios applied in the case study are detailed. It concludes with an outline of recently completed related research work.

INTRODUCTION

Contaminated land has become an area of increasing importance over recent years, as greater environmental awareness and growing pressure on land resources have brought about the protection of greenbelt and agricultural land. The government has stated that it requires the construction of 2.4 million new homes by the year 2016, 60% on brownfield sites (Jones & Hopkins, 1997). This has placed the onus firmly on the re-development of land that was originally used for industrial purposes. However, as a result of the land's past usage, increased levels of pollution might be found within the soil and groundwater which may preclude the site from immediate construction activity. Some type of ground remediation would, therefore, be required, the choice of which being almost certainly governed by performance, speed and economics. This in itself has fuelled the need for research into finding fast, effective and economical remediation techniques that could lead to commercialisation in the future.

In recent years solidification/stabilisation has steadily emerged as an in-situ remediation process for contaminated land (Wheeler, 1995). The system uses augers down which cement-based slurries are pumped. Mixing of the slurry takes place with contaminated soil remaining *in situ*, to produce soil-cement columns. The cementitious materials and additives used in the slurry prevent the mobility of contaminants by both chemical fixation and mass encapsulation processes. The technology was originally developed in the 1960's and used for groundwater cut-off and excavation support systems (Yang, 1994). Soil mixing is competitive with most conventional treatment, disposal and containment methods and has advantages in reduced health and safety risks, speed of construction, elimination of off-site disposal and low cost.

The success of the treatment relies on the slurry additives coming into direct contact with the contaminants in the ground. This is achieved by thorough mixing of the grout with the soil to produce a series of homogenous soil-grout columns. Stabilisation/Solidification techniques using cement-based materials have been practised for many years for the treatment of wastes containing inorganics and heavy metals (Conner, 1990). It is widely accepted, however, that some organic substances tend to interfere with the hydration of cementitious materials (Young, 1972). Special additives, such as organophilic clays, can be added to reduce these effects by acting as a link between the cement and the contaminant (Alther et al, 1988).

The paper begins with a summary of the findings of a research and development project on the in-situ solidification/stabilisation treatment of an Ministry of Defence owned site in West Drayton, Middlesex, contaminated with a cocktail of organic compounds and heavy metals. The project was funded by the Department of the Environment under its environmental technology innovation scheme and was carried out over a fifteen-month period in 1994-95. The project was carried out by the University of Birmingham, May Gurney (Technical services) Ltd and Envirotreat Ltd in collaboration. The project included a laboratory treatability study, a site trial and subsequent laboratory assessment of the properties of the in-situ treated ground according to a specified set of design criteria. This work led to the commercialisation of the treatment process and a case study in which the process was used commercially, for the first time in the UK in 1997, to control organic pollutants in contaminated soils and groundwater on an environmentally sensitive site also in West Drayton is then presented. It concludes with an outline of recent relevant research.

THE RESEARCH WORK

A former chemical works site located in West Drayton in Middlesex, owned by the Ministry of Defence Research Agency was selected for the research project. The ground conditions are given in Table 1 and show that made ground consisting of layers of sandy clay and clayey sand with fragments of bricks, metal and concrete was present to a depth of 1.7m. This was underlain by natural river terrace deposits of sands and gravels which in turn were underlain by London clay at a depth of 4-5m.   Due to the former use of the site a variety of contaminants were found to be present in the soils as detailed in Table 2 which included solvents, oils, petroleum hydrocarbons and heavy metals. The groundwater was also found to be contaminated with concentrations well above drinking water standards.

soil description	Depth
MADE GROUND: fine to coarse SAND	0.10 - 0.40
MADE GROUND: clayey silty fine to coarse SAND	0.40 - 0.87
MADE GROUND: very silty slightly sandy CLAY	0.87 - 1.62
MADE GROUND: spongy PEAT	1.62 - 1.72
Fine to coarse SAND with much fine to coarse flint gravel	1.72 - 2.00
Fine to coarse flint GRAVEL with much medium and coarse sand	2.00 - 2.30

Table 1. The ground conditions in the site trial area

As yet no criteria exist specifically for the solidification/stabilisation of contaminated land. In order to assess the performance of the treated material, a set of criteria based on the physical and chemical properties at 28 days after treatment, normally associated with solidified waste in the United States, were imposed. The reasoning behind the use of these criteria is given in detail in Al-Tabbaa & Evans (1998a). These are:

- unconfined compressive strength (UCS) of at least 350 kPa;
- durability to freeze-thaw and wet-dry cycles;
- permeability not greater than 1×10^{-9} m/s;
- low compressibility;

- leachability of up to 50 times drinking water standards using the Toxicity Characteristic Leaching Procedure (TCLP) (USEPA, 1986) and
- TCLP leachate pH of between 7 and 11.

sample (depth)	soil (0.7m)	soil (2.1m)	soil (2.6m)	Water (2.25m)
pH	8.2	7.2	7.8	7.3
Contaminant	mg/kg dry	mg/kg dry	mg/kg dry	mg/l
Arsenic	13.8	30.0	2.4	0.48
Lead	2801	2345	2785	438
Copper	1264	962	626	52.8
Nickel	105	232	24	10.5
Zinc	1589	1800	295	94.8
Mercury	15.5	3.7	1.8	0.46
Toluene extract	1700	8700	600	-
Coal tar	1400	6200	200	<1
Mineral oil	566	1900	325	30

Table 2. Contaminant concentrations for soil and groundwater

Treatability study
Extensive treatability study work was carried out using both model and site soils to arrive at suitable soil-grout mixes which contained different proportions of ordinary Portland cement (OPC), pulverised fuel ash (PFA), natural and modified bentonite clays and quicklime. Modified bentonite clays were developed as part of the project to stabilise a specific group of organic contaminants, namely polycyclic aromatic hydrocarbons (Lundie and Mcleod, 1997). The lime was mainly added to stabilise heavy metals and also to increase the pH. Parameter ratios were adjusted to influence the properties of the material. All the mixes had a sufficiently high water content to allow the grouts to be pumpable as required for the in-situ application process. One of the aims of the study was to investigate the use of the minimum amounts of cement so that the consistency of the resulting soil-grout material is similar to that of soft rock rather than concrete. This would reduce the cost of the treatment, minimise future excavation problems and would engineer the ground for future development purposes.

The mixes were all tested for the properties discussed above using standard tests: the unconfined compressive strength and durability tests using ASTM methods, the permeability and compressibility using BS methods and the leaching test using the TCLP method. Full details of the test procedures can be found in Evans (1998). At the end of the treatability study seven grout mixes, which showed sufficiently different behaviour were selected for use in a full-scale field trial. The range of additives used is shown in Figure 1. Two of the mixes contained cement and PFA only, three contained small percentages of lime in addition and the seventh contained cement and natural bentonite. A negligible amount of a modified bentonite clay was included in the grout. However, the amount added, up to 0.5g/kg of dry soil, was such that its effect on the properties measured, apart from leachability, was negligible.

Site Trial
The development of a suitable soil mixing auger was carried out by May Gurney (Technical Services) Ltd. Three prototypes in total were manufactured all based on a traditional continuous flight auger design. It was anticipated that the success of the treatment process would be largely dependent on the efficiency of the mixing process. The two main objectives were, therefore, to ensure that contaminants would come into contact with the treatment

additives and that arisings generated during mixing would be kept to a minimum. The auger used in the field trial can be seen in Fig. 2. It was 2.4 m in length and 0.6 m in diameter. Flights were restricted to two short sections near the toe on to which digging teeth were mounted. Mixing blades were also incorporated in the design to increase the mixing efficiency of the auger. Two grout ports were positioned on opposite sides of the auger at the toe under the leading flights for injection of grout in to the soil. The column installation was successful and resulted in well-mixed soil-grout material on excavation.

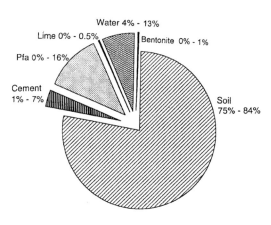

Figure 1. The range by percentage weight of the soil-grout mixes used in the site trial.

Figure 2. Prototype auger used for the field trial.

In the site trial, the seven grout mixes were used to produce a system of 23 overlapping soil-grout columns 2.4m deep and hence treating a volume of soil of 14m^3. The grout was mixed and fed to the rig by a small combination hydraulic grout pump and mixer. At each column position the auger was advanced clockwise at a rate of 30 rpm into the ground to the full treatment depth. The auger rotation was then reversed and grout was injected and mixed with the soil as the auger was withdrawn. Intermittently during withdrawal the auger was advanced again under clockwise rotation into the soil-grout column to aid compaction and to further homogenise the column of material. Only a relatively small volume increase of 7% was produced by the treatment. The treated ground was left to cure *in situ* for up to 45 days. Samples of diameter between 75 and 150mm, were then cored from all of the single soil-grout mix zones from the made ground only using conventional coring equipment.

Results

As it was not possible to test the samples at 28 days, testing took place 60 days after treatment. The seven soil-grout mixes used produced the following range of results:

- dry density: 1717 - 1809kg/m^3, which was consistent between the mixes,
- unconfined compressive strength: 990 - 1365kPa, showing less variation between the mixes compared to the treatability study results at 28 days,

- freeze-thaw durability: 1.1 - 28.5% average cumulative dry mass loss after 6 cycles at $-10^{\circ}C$ and 6 cycles at $-20^{\circ}C$, showing the largest variability between the mixes,
- wet-dry durability: -1.5 - 1.9% average cumulative dry mass loss after 12 cycles at $60^{\circ}C$, showing considerable resistance,
- permeability: 0.7 - 2.64 x 10^{-9} m/s, showing some values above the set criterion,
- coefficient of volume compressibility: 2.6 - 3.2 x 10^{-6} m^2/kN, similar to that of soft rock,
- TCLP Leachability: acceptable concentrations, some even lower than drinking water standards, for a number of polycyclic aromatic hydrocarbons and some heavy metals namely Copper, Lead, Zinc and Chromium (Lundie and McLeod, 1997), which were the only compounds assessed.
- TCLP leachate pH: 9.6 - 10.9, which is within the design range.

The properties of the samples obtained from the site trial at 60 days were far better than those from the treatability study at 28 days. In addition, the differences between the various mixes were generally not as pronounced in the site trial samples as in the laboratory treatability study samples. This could be attributed to a combination of different age, mixing and curing conditions. The cement-bentonite mix which contained cement:bentonite ratio of 10:1, water:dry grout ratio of 1.6:1 and soil:grout ratio of 3.7:1 produced the best results and satisfied all the design criteria set. These results were very encouraging and generally of the same order of magnitude as the set design criteria. The research project showed the importance of initial laboratory treatability study in the selection of suitable soil-grout mixes. It also validated the success of the in-situ treatment method for the site and the economical and environmental advantages associated with it.

Following this research and development work, May Gurney (Technical Services) Ltd together with Envirotreat Ltd, used the results to commercially treat organic contamination for the first time in the UK in 1997 using this in-situ solidification/stabilisation technique (Soudain, 1997). They also managed to advance the technology into the construction of active containment barrier systems.

THE CASE STUDY
The 7200m^2 site, formerly the location of a paint factory, again situated in West Drayton, was originally treated with soil vapour extraction to remove extensive hydrocarbon contamination, between depths of 3.5 and 4m in made ground and gravels. A local housing association required the site to be redeveloped with 70 housing units including terraced houses and flats. An examination of the site, however, prior to the start of the new development revealed that the original remedial treatment had not removed all of the pollution. Soil and groundwater, mainly in three areas, were found to be still contaminated with high levels of hydrocarbons.

The suitability of various remediation processes in the treatment of the site was evaluated. Disposal of the contaminated soil in a landfill was rejected due to the high costs, the environmental nuisance that would have been caused and also the impracticality of removing around 3,000 cubic metres of contaminated soil. On-site soil washing was considered and, although found to be cheaper than the 'dig and dump' scenario, was still rejected due to foreseen costs and duration. Bioremediation was also rejected on technical grounds as only partial success was envisaged. The OPC-based solidification/stabilisation soil mixing approach, using organophilic clay additives, as operated by May Gurney was considered to be the most cost effective treatment available.

Design criteria imposed were related to leaching test results of the treated material as compared with the commonly used Dutch Intervention Values for a number of contaminants including hydrocarbons and heavy metals. Given the proximity of this site to that used in the research work, together with similar contamination, the results from the research work were used in the development of a cement-based grout mix suitable for this site. Cement and bentonite based mixes were used as this type of mix produced the best performance in the research programme. However, as the treated material did not have to achieve a specific strength the following parameter ratios were used: OPC to bentonite in the range of 1-2.5:1; soil to grout in the range of 3-6:1; water to solids in the range of 3-6:1.

Two rigs were used on the site, fed from a centrally placed batching plant, which was imported from Italy specially for the contract. Dry powder materials are fed from 35 ton capacity silos by screw conveyors into a turbo mixer where water is added. Here the addition of the slurry ingredients is controlled by weight and a homogeneous mix is produced by high shear mixing. The slurry is fed into a $1.5m^3$ capacity holding tank, incorporating a paddle mixer. It is eventually pumped through grout hoses to open topped tanks, which are filled to the exact volume required for each soil-grout column. The slurry is then pumped under pressure to the rigs when required. The augers used were a new design consisting of a 300mm diameter leader connected to a shorter 600mm or 900mm diameter section depending on the required size of column, as shown in Figure 3. This design significantly reduced the possibility of any spoil being transported to the surface. An augering procedure similar to that used in the site trial was applied.

Three different treatment methodologies were applied on the site to cope with the different levels of contamination. The first was the treatment of three heavily contaminated areas which were block treated with the OPC-based slurry containing an organophilic bentonite mixed *in situ* with the soil down to the London clay, forming grids of overlapping columns. Then an OPC-bentonite slurry was used to construct a relatively impermeable cut-off wall, two columns thick around the site perimeter. This was to restrict the flow of groundwater to the nearby environmentally sensitive Fays river running along one edge of the site. Lastly, two relatively permeable sections known as 'active gates', consisting of an organophilic clay slurry, were installed in the cut-off perimeter barrier. These act as micro/chemical-sieves removing contaminants from groundwater as it passes through, therefore, in principle allowing only clean water to emerge on the other side. A total of 4,500 overlapping 600mm and 900mm diameter soil-cement columns were installed in eight weeks. The piles were installed at a rate of one every two to three minutes.

Throughout the treatment process leachability tests were carried out on samples of the soil-grout material to ensure that leachate concentrations were below Dutch Intervention Values. Leachate concentrations for Total Petroleum Hydrocarbons and the BTEX contaminants (Benzene, Toluene, Ethyl Benzene and Xylene) were well below Dutch Intervention Values. Subsequent groundwater monitoring carried out over a two-year period has also proved the treatment to be successful.

Figure 3. The commercial auger used.   Figure 4.   The batching plant.

FURTHER STUDIES

Further studies related to in-situ treatment of contaminated ground using soil mixing have recently been carried out. One study investigated the use of laboratory-scale augers and was able to produce soil-grout materials with very similar properties to those obtained from the site trial (Al-Tabbaa and Evans 1999a). Laboratory-scale soil mixing was also able to simulate certain in-situ conditions which were not possible with conventional treatability study manual mixing. This work showed that laboratory-scale soil mixing should be an integral part of the treatability study. Another study looked at the time-related properties of the in-situ treated ground from the site trial. Soil-grout samples were cured under laboratory conditions and tested at 14 and 28 months after treatment. The results showed that the properties listed earlier in the paper continually changed with time where some properties improved and some deteriorated (Al-Tabbaa and Evans, 1999b). This work emphasises the importance of taking time-related performance into account at design stage. Additional studies looked at soil mixing in heterogeneous soil and contaminant conditions (Al-Tabbaa et al, 1998) and at the application of soil mixing to the construction of biofilm active containment systems (Brough et al, 1998).

CONCLUSIONS

The research and development project was successful in the development and application of an in-situ treatment method for a specific site contaminated with a cocktail of organics and heavy metals. The treatment was solidification/stabilisation using cement-based grout which contained a modified bentonite and was applied using soil mixing. The project showed the success of the treatability study in assessing the suitability for the site conditions to the treatment and in the selection of soil-grout mixes for the in-situ application using soil mixing. The site trial showed the suitability of the in-situ application method to the site. Assessment of the resulting treated ground in terms of physical and chemical properties proved the success of the treatment. The case study showed the subsequent successful commercial application of the treatment. It also demonstrated the versatility of the system in the installation of two additional treatment and/or containment systems of a cut-off wall and an active barrier. The paper presents a clear picture of the technical effectiveness and versatility and commercial attractiveness of this treatment methodology. Recent research work has contributed to further advancing the technology, validating it for wider site conditions and for validating the application of soil mixing to other treatment methods.

REFERENCES

Alther, G.R., Evans, J.C. and Pancoski, S.E. (1988). Organically modified clays for stabilization of organic hazardous wastes. Superfund '88-Proc. 9th Nat. Conf., Washington D.C., 440-445.

Al-Tabbaa, A. and Evans, C.W. (1997). Medium-term performance of stabilised/solidified contaminated soil-grout material. Proc. 14th Int. Conf. Soil Mechanics and Foundation Engineering, Hamburg, Vol. 3, 1941-1946.

Al-Tabbaa, A. and Evans, C.W. (1998a). Pilot *in situ* auger mixing treatment of a contaminated site: Part 1: treatability study. Proc. Inst. Civ. Engrs., Geotech. Engng., Vol. 131, No. 1, 52-58.

Al-Tabbaa, A. and Evans, C.W. (1998b). Pilot *in situ* auger mixing treatment of a contaminated site: Part 2: results. Proc. Inst. Civ. Engrs., Geotech. Engng., Vol. 131, No. 2, 89-95.

Al-Tabbaa, A., Ayotamuno, J.M. and Martin, R.J. (1998). Contaminant transport and immobilisation in heterogeneous sands, Ground Engineering, Vol. 31, No. 11, 24-25.

Al-Tabbaa, A. and Evans, C.W. (1999a). Laboratory-scale soil mixing of a contaminated site. Accepted for publication in the Journal of Ground Improvement.

Al-Tabbaa, A. and Evans, C.W. (1999b). Pilot in-situ auger mixing treatment of a contaminated site: Part 3: time-related Performance. Submitted to the Journal of Geotechnical Engineering, ICE.

Brough, M.J., Martin, R.J. and Al-Tabbaa, A. (1998). *In situ* subsurface active biofilm barriers. Ground Engineering, Vol. 31, No. 3, pp 32.

Conner, J. R. (1990). Chemical Fixation and Solidification of Hazardous Wastes. Van Nostrand Reinhold Publishers.

Evans, C. W. (1998). Studies related to the in situ treatment of contaminated ground using soil mix technology. PhD Thesis, University of Birmingham, UK.

Jones, D. and Hopkins, M. (1997). Stabilization and solidification using cement: A study visit to the USA. Project Report, British Cement Association.

Lundie, P. and McLeod, N. (1997). Active containment systems incorporating modified pillared clays. Proc. Int. Containment Technology Conf. and Exhibition, Florida, 718-724.

Soudain, M. (1997). Fixed on site. Ground Engineering, 30, No.11, 19.

USEPA (United States Environmental Protection Agency). (1986). Toxicity Characteristic Leaching Procedure (TCLP). Federal Register, Vol. 51, No. 216, 40643-40652.

Wheeler, P. (1995). Leachate Repellent. Grnd. Engng., Vol. 28, No. 5, 20-22.

Yang, D.S. (1994). The Applications of Soil-Mix Walls in the United States. Geotechnical News, December, pp. 44-47.

Young, J.F. (1972). A Review of the Mechanisms of Set-Retardation in Portland Cement Pastes containing Organic Admixtures. Cement and Concrete Research, Vol. 2, 415-433.

Electroremediation: In situ Treatment of Chromate Contaminated Soil

R. Haus[1], R. Zorn[1], D. Aldenkortt[2],
[1]Department of Applied Geology, Karlsruhe University, D-76128 Karlsruhe, Germany
[2]Department of Applied Geology, Trier University, D-54286 Trier

ABSTRACT

High levels of chromate contamination were detected in a low permeability soil as a result of chromite ore processing residue (COPR) deposits. Hexavalent chromium is a class A carcinogen by inhalation. Laboratory experiments, pilot and field scale tests of chromium (VI) contaminated soils are presented. Bench scale results has shown that immobilising of chromium is possible. Hexavalent chromium migrates from the cathode to the anode, where it has been reduced to a geogenic trivalent form. The results have been verified in an ongoing field scale test at the "Hammerwerk Söllingen". After three months of operation the in-situ treatment has led to a significant reduction of Cr(VI) to stable Cr(III)-species.

INTRODUCTION

Electroremediation is an innovative in situ soil remediation technique well suited for contaminated fine grained sediments [1,9]. The extraction of pollutants is based on the two basic electrokinetic phenomena: Ion migration and electroosmosis [12,13].

Electromigration describes the transport of charged ions in the pore solution under the action of an electric field. In addition sediments composed of minerals with negative surface charge, such as clays, charge compensating cations of the diffuse double layer initiate an electroosmotic flow if a DC electric field is imposed. The drag force exerted by the hydration shell of the attracted double layer ions causes the pore water to flow. In contrast to hydraulic and microbiological treatment technologies the electroosmotic induced transport of contaminants is most effective in soils with pore sizes of micrometers and smaller.

Electroremediation is essentially a process of soil flushing, but has several advantages over the usual pressure-driven pumping technology. The transport rate induced by an electric field is not adversely affected by low soil permeability, and the path followed by the contaminants is confined by the electric field to the region between the electrodes. Electroremediation is therefore advantageous in soils of low or variable permeability and in situations where dispersion of the contaminants must be prevented.

Chromium is a major soil contaminant at numerous industrial sites. The extensive use of chromium e.g. in metallurgy, leather tanning, electroplating has led to its release into subsurface environment. Chromium occurs in two stable oxidation states in the soil, Cr(III) and

Geoenvironmental engineering, Thomas Telford, London, 1999, 384–391

Cr(IV). Cr(VI) occurs as oxyanions, $HCrO_4^-$ (bichromate ion), CrO_4^{2-} (chromate ion), and $Cr_2O_7^{2-}$ (dichromate ion). Due to the toxic and carcinogenic properties of Cr(VI) and its higher mobility in soils compared to the relatively immobile and non-toxic Cr(III), Cr(VI) oxyanions are of great concern [2,14]. Thus the remediation of chromium-contaminated sites is very important in order to protect public health and the environment.

Recent laboratory studies on Cr(VI) removal on soils and the in-situ treatment of chromium-contaminated soils with the electrokinetic phenomena have shown promising results [6,8,10,11,14]. HARAN et al. [4] have developed a mathematical model which neglects electroosmosis to simulate the removal of hexavalent chromate. Good agreements with laboratory experiments have been achieved. In this study laboratory experiments of chromium (VI) contaminated soils are presented and finally the in-situ treatment of a chromium-contaminated test site in Germany is introduced.

MATERIALS AND METHODS

In preliminary laboratory experiments loess loam samples at the "Hammerwerk Söllingen", an actual chromate contaminated site, were taken as undistributed soil cores. The loess loam had been contaminated with hexavalent chromium leached from the slag tailings of a chromium ore smelting plant. The hexavalent chromium was presented as the chromate ion, an anion which is highly mobile in the pore water and not significantly adsorbed on mineral surfaces [6]. The clay mineralogical composition of the loess loam is very important, because clay minerals together with the applied electric field control the electroosmotic permeability, removal efficiency and the transport mechanism of the electrokinetic remediation technology in fine grained sediments. The X-ray diffraction analysis of loess loam is given in Figure 1 and the mineralogical composition and soil parameters are given in Table 1.

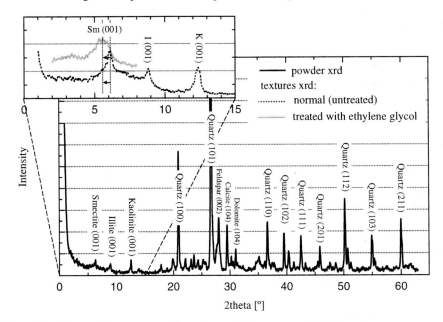

Figure 1.   X-ray diffractogram of loess loam at "Hammerwerk Söllingen".

Table 1.   Mineralogical composition (a) and soil parameters (b) of loess loam.

a)

Mineral		[%]
Quartz		43
Calcite		7,5
Dolomite		2
Smectite and Chlorite as		
irregular mixed layers: Smectite		5
Chlorite		11
Illite		15
Kaolinite		9
Feldspar		5
Organics		2,5

b)

Soil parameters		
liquid limit w_l [%]:		26,5
plastic limit w_p [%]:		20,1
plasticity index I_c[%]:		6,4
grain size distribution [%]:	Sand:	8-15
	Silt:	60-86
	Clay:	18-32
opt. moisture content w_{opt} [%]:		16,5
Proctor density δ_p [g/cm^3]:		1,69
skeletal density δ_s [g/cm^3]:		2,71
permeability k_f [m/s]		$2,9 \cdot 10^{-8}$

The water extractable Cr(VI)-concentrations were measured following the German regulations DIN 38 414-T4. Total chromium was determined by aqua regia extraction (DIN 38 414-T7) and sulfuric acid extraction. Sulfuric acid extraction was added because of the presence of chromium oxides after electrokinetic treatment [5,6].

ELECTROKINETIC LABORATORY EXPERIMENTS

The experimental cell was similar to those described in HAUS & ZORN [7] and consisted of a Plexiglas tube (10 cm in diameter and 12 cm long) fitted at each end with disk-shaped carbon electrodes.

HICKS & TONDORF [8] has shown that an oxidizing zone develops at the anode, where hexavalent chromium should be removed and a reducing environment develops at the cathode. To investigate the influence of changes of the Eh-pH conditions on the removal efficiency, different voltages were applied on the undisturbed loess loam samples. Three experiments (Run 1-3) were performed. A constant current of 2 mA and 16 mA was applied in Run 1 and Run 3 respectively, resulting in mean voltages of 3,6 V and 29,3 V respectively. To avoid pH changes and electrolysis reactions at the electrodes Run 2 was performed below the standard electrode potential of water (E^0=1,23 V) with constant voltage of 1 V.

The different voltages led to changes in the direction of chromate transport. At Run 2 the pH was constant across the sample, because of the lack of H$^+$ and OH$^-$ production at the electrodes. The chromate transport was in the direction of the electroosmotic induced water flow, thus an increase of the chromate concentration in the cathode effluent could be measured. The efficiency of chromium removal was up to 100 % (Figure 2). At higher voltages the electromigration of chromium to the anode became significant. Because of the production of H$^+$-Ions, the pH dropped near the anode. At the soil/reservoir interface a chromic oxide precipitate was formed, thus the chromium concentration in the pore water as well as the reservoir solution decreased progressively with time. A possible reaction to reduce Cr(VI) to Cr(III) is [3]:

$$CrO_4^{2-} + 8H^+ + 3e^- \rightarrow Cr^{3+} + 4H_2O \qquad (1)$$

The removal efficiency only reached 2,1 and 9 % respectively (Figure 2).

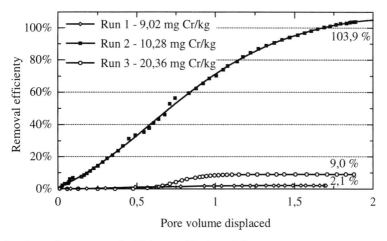

Figure 2.   Loess loam removal efficiency of test run 1-3.

The water soluble chromium in the different sample sections compared to the initial concentrations are given in Figure 3. The analyses of the soil sample used in Run 2 show, that most of the soluble chromate was removed. The difficulties in accurate balancing of the chromate concentration is based on the inhomogeneous distribution of the contaminant in the natural undisturbed soil core. In contrast most of the soluble chromium is still solubilized in the pore water of Run 1. Contaminant transport by electroosmosis and electromigration were detectable. 80% of the chromium removed has been collected at the cathode and 20% at the anode (Figure 2).

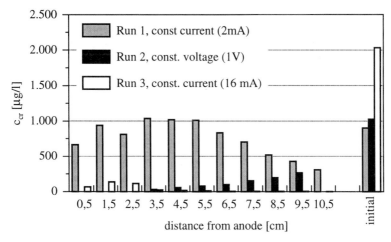

Figure 3.   Remaining water extractable chromium in soil compared to the initial concentrations.

In Run 3 the maximum water extractable chromium was found near the anode indicating the mechanism of ion migration of chromate (Figure 3). Only a small fraction of soluble chromate has been detected in the pore solution although only 9% of the chromate had been re-moved at the electrode reservoirs (Figure 2). The unbalanced chromium was extracted partly by aqua regia extraction (Figure 4), whereas sulfuric acid extraction indicates that chromate has been reduced to the trivalent form and precipitated as Cr(III)-oxides and –hydroxides by the pH drop near the anode.

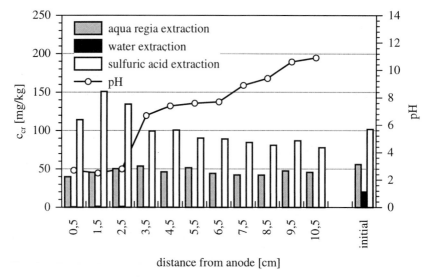

Figure 4.   Variations of chromium concentrations based on aqua regia, water and sulfuric extraction of Run 3.

FIELD SCALE RESULTS

Dependent on soil pH and electric field strength the results of the preliminary experiments indicated the transformation of Cr(VI) into Cr(III)-oxides. In this case the costs for water purification and the recovering of the contaminants at the electrode reservoirs could be ne-glected. The test site "Hammerwerk Söllingen" is situated in the unsaturated zone, so elec-tromigration should dominate electroosmotic processes. Steel electrodes were chosen on the test site, because the hexavalent chromium ions are reduced to the harmless trivalent form by chemical reaction with the anodic electrochemical release of Fe^{2+} [4]. The dissolution of Fe^{2+} has shown to enhance formation of chromium-iron hydroxide solid solution $[(Cr_xFe_{1-x})(OH)_3(ss)]$ which has a very low equilibrium solution activity, thereby relatively immobile in comparison to other species of chromium. The reduction mechanism of hexava-lent chromium in the presence of Fe^{2+} are shown below [11]:

$$HCrO_4^- + 7H^+ + 3Fe^{2+} \rightarrow Cr^{3+} + 4H_2O + 3Fe^{3+} \tag{2}$$
$$CrO_4^{2-} + 8H^+ + 3Fe^{2+} \rightarrow Cr^{3+} + 4H_2O + 3Fe^{3+} \tag{3}$$

The total volume of the test body is 112 m^3 with a soil mass of 190 t. The distance between the electrodes is 5 m, the depth of the electrode chambers measures 4,5 m. The electrode de-sign, consisting of a one-sided opened HDPE-electrode chamber, geomembrane, steel electrode and drainage tubes, was inserted in a narrow trench and adjacent filled with a filter gravel. The initial concentration of chromate on the test site was detected with up to 2,2 mg/l (Figure 5a). An electric field of 80 V was applied. The concentration distribution of water extractable chromium after three month of operation is given in Figure 5b. It can be shown that the chromate concentration on the test site had been reduced significantly.

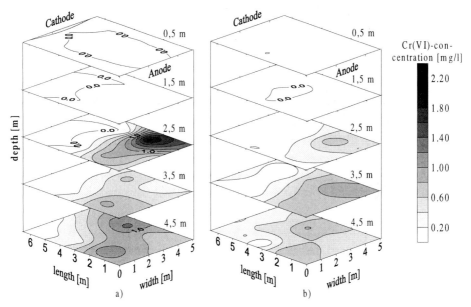

Figure 5.   Water extractable chromium a) before b) after three month of operation.

Due to the progressive movement of Cr(VI) to the anode the analyses of aqua regia extractable chromium show a decrease of total chromium in particular in the upper three meters of the test site (Figure 6). In the lower parts of the test site total chromium given by aqua regia extraction is almost similar before and after the test. This is not further surprising, because the water extractable chromium is still high in the lower two meters (Figure 5b). Furthermore most of the chromium in this depth is water insoluble and therefore does not involve in the electrokinetic processes. Mass transport by ionic migration will be approximately 10 times higher than the mass transport by electroosmotic advection [1]. For this reason only a light increase in chromium concentration has been detected in the electrode reservoirs. The chromium migrating towards the anode had to be found near the anode as a chromium-iron hydroxide precipitate, soluble only by sulfuric acid extraction. Sulfuric acid extraction analyses show increasing chromium concentrations in the soil samples near the anode (Figure 7a). In contrast cathode analyses of chromium concentrations by sulfuric acid and aqua regia extraction show almost equal results (Figure 7b). The electric field forced the Cr(VI)-ions from the cathode to the anode region, where they had been precipitated as Cr(III)-hydroxides. Finally a successful in-situ treatment of the chromium-contaminated test site "Hammerwerk Söllingen" could be proved.

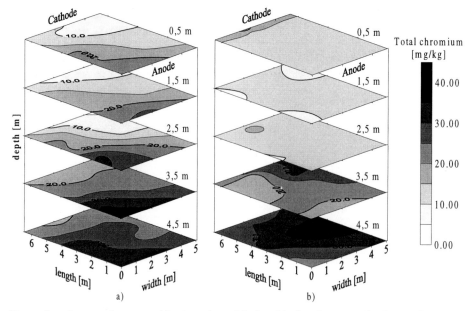

Figure 6.   Aqua regia extractable chromium a) before b) after three month of operation.

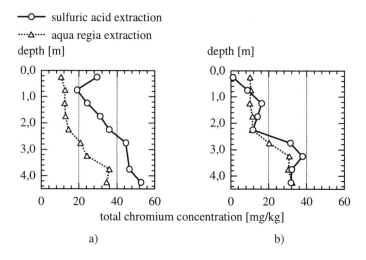

Figure 7.   Sulfuric acid extraction compared to aqua regia extraction nearby a) the anode
and b) the cathode.

CONCLUSIONS

The treatment of chromate contaminated loess loam from the test site "Hammerwerk Söllingen" demonstrates the direct impact of chromium speciation and the change of Eh-pH conditions on electroremediation. In preliminary laboratory experiments chromate has been reduced to the trivalent form and precipitated as Cr(III)-hydroxide near the anode due to a penetrating acid front. The formation of chromium-iron hydroxides by anodic release of Fe^{2+} has been verified in the field scale test. Toxic hexavalent chromium was reduced to a stable non-toxic trivalent chromium species. Electrochemical treatment may be engineered to immobilise the soluble species of cations and anions by electromigration without a need of water flushing and secondary treatments.

REFERENCES

[1] ACAR, Y.B. & ALSHAWABKEH, A.N. (1993): Principles of Electrokinetic Remediation.- Environ. Sci. Technol., **27** (13): 2638-2647.

[2] BARTLETT, R.J. & JAMES, B.R. (1988): Mobility and Bioavailability of Chromium in Soils. – In: NRIAGU, J.O., NIEBOER, E. (EDS.): Chromium in Natural and Human Environments, Wiley-Interscience, New York, pp. 267-304.

[3] DELTOMBE, E, DE ZOUBOV, N., POURBAIX, M. (1966): Chromium. – In: POURBAIX, M. [Ed]: Atlas of electrochemical equilibria in aqueous solutions; Pergamon, Oxford.

[4] HARAN, B.S., POPOV, B.N, ZHENG, G., WHITE, R.W. (1997): Mathematical modeling of hexavalent chromium decontamination from low surface charged soils. – Journal of Hazardous Materials, **55**:93-107.

[5] HAUS, R. (1998): Elektrokinetische Bodensanierung: Ein neues Verfahren für die in-situ Sanierung bebauter Altlaststandorte – Sch. Angew. Geol. Karlsruhe, 50, 350 S.; Karlsruhe

[6] HAUS, R. & CZURDA, K.A. (1998): Electrokinetic Remediation of Clays. – In: Clays for our Future - ICC 97, Ottawa: in press.

[7] HAUS, R., ZORN, R. (1998): Elektrokinetische in-Situ Sanierung kontaminierter Industriestandorte – Sch. Angew. Geol. Karlsruhe, 54, 200 S.; Karlsruhe

[8] HICKS, R.E. & TONDORF, S. (1994): Electrorestoration of Metal contaminated Soils.- Environ. Sci. Technol., **28**: 2203-2210.

[9] LAGEMANN, R. (1993): Electroreclamation - Applications in the Netherlands.- Environ. Sci. Technol., **27** (13): 2648-2650.

[10] MATTSON, E.D. & LINDGREN, E.R. (1995): Electrokinetic extraction of chromate from unsaturated soils.- In: TEDDER, D.W. & POHLAND, F.G.[EDS.]: ACS Symposium Series 607: Emerging Technologies in Hazardous Waste Management **V.**: 10-20; Washington/DC (American Chemical Society).

[11] PAMUKCU, S., WEEKS, A., WITTLE, J.K. (1997): Electrochemical extraction and stabilization of selected inorganic species in porous media. – Journal of Hazardous Materials, **55**:305-318.

[12] PROBSTEIN, R.F. & HICKS, R.E. (1993): Removal of Contaminants from Soil by Electric Fields.- Science, **260**: 498-503.

[13] PROBSTEIN, R.F. (1994): Physicochemical Hydrodynamics - An Introduction.: 2nd. ed.; New York (Wiley & Sons).

[14] REDDY, K.R., PARUPUDI, U.S., DEVULAPALLI, S.N., XU, C.Y. (1997): Effect of soil composition on removal of chromium by electrokinetics. – Journal of Hazardous Materials, **55**:135-158.

Optimising Contaminated Soil Removal Using a GIS Raster Model

E E HELLAWELL, University of Surrey, Guildford, UK,
A C KEMP and D J NANCARROW, WS Atkins Consultants Ltd., Epsom, UK

ABSTRACT:
A GIS raster model was used interactively by designers to evaluate the most efficient method of remediating a contaminant solvent plume by an appropriate combination of 'hot spot' excavation and *in situ* treatment. The plume was initially mapped using photoionisation detector (PID) readings taken from discrete soil samples. The optimum size and location of the excavation was determined, related to the cost of the operation.

ABBREVIATIONS
GIS   Geographic information system
PID   Photoionisation detector
TIN   Triangular irregular network
VOC   Volatile organic compound

INTRODUCTION
GIS has been used extensively within WS Atkins for the assessment and design of remedial and reclamation work for the development of former industrial sites. It has become a standard tool for such projects where large quantities of data need to be stored, linked and interrogated to produce visual maps of information.   Facilities within the main GIS package can also be utilised for calculating soil or waste volumes. This paper describes a technique using GIS raster and TIN models for locating a contaminant plume and evaluating remediation options in terms of efficiency and cost. The particular GIS software used for the model was ReGIS.

SITE
A site in Western Europe had been used for chemical storage and packaging. During the 1960s, eight above ground storage tanks were constructed that were used to store, pending bottling, a variety of acids (acetic, hydrochloric and sulphuric) and solvents (formaldehyde, trichloroethene, tetrachloroethene and chloroform). Occasional spillage or leakage or these solvents occurred. Acid spills were usually neutralised with soda ash and washed into the nearest foul sewer.

Geoenvironmental engineering, Thomas Telford, London, 1999, 392–399

Within the site, two distinct solvent plumes were identified.

- Plume 1 was centred near the northwestern corner of the site and was probably caused by spillage. Typical groundwater volatile organic compound (VOC) concentrations within the plume were 130 mg/l.

- Plume 2 was located in the vicinity of former tank filling operations and probably resulted from overfilling of these tanks and/or leakage of solvents from a damaged drain. Groundwater VOC concentrations were typically 70 mg/l.

The ground investigations did not identify free-phase chlorinated solvents. The chemicals were dissolved in the groundwater and had migrated in the general direction of groundwater flow. The geology of the site is made ground overlying gravels and a thick layer of clay. Solvent contamination was predominantly found within the gravels, with the clay acting as an impermeable barrier. There was little evidence either of contamination of the made ground or of a source of VOCs in the unsaturated zone.

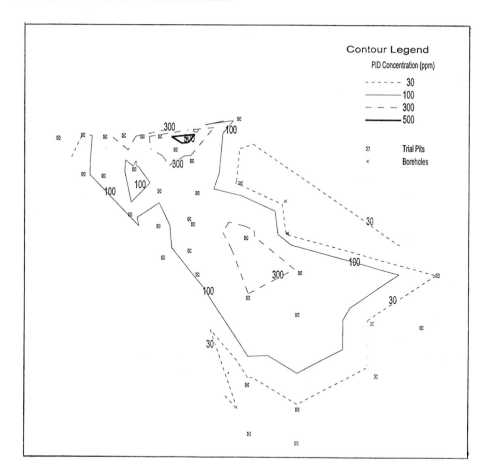

Figure 1 - Contour map of plume at 3.0m depth

CONTAMINATION MAPPING

A photoionisation detector (PID) was used to determine VOC concentrations from semi-quantitative headspace readings on soil samples, coupled with laboratory quantitative analysis. These were taken at a range of depths (at 0.5m separation) at various discrete locations. To understand the scale of the problem, a plan view of the plume at the different sampled depths was required (i.e. horizontal slices through the plume at each 0.5m depth). The PID readings from each depth were input into the GIS database and a TIN (Triangular Irregular Network) produced. Contour maps of the plume at each depth were then produced (Figure 1) and overlaid onto site maps.

These contaminant maps show two 'hot spots' of contamination with the plume, migrating in the direction of groundwater flow (towards the south east).

REMEDIATION DESIGN

The remediation strategy was developed based upon the results of site investigations, the groundwater monitoring and related risk assessment studies. The specific objectives of the works were to remove the majority of the solvent sources present and reduce groundwater contamination concentrations beyond the 'hot spot' centres of contamination. This was to be achieved by localised excavation of soil in the worst affected areas (with disposal of waste materials to licensed landfill) followed by implementation of an active groundwater remediation system. The issue to be explored was the optimum level of excavation, taking account mass removal and associated cost.

Five excavation options were proposed (figure 2). These were defined using professional judgement based upon the contamination mapping. All excavation zones were centred upon the known hot spots. Option 5 was the most expensive and was equal to the excavation of the entire contaminant plume. Options 1-3 comprised two discrete excavation zones. For option 4 these excavation zones were merged.

A method was required to evaluate these remediation proposals in terms of the amount of the plume removed and associated excavation costs. The following technique was developed using ReGIS' spatial analysis program ReSPAN.

OPTIMISING EXCAVATION EFFICIENCY

A raster model presents spatial data as a matrix of cells or pixels (AGI, 1998). Each cell is a discrete unit of data. In this problem, the requirement was for the value of each cell to represent the concentration of VOC at the cell's location. A spatial grid was therefore defined, made up of 150 x 100 rectangular nodes, covering the area of the contaminant plume. (Each node measured 0.4 x 0.35m.) The TIN models set up for each depth slice (produced to plot the contours of the contamination) were then rasterised forming a ReSPAN theme. Each node in the rasterised grid now had a PID value determined from the discrete points by linear interpolation.

The sum of the values of all the nodes in the grid was evaluated for each two-dimensional slice of the plume. To evaluate the total PID removed by each excavation option a spatial mask was required. This was produced by rasterising a CAD drawing of each proposed excavation area. All nodes within the excavation were set to 1, with all others set null. The mask was overlaid onto the plume leaving just those nodes within the excavation visible (Figure 3). The PID values of these nodes were then exported.

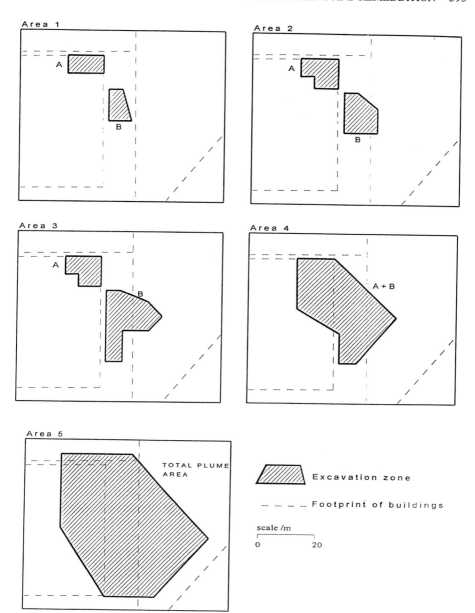

Figure 2 – Different excavation options

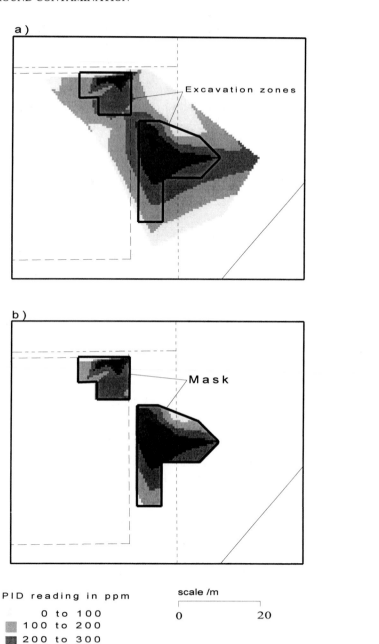

Figure 3 - Raster maps of the contaminant plume at 3m depth showing
a) all the plume and b) with the mask of excavation option 3 overlaid

The efficiency of an excavation was determined by adding the sum of the PID values of the nodes within the excavation divided by the total PID values within the grid for all the depth slices. This calculated the % of the plume removed by the excavation.

$$\% \text{ of plume removed} = \sum_{depth\ slice=1}^{n} \left(\frac{\sum PID\ values\ of\ nodes\ within\ the\ excavation\ boundary}{\sum PID\ values\ of\ raster\ model} \right) x100$$

The cost of the excavation was then calculated based upon the volume of soil removed and its disposal cost.

Table 1 - Calculation of excavation efficiency for option 3

Depth slice (m)	Total VOC concentration of plume (sum of all nodes) / ppm	VOC concentration in excavated nodes / ppm	% of plume excavated
0.5	54,764	29,078	53.1
1.0	324,278	180,727	55.7
1.5	132,086	85,141	64.5
2.0	348,725	213,435	61.2
2.5	559,744	310,388	55.5
3.0	824,219	395,101	47.9
3.5	689,958	321,898	46.7
4.0	574,794	271,186	47.2
Total	3,508,568	1,806,954	51.5

Table 1 shows the results for each 0.5m slice for excavation zone 3. This process was repeated for the five different excavation areas. The results are shown in table 2 and plotted on figure 4.

Table 2 - Results of excavation analysis

Option	% of total area	% VOCs removed	Cost £	Volume (m³)
1	8.4	24	30,450	580
2	14.5	39	52,920	1,008
3	22.1	52	80,325	1,530
4	40.4	78	147,000	2,800
5	100	99	364,014	6,934

Options were compared, taking into account the % of VOC removed, excavation areas/volumes and related costs to enable selection of the optimum excavation area. Results were plotted graphically to establish trends and aid in the decision making process (figure 4).

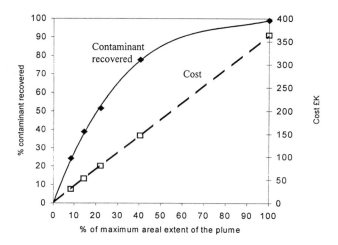

Figure 4 - Results of the excavation sensitivity analysis.

RESULTS
There is a linear relationship between the excavation area and cost (as would be expected based on a fixed depth assumption) as excavation/disposal rates will not (to a first approximation) vary as excavation volumes increase.

Contaminant recovery rises steeply up to approximately 30% of the total area of the plume. Recovery is therefore most economic below this 30% threshold. Above this threshold, recovery becomes increasingly less economic, with the contamination recovery rate steadily decreasing as the costs continue to rise in line with the excavation volume.

The plot reflects the concentrated nature of the contamination with approximately 70% of the contamination being held within two hot spot zones (occupying 30% of the total area of the plume.)

For remediation, the GIS analysis demonstrated that excavation option 1 was the most efficient; however, this was not considered to recover enough solvent. Option 2 was eventually chosen, combined with an in situ groundwater remediation system to remove the remaining groundwater contamination. The excavation would involve the removal of approximately 1000 m^3 or nearly 15% of the total area of the plume, recover almost 40% of the contamination at a cost of just over £50,000 ($80,000).

DISCUSSION
The above example demonstrates the use of spatial analysis in remediation design. Once the process had been developed it proved very quick to rerun the analysis for a modified excavation zone. The result is however only an estimate. There is some uncertainty in the analysis due to the spatial derivation and the two dimensional assumptions for the plume. Increasing the number of PID sampling points and using more sophisticated interpolation

techniques (e.g. Kriging) will reduce this error. However, the output represents a decision aid rather than a definitive answer.

Whilst the approach does not constitute a full cost-benefit analysis of remediation (since it focuses on the excavation component), it shows how the implications of different scenarios for excavation can be readily investigated. The output provides a rationale for deciding the extent of excavation that may be appropriate.

CONCLUSIONS

Spatial analysis and GIS were shown to be a useful aid in remediation design. The technology was used to locate a contaminant plume and determine the most efficient locations and extent for excavation and contaminant removal. The technique helped by targeting the worst areas of contamination and provided the client with a clear cost-benefit analysis of the remediation scheme.

The accuracy of the technique is affected by the number and spatial distribution of the sampling points in relation to the contaminant plume. Interpolation techniques also have considerable influence on the results. The technique is most appropriate for comparison of remediation schemes (excavation areas) rather than predicting absolute contamination volumes.

REFERENCES

Association for Geographic Information (1998)
AGI Dictionary
Web page: http:/www.geo.ed.ac.uk/agidict/

Automated Methods (1995), REGIS® 4.40 for Windows ®,
Automated Methods (Pty) Ltd. South Africa

Hydraulic Measures for the Treatment and Control of Groundwater Pollution

J. M. W. HOLDEN, M. A. JONES*, M. MORREY[+]
Scott Wilson Kirkpatrick & Co. Ltd., Bayheath House, Rose Hill West, Chesterfield, S40 1JF
[+]Aspinwall & Co. Ltd., Walford Manor, Baschurch, Shrewsbury, SY4 2HH
*Now at: Thames Water, Gainsborough House, Manor Farm Road, Reading, RG2 0JN

ABSTRACT
The technique of pump-and-treat has been used increasingly over the last 15 to 20 years in the USA and Europe for the treatment of polluted groundwater. It has proved to be successful as a means of achieving hydraulic control of contaminant sources and aqueous plumes and in preventing the spread of pollution. Furthermore, unlike other treatments, pump-and-treat is applicable to a wide range of contaminants and is normally the only treatment suitable for deep extensive plumes. However, problems have been experienced in using this technology as a sole treatment; e.g. remedial programmes can take much longer than anticipated and contaminant concentrations can rebound to unacceptable levels after pumping has stopped.

Where treatment of an entire contaminant plume to high standards cannot be achieved using pump-and-treat within reasonable limits of cost and time, application of alternative strategies and complementary techniques has proved successful. Consequently, pump-and-treat is becoming more successful as its limitations are more widely understood, and as appropriate strategies and complementary techniques assist in overcoming inherent difficulties. These strategies include hydraulic control, and partial treatment of contaminant plumes with natural attenuation, and complementary techniques such as free product recovery, soil vapour extraction, air sparging, in-ground barriers and soil treatment, all of which are potentially efficient and cost effective in treating or controlling sources and plumes.

These issues are addressed in a recently published CIRIA study entitled *Hydraulic measures for the control and treatment of groundwater pollution* (Holden et al., 1998). This paper summarises the findings of that study with reference to case histories, and summarises relevant legislation and regulations.

THE POTENTIAL OF PUMP-AND-TREAT SYSTEMS
A system involving abstraction of polluted groundwater with treatment at the surface prior to discharge is termed a pump-and-treat system (Figure 1). Such systems may be used for treatment of either sources or plumes of a wide range of dissolved contaminants. They are also being used increasingly for the containment of polluted groundwater, proving highly successful as a means of achieving hydraulic control of contaminant sources and plumes, and thus preventing the spread of pollution. Hydraulic control of groundwater pollution involves the use of pumped or gravity drainage from vertical and horizontal wells, sumps or drains, with or without injection of clean water, in order to control contaminant sources and plumes (see Box 1).

Geoenvironmental engineering, Thomas Telford, London, 1999, 400–407

Figure 1. Typical pump-and-treat system for groundwater remediation

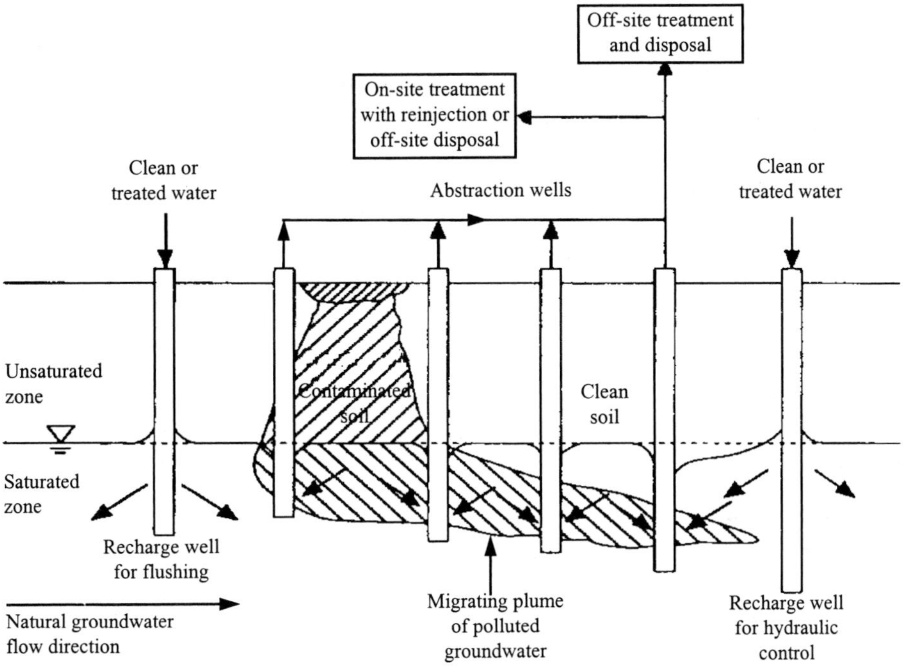

Box 1. Hydraulic Control: A Response to Potential Construction Impacts

Great Bridge Marl Pit (West Midlands, UK), infilled by assorted wastes including phenolic liquids, was to be crossed by a new road embankment. Although prior to construction polluted groundwater presented a minor risk to the adjacent River Tame, a significant increase in risk was expected as hydraulic gradients increased during ground consolidation arising from embankment construction. To control this risk, a simple pump-and-treat system was implemented to minimise migration of dissolved phenolic contaminants (Gore and Campbell, 1998). Groundwater monitoring during and after the 9 month operational period showed no evidence of additional contaminant migration.

This pump-and-treat system was not designed to treat groundwater pollution within the Made Ground or Coal Measures aquifer, but was instigated to provide hydraulic control and mitigate negative environmental impacts caused by road construction.

By implementing hydraulic control, the water table, hydraulic gradients, rates and directions of groundwater flow may be manipulated to:

- stabilise the size of a contaminant plume or source, i.e. plume or source containment
- lower the water table beneath sources of contamination either to reduce the amount of dissolved contamination feeding the aqueous plume, i.e. source control, or to allow the source to be treated or excavated, i.e. source treatment
- modify the direction of movement of a contaminant plume, thereby ensuring it does not affect sensitive receptors or minimises the impact, i.e. plume manipulation
- intercept contaminants immediately down-gradient from the source, preventing replenishment of the aqueous contaminant plume, i.e. source control.

THE PERFORMANCE OF PUMP-AND-TREAT SYSTEMS

As pump-and-treat systems can be applied to plumes of a diversity of dissolved contaminants which have affected extensive areas of an aquifer, and to great depths, it is clear that no other current technique is available which is so versatile. This is apparent from a database of pump-and-treat system performance, gathered from operational systems at hazardous waste sites in North America ranging in scale from 1 to over 200 wells (Table 1; US EPA, 1992).

Table 1. Summary of pump-and-treat system effectiveness from selected field studies

Geology	Contaminant	Start Date	Plume: area (ha)/ thick (m)	Maximum abstraction (l/min)	% decline to plateau	% above goal	Time (yrs)
Glacial silt, sand, gravel	PCE,TCE[1]	1987	4/35	1,000	91[2]	600[2]	1.7[1]
Glacial till & sandstone	TCE, DCE	1988	5/10	60	90	16000	1
Silt, sand, clay, gravel	TCE, DCE	1987	50/15	5,000	71	16000	0.4
Alluvium	VOC, pest	1985	15/10	700	-	-	-
Alluvium	TCA, DCE	1982	30/50	35,000	91	500	0.1
Glacial drift & bed rock	PCE, TCE	1985	45/10	1,500	92	370	0.5
Glacial sand, silt, gravel	VOC	1987	4/6	150	95	5000	3.5
Sand, silt & clay	VOC	1978	25/30	3,800	85	-	6
Alluvium	VOC, Oils	1982	300/75	23,000	-	-	-
Shale & sandstone	VOC	1988	1/30	250	90	2000	0.8
Sand, silt & clay	DCEE	1974	100/25	23,000	99	-	-
Silt, sand, clay, gravel	PCE, TCE	1984	9/25	7,500	90	1000	1
Sand, silt & clay	PCE, TCE	1985	400/50	2,000	71	-	-
Sand & limestone	VOC	1988	0.5/7	200	91	1000	-
Sand/gravel, sandstone	VOC	1983	3000/25	2,800	91	500	0.7
Shale & sandstone	PCP, Creos	1985	8/50	160	-	-	-
Alluvium	Pest	1982	300/60	2,300	95	-	6
Glacial sand, gravel	VOC, As	1981	6/30	1,100	41	345	0.7
Sand, clay & sandstone	TCP, Tol	1988	25/110	1,600	85	-	2
Alluvium	VOC	1988	6/20	750	62	10000	1

1   PCE = tetrachloroethane, TCE = trichloroethene, TCA = 1,1,1-trichloroethane, DCEE = dichloroethyl ether, DCE = trans 1,2-dichloroethylene, VOC = volatile organic compounds, PCP = pentachlorophenol, Pest = pesticides, TCP = 1,2,3-trichloropropane, Tol = toluene, Creos = creosote
2   Quoted with respect to contaminant concentrations; time refers to time to reach plateau

These systems generally decreased initial contaminant concentrations by 80 to 99% by continuous pumping over an average period of about 2 years. Subsequently however, the rate of contaminant removal slowed dramatically, and large volumes of groundwater needed to be abstracted to remove relatively modest amounts of contamination. In fact, concentrations either decreased slowly towards the treatment target or remained constant at levels typically ten times higher than treatment targets, features referred to as "tailing" and "apparent residual contamination levels" respectively. Where such behaviour was observed very long treatment periods are predicted. Such effects were not observed in all wells, but occurred typically in or adjacent to sources; concentrations in associated plumes typically continued to fall.

Protracted treatment periods and concentration rebound (i.e. increasing concentrations when pumping ceases) were generally attributed by US EPA (1992) to continued dissolution of light and/or dense non aqueous phase liquids (i.e. LNAPL and DNAPL) within source zones or diffusion of contaminants out of low permeability zones within the aquifer. Failure to deal with source zones was either due to the limited understanding of the importance of such action, or its impracticability in situations such as DNAPL penetration of heterogeneous soil or fractured rock aquifers and the presence of strongly adsorbed contaminants. Consequently, only at one of the 24 sites considered by US EPA (1992) had the treatment standard been achieved, although this also reflects the stringent treatment standards imposed (e.g. drinking water standards, even where groundwater was unlikely to be used for potable supply). This limited achievement of treatment standards reflects a treatment period typically less than 10 years at the time of the 1992 review. Predictions indicate that full aquifer restoration is unlikely to take less than 30 years, and at some sites it will take more than 100 years.

OPTIMISATION OF PUMP-AND-TREAT SYSTEMS
In the context of the inherent difficulties outlined above, it appears that pump-and-treat systems are now considered by some as a last resort for groundwater treatment. This is the case in North America according to Nyer and Fierro (1998). They emphasise that many practitioners have turned against pump-and-treat because of its high costs. However, one consequence of the findings of the US EPA, and later US NRC (1994), studies is the increasing acceptance in the USA that risk-based treatment standards are more appropriate; an approach developed and championed by the Environment Agency (EA) (e.g. Harris, 1997; see discussion in Holden et al., 1998). This, coupled with the development of alternative strategies and complementary techniques, has improved the efficiency of groundwater treatment and provided a continuing role for pump-and-treat.

Optimisation Using Complementary Techniques
In most of the cases reviewed by US NRC (1994) groundwater contaminant plumes appeared to have been successfully contained, and in many cases significantly reduced in size, by hydraulic control even though the source of contamination had not been treated or contained successfully. However, where treatment standards have been achieved, complementary techniques have usually been employed in conjunction with pump-and-treat, particularly in source areas and to deal with mixtures of contaminants (see Box 2).

Box 2. Synergistic Application of Complementary Remediation Techniques

Ten underground tanks containing petroleum, oils and paint products had leaked to contaminate groundwater in a sand-gravel aquifer (Dasch et al., 1997). A remediation system including 22 dual-extraction wells removed about 16,500 litres of contaminants over 21 months using: •   Free product recovery   •   Pump-and-treat   •   Soil vapour extraction (SVE)   •   Bioventing   Although SVE removed 59% of contaminants, synergism between techniques was clear. Pump-and-treat removed contaminants and hydraulically controlled the plume, while water table lowering facilitated air flow to optimise SVE. Conversely, on applying the SVE vacuum, which reduced air pressures in the unsaturated zone, increased groundwater abstraction rates were observed. Vacuum application also enhanced soil oxygenation levels and stimulated aerobic contaminant biodegradation. 92% of contamination was removed by synergistic application of these techniques.

In the UK there are fewer published examples of groundwater remediation schemes, but it is clear nevertheless, that complementary techniques are being used in conjunction with pump-and-treat. This is demonstrated by the treatment of TCE contaminated groundwater in the fractured Chalk aquifer of south-east England (Simonson and Clarke, 1993), comprising:

- SVE for removal of VOC in the unsaturated zone - source treatment
- Air sparging to optimise removal of VOC by SVE - source treatment
- Pump-and-treat for removal of dissolved TCE - source and plume treatment.

With such a wide range of complementary and alternative techniques being available it is useful to compare their effectiveness in remediating polluted groundwater. Table 2 presents such a comparison for contaminated groundwater in homogeneous multiple aquifers or heterogeneous single aquifers and, if its oversimplification is recognised, it may assist in selecting the most appropriate suite of remediation techniques.

Table 2. Effectiveness of various forms of treatment (adapted from US NRC, 1994)

Treatment	Dissolved contaminant		Non aqueous phase		Treatment period (yrs)		Status
	Mobile	Sorbed	LNAPL	DNAPL	Source	Plume	
Plume & source							
Pump-and-treat	1	2	2 to 3	4	>5	>5	P
Air sparging	2	2	2	3	<1 to 5	<1 to >5	P
Bioventing	2	2	2	3	<1 to 5	<1 to 5	P
Biological treatment							
-hydrocarbons	1	2	2 to 3	3 to 4	<1 to 5	1 to >5	P
-chlorinated solvents	1 to 2	3	NA	3 to 4	1 to >5	1 to >5	P
-metals	1 to 2	3	NA	NA	1 to >5	1 to >5	D
Permeable in situ treatment walls	2	NA	NA	NA	-	>5	D
Physical barriers	2	2	2	2	<5	>5	P
Intrinsic bioremediation	2	2	2 to 3	NA	<5	>5	P
Source							
SVE	NA	2	2	3	<1 to 5	-	P
In situ thermal technologies	NA	2	2	3	<1	-	D*
Soil flushing with chemicals	NA	NA	2 to 3	2 to 3	<1 to 5	-	D

Relative effectiveness: 1 is highly effective, 4 is least effective; NA means not applicable
Status: P means the treatment is proven, D means it is under development; * limited applications to date

Optimisation Using Alternative Strategies
In conjunction with the use of complementary techniques, alternative strategies provide the incentive for continued use of pump-and-treat. This has already been addressed in terms of its increasing use for hydraulic control of contaminant sources and plumes. However, as an extension of this hydraulic control strategy, pump-and-treat is increasingly being used to control, and in part treat, the zones of highest concentrations in contaminant plumes. This strategy can maximise the rate of contaminant mass removal, with conjunctive use of other treatment/control techniques for the remainder of the plume; there is particular potential for the use of natural attenuation including intrinsic bioremediation (see Figure 2 and Box 3).

Figure 2. Treatment and control of groundwater polluted by biodegradable compounds

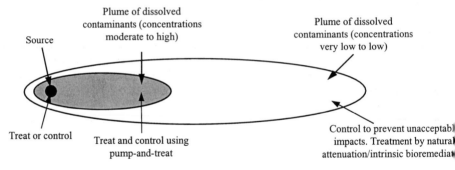

Box 3. Phased Remediation of a Petroleum Spillage (Holden & Tunstall-Pedoe, 1998)

Leakage from a single pipe at a storage terminal in New York released 4,000 m^3 of petroleum. This polluted the unsaturated zone, produced a LNAPL pool at the water table and a plume of dissolved contaminants in a glacial sand aquifer. The remediation strategy was phased as follows:

- Emergency Action - Free product recovery for source treatment recovered 1,800 m^3 in 2 years.
- Interim Remediation - For plume and source control/treatment, pump-and-treat of highest plume concentrations, plus SVE/air sparging.
- Final Remediation - Pump-and-treat and SVE to continue control and treatment, with intrinsic bioremediation for plume treatment.

The final phase acknowledged pump-and-treat would not achieve treatment standards. Instead it was used to control/treat highly contaminated parts of the plume, with intrinsic bioremediation treating the remainder within a rigorous monitoring framework, providing significant cost savings.

Applicability Of Pump-And-Treat: A Cautionary Note

The potential for the continued use of pump-and-treat as a key component of remediation schemes has been emphasised above, but its success even when used with complementary techniques is dependent on the characteristics of the hydrogeological regime and the type of contaminant (Table 3, adapted from US NRC, 1994). A specific example is noted by Nyer and Fierro (1998) in their consideration of chlorinated solvent source control. Application of pump-and-treat to such a source would appear to provide the potential to maximise the rate of contaminant mass removal, but source removal, and the flow of uncontaminated groundwater towards the abstraction well, will lead to the loss of the reducing environment created by the contaminant. As the reducing environment favours reductive dechlorination by natural bacterial action, its loss may reduce the efficiency of contaminant treatment. In this case, inclusion of pump-and-treat within the remediation strategy will depend on the homogeneity of the hydrogeological regime and the potential to optimise in situ bioremediation.

Table 3. Relative effectiveness of pump-and-treat (1 = most effective; 4 = least effective)

Aquifer type	Dissolved contaminant				Non aqueous phase	
	Mobile (degrades/ volatilises)	Mobile	Sorbed (degrades/ volatilises)	Sorbed	LNAPL	DNAPL
Homogeneous, single layer	1	1-2	2	2-3	2-3	3
Homogeneous, multiple layers	1	1-2	2	2-3	2-3	3
Heterogeneous, single layer	2	2	3	3	3	4
Heterogeneous, multiple layers	2	2	3	3	3	4
Fractured	3	3	3	3	4	4

LEGISLATION AND REGULATORY FRAMEWORK

To supplement the technical issues of applying hydraulic measures to groundwater remediation, there is a diversity of relevant legislation and regulation. Here, the important aspects in England and Wales are addressed (legislation in Scotland is generally analogous), although their complexity, and common law considerations, prevent a detailed presentation (see Holden et al., 1998, Harris, 1998, Mylrea, 1998 for further details). Nevertheless, it is clear that the Water Resources Act 1991, Environmental Protection Act 1990 and Environment Act 1995 define to the greatest extent current policy for contaminated land and groundwater. These acts, together with the Control of Pollution Act (1974) and Water Act (1989), are the driving force for groundwater remediation (see Box 4).

Box 4. Summary of Legislation Driving Groundwater Remediation

---
Water Resources Act 1991 (WRA 1991) - Responsible for:
- rendering it an offence to cause or knowingly permit pollution of controlled waters, including groundwater (Section 85),
- giving the EA powers to forestall or remedy pollution by carrying out necessary works and reclaiming costs from the polluter and/or site owner (Section 161), and
- providing for "works notices" to be served to cover groundwater pollution not directly linked to contaminated land (Section 161A).

Environmental Protection Act 1990 (EPA 1990) - Part IIA assigns liability for contamination to the "appropriate person" who caused or knowingly permitted the contamination, or, where such persons are unidentifiable, the current owner or occupier. The Local Authority or EA are empowered to serve "remediation notices" on appropriate persons.

Environment Act 1995 (EA 1995) - Introduced new provisions into EPA 1990, & defined contaminated land as land which by reason of substances in, on or under it, results in:
- significant harm being caused or a significant possibility of harm being caused; or
- pollution of controlled waters being caused or likely to be caused.

Local authorities have a duty to inspect their areas, identify contaminated land and ensure that risks posed are addressed; the EA will have executive responsibility for "special sites", to be confirmed by Statutory Guidance, including those polluting controlled waters.

Within this framework, there is a series of other legislation, permissions, consents and licences designed to control each stage of a groundwater remediation scheme (Table 4). Any potential owner/operator of a hydraulic system for control/treatment of groundwater pollution is strongly advised to contact the EA and Local Authority Planning Department at an early stage for specific advice regarding permissions required.

Table 4. Relevant legislation and regulation for pump-and-treat remediation

Stage	Legislation & regulatory controls
Initiation	• Health and safety - Health and Safety at Work Act (1974) & Construction (Design and Management) Regulations (1994).
	• Planning permissions and buildings regulations, e.g. Building Act (1984), Town & Country Planning Act (1990).
	• Boreholes to examine or abstract groundwater must have EA consent, and those >15m for exploration require notification to Natural Environment Research Council (WRA 1991).
	• All groundwater abstractions for pump-and-treat systems require an EA licence (WRA 1991), unless exempted by their small size, or the works are carried out by the EA or its agents under the remediation powers of WRA 1991.
Operation	• Treated water discharge to controlled waters must comply with an EA discharge consent (WRA 1991), & discharges to sewer will be subject to trade effluent consents. Restrictions on discharges to land or soakaway are covered by the Groundwater Regulations (1998).
	• Emissions to air are subject to the approval of local authority environmental health departments (EPA 1990); EA regulates some emissions with the greatest pollution potential.
	• For specified substances or processes, Integrated Pollution Control (IPC) authorisation may be required (EPA 1990) (until the requirements of IPPC come into effect in October 1999).
	• Solid wastes generated are subject to "duty of care" & must be disposed of to a facility licensed to receive that type of waste.
	• Contaminated groundwater requiring treatment is construed as a controlled waste, thus pump-and-treat systems may be subject to waste management licensing. Currently, a licence is unlikely to be required if abstraction licences & discharge consents have been obtained[1].
Shut-down	• Completion certificates under waste management licensing; licenses & consents surrendered
	• Implementation of monitoring programme agreed with EA
	• Boreholes & voids sealed to EA specification

1 The relationship between the waste management licensing regime and the remediation of contaminated land & groundwater is under review by the EA and the Department of the Environment, Transport & the Regions

CONCLUSION

Pump-and-treat is proven as a hydraulic control measure for groundwater pollution, and is applicable to a wide range of contaminants, however its use to restore groundwater to high quality standards can be very lengthy and costly.   Nevertheless, by using complementary treatment/control techniques (e.g. SVE, permeable reaction walls) especially in source zones, pump-and-treat effectiveness can be optimised. Similar benefits can be achieved by using a strategy of partial treatment and hydraulic control of the plume in conjunction with natural attenuation.   Pump-and-treat can continue to be effective for control and treatment if complementary techniques and alternative strategies are applied in well designed remediation systems based on comprehensive ground investigation information.

ACKNOWLEDGEMENTS

The authors thank David Banks (Holymoor Consultancy) and Jonathan Smith (EA Centre for Groundwater and Contaminated Land) for guidance on legislative and regulatory aspects.

REFERENCES

Dasch, J.M., Abdul, A.S., Rai, D.N., Gibson, T.L. & Grosvenor, N. (1997).   Synergistic application of four remedial techniques at an industrial site. *Ground Water Monitoring & Remediation*, Summer 1997, 194-209.

Gore, B.C. & Campbell, I.M. (1998).   Great Marl Pit: a case study in the prevention of contaminant migration. In: Mather, J. et al. (eds.) *Groundwater contaminants and their migration*. Geological Society Special Pub. No. 128, 313-331.

Harris, R.C. (1998). Protection of groundwater quality in the UK: present controls and future issues. In: Mather, J. et al. (eds.) *Groundwater contaminants and their migration*. Geological Society Special Pub. No. 128, 3-13.

Harris, R.C. (1997). Setting goals for the protection and remediation of groundwaters in industrial catchments. In: Chilton et al. (eds.) *Groundwater in the Urban Environment, Vol.1 Problems, Processes and Management*, 23-30. A. A. Balkema (Rotterdam).

Holden, J.M.W., Jones, M.A., Mirales-Wilhelm, F. & White, C. (1998).   *Hydraulic measures for the control and treatment of groundwater pollution.* Construction Industry Research and Information Association (CIRIA), Report 186, London, 576 pp.

Holden, J.M.W. & Tunstall-Pedoe, N. (1998).   Remediation of a petroleum spill to groundwater at a fuel distribution terminal (Long Island), USA) using pump-and-treat and complementary technologies. In: Mather, J. et al. (eds.) *Groundwater contaminants and their migration*. Geological Society Special Pub. No. 128, 165-180.

Mylrea, K. (1998).   Recent UK legal developments relating to pollution of water resources. In: Mather, J. et al. (eds.) *Groundwater contaminants and their migration*. Geological Society Special Pub. No. 128, 15-21.

Nyer, E.K. & Fierro, P. (1998). First, Do No Harm. *Ground Water Monitoring & Remediation*, Winter 1998, 65-68.

Simonson, J. & Clarke, A. (1993).   Practical implications of chlorinated solvent remediation in the Chalk Aquifer: A case study of TCE remediation in Southern England. In: *Proceedings of the third annual conference on groundwater pollution.* IBC Technical Services Ltd.

US Environmental Protection Agency (1992). *Evaluation of ground water extraction remedies: Phase II.* PB92-963346 (Summary Report) & PB92-963347 (Case Studies). US EPA Office of Emergency and Response (Washington DC).

US National Research Council (1994). *Alternatives for ground water clean up.* US National Academy Press (Washington DC).

Rapid ex situ bioremediation in the UK, the development of a commercial success.

A. A. KEAN, BSc, MSc, CGeol.
Bio-Logic Remediation Ltd, Glasgow, UK

INTRODUCTION

Ex situ bioremediation has, over the last few years, had an increasingly important role to play in the on-site treatment of hydrocarbon contamination in the UK. Bio-Logic Remediation Ltd (Bio-Logic) have developed a unique method of ex situ bioremediation. This development has largely been as a result of direct practical experience in large, commercial field projects. In addition to scientific and engineering breakthroughs, the company has gained considerable experience of how a commercial project can be shoehorned into a redevelopment site and its project schedule.

In this paper the development of the ex situ bioremediation system utilised by Bio-Logic will be discussed and a number of case studies will be presented. The process was devised to provide a viable, cheaper alternative to the traditional excavate and landfill approach to contaminated soil remediation when brownfield sites were being redeveloped.

HISTORY OF BIOREMEDIATION IN THE UK

Until the mid 1990's ex situ bioremediation in the UK was generally carried out using a technology known as landfarming. This process relied on the laying out of the hydrocarbon contaminated soils in extensive, shallow treatment beds. These beds were rotorvated with tractor towed agricultural harrows. Typically these landfarms were approximately 300mm deep and this meant that 10,000m^3 of soils would require a treatment area of at least 3ha. Further problems were caused by the fact that the beds only bioremediated effectively during warm summer weather and virtually ceased to operate during the winter months. This constrained landfarming to projects where the length of time was not an important issue. There were a number of well publicised projects where landfarming did not fulfil timescale expectations.

Other factors which made bioremediation by landfarming a less attractive option included the problem of disposing of leachate from heavy rainfall and the flushing of hydrocarbons to the base of the landfarm bed where they were least oxygenated and untreatable. Ten millimetres of rain on a saturated 1ha treatment bed would generate 100m^3 of oily leachate adding to the problems of trying to oxygenate a saturated, oxygen starved treatment bed. To prevent migration of the leachate the landfarms had to be lined with expensive synthetic liners which added to the cost of the product.

In 1994 it was recognised by Colin Grant, the founder of Bio-Logic, that if bioremediation was to become a widely practiced technology, it would have to become compatible with

Geoenvironmental engineering, Thomas Telford, London, 1999, 408–415

redevelopment programmes and to treat at lower costs than landfill. In addition such a process would have to address some of the environmental problems associated with landfarming such as leachate generation. The key aims were:

- short treatment times;
- to be able to treat throughout the year;
- to be able to deal with recalcitrant contaminants such as Polyaromatic Hydrocarbons.

EARLY TRIALS

An initial aim for the system was to devise a technique that would allow the soil in treatment to retain naturally generated heat and to keep water out. The challenge was to achieve this with a sophisticated system utilising added fungus as well as indigenous microbes, improved soil handling and aeration. In addition the reduced surface area to volume ratio would help to maintain any heat that was generated in the biopile. The problem with biopiles is how to keep the piles oxygenated as the centre and base of the piles is a long way from any circulating air. The two ways used to address this issue are to either construct the biopiles with air and water circulation pipework (known as engineered biopiles) or to turn the biopiles mechanically.

It was considered that engineered biopiles had several disadvantages over turned biopiles. For example, if further solid treatment products needed to be added to an engineered pile, then the whole pile along with the pipework would need to be demolished and reconstructed. Compaction and cracking would lead to areas of inconsistent remediation and sampling and verification of a large soil pile is problematic.

For these reasons turned biopiles were selected as being the most adaptable to innovative techniques of bioremediation. Early trials were conducted using small, static biopiles with very granular soils where efficient turning was less of a requirement. The trials were carried out at two sites in Scotland owned by Shell and Carless. The soils were selected for their high contamination levels of mineral oils and biopiles of approximately $15m^3$ were used.

The biological treatment of these hydrocarbons was to be brought about by two groups of organisms. Firstly the more simple hydrocarbons such as C_{10} to C_{30} alkanes and cyclo alkanes were to be degraded by the indigenous microbes. Laboratory studies carried out in conjunction with Strathclyde University microbiology department showed that the contaminated soils already had substantial microbial populations, often as high as 10^4 - 10^7 colony forming units per gram of soil. Secondly it was hoped that the more complex hydrocarbons could be degraded following an initial phase of fungal treatment using a variety of fungal species grown up on composted substrates. This was intended to be a more robust version of fungal technology already trialled by others (e.g. Holroyd, M. & Caunt, 1998).

TRIAL RESULTS

Results from these initial trials were very encouraging. 9,700mg/kg of diesel was remediated to 300mg/kg in 25 days during some of the worst December and January weather for years. The results were validated by the oil company's internal consultants and the Strathclyde Regional Analyst. These small trial biopiles lead to Bio-Logic receiving the John Logie Baird Award for innovative technology in 1994.

Importantly the piles had generated substantial internal heat from the efficacy of what can best be described as a low carbon content composting process. Temperatures averaging 25°C were consistently achieved in the windrows over a four week period despite the weather and the simple covers did help to keep the moisture out during heavy rainfall. It has been calculated that even in very heavy rain (>10mm/hour) typically the biopile moisture content will only rise by approximately 1% during turning, even assuming there was no surface run off.

The process had clearly coped with oxygenation purely by the natural flux through the biopiles. The next challenge was to find an efficient way of engineering a suitable soil turner to cope with the higher oxygen requirements of the anticipated volumes of soil during full scale site remediation. This would include soils with a smaller relative surface area, due to their larger size, and cohesive soils where the natural oxygen flux would be lower.

SOIL TURNERS

Bio-Logic required a machine capable of turning up to 2,000-3,000m^3 of soil per day to cope with the level of oxygenation required on projects of an anticipated size of 10,000 - 20,000m^3 of soil. After a number of false starts a relatively robust compost turner was identified in the South of England carrying out sewage sludge composting trials for a water company. The trials were nearing completion and the machine was surplus to requirements. The machine was an elevating face, tractor towed composter powered by a 100hp diesel engine, Figure 1. Close inspection revealed that the machine could be customised to cope with heavy soils and to this end heavy duty face paddles, drive chains and sprockets were fitted. By this time Bio-Logic had financial backing and the machine was purchased and named "Tina" Turner 1.

Figure 1

COMMERCIAL BEGINNINGS

After a small 300m^3 static project, the first full scale, turned biopile project was carried out in Bristol in the spring of 1996. A dockside tank farm was being redeveloped and the oily fill under this needed to be remediated. Bio-Logic's price was less than half the price of the landfill alternative and work commenced in June. The soils were anticipated to be granular fills but substantial contamination by diesel and lubricating oils was found in the underlying silty clays. To avoid damaging the turner the soils had to be screened to remove boulders and bricks. Approximately 5,000m^3 of soil were successfully laid out on a level treatment area and put into treatment. The Bristol project was successful completed with all of the soils being treated to less than the 2,000mg/kg target level within five weeks and within budget.

During this project the necessity of a sophisticated site management system became apparent due to the complexity of monitoring and maintaining what were effectively multiple soil bioreactors became apparent.

This system has been refined on subsequent projects but the crucial elements of successful field bioremediation remain:
- Full time site presence.
- A thorough knowledge of what hydrocarbons are present in each biopile.
- Daily monitoring of temperature, oxygen and carbon dioxide levels in the piles.
- Regular nutrient supplements based on monitoring the available concentrations not totals.
- Efficient moisture content analysis and control.
- Excellent quality systems to control data and ease handover.
- A high level of knowledge about what chemical tests are being specified and what the results are telling you.
- The ability to react quickly to requirements for additional nutrient, moisture or oxygen within the biopiles.

To this end each field team (a minimum of two full time site engineers/scientists) has been equipped with a computer and all of the monitoring data is logged in as soon as it is obtained and can be sent by modem to head office for discussion if required.

PROVING COMPLETION

It became very clear during the early projects and trials that robust and reproducible chemical testing was critical to the success of bioremediation projects. The shortcomings of laboratory analytical techniques became acutely apparent during trial projects where the same homogenised piles of soil were being analysed week after week. The level of homogenisation using the windrow turners was very good and it was a surprise to find results varying by as much as 50-70% from one week to the next with consistent trends being seen between sampling rounds that could not be explained by site variability. On occasions consistent reductions fitting the anticipated exponential degradation curves over four to five weeks were confounded by the next round producing results as high or higher than the starting concentration.

In depth reviews of these problems with experienced chemists at high quality laboratories has alleviated many of these problems. Many of the most intractable problems have been traced to inconsistencies in the hydrocarbon extraction procedures at the laboratories and the way that the samples were prepared for extraction. For example the samples should be:

- Taken as amalgamated samples, typically two or three per biopile per round. Check spot samples can be taken if deemed necessary.
- Given blind sample codes so that the laboratory will not anticipate a particular result.
- Sub sampled and air dried consistently.
- Always extracted by the same method with the same solvent for exactly the same time.
- Subjected to the most aggressive extraction available to minimise the effects of clay and humus binding.
- If possible analysed on the same, highly maintained gas chromatograph from one sample round to the next.
- Be subjected to the same integration method.

To illustrate this point, consider the uncertainty incurred in the anlalysis of an *in situ* soil sample taken and analysed without due care from an unhomogenised soil pile:
1. Field sampling ± 30%.
2. Sample preparation ±15%
3. Extraction ±30%
4. Analysis ±15%

$$\text{Uncertainty} = \sqrt{30^2 + 15^2 + 30^2 + 15^2}$$
$$= \pm47\%$$

Since resolving many of these problems and by homogenising the soil with the turning machines, Bio-Logic has been able to typically reduce sampling and testing variability to less than 20% as follows:
1. Field sampling ± 15%.
2. Sample preparation ±5%
3. Extraction ±10%
4. Analysis ±5%

$$\text{Uncertainty} = \sqrt{15^2 + 5^2 + 10^2 + 5^2}$$
$$= \pm19\% \qquad \text{(Risdon, 1998)}.$$

In addition to minimising variability the test methodology must be appropriate to the particular contaminants being remediated.

LARGER PROJECTS
Following on from the success of Bristol project the scale and complexity of the projects increased until, by the summer of 1997, the company was commencing work on a 34,000m^3 project in Norwich as part of a £26m redevelopment of the demolished Great Eastern Railway works, gasworks and aircraft factory that had previously occupied the site, Figure 2.

The contaminants included BTEX compounds, Polyaromatic Hydrocarbons and Mineral Oils. The primary site remediation targets were set at Total Petroleum Hydrocarbons 750mg/kg, Polyaromatic Hydrocarbons 20mg/kg, Benzene 0.5mg/kg, Toluene 25mg/kg, Ethyl Benzene 65mg/kg and xylene 12.5mg/kg. An on-site treatment area of 1.6ha was prepared by main contractor May Gurney. The project commenced well but record breaking rainfall in January virtually brought the project to a halt. Problems surfaced with the tarpaulins covering the biopiles blowing off, the site flooding and the soils piles consequently becoming too moist. In addition the additional weight in the soils meant that the tractor towed turners were now beginning to struggle. Clearly the system needed some improvement to cope with projects of this scale, particularly in mid winter.

Figure 2

The Norwich project was successfully completed within eight months but a lesson had been learned and improvements to the system were in design. In addition the Waste Management Licensing system was about to catch up with *ex situ* bioremediation.

WASTE MANAGEMENT LICENSING
Prior to 1998 most ex situ bioremediation projects in the UK had been carried out using either full Waste Management Licences or under exemptions to the Waste Management Regulations. The former took up to six months to process and left the site potentially blighted for redevelopment as the site was viewed by purchasers as little better than a landfill. The exemptions were being processed in an inconsistent manner by regional Environment Agency (EA) offices and by early 1998 certain projects were being landfilled because a suitable exemption could not be agreed.

As a consequence Bio-Logic had already begun lobbying the EA and the government with a lot of help from the Environmental Industries Commission for a solution to the problem. In early 1997 the idea of Mobile Plant Licences for ex situ bioremediation were first proposed. This process would allow the on-site remediation contractor to apply for a licence from their local EA or SEPA office and then use it to process soil on any site in the country. These sites would be controlled by the EA office closest to the site who would be responsible for approving a site specific Working Plan and adding any site specific conditions.

Bio-Logic took ownership of the first such Mobile Plant Licence in October 1998. The company now has several of these licences and they have proved to be the answer to many of the problems posed by the previous system.

RECENT DEVELOPMENTS

Following on from the experience gained on Norwich and other large scale projects, Bio-Logic has recently invested in improved soil turning and covering equipment. The new turners are called Straddlers as they stand over the top of the soil piles and rotorvate the piles with a 6m wide rotating drum. The machines, which weigh over 25 tonnes allow up to $0.7m^3/m^2$, a 40% improvement on the tractor system, and can rotorvate up to $1,500m^3$ in an hour. The first of these machines, called Goliath, is shown in Figure 3 below.

Figure 3

To be able to turn this volume of soil requires an improvement in the way that the soil pile covers are removed and replaced. Hand removed tarpaulins is no longer quick enough or cost effective. A concept from the composting industry was adapted to cope with the large, 5-6m wide biopiles being used. A tractor towed spool has been devised which can hold over the biopile a 50m length of polypropylene bioremediation felt. Two such lengths will cover a 100m long biopile. Using this device a biopile can be uncovered in five to ten minutes, up to ten times faster than the tarpaulin system. The principle of the machine can be seen in Figure 4 below. This shot was taken at Consett in County Durham where Bio-Logic were treating $21,000m^3$ of tarry soil on a former tar works.

Figure 4

The fabric becomes quickly saturated during heavy rainfall and further rain simply runs off the steep biopile profile. In addition the permeable nature of the fabric means that the soil can breathe more freely and less condensation builds up inside the cover when the piles are warm and there is cold air outside.

THE FUTURE OF *EX SITU* BIOREMEDIATION IN THE UK

It is clear that ex situ bioremediation is a technology that has gained widespread acceptance in the UK. More soil has been treated using *ex situ* processes than by any of the other innovative remedial technologies on the market.

One possible future development could be the licensing of some off site ex situ bioremediation centres to which contaminated soils could be taken for treatment. As yet the relatively low cost of landfill in the UK has made it cheaper to landfill than to transport and remediate. Landfill costs are rising and it is only a question of time before the first off site centres are opened.

Undoubtedly there will be other breakthroughs and innovations in what is a highly technical and young industry. It can be concluded that this is an efficient and cost effective technique for recycling polluted soils that is available as an off the shelf solution with a revised regulatory framework designed to allow it to operate quickly and effectively.

REFERENCES

Holroyd, M. and Caunt, P, 1998, Bioremediation, Principles and Practice Vol. 3, Technomic Publishing Co., ed. Sikdar & Irvine.

Risdon, G., 1998, TES Bretby Ltd., personal communication.

The Development of New Bentonite and Zeolite Based Products for Environmental Protection.

Cambridge, M. *KNIGHT PIESOLD LTD. Ashford, Kent, UK*
Headling, M. *KNIGHT PIESOLD LTD. Ashford, Kent, UK*
Kavanagh, D. *KNIGHT PIESOLD LTD. Ashford, Kent, UK*

INTRODUCTION

Horizontal and inclined barriers in the form of synthetic liners, compacted clay liners or combinations of the two have been extensively used in recent years for pollution containment. Synthetic liners have the advantage of negligible permeability, however their durability in the long term is not known. Clay liners act both as a low permeability barrier and as an adsorbent barrier that attenuates contaminants.

The current trend in industry is to seek for attenuating materials that will minimise both the advective and diffusive contaminant transport through clay liners. This paper is the second paper to present results of current research undertaken as part of BENPRO, an EC funded research project[a]. The main objective of this project was the development of new and modified bentonite products with enhanced properties for use in waste stabilisation and pollution control.

This paper describes civil engineering testing performed on modified bentonite and zeolite/base soil mixes to establish the engineering properties of the new materials, a cement-bentonite slurry trench cut-off field trial constructed in Kent, UK and a cost sensitivity analysis of using the new materials as additives in slurry walls.

CIVIL ENGINEERING LABORATORY TESTING OF BENTONITE-ZEOLITE SOIL MIXES

In the first BENPRO paper, Giannopoulou (1995) derived optimum properties for bentonite/soil mixes and methods of undertaking primary geotechnical screening.

Using the optimum bentonite sources identified by Giannopoulou, civil engineering tests were performed on bentonite/zeolite materials mixed with the base soil to establish the engineering properties of the new materials. Cone Penetrometer and Casagrande methods were used to determine Atterberg Limits and measurements were made of swelling pressure, coefficient of compressibility, permeability and quick undrained shear strength. Results are presented in Table 1, together with some results from Giannopoulou for comparison.

[a] The BENPRO project was an EC-DGXII funded research project under BRITE EURAM II, focusing on the development of new bentonite and zeolite based products and applications for environmental protection. The project leader was Silver and Baryte Ores Mining Company and the team members were the University of Karlsruhe-Applied Geology Department, the National Technical University of Athens, Ibeco Bentonit Technologie, Hermann Stumpp GmbH and Knight Piésold Limited.

Table 1 : Modified Bentonite/Zeolite Material Mixed with Base Soil Test Results

Test Results	Sample								
	Bentonite/Na-activated + base soil			Bentonite/Organophilic + base soil			Zeolite + base soil		
	3%	5%	7%	3%	5%	7%	3%	5%	7%
Liquid Limit (%) (Cone Method)	37	-	-	43	35	-	35	33	-
Plastic Limit (%)	21			22	24		22	22	
Liquid Limit (%) (Casagrande Method)	37	-	-	-	36	-	-	36	-
Plastic Limit (%)	21				24			22	
Swelling Pressure (kN/m²) (measured in Oedometer)	13	-	-	8	8	-	10	16	-
Coefficient of compressibility mv (m²/MN) (Consolidation test in Oedometer) for pressure 50-100 kN/m²	0.139			0.138	0.109		0.147	0.124	
Coefficient of compressibility mv (m²/MN) (Rowe Cell) for pressure 50-100 kN/m²	0.196 (Test 1) 0.084 (Test 2)	0.206	0.141	-	0.148	-	-	0.110	-
Permeability coefficient (m/s) (falling head test) γd = 1.79 Mg/m³, m/c = 16% (20°C)	5.1×10^{-12} (*)	4.7×10^{-12} (*)	3.7×10^{-13} (*)	1.9×10^{-11}	6.5×10^{-12}	4.8×10^{-12}	9.4×10^{-11}	9.6×10^{-11}	2.4×10^{-11}
Permeability co-efficient (m/s) (Rowe Cell) γd = 1.69 Mg/m³, m/c = 20% (20°C) Additional test γd = 1.81 Mg/m³	4.17×10^{-7} 5.82×10^{-7} 3.93×10^{-7} 2.12×10^{-7} 1.3×10^{-10}	3.78×10^{-9} 2.83×10^{-9} 2.03×10^{-9} 1.08×10^{-8} (*)	3.60×10^{-10} 3.27×10^{-10} 7.99×10^{-11} 2.18×10^{-10} (*)	-	2.78×10^{-9} 2.01×10^{-9} 1.61×10^{-9} 1.36×10^{-9}	-	-	3.98×10^{-9} 3.57×10^{-9} 2.46×10^{-9} 2.53×10^{-9}	(Stage 1) (Stage 2) (Stage 3) (Stage 4)
Triaxial Permeability Test (m/s) γd = 1.78, 1.7 Mg/m³, m/c = 17% (20°C)	8.5×10^{-10}	-	-	-	-	-	-	1.2×10^{-10}	-
Triaxial compressive strength (QU)	Cu = 66 kN/m² φ = 22° (*)	Cu = 68 kN/m² φ = 20° (*)	Cu = 64 kN/m² φ = 21° (*)	Cu = 79 kN/m² φ = 26°	Cu = 70 kN/m² φ = 33°	-	Cu = 80 kN/m² φ = 25°	Cu = 71 kN/m² φ = 28°	-

(*) Tests performed by Giannopoulou (1995) for comparison.

Results

A comparison of the Liquid Limit test results using the Cone Penetrometer and Casagrande methods indicated that there are only small variations between the two methods for tests performed on the Na-activated bentonite, organophillic and zeolite base soil mixes. Giannopoulou confirms this, for Liquid Limit values up to 100%.

Measurements of swelling pressure concurred with those previously produced by Giannopoulou .   The additives also caused a modest increase in the coefficient of compressibility when compared to the base soil alone.   Coefficients of compressibility measured in the oedometer were broadly similar to those measured in the Rowe cell apparatus.

The measurements of permeability in the Rowe cell for the various mixes showed marked variations in magnitude and trend when compared to the results of falling head tests. However, it has been observed that in general the initial dry density of samples prepared for falling head permeability tests was consistently higher than for samples prepared for the Rowe Cell. An additional test, undertaken to look at the variation of permeability with density, confirmed that the density of the initial sample can have a marked affect on measured permeability.

Additional permeability testing in a triaxial cell on samples 3% Na-activated bentonite/base soil and 5% zeolite/base soil were carried out. The results obtained tended to confirm the sensitivity of these samples to initial sample density but in addition also highlight the variation in measured values from different test methods.

The results of the quick undrained shear strength tests on organophillic bentonite/base soil and zeolite/base soil samples indicated that the undrained shear strength of these samples was higher than the strength of the Na-activated/base soil sample.

CEMENT-BENTONITE TESTING

Field Trial

The authors undertook the design of a pilot scale trial in Kent, UK to assess the permeability of cement-bentonite-zeolite/organoclay slurry trench cut-off walls in situ under field conditions. The arrangement of the field trial was such as to mimic the flow pattern through a wall that would occur under service conditions. To promote an adequate rate of flow it was proposed to apply a hydraulic gradient of four across the wall, achieved in part by suction through receptor panels fixed along the sides of each trench. The sides of each trench were lined with an impermeable membrane.

The field trial did not yield the anticipated results but a number of conclusions have been drawn from the site activities. The addition of modified bentonite and zeolite to the slurry mixes was shown not to have a deleterious effect on its mixing, placement or curing. Additionally, the construction of slurry trench walls with bentonite slurry containing additives is readily achieved using existing civil engineering plant. However, the accurate measurement of permeability of a slurry wall in the field is difficult. Consideration should be given to cutting full width block samples from walls and testing these in a specially manufactured test rig.

Laboratory Testing of Cement-Bentonite Samples

A suite of quality control testing was performed in the laboratory to investigate and record the strength and permeability characteristics of the new materials. Testing was performed on both laboratory prepared samples and samples taken during the construction of the field trial. Results of laboratory benchmark testing are presented in Table 2.

Table 2 : Laboratory Slurry Sample Test Results						
Sample	Marsh-Viscosity (sec)		Density (g/cm^3)		Fluid loss (ml at 2 bar after 30 min)	
	(1)	(2)	(1)	(2)	(1)	(2)
Bentonite/Cement/Water 35/200/918	32.5	58	1.025	1.15	-	112
Bentonite/Cement/Water/Zeolite 35/200/918/5	32	51	1.02	1.16	-	114
Bentonite/Cement/Water/Zeolite 35/200/918/20	33	-	1.03	1.15	18.5	129
Bentonite/Cement/Water/Organoclay 35/200/918/5	31	57	1.02	1.16	-	102
(1) Before cement addition						
(2) After cement addition						

Notes: Bentonite/water mixes prepared in a high shear mixer were allowed to stand for 24 hours. Cement was added to the bentonite suspensions with stirring.

12 No. field samples were tested after a minimum curing period of 21 days to determine their Unconfined Compressive Strength (UCS). Testing was performed at approximately 28 and 90 days in accordance with BS 1377 (1990). Failure deviator stresses are reported in Table 3.

The UCS test results show a high degree of variability in the strength of the field samples, whilst for the laboratory prepared samples less variation was evident. No trend of strength versus sample age could be determined.

Table 3 : Field Trial Quality Control Unconfined Compressive Strength Test Results						
Sample	Failure Deviator Stress (kN/m^2)					
	Field Samples		Laboratory Prepared Samples			
	28 day	90 day	46 day	68 day	291 day	354 day
Na$_2$-activated	197 440	279 473	-	694	-	-
Zeolite	454 614	353 572	739	670	-	-
Organophilic	236 673	246 283	638	647	698*	479*
* Contained air pockets						

Results of Triaxial Cell permeability testing, performed generally in accordance with BS1377, are presented in Table 4. The laboratory prepared samples consistently achieved coeffiecient of permeability values of 10^{-10}m/s^{-1} but the field samples failed to achieve the 1×10^{-9} m/s threshold in all tests, including an extended test with a duration of 170 days.

Variations in permeation time have been shown to affect test results (Mitchell, 1990). A review of the results of the extended test indicated that the shortest period over which a sample should be permeated is 48 hours. This is also the time period recommended in the UK Draft Specification for Slurry Trench Cuf-Off Walls. Permeability should then be calculated based on the flow rate at the end of the test. From the measure flowrates, the permeability at the end of the extended test was approximately 5×10^{-10} m/s.

Table 4 : Field Trial Quality Control Triaxial Permeability Test Results

Sample	Coefficient of Permeability k (m/s)			
	Field Samples			Laboratory Prepared Samples
	90 days		170 days	43 days
	10°C	20°C	20°C	20°C
Na$_2$-activated	5.3×10^{-9}	$6.9 \times 10^{-9(1)}$		6.4×10^{-10}
Zeolite	1.8×10^{-9}	$2.3 \times 10^{-9(1)}$	$5.0 \times 10^{-10(2)}$	9.9×10^{-10}
Organophilic	3.1×10^{-9}	4.0×10^{-9}		2.2×10^{-9}

(1)   short permeation time (less than 48 hours)
(2)   extended test period

Column Infiltration Tests

A simple laboratory model of the field trials was constructed, to simulate the flow of fluids through cement-bentonite mixes. Field samples were permeated, initially using tap water to promote hydration, and then organic, municipal and inorganic leachate solutions as in the field trials.

The laboratory test configuration of the nine systems is shown in Figure 1. The results of permeation with tap water and the leachates are given in Table 5 and indicate that the resistance of the cement-bentonite materials to the leachate solutions was generally good. This suggests that the condition of the corresponding materials in the field trial would also be generally good.

Table 5 : Laboratory Column Infiltration Tests

Sample	Coefficient of Permeability in m/s			
	Water	Inorganic Leachate	Municipal Leachate	Organic Leachate
Cement-Bentonite-Zeolite	7×10^{-11}	9×10^{-9} Rising[3]	4×10^{-11} Rising	1×10^{-10} Steady
Cement Bentonite	3×10^{-10}	1×10^{-8} Fluctuating[1]	1×10^{-8} Leak[2]	2×10^{-9} Steady
Cement-Bentonite-Organoclay	7×10^{-10}	2×10^{-11} Falling	1×10^{-7} Leak[2]	1×10^{-10} Steady

Notes: [1] Poor confidence in result, [2] Test aborted due to leakage, [3] Rising/Falling/Steady indicates trend in permeability at end of test

Immersion Testing

Samples of the cement-bentonite slurry walls were taken at the time of construction for chemical compatibility testing in the laboratory. After initial curing the samples were immersed in leachate solutions and monitored on a monthly basis over a period of 6 months, with a final inspection after 22 months. The samples were weighed, measured and photographed on each inspection and the relative performance of each cement-bentonite sample in resisting leachate attack assessed, with the results presented in Table 6.

These tests provided a simple visual indication of the compatibility of the samples to leachate attack. They must, however, be regarded as qualitative, due not only to their short term duration, but also due to the lack of confining pressures on the samples which would apply in the field. These pressures would reduce permeability and ingress of leachate, and therefore tend to improve the compatibility characteristics determined from the laboratory.

Table 6 : Immersion Test Results								
	Sample	Dec. 96	Jan. 97	Feb. 97	Mar. 97	Apr. 97	May 97	Sept 98
Municipal Leachate	ORG-300796-2-5	G	G	G	M	P	VP	VP
	ZE-100796-2-5	G	G	G	G	M	M/P	M/P
	NA-120796-2-6	G	G	G	G	G	M	M
Organic Leachate	ORG-300796-2-5	G	G	G	M	P	P	P
	ZE-100796-2-5	G	G	G	G	G	M	M
	NA-120796-2-6	G	G	G	G	G	G/M	M
Inorganic Leachate	ORG-300796-2-5	G	G	G	G	M	M	M
	ZE-100796-2-5	G	G	G	G	M	M	M
	NA-120796-2-6	G	G	G	G	M	M	M
G = Good, M = Moderate, P = Poor, VP = Very Poor								

COST SENSITIVITY ANALYSIS
Construction costs for specific applications were developed to assess the cost-effectiveness of the proposed new products, to enable direct comparisons with existing products and processes. The influence of using the new and modified materials on the cost of construction of slurry walls has been assessed, using data from a survey of UK contractors' rates for construction of similar small and large projects. The costs of a cement-modified bentonite cut-off wall has also been compared with the rates for a slurry trench cut-off incorporating a geomembrane.

Cement-Bentonite Slurry Trench Cut-Off Walls
The construction method costed was a single phase process, where the cement-bentonite slurry is left to self-harden in the trench to form a low permeability barrier. Rates for mobilisation and for cement-bentonite slurry for small projects (placement of up to 500 m^3 of slurry) and large projects (over 500 m^3) are presented in Table 7. These utilised a nominal price of bentonite of £4 per 25 kg bag, irrespective of the quantities used.

The analysis developed a costing formula including for the cost of establishment and production of varying volumes of cement-bentonite slurry, based upon the mix designs used in the field trial. The formula enabled the total construction cost for different volumes of slurry to be calculated and included a variable unit cost for the zeolite or the organoclay additives. Using the formula, the effect on construction cost of using specialist additives was investigated.

Results indicate that the price of the new and modified materials does not significantly affect the cost of construction of a cement-bentonite slurry cut-off wall.

Table 7 : Cut-Off Wall Construction Rates		
Item	Small	Large
Mobilisation (£)	23 000	31 000
Cement-Bentonite Slurry Rate (£/m^3)	50.5	34
Average values from survey of UK Contractors.		

Slurry Trench Cut-Offs With Geomembranes

Composite cut-off systems with a geomembrane positioned in the slurry trench are a hybrid technology currently used in environmental protection, when highly contaminated ground conditions or aggressive chemicals are encountered. In this case the cement-bentonite slurry is designed to support the geomembrane and enable it to be positioned during construction. It is the geomembrane which acts as the principal containment barrier.

The installation cost of a slurry trench cut-off with a geomembrane is estimated to be 40-50% more expensive than a trench without a geomembrane. This is due in part to the geomembrane material cost (assumed rate of £1.72/m^2 for 1.5 mm HDPE). However, mobilisation costs are higher as more equipment is required for geomembrane positioning and construction is slower.

Using a simple economic model, it is possible to determine cost multipliers for additives that would result in the same unit cost as a trench with a geomembrane. The values in Table 8 represent the multipliers by which the price of the new and modified materials would need to be increased before the cost of construction of a cement-bentonite-zeolite/organophilic slurry trench equals that of construction with a geomembrane.

Table 8 : Additive Cost Multipliers				
		Size of Project		
		Small (400m^3)	Large (5,000m^3)	V.Large (20,000m^3)
Additive	Zeolite	x 15	x 6	x 5
	Organoclay	x 60	x 30	x 30
Additive price multipliers for cost of cement-bentonite slurry trench to equal cost of construction with a geomembrane.				

Conclusion

The price of special bentonitic additives does not significantly affect the cost of construction of a cement-bentonite slurry trench. Until the size of a project becomes very large, the use of the new and modified materials will not be inhibited by cost, even if their price was to be three times the current price of bentonite.

Cement-bentonite slurry trenches with a geomembrane are more expensive than a trench containing a slurry with modified additives. The difference is dependent on the quantity of additive added to a mix. In circumstances where the application is other than for the prevention of gas migration the use of a slurry trench with modified additives to control migration of contaminants could be cost effective.

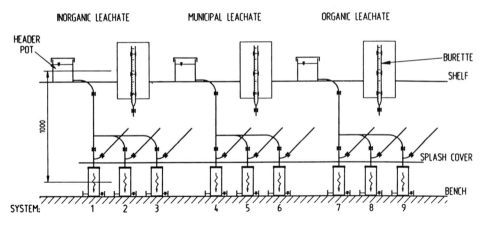

Acknowledgements
The authors gratefully acknowledge *Brett Gravel Limted,*for use of *Shelford Farm Quarry, Canterbury* and *Kvaerner Cementation* for assistance during construction of the field trial.

References

BRE (1994) Digest 395 "Slurry Trench Walls To Contain Contamination"

BS1377 (1990) "British Standard Methods of Test for Soils for Civil Engineering Purposes"

Cowland, J W, Leung, B N (1991) "A Field Trial of a Bentonite Landfill Liner" *Waste Management and Research 9, 277-291*

D'Appolonia, D J "Soil-Bentonite Slurry Trench Cut-Offs" *Journal of the Geotechnical Division, Proceedings of the American Society of Civil Engineers, Vol. 106, No. GT4, April 1980.*

Garvin, S, Tedd, P, Paul, V "Research on the Performance of Cement-Bentonite Containment Barriers in the United Kingdom".

Giannopoulou, M (1995) "The Geomechanical Characterisation of Bentonite/Soil Mixes For Use In The Construction of Barriers For Pollution Containment" *IX Conference of Young Geotechnical Engineers, Ghent, Belgium. 4-8th September 1995.*

Gill, S A, Christopher, B R "Laboratory Testing of Cement-Bentonite Mix for Proposed Plastic Diaphragm Wall for Complexe La Grande Reservoir Caniapiscau, James Bay, Canada" *Hydraulic Barriers in Soil and Rock, ASTM 874, A.I.Johnson, R.K.Frobel, N.J.Cavalli and Pettersson, Eds., American Society for Testing and Materials, Philadelphia, 1985, pp75-92.*

Head, K H (1990) "Manual of Soil Laboratory Testing" *Vols. 1, 2 and 3.*

Jefferis, S A "The Design of Cut-Off Walls for Waste Containment" *Land Disposal of Hazardous Waste : Engineering and Environmental Issues. Eds Gronow J.R., Schofield A.N., Ellis Horwood Ltd., Cichester 1989.*

Jefferis, S A "Contaminant-Grout Interaction" *Ground/Soil Improvement and Geosynthetics. Vol. 2 ASCE Geotechnical Special Publication No.30.*

Mitchel J K, Jaber, M "Factors Controlling the Long-term properties of Clay Liners" *Waste Containment Systems: Construction, Regulation and Performance.* Edited by Ruldolph Bonaparte, Geotechnical Special Publication, 26 ASCE 1990.

O'Sadnick, D L, Simpson, B E, Kasel, G K (1993) "Evaluation and Performance of a Sand/Bentonite Liner" *Geoenvironment 2000*

Tedd, P. (1996) "Investigation of the Performance of Cement-Bentonite Cut-Off Walls in Aggressive Ground at a Disused Gasworks Site" *Paper received from Tedd, P.*

Tedd, P, Paul, V, Lomax, C "Investigation of an Eight Year Old Slurry Trench Cut-Off Wall" *Waste Disposal By Landfill – GREEN '93*

Enhanced removal of copper from contaminated silt soil using bioelectrokinesis

GIACOMO MAINI and AJAY K. SHARMAN
(IBS Viridian Ltd., Whitstable, UK)
SIMON A. JACKMAN and CHRISTOPHER J. KNOWLES
(Oxford Centre for Environmental Biotechnology, Oxford, UK)
GARRY SUNDERLAND (EA Technology Ltd., Capenhurst, UK)

INTRODUCTION

The principle of electrokinesis is the application of low voltage DC current causing a pH gradient and an electroosmotic flow of water, which facilitates the directional movement/recovery of heavy metals (Yeung and Datla, 1995). Acidification of soils is obtained by the electrolysis of water at the electrodes. Under oxidative conditions at the anode hydrogen ions [H^+] are produced. These positively charged ions will move through the soil pore water towards the cathode lowering the pH of the soil thereby mobilising the metals species. Similarly electrolysis of water at the cathode produces an excess of [OH^-] which can produce alkaline conditions with the consequent precipitation of heavy metals into insoluble forms. In order to avoid heavy metals precipitation, acidic conditions can be successfully maintained at the cathode by adding a buffering agent such as acetic acid, carbon dioxide and hydrochloric acid (Hicks and Tondorf, 1994). Acetic acid has the advantages of both keeping the cathode under acid conditions and to increase the solubility of certain metals such as Cu, Zn, Pb and Ca in the cathode compartment. Nevertheless, the electrokinetic process alone may be ineffective in removing metals in soil with high buffering capacity where high concentration of [H^+] are required before achieving any significant drop in pH through the soil profile (Reed et al., 1996). Under these circumstances electrokinetic remediation may prove to be economically unacceptable (Will, 1995).

The presence of DC voltage in the soil has been shown to increase the activity of the indigenous bacterial population (Electorowicz and Boeva, 1996). Moreover, recent shake flask studies (Jackman et al., 1999) have shown that DC field can stimulate the activity of the naturally occurring Sulphur Oxidising Bacteria (SOBs). These bacteria have the capacity to use reduced sulphur as an electron acceptor producing H_2SO_4 as the end product, thereby 'naturally' acidifying the soil. Therefore, amending soil with sufficient elemental sulphur or sulphide can produce a soil acidification, which may result in the 'bioleaching' of certain metal species (White et al., 1998). The bioleaching of high value metals sulphides has now become a prominent method of recovery within the mining industry (Rawlings,1998). It is conceivable to assume that combining SOBs and electrokinesis may result in an enhanced the removal of heavy metals from difficult soils for the following reasons:

- It would reduce the energy costs which are usually associated with the electrokinetic process as SOBs activity would produce extra acid

Geoenvironmental engineering, Thomas Telford, London, 1999, 424–431

- It could increase the rate of acidification of the indigenous SOBs due to the beneficial presence of a DC field in the soil.

This paper investigates the potential advantages of using the integrated process combining the bioleaching of the SOBs with electrokinesis to a well-characterised silt soil spiked with 1000 mg/kg of Cu.

MATERIALS AND METHODS
A silt soil was used. The characteristics of the silt soil are presented in Table 1.

Table 1.   Silt soil characteristics.

Parameter	Value
pH	8.1
Extractable phosphorus (mg/L)	24
Cationic exchange capacity (me/100g)	21.7
Soil buffering capacity (g/kg) *(calculated as the amount of [H⁺] from sulphuric acid required to acidify the soil to pH 2)*	0.52
Particle size 2000-600 μm (%)	1
Particle size 600-212 μm (%)	4
Particle size 212-63 μm (%)	7
Particle size 63-20 μm (%)	43
Particle size 20-2 μm (%)	19
Particle size <2 μm (%)	26
Calcium carbonate (%)	2
Total Sulphur (%)	0.08
Total nitrogen (%)	0.25
Organic matter (5)	3.40
Organic carbon (%)	1.97
C:N ratio	7.9:1

Silt soil was spiked with a solution containing $Cu(NO_3)_2$ in order to provide a concentration of 1000 mg/kg as Cu. Approximately 1 kg of soil was tightly packed in the reactor prior to starting the experiment. A diagram of the electrokinetic rig is presented in figure. 1. The anode consisted of a carbon felt while the cathode was composed of a stainless steel electrode, housed in a semiporous membrane, which allowed the recirculation of the catholyte.

The reactor vessel comprises of six cells allowing a modular design to be used for a time course sampling of the soil profile. This was achieved by further subdividing each cell into 5 subsections at 1 cm intervals with 2 sampling cores per section.

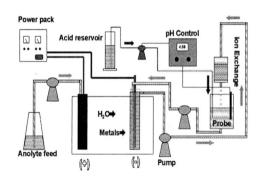

Indigenous SOBs present in the spiked silt soil were stimulated by the addition of 5% (w/w) elemental sulphur and soil was maintained at 30% (w/w) soil moisture. Soil was incubated at room temperature (~20°C) and mixed once a week in order to ensure the oxygen necessary for the oxidation process. Sulphate concentration in soil, pH and moisture were monitored weekly. Once the soil profile showed no further decrease in pH, the soil was transferred into the electrokinetic reactor in order to commence the electrokinetic experiment. The experimental conditions used during the two experiments are presented in Table 2.

Water was added at the anode to prevent the soil drying out during the electrokinetic runs. The pH control unit was set to maintain the catholyte at pH ≤ 4.5 using a concentrated acetic acid solution. Soil samples were taken from the electrokinetic reactor during the experiment at time intervals.

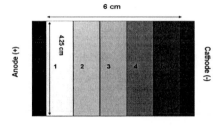

Figure 1.   Electrokinetic reactor set up and cells sampling diagram.

Table 2.   Condition imposed to the two electrokinetic experiments.

Parameter	Run 1	Run 2
Soil Type	Silt soil	Silt + 5% w/w sulphur
Permeability (K)	1.81×10^{-11} m/s	1.81×10^{-11} m/s
Contamination	Cu 1000 mg/kg	Cu 1000 mg/kg
Current Density	3.72 A/m^2	3.72 A/m^2
Catholyte addition	4 M Acetic acid	4 M Acetic acid
Anolyte addition	0.05M H_2SO_4	Water
Metal recovery device @ the cathode	Ion exchange	Ion exchange
Time of run	15 days	16 days

At the end of the experiment soil cores were taken from each cell and analysed for soil moisture, exchangeable sulphate and pH. Analyses of Cu, Ca, Fe, Mn and Mg in the soil cores were performed by Electron-Diffraction X-ray Fluorescence Spectrometry (ED-XRF). Concentrations of Cu, Ca, Mn, Mg and Fe in the ion exchange elution were measured using Atomic Absorption Spectrometry (AAS).

RESULTS

The result of Run 1 showed that approximately 92% of the Cu in the silt soil was removed after 15 days operation. In addition, most of the solubilised Cu, which migrated in the catholyte was recovered in the ion-exchange column as shown in the mass balance (Table 3).

Table 3.   Calculation of the Cu mass balance in the two electrokinetic experiments.

Run	Initial Cu concentration (mg/kg)	Total Cu in the reactor (mg)	Final Cu concentration (mg/kg)	Total removed (%)	Total recovered (%)
1	1080	1086.76	89	92	100.1
2	920	901.35	137	86	108.0

The soil pH decreased from 7.5 to 2.8 during the 15 days electrokinetic run. The addition of $[H^+]$ from the sulphuric acid added (~0.086 g/kg soil) at the anode was well below the buffering capacity of the soil, thus suggesting that the electrokinetic process alone can produce enough acid to decrease the pH (Table 4). The mobilisation of the metal present in soil was Ca>Mn>Cu and was affected by the decrease in the soil pH (Fig.2). In contrast, Fe and Mg were poorly mobilised and low concentrations were found to be dissolved in the catholyte. Metal banding across the soil cell during electrokinetic remediation was found to be caused by the change in the pH in the soil with the highest concentration measured in the proximity of the pH 'jump' (Table 4).

Table 4.   Cu distribution, SO_4^{2-} and pH profile during electrokinetic remediation in Run 1. Highlighted values showed the greatest Cu and SO_4^{2-} concentrations in soil compared to the highest pH values.

Parameter	Time (d)	Cu removal (%)	Distance from the anode (cm)				
			1	2	3	4	5
Cu (mg/kg)			231	2323	1206	1166	1132
SO_4^{2-} (mg/kg)	2	0	2786	1464	0	0	0
pH			2.03	3.65	7.24	7.61	6.69
Cu (mg/kg)			73	105	313	466	906
SO_4^{2-} (mg/kg)	9	65	2909	2446	82	62	41
pH			2.38	2.61	3.27	3.35	4.08
Cu (mg/kg)			40	69	107	156	196
SO_4^{2-} (mg/kg)	15	92	2407	2174	3805	79	123
pH			2.12	2.40	2.89	3.32	3.42

Silt soil amended with sulphur (5%, w/w) was packed into the electrokinetic reactor when it reached a pH value of 5.4 and an average sulphate level of 3101 mg/kg. Once the electrokinetic remediation was initiated more than 80% of Cu was removed after 10 days. This was partly caused by the lower initial pH of the silt soil (i.e. pH 5.4), which was already acidified by the SOBs. Again, most of the Cu removed from the soil was recovered in the ion-exchange column (Table 3) and Cu metal banding was observed in a similar manner as reported for Run 1 (Table 5).

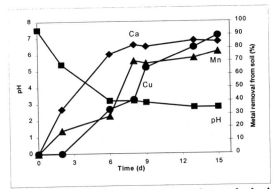

The final removal of Cu from the soil after 16 days treatment was approximately 86%. Ca and Mn were found to be readily mobilised in the presence of the SOBs while Fe and Mg showed almost no movement. Sulphate appeared to accumulate as a band in the middle of the reactor cell (Table 5) and its average concentration across the cell increased significantly during the electrokinetic run.

FIGURE 2.   Removal of Cu and earth metals during electrokinetic remediation in relationship to the soil pH.

TABLE 5.   Cu distribution, SO_4^{2-} and pH profile during electrokinetic remediation in Run 2. Highlighted values showed the greatest Cu and SO_4^{2-} concentrations in soil compared to the highest pH values.

Parameter	Time (d)	Cu removal (%)	Distance from the anode (cm)				
			1	2	3	4	5
Cu (mg/kg)			384	836	2680	1360	1160
SO_4^{2-}(mg/kg)	3	0	8610	11705	2871	4011	2136
pH			1.83	2.4	4.08	5.34	5.27
Cu (mg/kg)			103	124	168	225	310
SO_4^{2-}(mg/kg)	10	80	5493	17313	16635	19981	5483
pH			2.19	2.11	2.10	2.15	2.66
Cu (mg/kg)			103	100	132	149	202
SO_4^{2-} (mg/kg)	16	86	6335	7011	7399	4689	6876
pH			1.97	2.05	2.18	2.02	2.31

Sulphate production rate reached a maximum of 1131 mg/kg/day between day 7 and 10, whereas in previous static soil, using uncontaminated silt SOBs averaged only 328 mg/kg/day). Plating techniques identified that at pH <4.0, the predominant bacteria of the SOB population present in the silt soil were similar to the acidophilic *Thiobacillus thiooxidans*. A decrease in the total sulphate concentration within the soil profile was recorded after day 14 as the negatively charged sulphate molecules started to migrate into the carbon anodic felt.

DC voltage across the soil cell was measured during the two electrokinetic runs; this was used to calculate the power consumption required to remove Cu from soil. There was a significant decrease in the power consumption in Run 2 where the initial step of the acidification was carried out by the SOBs (Table 6).

TABLE 6.   Power requirements and associated costs calculated for the two electrokinetic runs

Run	Run time (Days)	Current applied (A)	Current density (A/m²)	Power Input (Kwh/m³)	Cost ($/Tonne)* @ $0.32 kWh
1	15	0.100	3.72	490	26.16
2	16	0.100	3.72	173.95	9.26

1 m³ of soil is approx. 1.5 tonne

DISCUSSION

These experiments highlight the potential of electrokinesis alone and in combination with SOBs in remediating soils contaminated with heavy metals. The data of both Run 1 and 2 showed that soil acidification must be achieved prior to mobilising Cu. Cu started to be removed once the soil had achieved a level < pH 4.0. Under these conditions Cu is present in the pore water as the ionic form Cu^{2+} and electromigrated towards the cathode. However, acidification of the soil could be achieved only when the carbonate buffering system could be overcome. The $CaCO_3$ in the silt soil was ~2% therefore large quantities of Ca were removed and recovered at the cathode. Ca did not precipitated at the cathode, as the solubility of calcium acetate is greater than 1g/L. Mn showed similar behaviour, while Mg showed almost no mobility. Fe is present in the silt soil at concentration of 30,000 mg/kg, but seemed to be locked up as insoluble ferric hydroxide species, which becomes mobile only at pH values < 3.0 (Ulrich, 1991).

In Run 2, the removal of Cu and earth metals showed a similar trend to Run 1, although the process was accelerated. The increase in the SO_4^{2-} production in the cell inoculated with SOBs clearly indicated an enhanced microbial activity. Previous experiments conducted by Jackman *et al.* (1999) in shaken flasks showed that indigenous SOBs in similar sulphur amended silt soil could not achieve acidification < pH 4.0, as the mobile Cu^{2+} inhibited any further bacterial activity. Similarly, pre-acidified soil did not reach acidification < pH 5.4 prior to starting the electrokinetic remediation, even after 90 days incubation. It is possible that the 'switch on' of the electrokinetic process, with the consequent migration of the Cu^{2+} towards the anode, removed the inhibition caused by this heavy metal upon the indigenous SOBs population. A previous study (Tuovinen *et al.*, 1971) has reported that high concentrations of ionic Cu^{2+} (i.e. >1000 mg/kg) may prolong the lag phase and partly inhibit the metabolism of *Thiobacilli*.

Data for SO_4^{2-} movement in soil were similar to those reported by Acar *et al.* (1997). SO_4^{2-} moved from the cathode toward the anode as result of electromigration, however this was in part counteracted by the electroosmotic flux of the water towards the cathode. For instance, during Run 1 the H_2SO_4 was added at the anode and the SO_4^{2-} remained confined near the anode although some migration caused by the water flux towards the cathode was recorded towards the end of the run. On the other hand, in Run 2 the SO_4^{2-} concentration was evenly distributed across the soil profile at T=0 as the result of SOBs acidification. As electrokinetic remediation proceeded a band in the middle of the soil cell was observed from day 3 due to the counteracting effects of electroosmosis and electromigration. This was also in combination with a general increase in SO_4^{2-} concentration caused by the SOBs activity. The permeability coefficient k measured for the silt soil used in this experiment (1.81×10^{-11} m/s) suggested that electroosmosis is likely to be very significant during the electrokinetic process. Nevertheless, the electromigration effect has been reported to be more pronounced (up to 1 order of magnitude) than electroosmosis (Probstein and Hicks, 1993) and by day 16 most of the SO_4^{2-} was confined near or within the anode.

The electrokinetic cell inoculated with SOBs showed faster initial removal rates, however the overall level of removal was similar to the experiment without SOBs activity. Nevertheless, the electrokinetic process coupled with bioleaching showed significantly lower power requirements. Another advantage in combining the two processes together is the fact that electrokinesis stimulated the activity of the SOBs and at the same time selectively removed SO_4^{2-} from soil. This is a major step forward in the application of the bioleaching process to soil as acidification due to the production of H_2SO_4 from the SOBs is likely to produce levels of SO_4^{-2} above most national guidelines for contaminated soil. Therefore, these initial results suggest that an integrated electrokinetic and microbial bioleaching system (Bioelectrokinesis) may have substantial advantages over the single electrokinetic or bioleaching process. A detailed consideration of a number of operational parameters such as electrode costs, applicability to soil with high buffering capacity, presence of weathered metals will be required to determine the commercial viability of the process.

REFERENCES

Acar, Y. B., M. F. Rabbi, and E. E Ozsu. 1997 "Electrokinetic Injection of Ammonium and Sulphate Ions Into Sand and Kaolinite Beds." *Journal of Geotechnical and Geoenvironmental Engineering* 123:239-249

Electorowicz, M and V. Boeva. 1996. " Electrokinetic Supply of Nutrients in Soil Bioremediation." *Environmental Technology*. 17:1339-1349

Hicks. R.E. and S. Tondorf. 1994. "Electrorestoration of Metal Contaminated Soils." *Environ. Sci. Technol.* 28: 2203-2210

Jackman, S. A, G. Maini, A. K Sharman and C. J. Knowles. 1999 "The effects of Direct Electric Current Upon the Viability and Metabolism of Acidophilic Bacteria." *Enzyme Microbial Technology* 24: 316-324

Probstein, R. F. and R. E. Hicks. 1993 "Removal of Contaminants from Soil by Electric Fields." *Science*, 260: 498-503

Rawlings, D. E. 1998 "The Industrial Practice and the Biology of Leaching of Metals from Ores" *Journal of Industrial Microbiology and Biotechnology'* 20:268-274

Reed, B. E., P. C. Carriere, J.C. Thompson and J. H. Hatfield. 1996 " Electronic (EK) Remediation of a Contaminated Soil at Several Pb Concentrations and Applied Voltages." *Journal of Soil Contamination.* 5 (2): 95-120

Tuovinen, O. H., S. I. Niemelä and H. G. Gyllenberg. 1971 "Tolerance of *Thiobacillus ferrooxidans* to Some Metals" *Antonie van Leeuwenhoek,* 37: 489-496

Ulrich B. 1991 "Deposition of Acids and Metal Compounds." In E. Merian Eds. *Metals and Their Compounds in the Environment.* VCH Publishers, Inc., New York, N.Y.

Yeung, A. T. and S. Datla. 1995 "Fundamental Formulation of Electrokinetic Extraction of Contaminants from Soil" *Can. Geotech. J.* 32: 569-583

White, C., A. K. Sharman and G. M. Gadd. 1998 "An Integrated Microbial Process for the Bioremediation of Soil Contaminated with Heavy Metals." *Nature Biotechnology* 16: 572-575

Will, F. G. 1995 "Removing Toxic Substances from the Soil Using Electrochemistry." *Chemistry & Industry* 5:376-379

An evaluation of remediation technologies for metal-contaminated soils and sediments

Catherine N. Mulligan, Concordia University, Montreal, Canada, Raymond N. Yong, Cardiff University, Cardiff, Wales and Bernard F. Gibbs, Bivan Consultants, Montreal Canada

INTRODUCTION

In the United States, 1200 sites are on the National Priority List (NPL) for the treatment of contaminated soils, indicating the extensiveness of this problem. Approximately 63% of sites include contamination from toxic heavy metals (Hazardous Waste Consultant, 1996). For example, lead was found at 15% of the sites, followed by chromium, cadmium and copper at 11, 8 and 7 % of the sites respectively. Therefore, metal contamination is a major problem.

Cadmium, copper, lead, mercury, nickel and zinc are considered the most hazardous and are included on the Environmental Protection Agency's (EPA) list of priority pollutants (Cameron, 1992). Sources of metals include, domestic and industrial effluents, the atmosphere, runoff and lithosphere. Once metals are allowed to pass through the municipal waste treatment facility, the heavy metals return to the environment where they are persistent, cannot be biodegraded and can thus follow a number of different pathways. Exposure to the heavy metals through drinking water (particularly where water is reused) and foods can lead to accumulation in both animals, plants and humans. Metal accumulation can lead to extinction or mutation of plants and animals. Levels of metals can accumulate in the following order: river sediments, bacteria, tubicids and then fish and man if he eats these fish.

Over the past years, use of metals such as copper, cadmium and zinc have increased substantially (Table.1). Copper is produced more than any other metal, whereas more zinc reaches the soil than any other metal. Lead use has decreased due to toxicity concerns. In Canada, according to the National Pollutant Release Inventory, approximately 13,300 tonnes of copper, 9,500 tonnes of zinc, 1300 tonnes of lead and 33 tonnes of cadmium were released to the air, water and soil (NPRI, 1995).

MOBILITY OF VARIOUS METALS

In its natural form, cadmium is relatively rare and concentrated in argillaceous and shale deposits as greenockite (CdS) or otavite ($CdCO_3$) and is usually associated with zinc, lead or copper in sulfide form (Cameron, 1992). It is a bluish-white soft metal or grayish powder. It is more mobile, though, than zinc at low pH values, particularly at pH values between 4.5 and 5.5. Above pH 7.5, cadmium is not very mobile. Its divalent form is soluble but it can also complex with organics and oxides. Volcanoes release cadmium into the air. It is only in the last twenty years that cadmium has become a concern due to the extensive extraction for use in steel plating, pigment stabilization and nickel-cadmium batteries (Fasset, 1980).

Geoenvironmental engineering, Thomas Telford, London, 1999, 432–440

Sources of cadmium include alloys, polyvinyl chloride plastic manufacture (PVC), solders, fungicides, enamels, motor oil, textile manufacturing, electroplating and rubber, sewage sludges and phosphate fertilizers (Matthews, 1984). It enters the environment through industrial effluents and landfill leaching, spills and leaks at hazardous waste sites, mining and household wastes.

Table 1   Global production of metals and the rate of metals reaching the soil (10^3 t/year)

Metal	1975	1980	1985	1990	Emissions to the soil in 1980s
Cd	15.2	18.2	19.1	20.2	22
Cu	6739.0	7204.0	7870.0	8814.0	954
Pb	3432.2	3448.2	3431.2	3367.2	796
Zn	3975.4	4030.3	4723.1	5570.9	1372

Adapted from World Resources Institute (1992/1993) and Nriagu and Pacyna (1988)

Copper is found naturally in sandstones and in minerals such as malachite and chalcopyrite. It is a reddish-brown metal that binds strongly to organic matter and clay minerals, thus decreasing its mobility. The organic matter, however, can be degraded through anaerobic or aerobic means releasing the copper in its monovalent or divalent states, respectively. The average content of copper in soils is 2 to 100 ppm. Plants can accumulate with copper average contents are in the range of 5 to 30 ppm. Toxic levels in plants are in the range of 20 to 100 ppm. Increased levels of copper are due to uses in fertilizers, building materials, rayon manufacture, pesticide sprays, agricultural and municipal wastes and industrial emissions (Cameron, 1992).

Lead is found naturally in soils. It is most commonly found in the ore gelena (PdS) and in smaller quantities in cerussite ($PbCO_3$), anglesite ($PbSO_4$) and crocoite ($PbCrO_4$). It is a bluish-white, silvery or grey metal with a high density of 11.4 g/cm^3. Lead can be found in soils at the surface and organic matter in higher quantities. Sources of lead include lead-zinc smelters, ammunition, solder, glass, piping, insecticides, paints and batteries (Jaworsky, 1978). The divalent form is the most common and is capable of replacing calcium, strontium, barium and potassium in soils. In general, levels less than 10 ppm are found and mobility of lead in soils is low. Organic matter can adsorb substantial quantities of lead. Lead is released into the air from burning of wastes and fossil fuels and subsequently lands onto the soil. It also reaches the soil from landfills and paints.

Although not as toxic as cadmium, zinc is quite often associated with this metal. Zinc is a soft, white metal with bluish tinge. Soil texture, pH, nature of the parent rocks and organic content all affect the natural content of zinc in the soil. Under acidic conditions, zinc is usually divalent and quite mobile. At high pH, zinc is bioavailable due to the solubility of its organic and mineral colloids. Zinc hydrolyzes at pH 7.0 to 7.5, forming $Zn(OH)_2$ at pH values higher than 8. Under anoxic conditions, ZnS can form upon precipitation, whereas the unprecipitated zinc can form $ZnOH^+$, $ZnCO_3$ and $ZnCl^+$. Natural levels of zinc in soils are 30 to 150 ppm. Levels of 10 to 150 ppm are normal in plants while 400 ppm is toxic. Sources of zinc include brass and bronze alloys, galvanized products, rubber, copying paper, cosmetics, pharmaceuticals, batteries, televisions, tires, metal coatings, glass, paints and zinc-based alloys (Cameron, 1992). It can enter the environment from galvanizing plant effluents, coal and waste burning, leachates from galvanized structures, natural ores and municipal waste treatment plant discharge. Zinc is commonly found in wastes as zinc chloride, zinc oxide, zinc sulfate and zinc sulfide (Agency for Toxic Substances and Disease Registry, 1995).

REMEDIATION TECHNIQUES
Isolation and containment
Contaminants can be isolated and contained to prevent further movement, to reduce the permeability of the waste to less than 1×10^{-6} m/sec and to increase the strength or bearing capacity of the waste (USEPA, 1994). Physical barriers made of steel, cement, bentonite and grout walls can be used. Another method is encapsulation which contains the contaminants in the area. This process is also called solidification or stabilization. Some metals such as arsenic, chromium (VI) and mercury are suitable for this type of treatment. Liquid monomers that polymerize and cement are injected to encapsulate the soils. Leaching of the contaminants must, however, be carefully monitored as is the case for vitrification - the formation of a glassy solid. Vitrification involves the insertion of electrodes into the soil which must be able to carry a current. The soil solidifies as it cools. Toxic gases can also be produced during vitrification.

Soils can be treated *in situ* or after excavation. The main types of remediation technologies include biodegradation, desorption, phase separation such as froth flotation and hydrocyclones, chemical or thermal destruction. Containment and treatment methods are summarized in Table 2.

Mechanical separation
The aim of size selection processes is to remove the larger, cleaner particles from the smaller more polluted ones. To accomplish this, several processes are used. They include: hydrocyclones which separate the larger particles greater than 10 to 20 microns by centrifugal force from the smaller particles, fluidized bed separation which remove smaller particles at the top (less than 50 microns) in the countercurrent overflow in a vertical column by gravimetric settling and flotation which is based on the different surface characteristics of contaminated particles. Addition of special chemicals and aeration in the latter case causes the contaminated particles to float.

Thermal treatment
Temperatures of 200 to 700°C are used to evaporate contaminants. Organic pollutants and organic soil components such as humic acids can also be transformed. Higher temperatures (900 to 1100°C) can be used to oxidize the components and prevent dioxin emissions in a separate incineration step (van de Leur, 1990). Petrol, diesel oil, polycyclic aromatics, iron cyanides and halogenated organic compounds can all be removed. In terms of heavy metal treatment, mercury, arsenic and cadmium and its compounds can be evaporated at 800°C with the appropriate air pollution control system. Some of the metals remain in the solid residues which will have to be properly disposed.

Pyrometallurgical separation
Pyrometallurgical processes are most applicable for mercury since is easily converted to its metallic form at high temperatures. Other metals including lead, arsenic, cadmium and chromium may require pretreatment with reducing or fluxing agents to assist melting. This type of treatment is usually performed off site due to a lack of mobile units and is most applicable to highly contaminated soils (5-20%) where metal recovery is profitable. Mercury, however, can be easily recovered at lower concentrations.

Chemical treatment
Chemical treatment is used to decrease the toxicity or mobility of metal contaminants (Evanko and Dzombak, 1997). Oxidization reactions involve addition of potassium permanganate, hydrogen peroxide, hypochlorite or chlorine gas. Reduction reactions are induced through the addition of alkali metals such as sodium, sulfur dioxide, sulfite salts and ferrous sulfate. Neutralization reactions are performed to adjust the pH of acidic or basic soils. Sometimes chemical treatment is used to pretreat the soil for solidification or other treatments.

Permeable treatment walls
Permeable barriers that contain a reactive substance (physical, chemical or biological or a combination) are being evaluated for reduce the mobilization of metals in groundwater at contaminated sites. Various materials have been investigated and include zeolite, hydroxyapatite, elemental iron and limestone (Vidic and Pohland, 1996). Preliminary results have shown that elemental iron can be used for chromium reduction and limestone for lead precipitation.

Electrokinetics
Electrokinetic processes involve passing a low intensity electric current between a cathode and an anode imbedded in the contaminated soil. Ions and small charged particles, in addition to water, are transported between the electrodes. Anions move towards the positive electrode and cations towards the negative. An electric gradient initiates movement by electromigration (charged chemicals movement), electro-osmosis (movement of fluid), electrophoresis (charged particle movement) and electrolysis (chemical reactions due to electric field) (Rodsand and Acar, 1995). The process can be used in situ or with excavated soil. Metals as soluble ions and bound to soils as oxides, hydroxides and carbonates are removed by this method. Other non-ionic components can also be transported due to the flow. Large metal objects can interfere with the process. Unlike soil washing, this process is effective with clay soils. Demonstrations of this technology have been performed but are limited. In Europe. this technology is used for copper, zinc, lead, arsenic, cadmium, chromium and nickel. Electrode duration and spacing is site specific.

Biochemical processes
Techniques for the extraction of metals by biological means are rather limited at this point. Bioleaching involves *Thiobacillus sp.* bacteria which can reduce sulphur compounds under aerobic and acidic conditions (pH 4) at temperatures between 15 and 55°C, depending on the strain. Leaching can be performed by indirect means, acidification of sulfur compounds to produce sulfuric acid which then can desorb the metals on the soil by substitution of protons. Direct leaching solubilizes metal sulfides by oxidation to metal sulfates.

Several options are available for bioleaching including heap leaching, bioslurry reactors and in situ. Anoxic sediments are more suitable for treatment since the bacteria can solubilize the metal compounds without substantially decreasing the pH. Soils require lower pH values to extract the metals since they have already been exposed to oxidizing conditions. For both heap leaching and reactors, the bacteria and sulfur compounds are added. In the reactor, mixing is used and pH can be controlled more easily, Leachate is recycled during heap leaching. Copper, zinc, uranium and gold have been removed by *Thiobacillus sp.* in biohydrometallurgical processes (Karavaiko et al, 1988). Fungus such as *Aspergillus niger* can produce citric and gluconic acids which can act as chelating agents for the removal of metals such as copper from oxide mining residues (Mulligan et al., 1999a). Several

Table 2. Summary of remedial technologies

Technology	Description	Applicability	Costs ($US/metric ton)
Contaminant and isolation			
Physical	Prevent movement by preventing fluid flow	Landfill covers and slurry walls	10 to 90
Encapsulation	Creation of an inert waste	Injection of solidifying chemicals	60 to 290
Vitrification	Application of electrical energy to vitrify contaminate	Shallow metal-contaminated soil, low volatility metals	400 to 870
Ex situ treatment			
Physical separation	Includes, froth flotation, gravity separation, screening,etc.	For high metal concentrations	170 to 270
Soil washing	Addition of surfactants and other additives to solubilize	For water soluble contaminants	60 to 245
Pyrometallurgical	Elevated temperature extraction and processing for metal removal	Highly-contaminated soils (5-20%)	250 to 9,000
In situ			
Reactive barriers	Creation of a permeable barrier	Sorption or degradation of contaminants in barrier	Little info
Soil flushing	Water flushing to leach contaminants	For soluble contaminants	60 to 245
Electrokinetic	Application of electrical current	Applicable for saturated soils with low groundwater flow	100 to 200
Phytoremediation	Use of plants for metal extraction	Shallow soils and water	Good ($50,000 to 200,000/acre)

feasibility studies have indicated that contaminated soils can be remediated (Tichy et al., 1992). Sludges from anaerobic processes that contain metal sulfides could be treated in this manner (Blais et al., 1992). Biosorption is a biological treatment method which involves the adsorption of metals into biomass such as algae. There is, however, strong competition from ion exchange resins. If large scale inexpensive production techniques for the biomass are developed, this heavy metal treatment is promising (Hazardous Waste Consultant, 1996). Microorganisms are also known to oxidize and reduce metal contaminants. Mercury and cadmium can be oxidized while arsenic and iron can be reduced by microorganisms. This process (called "mercrobes") has been developed and tested in Germany at concentrations greater than 100 ppm. Since the mobility is influenced by its oxidation state, these reactions can affect the contaminant mobility.

Phytoremediation

Plants such as *Thlaspi, Urtica, Chenopodium, Polygonum sachalase* and *Alyssim* have the capability to accumulate cadmium, copper, lead, nickel and zinc and can be therefore be considered as an indirect method of treating contaminated soils (Baker et al., 1991). This method is called phytoremediation and is limited to shallow depths of contamination.

Phytoextraction involves uptake of metals by trees, herbs, grasses and crops. Phytostabilization is a process to excrete components from the plants to decrease the soil pH and form metal complexes. The plants will have to be isolated from wildlife and agricultural lands. The climatic conditions and bioavailability of the metals must be taken into consideration when using this method. Once contaminated, the plants will have to be disposed in an appropriate fashion. Some techniques include drying, incineration, gasification, pyrolysis, acid extractions, anaerobic digestion, extraction of the oil, chlorophyll fibers from the plants (Bolenz et al., 1990) or disposal since plants are easier to dispose of than soil. Phytoremediation will be most applicable for shallow soils with low levels of contamination (2.5 to 100 mg/kg)

In situ treatment (soil flushing)

Extracting solutions are infiltrated into soil using surface trenches, horizontal drains or vertical drains. Water with or without additives are employed to solubilize contaminants. The efficiency of the extraction depends on the hydraulic conductivity of the soil. High permeability gives better results (greater than 1×10^{-3} cm/s). The solubility of pollutants and if the pollutant was originally solubilized in water or not affects removal efficiencies. Prior mechanical mixing of the soil can disturb the infiltration of the extractant. Understanding the chemistry of the binding of the contaminant and the hydrogeology of the site are very important (USEPA, 1987).

Since water solubility is the controlling removing mechanism, additives are used to enhance efficiencies. In an analysis of a test site, it was determined that 400 years would be required to treat a site with water alone compared to 4 years with chemical enhanced flushing (AAEE, 1993). The research in this area is still quite limited, particularly where metal removal is concerning (USEPA, 1987). They include organic and inorganic acids, sodium hydroxide which can dissolve organic soil matter, water soluble solvents such as methanol, displacement of toxic cations with nontoxic cations, complexing agents such as EDTA, acids in combination with complexation agents or oxidizing/reducing agents. Soil pH, soil type, cation exchange capacity (CEC), particle size, permeabilities and contaminants all affect removal efficiencies. High clay and organic matter contents are particularly detrimental. Once the water is pumped from the soil, it must be extracted and then treated to remove the metals. Several technologies exist such as sodium hydroxide or sodium sulfide precipitation, ion exchange, activated carbon adsorption, ultrafiltration, reverse osmosis, electrolysis/ electrodialysis and biological means (Patterson, 1985).

Large-scale treatment has been done mostly for organic removal. A 30,000 m³ volume has been successfully treated in the Netherlands to decrease the cadmium content in 90% of the soil from 10 mg/kg to less than 1 mg/kg with dilute hydrochloric acid (pH 3 (Urlings, 1990). Another full scale process was used at a electroplating shop with high chromium levels. Two infiltration basins were used to flush the low permeability soils. Treatment is still ongoing. (USEPA, 1996). Chromium concentrations have decreased from 2,000 mg/L to 18 mg/L.

Treatment of sediments

Since sediments contain large quantities of water, dewatering is frequently necessary after dredging to enable treatment. Methods include draining of the water in lagoons with or without coagulants and flocculants or using presses or centrifuges. Treatment methods are similar to those used for soil such as hydrocyclone pretreatment, biological decontamination (landfarming for organic contaminants in particular) and chemical extraction (acids). Very few of these techniques in comparison to soil treatment have been used commercially.

Soil washing (chemical leaching)

Heavy metals can be removed from soils using various agents added to the soil. This can be done in reactors or as heap leaching. These agents are: inorganic acids such as sulfuric and hydrochloric acids with pH less than 2, organic acids including acetic and citric acids (pH not less than 4), chelating agents such as ethylenediaminetetraacetic acid (EDTA) and nitrilotriacetate (NTA), and various combinations of the above (USEPA, 1991a). The cleaned soil can then be returned to the original site. Soils with less than 10 to 20% clay and organic content (ie.,sandy soils) are most effectively remediated with these extractants. Both organics and metals are removed. However, modifications to the process which is commercially used have to be made for each type of soil (Hinsenveld et al., 1990). In general, soils with low contents of cyanide, fluoride and sulfide, CEC of 50 to 100 meq/kg and particle sizes of 0.25 to 2 mm, with contaminant solubility in water of greater than 1,000 mg L, can be most effectively cleaned by soil washing (Hazardous Waste Consultant, 1996).

Soil washing with biosurfactants

The feasibility of using biodegradability biosurfactants to remove heavy metals from an oil-contaminated soil was recently demonstrated by batch washes with surfactin, a rhamnolipid and a sophorolipid (Mulligan et al., 1999b). The soil contained 890 mg/kg of zinc, 420 mg/kg of copper with a 12.6% oil and grease content. A series of five batch washes removed 70% of the copper with 0.1% surfactin/1% NaOH while 4% sophorolipid/0.7% HCl was able to remove 100% of the zinc (Figure 1). The results clearly indicated the feasibility of removing metals with the anionic biosurfactants tested even though the exchangeable metal fractions were very low. These biosurfactants were also able to remove metals from sediments (Mulligan et al., 1999c). Since these agents are biodegradable, can enhance hydrocarbon removal and can potentially be produced *in situ*, they have a great potential for soil washing and soil flushing applications.

A   B

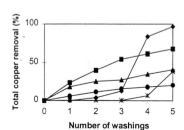 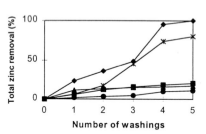

Figure 1. Cumulative copper (A) and zinc (B) removal after five washings using different surfactants and controls. (●) 1% NaOH ; (■) 0.1% surfactin/1% NaOH; (▲) 0.1% rhamnolipid/1% NaOH; (*) 0.7% HCl and (♦) 4% sophorolipid/0.7% HCl.

CONCLUSIONS

Since metals are considered relatively immobile, methods for metal-contamination have focused on solid-phase processes such as solidification/stabilization and vitrification. Electrokinetics and in situ extraction have been used in a few sites but results are promising. Phytoremediation and bioleaching are not as well developed but could be useful for areas of low contaminations. Treatment walls are low cost, passive treatment methods that are promising Our research has indicated that biosurfactants, biologically produced surfactants, may be promising agents for the enhanced removal of metals from contaminated soils and sediments.

REFERENCES

American Academy of Environmental Engineers (AAEE) (1993) *Soil Washing / Soil Flushing, Innovative Site Remediation* (Anderson W,C, ed.), Volume 3, WASTEC.

Agency for Toxic Substances and Disease Registry (1995) Division of Toxicology, Atlanta.

Blais, J.F., Tyagi, R.D. and Auclair, J.C. (1992) Bioleaching of metals from sewage sludge by sulfur-oxidizing bacteria. *J. Environ. Eng.* 118/5, 690-707

Bolenz, S., Omran, H., Gierschner, K. (1990) Treatments of water hyancinth tissue to obtain useful products. *Biol. Wastes* 22, 263-274.

Cameron, R.E. (1992) *Guide to Site and Soil Description for Hazardous Waste Site Characterization. Volume 1: Metals.* Environmental Protection Agency EPA/600/4-91/029.

Evanko, C.R. and Dzombak, D.A. (1997) *Remediation of Metals-Contaminated Soils and Groundwater.* Technology Evaluation Report, TE-97-01,Gound-Water Remediation Technologies Analysis Center, Pittsburgh, PA.

Fassett, D.W. (1980) Cadmium. In *Metals in the Environment* (Waldron, H.A., ed.) Academic Press, pp. 61-110.

Hazardous Waste Consultant (1996) Remediating soil and sediment contaminated with heavy metals. Nov/Dec, Elsevier Science.

Hinsenveld, M., Soczo, E.R., van de Leur, G.J., Versluijs, C.W., Groenedijk, E. (1990) Alternative physico-chemical and thermal cleaning technologies for contaminated soil. In: *Proc. Int. Conf. Contaminated Soils.'90*, (Arendt, F., Hinsenveld, M. and van der Brink (eds.) Kluwer Academic Publishers, pp. 873-881.

Jawarsky, J. (1978) *Effects of Lead in the Environment I & II.* National Research Council Canada.

Karavaiko, G.I., Rossi, G., Agates, A.D., Groudev, S.N., and Avakyan, Z.A (1988) *Biogeotechnology of metals: Manual.* Center for International Projects GKNT, Moscow, Soviet Union.

Matthews, P.J.(1984) Control of metal application rates from sewage sludge utilization in agriculture. *Crit. Rev. Environ. Control*, 14, 199.

Mulligan, C.N., Galvez-Cloutier, R. and Renaud, N. (1999a) Biological leaching of copper mine residues by *Aspergillus niger*, Presented at AMERICANA 1999, Pan-American Environmental Trade Show and Conference, Montreal, Canada, March 24-26, 1999.

Mulligan, C.N., Yong, R.N. and Gibbs, B.F. (1999b) On the use of biosurfactants for the removal of heavy metals from oil-contaminated soil. *Environmental Progress* 18(1), 1-5.

Mulligan, C.N., Yong, R.N. and Gibbs, B.F. (1999c) Removal of heavy metals from contaminated soil and sediments using the biosurfactant surfactin. *J. Soil Contamination*, 8, 231-254.

National Pollutant Release Inventory (NPRI) (1995) Environment Canada.

Nriagu, J.O. and Pacyna, J.M. (1988) *Nature* (London), 333, 134-139.

Patterson, J.W. (1985) *Industrial wastewater treatment technology* (Second edition), Butterworth Publishers, Boston.

Rodsand, T. and Acar, Y.B. (1995) Electrokinetic extraction of lead from spiked Norwegian marine clay. *Geoenvironment 2000*, Vol. 2, 1518-1534.

Tichy, R., Grotenhuis, J.T.C., and Rulkens, W.H. (1992) Bioleaching of zinc- contaminated soil with thiobacilli. In: *Proc. Int. Conf. Eurosol.* 92, Sept., 1992 Masstricht, The Netherlands.

Urlings, L.G.C.M. (1990) In situ cadmium removal-full scale remedial action of contaminated soil. *International Symposium on Hazardous Waste Treatment: Treatment of Contaminated Soil. Air & Waste Association and U.S. EPA Risk Education Laboratory, Cincinatti, Ohio, February, 5-8, 1990.*

USEPA (1987) Treatability studies under CERCLA: An Overview. OSWER Directive 9380.3-02FS.

USEPA (1991) Innovative treatment technologies. Semi-annual status report (Third edition) EPA/540/2-91/001, U.S. EPA Office of Solid Waste and Emergency Response, Washington, D.C.

USEPA (1996) *Superfund Fact Sheet. United Chrome Products Inc. Corvallis Oregon, June 10.*

USEPA (1994) *Selection of Control Technologies for Remediation of Soil Contaminated with Arsenic, Cadmium, Chromium, Lead or Mercury. Revised Draft Engineering Bulletin, January 31.*

USEPA Office of Solid Waste and Emergency Response (1997) *Recent Development for In Situ Treatment of Metal Contaminated Soils.*

van de Leur, G.J. (1990) Study on alternative physico-chemical and thermal treatment of contaminated soil. RIVM report no. 736102003, Bilthoven, The Netherlands.

Vidac, R.D. and Pohland, F.G. (1996) Treatment Walls. Technology Evaluation Report TE-96-01, Groundwater Remediation Technologies Analysis Center, Pittsburgh, PA.

World Resources Institute (1992), *World Resources 1992/93*, Oxford University Press, New York.

A review of surfactant-enhanced remediation of contaminated soil

Catherine N. Mulligan, Concordia University, Montreal, Canada, Raymond N. Yong, Cardiff University, Cardiff, Wales and Bernard F. Gibbs, Bivan Consultants, Montreal, Canada

INTRODUCTION

The injection or infiltration of solutions into soil using surface trenches, horizontal drains or vertical drains is called *in situ* flushing. Water with or without additives are employed to solubilize contaminants. The efficiency of the extraction depends on the hydraulic conductivity of the soil. High permeability gives better results (greater than 1×10^{-3} cm/s). The solubility of pollutants and if the pollutant was originally solubilized in water or not affect removal efficiencies. Prior mechanical mixing of the soil can disturb the infiltration of the extractant. Understanding the chemistry of the binding of the contaminant and the hydrogeology of the site are very important (USEPA, 1987).

Since water solubility of many organic contaminants is the controlling removing mechanism, additives are used to enhance efficiencies. Those contaminants that remain as a separate phase are called nonaqueous phase liquids (NAPL). NAPLs that sink below the water table are denser than water and are called (DNAPLs) and those that are lighter are called (LNAPLs). The former are thus particularly difficult to remediate (Pankow and Cherry, 1996). Some examples include chlorinated solvents such as trichloroethylene (TCE), polycyclic aromatic hydrocarbons (PAHs) including phenanthrene, naphthalene, among others that are found in coal tar and creosote and polychlorinated biphenyls (PCBs) such as Arochlor 1242. Matters become even more complicated by the fact that these contaminants are often mixed with metals or radionuclides.

BACKGROUND ON SURFACTANTS

Cationic, anionic and nonionic surfactants can be used for soil washing or flushing. They are useful in displacing DNAPL by reducing interfacial tension between DNAPL and groundwater. It is these capillary forces that restrict the mobility of the DNAPL. The mobilized contaminant can then be recovered in extraction wells (Figure 1). Surfactants can be used in mixtures, or with additives such as alcohol and/or salts such as sodium chloride. Polymers or foams can also be added to control the mobility of the contaminants. The surfactants must be recovered and reused for the process to be economic.

Surfactants are amphiphilic compounds (containing hydrophobic and hydrophilic portions) that reduce the free energy of the system by replacing the bulk molecules of higher energy at an interface. They contain a hydrophobic portion with little affinity for the bulk medium and a hydrophilic group that is attracted to the bulk medium. Surfactants have been used industrially as adhesives, flocculating, wetting and foaming agents, de-emulsifiers and penetrants (Mulligan and Gibbs, 1993). Petroleum users have

Geoenvironmental engineering, Thomas Telford, London, 1999, 441–449

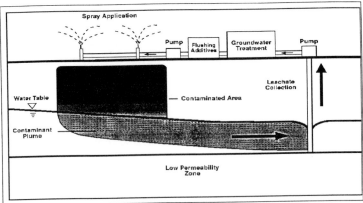

Figure 1.   Typical Soil Flushing System (Surface Sprinklers)

traditionally been the major users, as in enhanced oil removal applications. In this application, surfactants increase the solubility of petroleum components (Falatko, 1991). They have also been used for mineral flotation and in the pharmaceutical industries. Typical desirable properties include solubility enhancement, surface tension reduction, the critical micelle concentrations, wettability and foaming capacity.

FIELD STUDIES

Numerous batch and column studies have indicated that surfactants enhance recoveries of NAPL (Ellis et al., 1985; Abdul, et al., 1990; Kan et al, 1992; Soerens et al., 1992). There have also been indications that pretreatment of a soil with surfactant washing (Igepal CA-720) to solubilize PAHs enhanced biodegradation of these contaminants (Joshi and Lee, 1995).

Several field studies have also been performed with surfactant *in situ* flushing. In 1988, a wood treating site was used to test surfactant washing (Sale, 1989). Two blends were tested, one as a prewash and the other to lower oil levels even further. Blend 1 consisted of 1% Polystep A, 0.7% $NaHCO_3$, 0.1% Na_2CO_3 and 1,000 mg/L xanthan gum and second blend was comprised of 1.4% Makon-10 surfactant, 0.7% $NaHCO_3$, 0.8% Na_2CO_3 and 1000 mg/L xanthan gum. 95% of the oil was reduced and 99% of the surfactants were recovered. The combination of surfactants and sodium hydroxide has been used in enhanced oil recovery to reduce interfacial tensions. Alkaline agents can also react with hydrocarbons to form surfactants. The combination of these surfactants and the added surfactants can effectively reduce interfacial tension and enhance oil recovery (Sale and Pitts, 1989). This same mechanism then can be responsible for removal of hydrocarbons from a contaminated soil. The addition of polymer enhances the mobility ratio (i.e., enables the contaminant to be pushed instead of flow passing around it).

At a Canadian Forces Base Borden in Ontario, a test was performed in 1990-1991. The hydraulic conductivity at the site was 1×10^{-4} cm/sec with a low CEC and organic matter content. Based on the data at this site, approximately 21 pore volumes of 1% surfactant solution would be needed to remove trichloroethylene (TCE) over a 4 year period, whereas pumping and treating groundwater would require 2,000 pore volumes over 400 years to decontaminate the site (AAEE, 1993). Other examples are shown in Table 1.

Group	Process	Contaminant removed	Reference
SUNY-Buffalo with Dupont at Corpus Christi Site	Addition of 1% surfactant (approved in food preparation) Witconol 2722 and Tergitol 15-S012	73 gal carbon tetrachloride successfully removed in comparison to pump and treat	Fountain (1993)
Canadian Air Forces Base at Borden (SUNY)	2% mixture of surfactant added to sandy aquifer	80% of PCE recovered	Fountain and Hodge (1992)
General Motors NAO Research and Development Center	0.75% nonionic ethoxylated surfactant washing with recovery	10% of PCBs, oils removed in first test and 14% in second exceeding expectations	Ang and Abdul (1991)
University of Michigan	Injection of 4% Witconol and Tween 80 and modelling	Removal of 10% dodecane required 0.7L surfactant and 130,000L water	Abriola (1993)
Eckenfelder, Inc.	Injection of 2.5% SDS and recycle/reuse of surfactant	Injection of 7.7 pore volumes of 2.5% SDS removed as much PCB as 20 to 40 pore volumes of water (90% waste volume reduction)	Underwood et al. (1993)
U.S. DOE Gaseous Diffusion Site Portmouth, OH	4% sodium dihexyl sulfosuccinate surfactant / 4% isopropyl alcohol as cosolvent and 2% electrolyte (1:1 NaCl and $CaCl_2$)	DNAPLs, TCE with some PCBs and other chlorinated solvents, >90%removed by solubilization	Jafvert (1996)
Dover AFB, Dover DE (single phase microemulsion)	Surfactant	DNAPL (PCE or perchloroethylene)	Internet report: www.epa.gov/ Superfund/index.htm
Ecosites Inc., Estrie Reg. Shop, Quebec, Canada	Biodegradable nontoxic surfactant	Hydrocarbons, LNAPLs, "Cutting Oil", 160,000 kg hydrocarbon recovered in 12 months, $1.2 million Cdn for full scale, commercial completed project	Jafvert (1996)
Fredicksburg, VA Wood treating site	Alkaline agent (0.5% Na_2CO_3), nonyl phenol surfactant with 10 mol ethylene oxide (0.1% wt Makon-10) and polymer (1,500 mg/L xanthan gum)	Creosote-based wood treating oil (DNAPL), unable to inject fluid, poor site evaluation	Rice University (1997)
TWCC Biosolve Group	Patented non-hazardous, biodegradable surfactant (1 to 2% concentration)	80% removal of NAPL, 90% removal of transmission fluids in Northern California , increases biodegradation by 30%. Cost $0.35 to 0.85 per gallon.	VISITT 6.0

Group and/or Location	Process	Contaminant removed	Reference
Hill Air Force Base, Layton, UT Cell 5, Surfactant Mobilization	6.6 PV of surfactant (2.2% Aerosol OT/ 2.1%Tween) and Electrolyte (CaCl₂) injected	LNAPL (JP-4 jet fuel, chlorinated and nonchlorinated VOCs, naphthalene, pesticides, PCBs, dioxins)	Jafvert (1996)
Cell 6, Surfactant Solubilization	10 PV of 4.3% surfactant (Dowfax)	Mixture of VOCs, naphthalene, pesticides, PCBs, dioxins, JP-4 fuel	Jafvert (1996)
Cell 8, Surfactant/cosolvent solubilization	Surfactant (3.5% wt Brij 91) and cosolvent (2.5% wt n-pentanol), < 10 PV injected	LNAPL (JP-4 jet fuel, chlorinated and nonchlorinated solvents, PCBs), 72% average reduction	AATDF (1998)
OU2 - Micellar Flood	0.6 PV of surfactant (7.5% sodium dihexyl sulfosuccinate), cosolvent (3.75% isopropyl alcohol) and electrolyte (7,000 mg/L NaCl)	Chlorinated solvents (TCE, TCA, PCE, and TCET), petroleum hydrocarbons, DNAPL (70% TCE), 99% recovery of DNAPL	Jafvert (1996)
Picatinny Arsenal, NJ	400 mg/L of Triton X-100	TCE, rate of desorption increased by 30% by surfactant addition	Jafvert (1996)
S.S. Papadopoulos & Assoc. (DeNAPL process) Delmont, PA	Non-toxic biodegradable surfactant (91% Witconol SN-70, 9% Mirataine BET C-30), maximum concentration of 13,000 mg/L	PCBs (solubility increased by 50 times) in fractured rock at a cost of $40 to 100 per square ft	VISITT 6.0
Serrener/Varisco Consortium, Quebec, Canada	Surfactants	BTEX, aliphatic hydrocarbons (2,500 kg/day of hydrocarbons recovered, 85% extraction efficiency, 1000 m³/week) Cost $50 to 250/m³	VISITT 6.0 GSI Environment Marketing Information
Surtek, Inc. Mobility controlled Surfactant Flushing	Low concentration of EPA and FDA approved surfactants	23,000 gallons of residual wood treating oil (DNAPL) recovered (89%) Cost of $50 to 125 per cubic yard	VISITT 6.0
Thouin Sand Quarry, Quebec, Canada, Laval University	Surfactant/cosolvent (n-butanol, hostapur (SAS), d-limonene)	Oil waste and chlorinated solvents in the form of DNAPL (density-1.02, viscosity, 18 cp), 86% recovery of DNAPL.	Jafvert (1996)
U.S. Coast Guard, Traverse City, MI	2100 L of 36,000 mg/L Dowfax 8390 injected	PCE, TCE and reacalcitrant jet fuel, PCE and LNAPL concentrations increased 40 and 90 fold, respectively	Jafvert (1996)
U.S. DOE Gaseous Diffusion Pint, Paducah, NY	1% food grade sorbitan monooleate, (3.8 L/min over 3.8 days)	TCE (DNAPL)	Rice University (1997)
Volk Air National Guard Base, WI	9 to 14 PV of Adsee 799 and Hyonic PE-90 (50:50) blend	Hydrocarbons, chlorinated hydrocarbons (Dichloromethane, chloroform, TCA, TCE)	Nash (1988)

PV, Pore volume; BTEX, benzene, toluene, ethylbenzene, xylene ; TCE, trichloroethylene; TCA, trichlorethane; TCET, tetrachloroethane; PCE, perchloroethylene: PCB, polychlorinated biphenyls

Abdul et al. (1992) examined treatment of PCB-contaminated soils. The hydraulic conductivity was 1 x 10^{-3} cm/sec. A 0.75% solution of Witconol was applied to the surface. During the test, 1.6 kg of PCBs and 16.9 kg of carrier oil were recovered.

Anionic and nonionic surfactants are less likely to be absorbed to the soil. Cationic surfactants have been used to lower aquifer permeabilities by sorption on to the aquifer materials (Westall et al, 1992) . Surfactant soil washing was originally developed in petroleum recovery operations. Surfactants have potential for use in aquifer remediation of DNAPLs.

Several factors can influence the efficiency of soil flushing with surfactants, Groundwater that is too hard may be detrimental to the effectiveness of a surfactant (AAEE, 1993). Surfactants can adsorb onto clay fractions, reducing their availability. Too quick biodegradation can inactivate the surfactant while some degradability is required to avoid accumulation. Removal of the surfactant from the recovered water from flushing can be difficult and lead to high consumption rates. In summary, desirable surfactant characteristics include biodegradability, low toxicity, solubility at groundwater temperatures, low adsorption to soil, effective at concentrations lower than 3%, low soil dispersion, low surface tensions and low CMC (Kimball, 1992). Anionic surfactants may precipitate. However, coinjection with a nonionic surfactant can reduce precipitation and also CMC values (Sabatini et al., 1995). Biosurfactants may be more biodegradable, more tolerant to pH, salt and temperature variation and in some cases less expensive to use (West and Harwell, 1992).

To reduce risk, food grade surfactants (T-MAZ 28, T-MAZ 20 and T-MAZ 60) which have been approved by the Food and Drug administration have been examined (Shiau et al., 1995b). These surfactants were able to remove one to two orders of magnitude more chlorinated organics such as PCE, TCE and 1,2 DCE than water alone by formation of microemulsions. Other surfactants with indirect food additive status such as alkyl diphenyl disulfonate (DOWFAX) indicated lower levels of losses via sorption and precipitation while substantially solubilizing naphthalene and other PAHs. Recently, a plant-based surfactant from the fruit pericarp of *Sapindus mukurossi*, a plant from the tropical regions of Asia has shown potential for the removal of hexachlorobenzene (Roy et al., 1997).

In 1996, Intera, Radian and the University of Texas conducted a study where 2.5 pore volumes of an 8% surfactant solution, 4% isopropanol and sodium chlorides were used to remove mainly TCE. No confining walls were used. Approximately 99% of the DNAPL was removed to reach as final groundwater concentration of 10 mg/L (Brown et al., 1997).

In 1996, the University of Okalahoma and the U.S. EPA R.S. Kerr Laboratory conducted a surfactant flood by pumping 6.5 pore volumes of a 4.3% surfactant through a 3 x 5 meter sheet piling cell at Hill, AFB, Utah. In this case, the contamination was mainly a 8.5% saturation of LNAPL consisting of weathered jet fuel and other components. The surfactant was able to remove 90% of the LNAPL by mobilization. In the same year at the same site, the University of Florida conducted a test by injecting 9 pore volumes of 3% surfactant and 2.5% pentanol followed by a pore volume of 3% surfactant and then 6.5 pore volumes of water. Core data indicated that 90% of the NAPL was removed while partitioning tracer data indicated that 72% removal was achieved (Jawitz et al., 1998). In this case, solubilization was the main mechanism of removal.

METAL REMOVAL BY SURFACTANTS

The research in the area for metal removal is still quite limited, particularly where metal removal is concerning (USEPA, 1987). They include organic and inorganic acids, sodium hydroxide which can dissolve organic soil matter, water soluble solvents such as methanol, displacement of toxic cations with nontoxic cations, complexing agents such as EDTA, acids in combination with complexation agents or oxidizing/reducing agents. Soil pH, soil type, cation exchange capacity (CEC), particle size, permeabilities and contaminants all affect removal efficiencies. High clay and organic matter contents are particularly detrimental.

At sites with signed Records of Decision (ROD), the combination of hydrocarbons and metals is found at 49% of the sites (USEPA, 1997). Even though organic and metal contamination is a major concern, very few technologies are capable of dealing with both types of contaminants. As previously mentioned, surfactants can be used to assist in the remediation of numerous types of hydrocarbon contaminants. Only recently, has it been shown that surfactants can be used to enhance metal removal (Mulligan et al., 1999a; Mulligan et al., 1999b). Biologically produced surfactants, surfactin, rhamnolipids and sophorolipids (Mulligan et al., 1999a) were able to remove copper and zinc from a hydrocarbon-contaminated soil. This is due to the anionic character of these surfactants. A series of washings was performed with surfactin and compared to a control. Initial concentrations of oil and grease were 12.6% and the initial copper content of the soil was 550 mg/kg). Five consecutive washings were performed each lasting 24 h (Figure 2). For copper, the control showed a final cumulative removal of 20% while approximately 70% was removed by the surfactin. At the same time, approximately 50% of the hydrocarbons were removed by surfactin compared to 30% by the control. Therefore, these results seem very promising. Other advantages of these biosurfactants are that they are low in toxicity and biodegradable. In addition, they potentially can be produced in situ using the organic contaminants as substrates for their production. Larger scale studies, however, are needed.

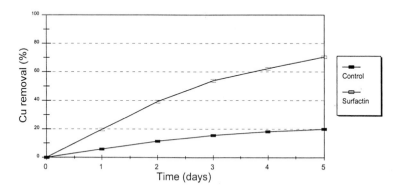

Figure 2. Series of washings for copper by 0.25% surfactin with 1% NaOH and the control (1% NaOH) according to Mulligan et al. (1999b).

CONCLUSIONS

It has been demonstrated that high contaminant removal rates are possible by using surfactants in *in situ* field tests. Large quantities of soil do not need to be excavated and handled. The addition of the surfactants can accelerate remediation work and potentially a wide variety of contaminants in the vadose and saturated zones can be remediated. Only a few tests, however, have been performed over a long period of time to determine the levels of residual contamination that are achievable at full scale. Some of the limitations may include low soil permeability, heterogeneity and extremely insoluble contaminants. For example, low permeability clays with hydraulic conductivities in the order of 10^{-4} cm/sec may significantly increase the time for the surfactant to permeate through the contaminated zone. Heterogeneities could cause some sections of the treatment zone to receive more solution than others and thus increase solution requirements. Control of mobility by the addition of polymers or foam can reduce this problem. Another potential factor in surfactant flushing that must be taken into consideration is that since the NAPLs are mobilized by the surfactant, there is the potential for downward, or horizontal movement of the contaminant and subsequent groundwater contamination. Research is underway to minimize this possibility (Fountain, 1998). Another concern of the regulators is the toxicity of the flushing solutions. Therefore, it is very important to develop the use of non-toxic biodegradable surfactants to enhance the acceptability of this technology. The development of the use of biosurfactants both for hydrocarbon and metal removal should help alleviate this concern.

REFERENCES

American Academy of Environmental Engineers (AAEE) (1993) *Soil Washing / Soil Flushing, Innovative Site Remediation* (Anderson W,C, ed.), Volume 3, WASTEC.

AATDF (1998) Surfactants and Cosolvents for NAPL Remediation. Technical Practices Manual, Ann Arbor Press.

Abdul, A., Gibson, T., Ang, C., Smith, J. and Sobczynski, R. (1992) In-situ surfactant washing of polychlorinated biphenyls and oils from a contaminated site. *Ground Water*, **30**, 219-231.

Abriola, L.M, Pennell, K.D., Dekker, T.J. and Weber, W.J., Jr. (1993) Laboratory and modeling investigations of surfactant enhanced aquifer remediation. *19th Annual Risk Reduction Engineering Laboratory Hazardous Waste Research Symposium*. Cincinnati, Ohio, April 13-15, pp. 173-176.

Ang, C.C. and Abdul, A.S. (1991) Aqueous surfactant washing of residual oil contamination from sandy soil. *Ground Water Monit. Rev*, **11**, 121-127.

Brown, C.L., Delshad, M., Dwarakanath, V., Jackson, R.E., Londergan, J.T., Meinardus, H.W., McKinney, D.C., Oolman, T., Pope, G.A. and Wade, W.H. (1997) A successful demonstration of surfactant flooding of an alluvial aquifer contaminated with DNAPL, *Submitted as a Research Communication to Environmental Science and Technology*.

Ellis, W.D., Payne, J.R. and McNabb, G.D. (1985) *Treatment of Contaminated Soils with Aqueous Surfactants*, U.S. EPA No. EPA/600/2-85/129.

Falatko, D.M. (1991) Effects of biologically reduced surfactants on the mobility and biodegradation of petroleum hydrocarbons. M.S. Thesis. Virginia Polytechnic Institute and State University, Blackburg, VA.

Fountain, J.C. (1993) A pilot scale test of surfactant enhanced pump and treat. *Proceedings of Air and Waste Management Association 85th Annual Meeting* in Denver, Colorado, June 13-18, 1993.

Fountain, J.C. and Hodges, D. (1992) *Project Summary: Extraction of Organic Pollutants Using Enhanced Surfactant Flushing-Initial Field Test (Part 1)*. NY State Center for Hazardous Waste Management, February 1992.

Fountain, J.C. (1998) Report: Technologies for Dense Nonaqueous Phase Liquid Source Zone Remediation. Ground-Water Remediation Technologies Analysis Center, Document TE-98-02 [gwrtac.org].

Jafvert, C.T. (1996) Report: *Surfactants/Cosolvent*. Ground-Water Remediation Technologies Analysis Center, Document TE-96-02.

Jawitz, J.W., Annable, M.D., Rao, P.S.C. and Rhue, R.D. (1998) Field Implementation of a Winsor Type I Surfactant/Alcohol Mixture for in Situ Solubilization of a Complex LNAPL as a Single-Phase Microemulsion. *Environmental Science and Technology*, 32, 523-530.

Joshi, M.M. and Lee, S. (1995) A novel treatment train for remediation of PAH contaminated soils. *Fresenius Envir. Bull.* **4**, 617-623.

Kimball, S.L. (1992) Surfactant-enhanced soil flushing: An overview of an in situ remedial technology for soils contaminated with hydrophobic hydrocarbons. In: *Hydrocarbon Contaminated Soils. Vol. II., (Kostecki, P.T., Calabrese, E.J., Bonazountas, M., eds) Lewis Publishers, Boca Raton.*

Mulligan, C.N. and Gibbs, B.F. (1993) Factors influencing the economics of biosurfactants, *Biosurfactants, Production, Properties, Applications*, (N. Kosaric, ed.) Marcel Dekker, New York, pp. 329-371.

Mulligan, C.N., Yong, R.N. and Gibbs, B.F. (1999a) On the use of biosurfactants for the removal of heavy metals from oil-contaminated soil. *Environmental Progress*, 18(1), 1-5.

Mulligan, C.N., Yong, R.N. and Gibbs, B.F. (1999b) Removal of heavy metals from contaminated soil and sediments using the biosurfactant surfactin. *J. Soil Contamination.* 8, 231-254.

Nash, J.H. (1998) *Project Summary Field Studies of In situ Soil Washing*. EPA/600/S2-87/1110. U.S. Environmental Protection Agency. February 1988.

Pankow, J.F. and Cherry, J.A. (1996) *Dense Chlorinated Solvents and other DNAPLs in Ground-Water*, Waterloo Press, Portland.

Rice University (1997) *Technology Practice Manual for Surfactants and Cosolvent*. Rice University, Houston, TX, February 1997.

Roy, D., Kommalapati, R.R., Mandava, S.S., Valsarai, K.T. and Constant, W.D. (1997) Soil washing potential of a natural surfactant. *Environ. Sci. Technol.* 31, 670-675.

Sabatini, D.A., Knox, R.C. and Harwell (1995) Emerging technologies in surfactant-enhanced subsurface remediation. In: *Surfactant-Enhanced Subsurface Remediation, Emerging Technologies* (Sabitini, D.A., Knox, R.C. and Harwell, J.H. eds). ACS Symposium Series 594, American Chemical Society, Washington, pp. 1-9.

Sale, T. and Pitts, M. (1988) Chemically enhanced in situ soil washing. In *Proc. Conference on Petroleum Hydrocarbons and Organic Chemical in Ground Water: Prevention, Detection and Restoration.* 487. Houston. Dublin, Ohio: National Water Well Association.

Sale, T., Pionek, K. and Pitts, M. (1989) Chemically enhanced in situ washing, In: *Proceedings of the 1989 NWWA/API Conference on Petroleum Hydrocarbons and Organic Chemicals in Ground Water-Prevention, Detection and Restoration.* Houston, TX, November 15-17, 1989.

Shiau, B.-J., Sabatini, D.A. and Harwell, J.H. (1995) Properties of food grade (edible) surfactants affecting subsurface remediation of chlorinated surfactants, *Environ. Sci. Tech.* 29, 2929-2935.

Soerens. T., Sabatini, D.and Harwell, J. (1992) Surfactant enhanced solubilization of residual DNAPL: Column studies. *Subsurface Restoration Conference*, Dallas, TX, June 21-24.

Underwood, J.L., Debelak, K.A., Wilson, D.J., and Means, J.M. (1993) Soil cleanup by in- situ surfactant flushing, V. *Sep. Sci. Technol.* 28(8), pp. 1527-1537.

USEPA (1987) Treatability studies under CERCLA: An Overview. OSWER Directive 9380.3-02FS.

USEPA Office of Solid Waste and Emergency Response (1997) *Recent Development for In Situ Treatment of Metal Contaminated Soils.*

Vendor Information System for Innovative Treatment Technologies (VISITT) version 6.0 Database, EPA 542/C-98-001.

West, C.C. and Harwell, J.H. (1992) Surfactant and subsurface remediation. *Environ. Sci. Technol.* 26, 2324-2330.

Westall, J., Hatfield, J. and Chen, H. (1992) The use of cationic surfactants to modify aquifer materials to reduce the mobility of hydrophobic organic compounds. A study of equilibrium and kinetics. *Subsurface Restoration Conference*, Dallas, TX, June 21-24.

Remedial Treatment of Contaminated Materials with Inorganic Cementitious Agents: Case Studies

DR J. M. REID and G. T. CLARK,
Transport Research Laboratory, Crowthorne, UK

ABSTRACT

Two case studies of the field behaviour of lightly contaminated fine-grained materials treated with lime are described. The first is a small-scale outdoor test bed at TRL, which was monitored over a period of 15 months. The results are compared with dynamic leaching tests on the same material conducted in the laboratory. The strength of the material varied in relation to the weather, but was not significantly different from that of samples stored in the laboratory. The pattern of leaching of contaminants was similar to that observed in laboratory tests. However the pH decreased to near neutral over the period, whereas in laboratory tests the leachate pH remained greater than 12. The second study consisted of sampling an embankment of lime-modified material three years after construction. The shear strength of the material had not changed, and the pattern of leaching of contaminants was very similar to that found in laboratory tests prior to construction. As with the test bed, the pH of the leachate was near neutral, whereas it had been strongly alkaline in tests prior to construction. It is concluded that both cases are examples of *solidification* rather than *stabilisation*. The studies indicate that both materials have retained their strength, and have not caused significant pollution of controlled waters. Environmental effects can be predicted from simple laboratory leaching tests.

INTRODUCTION

The remedial treatment of contaminated materials with inorganic cementitious agents such as cement, lime and pulverised fuel ash (pfa) has become increasingly common in recent years. In the United States, it is estimated that stabilisation with cement is used in about 30% of site clean-up cases (Jones and Hopkins, 1997). Recent cases in the UK include the use of lime and pfa to transform 100,000 m^3 of lightly contaminated silt dredgings into acceptable earthworks fill on a major highway scheme (Nettleton *et al.*, 1996). The method is particularly suitable for inorganic contaminants and for wet, fine-grained materials such as sludges and dredgings, where the addition of cementitious agents produces a considerable improvement in the physical properties of the materials.

DEFINITIONS

The processes by which cementitious agents act on contaminated materials are *stabilisation* - changing the chemical form of the contaminants to render them less mobile - and *solidification* - changing the physical properties of the material (Harris *et al.*, 1995). The processes are referred to as stabilisation/solidification, or s/s methods, and most applications involve a combination of both effects. There is potential confusion with the terms *stabilisation* and *modification* which are used in civil engineering in regard to the improvement of the physical properties of soil by the addition of lime or cement (MCHW 1). In this terminology, *modification* refers to the reduction in moisture content and plasticity index which occurs immediately on addition of the cementitious agent, whereas *stabilisation* refers to the formation of cementitious compounds, which develop over time in the material (Sherwood, 1993). However, the formation of cementitious compounds does not mean that the chemical form of the contaminants has been altered, which the use of the term *stabilisation* implies, and *modification* is essentially an example of a *solidification* process.

One of the advantages of s/s is that it is familiar to contractors and consultants from civil engineering works. However, this familiarity may lead to an incorrect understanding of what s/s technology does and how it should be implemented. Each case has to be considered individually to see what effects the addition of cementitious agents will have for the particular material under consideration.

EARLIER WORK AT TRL

Research has been carried out by the Transport Research Laboratory (TRL) in recent years into treatment with lime for immobilising contaminants and improving the geotechnical properties of weak contaminated materials (Reid and Brookes, 1997, 1999). The work has been carried out in the context of the remediation and beneficial use of contaminated materials where they are encountered in highway schemes, but has applications to the wider use of cementitious agents for remediating contaminated land.

The work reported earlier dealt with the results of long-term dynamic leaching tests on fine-grained contaminated material treated with lime. The results showed that the shear strength and physical integrity of the materials were not affected by prolonged leaching under laboratory conditions. Most of the metals were effectively immobilised by the treatment with lime. However, concentrations of copper, nickel and phenol were significantly higher in the initial leachate from the treated material than from the untreated material. This initial peak was not detected by the NRA interim leaching test (Lewin *et al.*, 1994), which is a single stage batch test, but would have been detected by a multi-stage test such as the draft CEN two stage test (CEN, 1996).

A programme of detailed chemical and mineralogical analysis was carried out on the test materials to establish the reason for this behaviour. This work was carried out by the Geoenvironmental Engineering Research Centre at Cardiff University and is reported elsewhere in this volume (McKinley *et al.*, 1999). It was concluded that the copper and nickel were largely bound to organic matter in the untreated material, and that the addition of lime had caused the organic compounds to break down to soluble organo-metal complexes.

It was recognised that the laboratory leaching tests did not simulate the field conditions for highway earthworks. Unlike the constant temperature and full saturation of the laboratory tests, in the field the materials will be subject to variable temperatures and degrees of saturation. This could lead to differences in their physical and chemical performance. Two approaches were taken to address this problem: a pilot scale outdoor test bed was constructed at TRL and monitored for 15 months; and samples were taken from an existing highway embankment of material treated with lime. These tests are described in this paper.

TEST BED AT TRL

The materials were obtained from a site on the A13 improvement scheme at Rainham Marsh, Essex where modification with lime was successfully used to treat lightly contaminated silt dredgings to which pfa had been added (Nettleton *et al.,* 1996). Samples used in the TRL tests consisted of a mixture of the silt dredgings and pfa: to this was added 5% w/w of sewage sludge, which contained higher levels of metals and so produced a more heavily contaminated material.

The criteria used to determine a suitable percentage addition of lime were those used in the assessment of materials for general earthworks fill in highway works (MCHW 1). Acceptability was defined as an MCV between 8.5 and 12 prior to final compaction, immediate CBR greater than 3%, rising to greater than 5% after 7 days and swell of less than 5mm after 28 days (Reid and Brookes, 1997). Extensive trial mixes indicated that, for the test materials, a sample with 5% w/w lime yielded an acceptable earthworks fill material.

A pilot scale outdoor test bed (5.0 m x 1.7 m x 0.2m) of the material was constructed at TRL in July 1997. The test bed consisted of a 200 mm deep bay with concrete base and surround. The base of the bay, which was constructed at a slight gradient, was lined with an impermeable rubber membrane to allow any water that percolates through the soil to collect and run out towards one end. The treated material was compacted in thin layers (50 mm to 100 mm) with the aid of a JCB and a small vibrating plate compactor. The final, compacted thickness of the soil was approximately 200 mm. After monitoring for 15 months, the bed was excavated in December 1998.

Samples have been taken from the test bed for shear strength testing at specified time intervals. Samples of the material (taken before compaction in the test bed) were compacted in the laboratory in accordance with BS 1924: Part 2 (British Standards Institution, 1990) and stored in a curing tank under water. In order to compare the field and laboratory test results, testing of both materials was carried out at the same time intervals.

Some key properties of the materials are summarised in Table 1. The untreated material was a silty clay, with Liquid Limit 63%, Plastic Limit 27% and Plasticity Index 36%. The addition of lime had very little effect on the Atterberg Limits, though it had a major effect on the moisture content and geotechnical properties. Frost heave tests were carried out on the treated material according to BS 812: Part 124 (British Standards Institution, 1989), and it was found to be non-frost susceptible.

Table 1. Geotechnical properties of test bed material

Material	mc (%)	Dry Density (Mg/m^3)	MCV	CBR (%)	Shear Strength (kPa)	pH
Untreated	53	1.03	6.7	0.8	25	8.1
Treated, laboratory	44	1.10	10.0	4.5	129	12.2
Test bed as placed	44	0.98	11.7	7.4	98	12.5
Test bed after 1 year	39	1.07	11.4	10.7	232	10.6

The development of strength in the test bed compared to the samples allowed to cure in the laboratory is shown on Figure 1. The test bed showed a rapid gain in strength over the first two months, corresponding to the warmest summer weather. The shear strength decreased during the winter months, and remained variable during the spring, before increasing to a maximum in July 1998. By contrast, the saturated samples were slow to increase in strength, reached a maximum after six months and then decreased. The influence of atmospheric conditions on shear strength is evident. However, despite being open to the weather, the shear strength of the test bed did not drop below 140 kPa.

Figure 1. Development of shear strength in the test bed

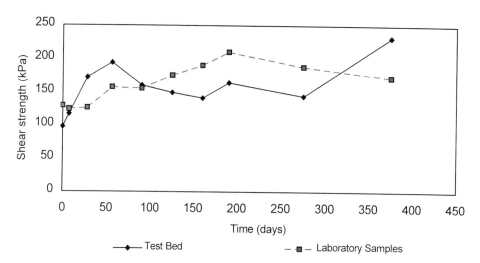

The drainage from the test bed was collected and samples taken for chemical analysis at regular intervals over a period of 15 months. Most of the water collected was probably runoff rather than percolation, given the low permeability of the material (2×10^{-9} m/s in laboratory tests; Reid and Brookes, 1997). The results for selected species are shown on Figures 2 and 3.

The most immediately noticeable feature about the test bed drainage compared to the laboratory tests is that the pH decreases with time from 12.2 at the start of the test to 7.5 after one year. In the laboratory tests, the pH remained greater than 12 throughout (Reid and Brookes, 1997). The concentration of calcium in the drainage rapidly drops to much lower levels than in the laboratory tests. This is attributed to uptake of carbon dioxide by the drainage, and white precipitates of calcium carbonate were found in the collection vessel each month. The pH of the test bed material decreased over the period (Table 1) from 12.5 to 10.6. This suggests that free lime has been leached out of the material or converted to calcite by carbonation.

Of the major ions, sodium, chloride and potassium show broad peaks between 3 and 6 months with a significant decrease after one year. This parallels the behaviour in the laboratory tests, where concentrations fell from high initial values to background levels as the test progressed (Reid and Brookes, 1999). Copper and nickel show similar trends, again paralleling the behaviour in laboratory tests. Aluminium and cyanide are higher than in the laboratory tests and phenols lower. The high value of 8 mg/l aluminium recorded after 300 days is probably due to suspended material rather than dissolved aluminium.

The significance of the results of the chemical tests needs to be seen in the context of the chemistry, drainage and low permeability of the material. Because of the low permeability, the amount of leachate that would be generated is very small. It would thus be diluted to acceptable levels before it could cause pollution of controlled waters.

Overall, the lime-treated material in the test bed behaves in a similar fashion to saturated samples in the laboratory, both in terms of shear strength and leaching of contaminants. The main difference is the reduction in pH of the test bed material and drainage compared to the laboratory leaching tests. This implies rapid carbonation of free lime in the test bed material.

Figure 2. Test bed drainage: major ions and pH

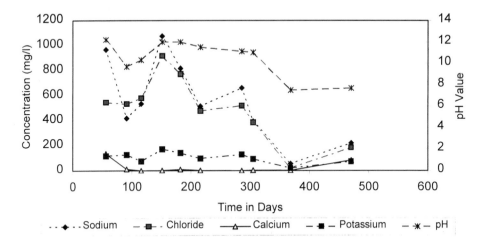

Figure 3. Test Bed Drainage: Major Contaminants

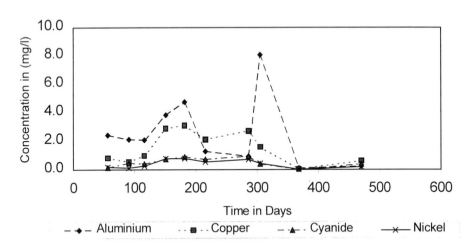

SAMPLES FROM EXISTING CONSTRUCTION

In 1995 about 100,000m^3 of lightly contaminated silt dredgings mixed with pfa were treated with lime and incorporated in the highway embankment on the A13: Contract 2 highway scheme at Rainham Marsh, Essex (Nettleton et al., 1996). A lime content of 3% w/w was used. The lime was used to *modify* the silt, that is, to dry it out so that it formed an acceptable earthworks fill. There was no intention to *stabilise* the material, in the sense of forming cementitious compounds.

In 1998, an opportunity arose to sample some of the lime-modified material. The material is surrounded by a 4m zone of pfa on the embankment shoulders, but during construction of Contract 1 of the scheme, some of the material was exposed during excavations for a viaduct abutment at the junction of Contract 1 and Contract 2. Undisturbed block samples were taken using the method of Barton *et al.* (1986). Small (60mm) shear box tests were carried out on the material to compare with similar tests carried out during construction. All tests were carried out in accordance with BS 1377 (British Standards Institution, 1990).

A comparison of the results from 1995 (from Nettleton *et al.*, 1996) and 1998 are given in Table 2. Very little, if any change has occurred in the strength of the material since 1995. The marginal differences recorded are nearly all within the range of values obtained in 1995. The only exception was the lower apparent residual cohesion value obtained from the single series of residual strength tests carried out in 1998. This may be attributable to the different test procedure (multi-reversal) used in 1998, which gives a more accurate figure for the residual cohesion.

Table 2. Engineering properties of lime modified silt, A13 Rainham Marsh, 1995 and 1998.

Soil property	Units	Mean value		Range of values	
		1995	1998	1995	1998
Bulk density	Mg/m^3	1.55	1.70	1.41-1.64	1.61-1.76
Dry density	Mg/m^3	1.14	1.08	1.03-1.25	1.00-1.14
Void ratio e		1.02	1.20	nd	1.06-1.46
Apparent cohesion peak c'	kPa	24	26	13-38	23-27
Apparent cohesion residual c'	kPa	25	7	16-32	*
Angle of internal friction peak ϕ'	^{o}	34	32	30-38	30-34
Angle of internal friction residual ϕ'	^{o}	31	31	30-32	*

* Single series of tests
nd: not determined

Leaching tests were carried out on the treated silt prior to construction in 1995. These showed that the leachate met the Water Quality Standards for irrigation (Nettleton *et al.*, 1996). The NRA leaching test was conducted on a sample of the material in 1998 and chemical analysis of the soil and leachate carried out. The leachate was relatively innocuous. Only potassium, boron, sulphate and nickel concentrations exceeded the UK drinking water standards, and then only marginally. The only significant change in leachate quality since 1995 was in the pH. In the 1998 tests, the leachate had a pH of 7.6 and the soil pH was 8.2. This compares with values of 11.75 for leachate and 10.7 for soil in 1995. This suggests that all free lime has been leached out of the material or been carbonated between 1995 and 1998, and confirms that the effect of the lime has been to *modify* the soil rather than *stabilise* it. In remediation terminology, the process has been *solidification*, not *stabilisation*. The process has been very satisfactory, as the formerly unacceptable material has been rendered into an acceptable general earthworks fill which has retained its geotechnical properties and is unlikely to cause significant pollution of controlled waters.

CONCLUSIONS

The two case studies presented are examples of *solidification*, or *modification* rather than *stabilisation*. In order to achieve *stabilisation*, in the sense of changing the chemical form of contaminants to reduce their mobility, it would be necessary to use much higher levels of cementitious agents than the 3% to 5% w/w used in these examples. However, this would have resulted in a material with much higher strength and less flexibility, which would have exceeded the geotechnical requirements for general earthworks fill. It would also have been much more expensive, and might have rendered the work uneconomic. The evidence from the A13 site is that the modification with lime has successfully achieved its aim, with no indication of significant loss of strength or release of contaminants in the three years since construction. Similarly the test bed, exposed to more extreme conditions of temperature and humidity over a period of one year, shows no loss of strength and the leaching of contaminants follows a similar pattern to that observed in laboratory leaching tests.

The major difference found between laboratory tests and field performance is the pH of the leachate. This is strongly alkaline in the laboratory tests, but in the field uptake of carbon dioxide takes place and the pH drops to near neutral. This is much less harmful to controlled waters and construction materials. However, in the test bed drainage this was accompanied by precipitation of calcium carbonate in the collection vessel, and there is a risk that this could occur in any drains adjacent to lime-modified material. No calcium carbonate was observed on the A13 site, and it may be that under natural conditions the carbonate redissolves as the pH falls to near neutral. This is predicted by speciation modelling carried out by McKinley *et al.* (1999).

The case studies indicate that treatment with lime can be an effective way of dealing with contaminated fine-grained material for earthwork purposes. However, at the low levels of cementitious agents normally used for earthwork purposes it should be appreciated that the effects are primarily *solidification* or *modification* rather than *stabilisation*. Thus it may be found that some contaminants are more mobile in leaching tests on the modified material than on the original material, as with copper and nickel in the example above. This effect will generally be negated by the low permeability of the modified material. Further precautions against pollution of watercourses can be taken by installing a break layer of inert material around the modified material to limit contact with percolating water.

Each material will react differently when treated with inorganic cementitious agents. The simplistic assumption that raising the pH will automatically reduce the mobility of contaminants is unsustainable. However, the leaching behaviour of a material can be predicted by simple multi-stage leaching tests such as the draft CEN test (CEN, 1998). The evidence from the studies reported is that this will give a generally accurate picture of leaching behaviour in the field, but that the high pH values found in the leaching tests will not occur in the field. Tests of this sort should be carried out whenever it is proposed to use s/s technology for contaminated materials.

ACKNOWLEDGEMENTS
This paper is based on research carried out by TRL under a contract with the Department of the Environment, Transport and the Regions. Any views expressed are those of the authors and not necessarily those of the Department.

The authors would like to thank Mr Geoff Turpin and Mr Dave Purves of Hyder Consulting Ltd for arranging access for sampling the lime-modified material on the A13, Mr Andrew Brookes for carrying out the sampling and the Department of Civil Engineering of the University of Southampton for the use of their block sampling equipment and for carrying out the shear box tests.

REFERENCES
Barton M E, A H Brookes, S N Palmer and Y L Wong (1986). A collapsible sampling box for the collection and transport of intact block samples of friable uncemented sands. *Journal of Sedimentary Petrology,* Vol 56, pp 540-541.
British Standards Institution (1989). Method for the determination of frost heave. *British Standard BS 812: Part 124.* London, British Standards Institution.
British Standards Institution (1990). Methods of test for soils for civil engineering purposes. *British Standard BS 1377.* London, British Standards Institution.
British Standards Institution (1990). Methods of test for cement-stabilized and lime-stabilized materials. *British Standard BS 1924: Part 2.* London, British Standards Institution.
CEN (1996). Characterisation of waste. Leaching. Compliance test for leaching of granular waste materials. Determination of the leaching of constituents from granular waste materials and sludges. Draft European Standard prEN 12457, CEN/TC 292.
Harris M R, S M Herbert and M A Smith (1995). Remedial treatment for contaminated land: Volume VII, Ex-situ remedial methods for soils, sludges and sediments. *CIRIA Special Publication SP107.* London: Construction Industry Research and Information Association.
Houses of Parliament (1989). Water Supply (Water Quality) Regulations 1989. HMSO, London.
Jones D and M Hopkins (1997). Stabilization and solidification using cement: a study visit to the USA. *British Cement Association Project Report.* British Cement Association, Crowthorne.
Lewin K, K Bradshaw, N C Blakey, J Turrell, S M Hennings and R J Fleming (1994). Leaching tests for assessment of contaminated land: Interim NRA Guidance. *National Rivers Authority R&D Note 301.* National Rivers Authority, Bristol UK.
Manual of Contract Documents for Highway Works. The Stationery Office, London.
  Volume 1: Specification for highway works (MCHW 1).
McKinley J, *et al.* (1999). *Ibid.*
Nettleton A, I Robertson and J H Smith (1996). Treatment of silt using lime and pfa to form embankment fill for the new A13. *Lime stabilisation,* CDF Rogers, S Glendinning and N Dixon, Editors. London: Thomas Telford, pp 159-175.
Reid J M and A H Brookes (1997). Stabilisation of contaminated material using lime. In *Geoenvironmental Engineering Contaminated ground: fate of pollutants and remediation,* edited by R N Yong and H R Thomas. Thomas Telford, London, pp 409-414.
Reid J M and A H Brookes (1999). Investigation of lime stabilised contaminated material. *Engineering Geology* (in press).
Sherwood P T (1993). Soil stabilization with cement and lime. *TRL State-of-the-Art Review.* London: The Stationery Office.

Comparison of methods for the determination of surface area and cation exchange capacity of clay soils.

S. RICHARDS AND A. BOUAZZA
Department of Civil Engineering, Monash University, Clayton, Victoria, Australia

ABSTRACT

In this paper, the generally accepted laboratory methods of surface area and cation exchange capacity (CEC) determination have been compared with more basic tests that could be performed in a soils laboratory. The application of the BET equation (Brunauer et al., 1938) to a nitrogen adsorption isotherm constructed at 77.35 K was used to measure the external surface area of four natural clay soils. Ethylene glycol monoethyl ether (EGME) retention at room temperature was used to estimate the total surface area of the same four materials. Cation exchange capacity was measured using the ammonium acetate method. Both the sum of the exchanged cations and the amount of ammonium ions later exchanged were considered. Methylene Blue titration was also used to estimate CEC. It is considered unlikely that the simple test methods used, reliably measure either surface area or cation exchange capacity. However, the results do provide an indication of the material interaction with cationic and polar compounds. This information may be useful in the assessment of contaminant interaction with natural soils.

INTRODUCTION

Cation exchange capacity and surface area are fundamental clay characteristics which affect soil – contaminant interaction and are considered to influence plasticity and hydraulic conductivity. An estimate of surface area and cation exchange capacity of site soils may provide valuable information on contaminant transport for the design of landfill liners or assessment of contaminated sites. The currently accepted methods of assessment of surface area and cation exchange capacity involve time consuming laboratory methods requiring purpose built equipment or chemical analysis.

The most commonly used method for analysis of the surface area of clays is application of the BET theory to a nitrogen adsorption isotherm (van Olphen, 1969). Nitrogen only bonds to easily accessible sites, and therefore this method is considered to measure the total surface area for non-expanding clays and the external surface area for expanding clays. This method requires specific apparatus and mathematical analysis. Several researchers have proposed alternatives based on adsorption of a monolayer of a polar liquid, such as glycerol, ethylene glycol or ethylene glycol monoethyl ether (Dyal and Hendricks, 1950; Carter et al., 1965). These methods are considered to measure internal and external surface area. These tests require simple equipment and several samples can be tested simultaneously.

There are many accepted methods for analysing cation exchange capacity. Adsorption of any cation could be observed, however, the ideal reagent would replace all exchangeable cations, yet be readily exchanged itself. Due to the nature of the net negative charge, which generates

Geoenvironmental engineering, Thomas Telford, London, 1999, 458–464

the cation exchange capacity, the measured capacity is sensitive to test conditions such as pH and ionic strength. Analysis of cations displaced by ammonium ions and subsequent displacement and analysis of the ammonium ions is a popular method.

Four clay materials were chosen for assessment. All are natural clay soils which have undergone little or no treatment. Tru-bond and Albion clay contained some smectite minerals. The Saponite clay contains saponite and calcite. Kaolin is considered to contain mostly kaolinite.

By estimating surface area and cation exchange capacity, along with the more traditional engineering properties of soil, an assessment of the likely soil – contaminant interaction can be made. This improves the accuracy of transport models and can provide an indication of where further assessment, such as column diffusion tests, may be required.

MATERIALS AND METHODS
Materials
The two commercially available materials, Tru-bond and Kaolin, were supplied by Commercial Minerals. The Albion clay was obtained from Albion, Victoria, Australia and is the result of weathering of Quaternary age basalt. The saponite clay was obtained from a farming property near Perth, Western Australia.

Surface Area
External surface area was assessed by applying BET theory to an N_2 adsorption isotherm. A Micromeritics ASAP 2010 instrument was used. The Albion clay and Tru-bond samples were degassed at 130°C, Saponite at 170°C and Kaolin at 140°C. The adsorption was conducted at the temperature of liquid nitrogen, 77.35 K. The BET analysis was conducted by the equipment interface.

Total surface area was assessed by ethylene glycol monoethyl ether retention on dried clay samples. These tests were conducted in triplicate. The procedure presented by Eltantawy and Arnold, (1973), was followed. Approximately 1g of soil sample, which had been dried in an oven for 24 hours was placed in a shallow aluminium tin. Approximately 3ml of EGME was added to each dried sample to form a slurry. The slurry was left to cure for at least one hour. Samples were then placed in a desiccator over dry $CaCl_2$. A dish containing liquid EGME was also placed in the desiccator. The desiccator was evacuated with a vacuum pump for 45 minutes. The aluminium tins and their contents were weighed every 2 hours until a near constant weight was achieved. The EGME retention test produces a result in terms of grams of EGME per gram of soil. To convert this to surface area per gram some correlation is required between the weight of EGME and surface coverage.
By comparing EGME retention on a montmorillonite of known surface area, Eltantawy and Arnold (1973) established that 3.78×10^{-4} g of EGME was required to cover 1 m^2. Later work by Nguyen *et al* (1987) investigated the coverage of EGME molecules on montmorillonite and kaolinite minerals. They suggested that, based on their Fourier-transform infra-red study, each molecule of EGME covered 40.2 A^2 (3.72×10^{-4} g/m^2) on montmorillonite and 20 A^2 (7.48×10^{-4} g/m^2) on kaolinite. These values compared well with the empirical values obtained on the same materials.

Cation Exchange Capacity
The ammonium acetate method as developed by Lavkulich, (1981), was used. Approximately 4g of soil was saturated with 1M ammonium acetate solution with a pH of 7. The liquid and

solid phases were separated and the liquid analysed for common earth and metal cations by atomic adsorption spectrophotometry. The solids were rinsed several times with isopropanol then saturated with 1M potassium chloride solution. Again the liquid was separated and analysed for ammonium. Summation of the displaced earth and metal cations should approximately equal the displaced ammonium. These tests were performed in duplicate.

The methylene blue titration method proposed by Cokca and Birand, (1993), was generally followed, the methylene blue solution and sample slurry concentrations were varied slightly. Approximately 1g of clay was mixed in 100ml distilled water. Drops of 0.1M nitric acid were added until the slurry just turned blue litmus paper pink. A 0.01 M solution of methylene blue was added in stages. After each addition of dye the slurry was mixed well and a drop of slurry was placed on filter paper. When a pale blue ring was visible around the clay particles, the suspension was mixed for a further two minutes. If a drop placed after this additional two minutes also had a visible blue ring then the end point of the titration was reached. This end point approximately represents the methylene blue adsorption capacity of the clay. In fact, if you consider the shape of an adsorption isotherm, it represents the point at which the curve departs from the straight line of amount added equals amount adsorbed. This test was performed in triplicate. The CEC is calculated from this volume by the following equation.

$$CEC = \frac{100}{W_{clay}} \times V_{MB} \times \left(\frac{W_{MB}}{320} \times \frac{100 - mc_{MB}}{100} \right) \qquad \text{Eq. 1}$$

W_{clay} is the weight of dry clay tested
V_{MB} is the volume of methylene blue solution
W_{MB} is the weight of methylene blue powder added to made 1L of solution
mc_{MB} is the moisture content of the methylene blue powder.

RESULTS AND DISCUSSION
Surface Area
The results of the BET analysis of the Nitrogen adsorption on the four clay soils are presented in Table 1. The EGME retention obtained was multiplied by the theoretical values developed by Nguyen et al (1987) of 40.2 Å^2 (3.72x10^{-4} g/m^2) for montmorillonite and 20 Å^2 (7.48x10^{-4} g/m^2) for kaolinite. An estimated surface area coverage of Methylene Blue has also been included. These values are based on a surface coverage of 130 Å^2, which reflects a flat orientation.

Table 1. Results of Surface Area tests.

Material	BET surface area (m^2/g)	EGME surface area (m^2/g)	Estimated surface area coverage * (m^2/g)
Saponite	78.2	270.4	251.3
Basaltic	85.9	252.4	354.6
Bentonite	43.3	410.2	497.2
Kaolin	23.0	27.8	47.8

* based on a molecular area of 130 Å^2 and flat orientation (Hang and Brindley, 1970).

The results for Kaolin show a good comparison between the Nitrogen surface area and the result obtained by EGME retention. The other results can not be compared as they are measuring two different areas. The two sets of results do however provide some insight into the structure of the clay in terms of internal / external surface areas.

The surface area values estimated by Methylene adsorption compare favourably with the three expandable clays, where the same adsorbed molecule orientation has been assumed. However, for the Kaolin the methylene blue estimated surface area is approximately double the value obtained through nitrogen adsorption and EGME retention. This is due to a different assumption of adsorbed molecule orientation. Early work on methylene blue, ethylene glycol and EGME adsorption assumed the formation of a monolayer on each surface, internal and external. The infra-red study conducted by Nguyen *et al* (1987) showed that the EGME molecules adsorb to the mineral surface in an angled overlapping pattern. On internal surfaces this orientation results in coverage roughly equivalent to two monolayers (see Figure 1). On external surfaces the coverage is approximately equivalent to a double layer.

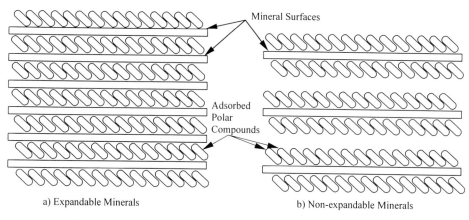

Figure 1. Schematic of interpreted adsorption of polar compounds.

Due to the high ratio of internal to external surfaces on expanding minerals, such as montmorillonite, many researchers, based on empirical data from montmorillonite, have suggested that the coverage can be approximated to a monolayer covering all surfaces. This may be a suitable approximation for layer silicates with high internal to external surface ratio. However, as the ratio gets lower and the structure approaches a non-expandable mineral such as kaolinite, this approximation produces a surface area that is overestimated by a factor of 2. If we apply this principle to the methylene blue adsorption, then we obtain an estimated surface area of 23.9 m^2/g. This value compares well with the values obtained by nitrogen adsorption and EGME retention. Unfortunately, to achieve this good relationship a prior knowledge of the material mineralogy or external surface area was required.

Cation Exchange Capacity
The results of the ammonium acetate replacement method, including both summation of displaced cations and subsequently displaced ammonium, and the methylene blue titration method are presented in Table 2.

Table 2. Results of cation exchange capacity tests.

Material	Ammonium acetate method CEC (meq/100g)		Methylene Blue titration CEC (meq/100g)
	Summation of exchange cations	Displaced ammonium	
Saponite	96.4	46.5	32.1
Basaltic	50.3	63.5	45.3
Bentonite	62.9	86.8	63.5
Kaolin	11.8	33.7	6.1

The estimated cation exchange capacity using the summation of exchange cations and the methylene blue adsorption methods compare reasonably well for the Albion clay and Trubond materials. The ammonium displacement produces a value approximately 40% higher. This may be due to adsorption of ammonium in excess of the CEC. The estimated CECs produced by the three methods were all within the range expected for materials containing montmorillonite.

The CEC value obtained by the summation of exchange cations for the Saponite clay does not compare well with the results of the other methods. The material contains high concentrations of calcite and dolomite, which produce high concentrations of calcium and magnesium in the analyte. If the concentrations of calcium and magnesium are not included in the total concentration then the estimated CEC becomes 38.5meq/100g. This value compares much more favourably with the estimate by the methylene blue adsorption method. Again the CEC obtained by the ammonium displacement method is approximately 40% higher than that obtained by methylene blue adsorption. These results are within the range expected for a material containing saponite minerals.

The three methods of estimating CEC for Kaolin produced three considerably different results. Based on the documented range of cation exchange capacity for kaolinite, it is unlikely that the CEC is as high as 33.7 meq/100g. The reasons for the discrepancies are unclear. However, methylene blue adsorption may have been limited by the available surface area.

General
The EGME retention and methylene blue adsorption tests are in effect quite similar. Both tests assess the adsorption capacity of a polar liquid to the material under some particular condition. Therefore, it is not surprising that the results of both tests are quite similar. This raises the question, "Are the tests really estimating two different characteristics?" There are some examples of work that have been carried out that compare a theoretical surface area coverage per molecule with empirical results (Nguyen et al., 1987). The results compared favourably. However, the theoretical model was based on a known adsorption orientation to a mineral of known area. It is rare to have that much information for a mineral, even less likely for a soil.

It would appear that the adsorption mechanisms of the methylene blue adsorption and EGME retention tests do not vary significantly. The adsorbed molecules are either packed evenly on the surfaces of the minerals, bound to the surface charge sites or some combination of the two. It is only the interpretation of results after the test, which determines whether surface area or CEC is being measured. Obviously the empirical or theoretical relationships used during the interpretation greatly affect the results. For surface area assessment, applying the wrong relationship can over or under estimate the area by a factor of two. A more accurate measurement of surface area could be obtained by using the EGME retention test in conjunction with nitrogen adsorption. If the area per gram of EGME relationship for external surfaces is applied to the area obtained from nitrogen adsorption and this value is subtracted from the total EGME retained, then the remainder can be assumed to have been adsorbed onto internal surfaces. The area per gram of EGME relationship for internal surfaces can then be applied. This technique removes the requirement of having some knowledge of the sample mineralogy. However, it also reduces the simplistic appeal of the method. Some researchers have had some success in measuring the external surface area using the EGME retention method by heating the sample to 600° C prior to testing (Carter et al., 1986). However kaolinitic minerals are destroyed at this temperature.

The variable results obtained by the three methods of cation exchange capacity estimation indicate that there are several complicating factors. Soluble or fine (sub 0.45 μm) materials which contain the exchange cations being analysed will result in an over estimate of the CEC. Some knowledge of the composition of the material may be used to compensate. Surface area may affect the results in some way for some materials. In 1982, Rhodes noted that results of cation exchange capacity measurement, using any displacement methods, are doubtful in the presence of calcite, gypsum, mica, vermiculite, zeolites or feldspathoids.

CONCLUSIONS

The results of this study indicate that the surface area and CEC estimations obtained by EGME retention, methylene blue titration or ammonium acetate replacement should not be applied as absolute values. Some knowledge of the mineralogy of the samples is required to assess the accuracy of the method used. In particular, the EGME retention test for surface area, which requires the application of a relationship between EGME adsorption and unit area. This relationship varies by a factor of 2 for internal and external surfaces. Some knowledge of mineralogy, internal to external ratio or external surface area is required to accurately apply the relationships.

It was found that the three methods of cation exchange capacity estimation produce varying results. Again some prior knowledge of material mineralogy can be used in the interpretation of the results. However, this is of little use when the tests are being used to characterise a soil material.

Due to the limitations of the simple test methods demonstrated, the results obtained from these tests should not be considered to give the absolute values of surface area or cation exchange capacity. This does not mean that the tests do not provide a valuable insight into interactions between the soil and polar or cationic environmental contaminants. In reality this is the only thing we can be sure that the tests reflect.

ACKNOWLEDGEMENTS

The authors would like to thank the Australian Research Council and ENRID Ltd for their financial support of the primary author's doctoral research.

REFERENCES

Brunauer, S., P. H. Emmett and E. Teller (1938). "Adsorption of gases in multimolecular layers." Journal of the American Chemical Society **60**: 309-319.

Carter, D. L., M. D. Heilman and C. L. Gonzalez (1965). "Ethylene glycol monoethyl ether for determining surface area of silicate minerals." Soil Science **100**(5): 356-360.

Carter, D. L., M. M. Mortland and W. D. Kemper (1986). Specific Surface. Methods of Soil Analysis, Part 1, Physical and Mineralogical Methods. Madison, WI, USA, American Society of Agronomy-Soil Science Society of America: 413-423.

Cokca, E. and A. Birand (1993). "Determination of cation exchange capacity of clayey soils by the methylene blue test." Geotechnical testing journal **16**(4, December 1993): 518-524.

Dyal, R. S. and S. B. Hendricks (1950). "Total surface of clays in polar liquids as a characteristic index." Soil Science **69**: 421-432.

Eltantawy, I. M. and P. W. Arnold (1973). "Reappraisal of ethylene glycol mono-ethyl ether (EGME) method for surface area estimations of clays." Journal of Soil Science **24**: 232-238.

Hang, P. T. and G. W. Brindley (1970). "Methylene blue absorption by clay minerals. Determination od surface areas and cation exchange capacities (Clay-organic studies XVIII)." Clays and Clay Minerals **18**: 203-212.

Lavkulich, L. M. (1981). Methods manual, pedology laboratory, University of British Columbia, Vancouver.

Nguyen, T. T., M. Raupach and L. J. Janik (1987). "Fourier-transform infrared study of ethylene glycol monoethyl ether adsorbed on montmorillonite: Implications for surface area measurements of clays." Clays and Clay Minerals **35**(1): 60-67.

Rhodes, J. D. (1982). Cation Exchange Capacity. Methods of Soil Analysis, Part 2. Chemical and Microbiological Properties. Madison, WI, USA, ASA-SSSA: 149-158.

van Olphen, H. (1969). Determination of surface areas of clays - Evaluation of Methods. International Symposium on Surface Area determination, Bristol, UK, Butterworths, London.

Use Of Shredded Tyre Chips In Permeable Reactive Barriers

C.C. SMITH, W.F. ANDERSON, and R.J. BURGESS
Department of Civil and Structural Engineering, University of Sheffield, UK

INTRODUCTION

The research reported here forms part of a programme of work looking at the feasibility of using tyre chips in permeable reactive barriers and other geo-environmental applications. Tyre chips have significant sorption capacity due to their principal constituents, rubber (~63%), and carbon black (~31%). This work addresses the scope of such materials for attenuating soluble organic and inorganic (such as heavy metal) contaminants, free product (LNAPL and DNAPL), and organic vapours. Relevant engineering properties such as compressibility and permeability of tyre chips and tyre chip/gravel mixtures have also been studied as well as the potential leaching of heavy metals inherent in the tyre. Various tyre chip size ranges (1-3mm, 2-8mm, and 30mm, sourced from the tread and side walls of truck tyres) have been investigated. Due to space restrictions, only organic sorption characteristics and compressibility behaviour will be dealt with in this paper.

ORGANIC SORPTION CHARACTERISTICS

The sorption capacity of tyre chips for aqueous phase organic contaminants has been studied by a number of authors, e.g. Kim et al. (1997), Park et al. (1996), Park et al. (1997), Kershaw et al. (1997), in both batch and column tests. Kim et al. (1997) investigated the relative contributions of the carbon black and rubber to the overall sorption capacity and found that that the carbon black provided 5 - 15% of the sorption capacity of tyre chips for the organics that they investigated. The sorption characteristics of hydrophobic compounds (e.g. toluene, xylene) have been relatively well investigated, however there is less data on desorption characteristics and on sorption/desorption of hydrophilic organics. Some researchers have investigated the sorption of free product by tyre chips e.g. Al Tabbaa & Aravinthan (1998), and Baykal et al. (1993), but this is primarily related to mechanical effects. Data on desorption is again limited. In this paper the results so far available from sorption/desorption tests on free product petrol and paraffin, and on aqueous phase phenol and cresols are reported and interpreted together with data in the literature.

The sorption process is a combination of two different mechanisms: adsorption and absorption. In the former a sorbate molecule attaches itself to a sorbent, due to one of a number of different attracting forces, for example organic compounds such benzene, toluene, show non-polar hydrophobic adsorption; the non-polar solute in aqueous phase is attracted toward the other non-polar sorbent due to the slightly polar nature of water molecules. Absorption then follows adsorption in that the sorbate molecule then penetrates the sorbent Absorption is often the predominant form of sorption of organic molecules onto polymeric materials in a binary aqueous solution due to the free volume between the long polymer

Geoenvironmental engineering, Thomas Telford, London, 1999, 465–472

chains. The absorption rate is controlled by the process of diffusion of the organic molecules into the polymer structure.

Liquid phase free product

The sorption and desorption characteristics of petrol, paraffin, and water were examined. Simple batch tests on 2-8mm and 30mm tyre chips were carried out in chemically resistant reaction vessels. For the 2-8mm tyre chips, 30g of tyre chips were immersed in 150ml of free product to examine sorption, and subsequently either in 150ml of deionised water, or exposed to the air to examine desorption. For 30mm tyre chips, 60g of chips were used with 300ml of fluid. The rubber mass was chosen such that enough head space was available in the reaction vessels to allow expansion of the rubber, and the volume of free product used was sufficient to cover the rubber tyre chips. At various time intervals from the initial contact of the free product with the rubber the free product was drained from the reaction vessel and the mass of rubber determined. This was achieved by using a filter connector with 1mm filter holes to drain the excess free product for a 1 minute duration. After recording the weight of the chips the free product was poured back into the reaction vessels. Once the mass had stabilised, desorption tests were carried out by exposing the tyre chips to either deionised water or air.

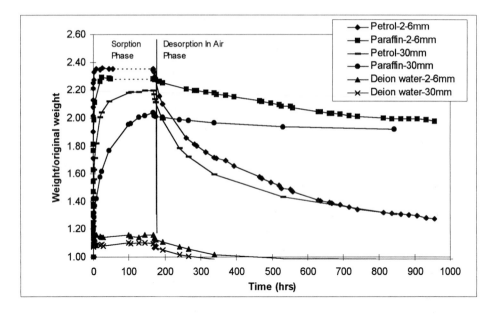

Figure 1: Free product sorption and desorption in air

The results are presented in Figure 1. All masses are subject to a small error, since residual surface fluid would be present on the tyre chips following drainage of the free product. This may explain the small weight gain of the tyre chips immersed in deionised water.

Tyre chips immersed in organic free product absorbed more than their own weight of fluid, the petrol being absorbed faster than the paraffin, presumably due to the smaller molecular weight of its components. The overall capacity was however similar. Likewise the sorption rate in the larger 30mm tyre chips was slower presumably due to the longer diffusion pathways within the larger tyre chips. No subsequent desorption in water could be measured by this technique. This is consistent with the solubility of the free product in water being negligible compared to that in the rubber. In contrast the lighter components of the free products can partition in the air fairly readily.   Tyre chips soaked in petrol lose most of their sorbed organics by evaporation, however it is likely that the heavier components of the paraffin are not vaporising at a significant rate leading to a reduced mass loss. This was confirmed by parallel tyre chip - vapour sorption/desorption tests, in which almost all weight gain during sorption from a petrol or paraffin vapour saturated atmosphere could be reversed upon desorption in air.

Aqueous phase organics

Several researchers have reported sorption test results for a range of organic compounds *e.g.* chloroform, ethylbenzene, methylene chloride (MC), toluene, 1,1,1-trichloroethane, trichloroethylene (TCE), *m*-xylene, (Kim et al., 1997), benzene, toluene, ethylbenzene, *o*-xylene, (Kershaw et al., 1997), MC, TCE, toluene, m-xylene (Park et al., 1996).   Kim et al. found that the tyre chip-water partition coefficients correlated well with the ocatanol-water partition coefficient, and showed that the presence of other organics does not appear to have a significant influence on the partition coefficients. Park et al. (1996) reported some desorption data which indicated that the partition coefficient was reduced by approximately 2 for the TCE, toluene, and *m*-xylene, but that that for MC was reduced by a factor of 15.   The implication is that there is some permanent sorption, and some retardation of these organic compounds. However column tests are essential to investigate this thoroughly. Their results also indicated that the partition coefficient was affected by the batch test tyre-liquid volume ratio. MC was particularly sensitive relative to TCE, toluene and *m*-xylene, and did not seem to follow the same pattern of behaviour as the other compounds investigated. This may be due to its high solubility in water (hydrophilic nature) and low octanol-water partition coefficient. Most researchers report that pH and temperature have little influence on organics sorption though Kim et al. note that this may not be the case for MC.

In the current work, additional hydrophilic compounds, phenol, *p*- and *o*-cresol were examined: a stock solution was prepared by adding 0.05% sodium azide by weight to deionised water in order to inhibit biodegradation of the solute. Analar grade reagents were used to prepare 1000ppm stock multi-solute (phenol, p-cresol, and o-cresol), and single species (phenol) solutions. Solute concentrations of approximately 1000ppm, 500ppm, 200 ppm, 100ppm, 10ppm, and 0ppm were used in the batch tests in order to provide equilibrium isotherm plots. The batch tests were carried out in 250ml Pyrex volumetric flasks with nitrile bungs into which 200ml of test solution was added to 20g of air-dried tyre chips. The flasks were then shaken on an orbital shaker at 20°C for 7 days. At the end of the 7 day test, 1ml of solution was removed from the batch test for analysis by HPLC and the solution pH determined.   The excess solution was then drained from the reaction vessels and 200ml of deionised water containing 0.05% sodium azide was added to the tyre chips for the desorption phase of the test.   The batch reactors were again agitated at 20°C for 7 days and the concentration of solute in solution determined.

Sorption and desorption isotherms are presented in Figure 2. It may be observed that for single species phenol, the desorption isotherm is approximately twice as steep as the sorption isotherm (in line with some of the observations discussed above), indicating that sorption is not fully reversible. However when the multi-solute results are examined, it can be seen that the cresol sorption-desorption appears to be fully reversible and the phenol sorption does not appear to be affected by the presence of the cresols (this follows the results of other researchers who report little competitive sorption between organics). However the multi-solute desorption results imply that all phenol is re-released. This is being investigated further. Some of the results imply that more of a compound is desorbed than was initially present. This may due to other compounds in the tyre rubber giving false readings, however it appears to be a relatively small effect.

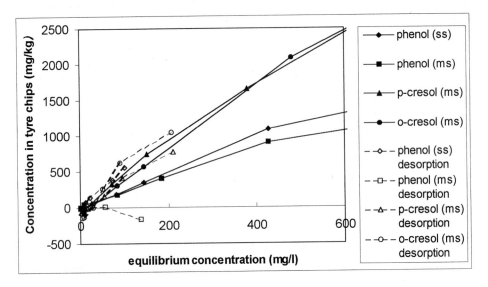

Figure 2: Single solute (ss) and multi-solute (ms) phenol and cresol batch test isotherms for 1-3mm tyre chips

Based on the sorption tests, the ratios of the partition coefficient for cresols (~5 L/kg) and phenol (~2 L/kg) appear to correlate reasonably well with the ratios of their respective octanol water coefficients ($K_{ow\ phenol}$ = 29, $K_{ow\ cresol}$ = 89). However the actual values are lower than would be predicted by comparison with other organics data by Park et al. (1996).

Discussion
The use of tyre chips in reactive barriers would appear to be an attractive proposition for a selected range of scenarios. For example tyre chip barriers could be used to sorb residual free product present following *e.g.* well pumping extraction of floating LNAPL, or in circumstances where spillages are possible. Tyre chips appear able to absorb their own weight in free product, but will re-release negligible quantities to water. Larger amounts of products might be released to the to air in the form of vapour.

In terms of aqueous phase organics, current data would appear to be inconclusive regarding whether tyre chips result in permanent sorption or simply retardation. Column tests are needed to fully ascertain the response. However based on the magnitudes of sorption partition coefficients derived above and in the literature (e.g. ~1000 L/kg for *m*-xylene, Kim et al.,1997) retardation factors of around 5-10 for phenols and cresols and 1000 - 2000 for *m*-xylene, might be expected for pure tyre chip barriers. In practice flow has to be very slow to allow complete sorption; column tests by Kershaw et al. (1997) indicated utilisation efficiencies of ~40% for hydraulic residence times of 15 minutes in ground rubber, though only the sorption phase was studied. Retardation factors of ~50 for benzene and >350 for *o*-xylene can be estimated from their data. This implies that a 1m thick tyre chip barrier might be used to treat approximately ~350m of groundwater contaminated with *o*-xylene, but perhaps only 5m of phenol contaminated groundwater. It might be necessary to remove the chips (and pore fluid) after a certain period and replace or regenerate if re-release is likely. Since this period (a function of retardation coefficient) is different for each of the contaminants in any particular scenario, the removal time may well be determined by the least sorbent compounds.

Tyre chip barriers could also have a role to pay in enhancing natural attenuation; they may be able to act like a capacitor, soaking up free product or smoothing out high concentration pulses of contaminant which might inhibit microbiological activity, and then rereleasing them in biologically favourable concentrations.

MECHANICAL BEHAVIOUR
The chemical testing suggests that tyre chips have the potential to be used as reactive barriers, either on their own or mixed with gravel where additional strength and stiffness is required, but a high permeability is to be maintained. Where they are to be used to sorb organics (either from solution or as free product) then the observed swelling may have some implications for mechanical stability. Initial tests examined the ability of different sizes of tyre chip and gravel to be mixed without segregation or significant loss of material during vibratory compaction. On the basis of these tests the study of mechanical behaviour has concentrated on 2mm - 6mm tyre chips on their own (dry and soaked in paraffin) and mixed with 20mm single size washed river gravel.

Tyre chips only
Initial studies of one dimensional compressibility were made in a 150mm diameter CBR mould, with the specimen being loaded via a dead load and lever arm system. The 130 mm high specimen was prepared by filling the greased mould in three layers with dry tyre chips and compacting each layer with a vibrating hammer for one minute. During the compression test each load increment was applied for 15 minutes and it was noted that the majority of the settlement occurred in the first couple of minutes. Figure 3 shows the settlement - applied stress relationship for a specimen consisting of the 2 - 6mm tyre chips as supplied and a specimen made up of 2mm - 6mm tyre chips which had been soaked overnight in paraffin and surface dried before compaction in the CBR mould. The deformation of the specimen is expressed as a percentage of the original height of the specimen

These compression tests in the CBR mould were complemented by tests carried out on tyre chips in a 250mm diameter Rowe hydraulic consolidation cell (RC) which had been modified to ensure no ill effects from the paraffin fluids used and to allow high flow permeability testing of the specimen under an applied stress. The specimens for the Rowe cell tests were compacted in a similar manner to those for testing in the CBR mould and taken through a similar load and unload sequence, although each load increment was applied for 30 minutes instead of the 15 minutes. In the first test the tyre chips were initially loaded and unloaded in a dry condition. The cell was then flooded with paraffin until the specimen was fully saturated. The swelling due to absorption of the paraffin was monitored until it stabilised after about 16 hours. When swelling was complete the excess paraffin was allowed to drain off through the modified Rowe cell base and a further loading/unloading sequence was applied to the specimen. The results of these tests are also plotted in Figure 3. It can be seen that significant settlements occur even under relatively low stresses and the results of tests on 'dry' chips in the CBR mould and Rowe cell are very similar. There also appears to be little difference in the behaviour between the 'dry' and 'soaked' tyre chips when the latter are compacted in the CBR mould after soaking.

When the 'dry' specimen in the Rowe cell was saturated with paraffin it underwent significant swelling as shown in Figure 3, but yielded a similar shape loading curve to the 'dry' specimen, and on unloading returned to almost its preloading void ratio. This suggests that absorption of paraffin by the tyre chips has little or no effect on the compressibility of the compacted tyre chips.

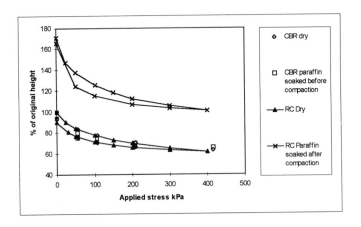

Figure 3: CBR and Rowe Cell tests on tyre chips only

Tyre chip – gravel mixtures
The combination of tyre chips with gravel will lead to a more stable barrier than tyre chips alone. A series of tests was therefore carried out in the CBR mould on dry tyre chip/gravel mixtures, the percentage of tyre chips by weight being varied in each test. Specimens were prepared by thoroughly mixing the constituents on a tray before compacting as in the tyre chip only tests. Figure 4 shows the settlement - applied stress curves. It can be seen that there is very little settlement until the percentage of tyre chips by weight exceeds 10%, above which the settlement increases significantly with increasing percentage of tyre chips.

Rowe cell tests were carried out on 10% tyre chip and 15% tyre chip specimens, but the maximum applied stress in the former test had to be limited to 200 kPa as the cell had not been fully modified at this stage. The results of two load - unload stages, i.e. 'dry' and 'paraffin soaked' are shown in Figure 5. As in the CBR mould tests on low percentage tyre chips mixtures the compression of the 'dry' chip specimens was relatively small. It can be seen that on soaking the 10% tyre chip swelled by 21% of its original height and the 15% tyre chip specimen by 42%. Significant compression occurred during the loading of the 'soaked' mixtures but on unloading they returned to almost their preloading thickness, similar to the tyre chip only tests. Rowe cell tests on mixtures which had been water saturated before paraffin soaking gave almost identical data to 'dry - paraffin soaked' tests indicated that uptake of paraffin is not influenced by whether or not their surface is dry or wet.

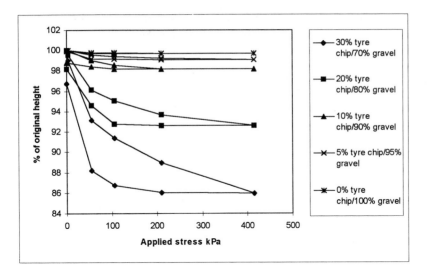

Figure 4: Tests on dry tyre chip/gravel mixtures carried out in the CBR mould

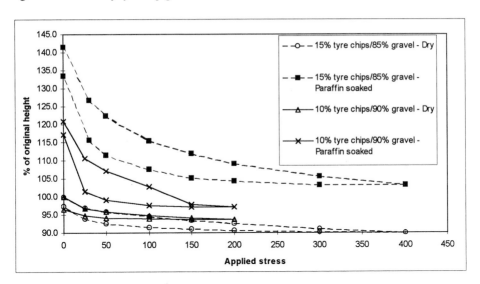

Figure 5: Rowe Cell tests on tyre chip/gravel mixtures

An insight into the compression behaviour of the tyre chip/gravel mixes may be gained by considering the structure of the specimens. The vibratory compaction of gravel on its own (0% tyre chips) will result in it having a dense granular packing. When small amounts of tyre chips are added they will fit into the void spaces in this dense granular packing. However, at a certain stage the percentage of tyre chips will have reached a level at which they can only be included in the specimen if they disrupt the dense packing moving the gravel particles apart, thus lowering the bulk density of the specimens and resulting in increased settlement when loaded. For the materials reported in this study this value this would appear to lie between 5-10% tyre chips by weight, implying that for any application where geotechnical stability is required, percentages of between 5 and 10% tyre chips in gravel will be required. This may need be reduced where significant organic sorption and associated swelling is likely, but will adversely affect the reactive performance of the barrier.

Attempts to determine the hydraulic conductivity of untreated tyre chips alone and in mixtures using a constant head technique in the Rowe cell have not yet been completely successful. However, the results indicate that the flow rate is not significantly influenced by the applied vertical stress level up to 400 kPa and values are likely to be in excess of $2x10^{-5}$ m/s.

CONCLUSIONS
Tyre chips show significant promise as a reactive barrier, particularly for aqueous phase hydrophobic organic contaminants, and residual free product. However more data on desorption characteristics of hydrophobic contaminants and on sorption and desorption characteristics of hydrophilic contaminants is needed, particularly from column tests. It is not clear whether a tyre chip barrier will permanently absorb or simply retard contaminants. Where geotechnical stability is an issue then it may be necessary to mix the tyre chips with gravel. Compressibility data for tyre chips on their own and mixed with gravel have been presented. A 5-10% mixture of 2-6mm tyre chips in 20mm gravel should achieve a relatively stable material.

ACKNOWLEDGEMENTS
This research has been supported by an EPSRC grant under the Waste Minimisation through Recycling, Re-use and Recovery in Industry programme, and by Charles Lawrence Surfaces, PLC, Newark, through provision of materials. The assistance of P. Carpenter and G. Devany in obtaining some of the test data is gratefully acknowledged.

REFERENCES
Al-tabba, A., Aravinthan, T. (1998) 'Natural clay-shredded tire mixtures as landfill barrier materials' Waste Management, 18, pp 9-16.

Baykal, G., Kavak, A., Alpatli, M., (1993) 'Rubber-kaolinite and rubber-bentonite liners' Waste Disposal by Landfill, Proc. of Symposium: Green '93. Bolton.

Kershaw, D.S., Kulik, B.C., Pamukcu, S. (1997) 'Ground rubber: Sorption media for ground water containing benzene and o-xylene' ASCE Journal of Geotechnical and Geoenvironmental Engineering, Vol 123, No. 4, pp324-334.

Kim, J.Y., Park, J.K. Edil, T.B. (1997) 'Sorption of organic compounds in the aqueous phase onto tire rubber' ASCE Journal of Environmental Engineering, Vol. 123, No. 9, pp 827-835.

Park, J.K., Kim, J.Y., Edil, T.B. (1996) 'Mitigation of organic compound movement in landfills by shredded tires' Water Environment Research, Vol. 68, No. 1, pp 4-10.

Park J.K, Kim J.Y, Madsen, C.D., Edil, T.B. (1997). 'Retardation of volatile organic compound movement by a soil-bentonite slurry cut-off wall amended with ground tyres'. Water Environment Research, Vol. 69, No.5, pp1022-1031.

Risk Assessment and Management

Abandoned mines drainage: impact assessment and mitigation of discharges from coal mines in the UK

SHEILA B BANKS[1] and DAVID BANKS[2]
[1]Scott Wilson Kirkpatrick & Co Ltd, Chesterfield, UK
[2]Holymoor Consultancy, Chesterfield, UK

BACKGROUND

The UK has a legacy of pollution caused by discharges from abandoned coal mines. Since the autumn of 1992, when the Government announced the closure of half of the remaining deep coal mines in Britain within six months, public awareness of the potential problems associated with rising mine water levels and the environmental impacts of existing mine water discharges has increased dramatically. General awareness has been further heightened by incidents such as the overflow of the Wheal Jane Mine, even though this was not associated with coal mining (Banks et al. 1997b, Bowen et al. 1998).

The chemistry of coal mine discharges has been described and discussed in numerous publications (Younger 1995, Banks et al. 1997a, Wood et al. 1999). They are characterised by elevated iron and sulphate concentrations, arising from the oxidation of pyrite:

$$2FeS_2 + 2H_2O + 7O_2 = 2Fe^{++} + 4SO_4^{=} + 4H^{+}$$

The ferrous iron often subsequently oxidises and hydrolyses, precipitating "ochre", which smothers stream beds, devastating benthic fauna/flora:

$$4Fe^{++} + 10H_2O + O_2 = 4Fe(OH)_3 + 8H^{+}$$

The chemical reactions involved also generate protons, which may give rise to low pH values, unless there is a sufficient rate of alkalinity production (Strömberg & Banwart 1994) from weathering of basic mineral phases (carbonate / silicate). Where alkalinity generated exceeds the potential acidity of the water (sum of protons and equivalent hydrolysable metals), discharges will tend to be *net alkaline*, and treatment may be accomplished by simple aeration and precipitation by active or passive (aerobic wetland) methods (Hedin et al. 1994). This is most commonly (though not exclusively) the case for minewater discharges from long-abandoned UK collieries. In contrast, where potential acidity cannot be neutralised by available alkalinity, the discharge is *net acidic* and will often have a pH below c. 5.5, leading to the dissolution of other metals, notably aluminium. Treatment of such water involves either (i) addition of alkalinity by anaerobic limestone units or active dosing prior to metal hydroxide precipitation, or (ii) anaerobic compost wetlands where alkalinity is generated by sulphate reduction and metals are to a large extent removed as sulphides. Acidic discharges are typical of spoil tips, new mine discharges or highly aerobic, under-drained mines.

Geoenvironmental engineering, Thomas Telford, London, 1999, 475–482

Regulatory authorities with responsibility for the quality of controlled waters (the former National Rivers Authority - NRA [now part of the Environment Agency -EA] and the former Scottish River Purification Boards - SRPBs [now Scottish Environment Protection Agency - SEPA]), recognised the existing problems and initiated their own assessments of the impacts of these uncontrolled minewater discharges.

The first major study, undertaken in 1992 and 1993, was an assessment of the impact of discharges in Wales and was jointly funded by the Welsh office and the NRA (Welsh Region). The subsequent report (Butler et al. 1994) presented a ranked list of all the known abandoned minewaters in Wales. The ranking process was carried out in two stages. In the first stage, the physical and chemical impact of 90 discharges on receiving watercourses were compared, the most important criterion being area of watercourse affected. In the second stage, the chemical, biological and fisheries impacts were assessed in more detail for 33 of the highest-ranked sites from stage 1.

This provided a basic methodology for subsequent studies and, in 1995, the NRA North-East Region (Northumbria and Yorkshire) applied a slightly modified version (Fig. 1) to assess their minewater discharges (Firth et al. 1997). As in the Welsh study, discharges were preliminarily ranked according to the severity of their physicochemical impact on the receiving watercourses. Thirty eight of the 120 minewaters assessed in the first stage were then selected to proceed to an ecotoxicological assessment and a ranking list was produced on the basis of chemical, physicochemical and biological impacts. The study subsequently expanded to become a nationwide EA scheme, while a similar scheme has been adopted by SEPA in Scotland.

The current, periodically-updated EA priority list for England and Wales, is determined using a scoring algorithm called the Multi-Attribute Technique (MAT - Fig. 1), which enables the biological, physical and chemical impacts to be combined with socio-economic factors and the potential uses of the receiving waters (Environment Agency, 1997). This methodology implicitly retains many of the factors assessed in Firth et al.'s (1997) methodology, but places increased emphasis on human / aesthetic impacts (recreation / visibility).

THE COAL AUTHORITY SCOPING STUDIES

Mine owners/operators will become liable, under the Environment Act (1995) (codified by the Mines (Notice of Abandonment) Regulations 1998) for permitting pollution of controlled waters by discharges from mines abandoned after 31st December 1999. Gravity discharges from mines abandoned prior to that date will, however, remain exempt from liability by dint of Section 89(3) of the Water Resources Act 1991: "*A person shall not be guilty under section 85.... by reason only of his permitting water from an abandoned mine ..... to enter controlled waters.*"

Despite the fact that the Coal Authority has very few obligations in law for discharges from abandoned mines, Lord Strathclyde indicated (in 1994) that the Coal Authority "*would be expected to go beyond the minimum standards of environmental responsibility which are set by its legal duties and to seek the best environmental result which can be secured from the resources available to it for these purposes.*" Thus, the Coal Authority commissioned Scott Wilson CDM (Scott Wilson) to undertake a scoping study of 30 discharges, prioritised by the NRA and the Scottish River Purification Boards, over a 5 month period from October 1995 to

February 1996 (Scott Wilson CDM 1996). The study considered fourteen minewaters from the 1995 NRA study of the NE Region, and 8 discharges each from Scotland and the North West (Table 1).

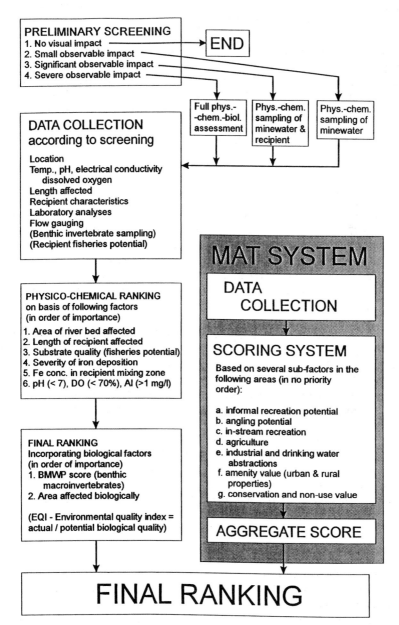

Figure 1. The ranking system applied by the Environment Agency / NRA North East Region (Firth et al. 1997) on the left hand-side of the diagram, compared with the current EA Multi-attribute technique (MAT) for assessing surface water pollution (shaded boxes on right).

The scoping studies comprised assessments (Fig. 2) of:
- the physico-chemical characteristics of each discharge,
- the mining system , related geology and hydrogeology,
- treatment requirements (based on loading to the recipient watercourse and the required quality of that watercourse),
- potential treatment options,
- outline environmental and socio-economic impacts of the discharge, and therefore benefits to be gained by treatment,
- likely costs associated with selected treatment methods.

Table 1. Minewater discharges included in scoping studies (Fe in mg/l, Q = discharge in l/s)

Name, Location	1	2	3	4	5	6	7	Type	Fe	Q
Acomb, Northumberland	*	*						Ak	45	5-10
Aspen V., Ostwaldtwistle, Lancs	*	*						?	4-12	15
Aspull Sough, Wigan, Lancs.	*	*						Ak	23	8-58
Bullhouse, nr. Penistone, Yorks	*	*	*	*	*	*	*	Ac	50	>25
Clough Foot, Todmorden, Yorks.	*	*	*	*				M/Ak	37	7-16
Craggs Moor, Bacup, Yorks.	*	*						Ac	110	2-5
Cuthill, Addiewell, Lothian[+]	*	*	*					Ak	190	10
Deerplay, Burnley, Lancs.[+]	*	*	*	*	*			M	50	10
Douglas, Strathclyde	*	*						Ak	55	60
Edmondsley, Co. Durham	*	*	*	*	*	*		Ak	27	3-5
Elginhaugh, Lothian[+]	*	*	*					M/Ac	95	30
Fauldheads, Harthill, Strathclyde	*	*						Ac	400#	<2#
Fender, Chesterfield, Derbys.	*	*	*	*	*			Ak	10	26
Fordell Castle, Fife	*	*	*					Ak	16	>100
Hapton Valley, Burnley, Lancs.$	*	*						Ak	0.1	>9
Haydock Sough, Lancs.	*	*						Ak	4	>20
Helmington Row, Co. Durham[++]	*	*						Ac	80	<4
Jackson Br., Holmfirth, Yorks.	*	*	*					Ak	24	31
Kames 1 & 2, Muirkirk, Strathcl.[+]	*	*	*	*				Ak	15	30-50
Lambley, Haltwhistle, N'land[+]	*	*						Ak	5	120
Loxley Bottom, Sheffield[+]	*	*						Ac	50	13
Minto, Lochgelly, Fife[+]	*	*	*	*	*	*	*	Ak	13	30-60
Oatlands, Cumbria[++]	*	*						Ac	290	3-6
Old Meadows, Bacup, Yorks.	*	*	*	*	*	*		M/Ak	25	40
Pool Farm, Auchingray, Strathcl.	*	*						Ak	8	45
Sheephouse Wood, nr. Langsett	*	*						M	30	6
Stony Heap, Co. Durham[+]	*	*						Ak	17	8
Summerley 1, Dronfield, Derbys.	*	*						Ak	14	>15
Summersales, Highfield, Lancs.[+]	*	*	*					M/Ak	75	>30
Unstone 1, Dronfield, Derbys.	*	*						Ak	12	14

No.	Current Status	Symbol	Explanation
1	Priority monitoring site	Ac	Net acidic
2	Scoping study complete	Ak	Net alkaline
3	Feasibility study complete	M	Marginal net acidity/alkalinity
4	Post feasibility programme	+	Several discharges
5	Implementation programme commenced	++	Spoil tip run-off
6	Construction completed	#	Affected by opencasting/backfill
7	Fully operational site	$	V. alkaline, H_2S discharge

Note that Fe and Q are "typical" values and may be very variable.

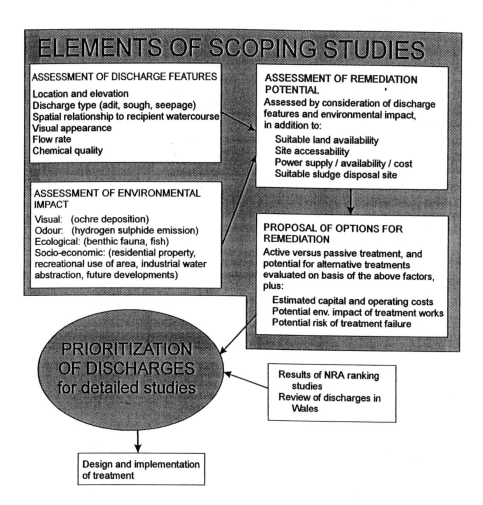

Figure 2. The structure of the Scott Wilson CDM (1996) scoping studies (in shaded box), and their subsequent consequences (detailed studies, remediation - outside shaded box).

The scoping studies recognised that selection of discharges for treatment will be dependent not only on their ranking, but also on factors such as land availability (negotiations to purchase land may be complex and can affect whether work should go ahead), the ease with which planning permission could be achieved, the status of other river improvement schemes (for example, alleviation of storm or foul sewage impacts), current public concerns, plans for other developments in the area (which may necessitate treatment or may provide a means of funding treatment - e.g. opencast schemes) and economics (available budget versus probable cost). Consideration of all these factors has resulted in a practical programme of mine water remediation which covers all the environmentally significant discharges identified throughout the UK. Progression of the schemes will not necessarily be in order of environmental impact and hence the Environment Agency list cites discharges in alphabetical order only.

MITIGATION MEASURES FOR EXISTING DISCHARGES

Subsequent to completion of the scoping studies and following discussions with the EA and SEPA, the Coal Authority, selected fourteen discharges (ten from the scoping studies, plus four identified by similar studies in Wales for the NRA), for more detailed feasibility studies. The feasibility studies were undertaken, as part of the Coal Authority Minewater Treatment Programme, by several consultancy firms, with Scott Wilson responsible for programme management. The first feasibility studies were completed by the end of 1996 and site works on the first of the treatment works, at Minto, began in November 1997. To date (April 1999), nine minewater treatment schemes have been, or are near to, completion (Table 2) and feasibility studies have been completed for a further eighteen. The status of the discharges studied during the Coal Authority scoping exercise is shown in Table 1. The Coal Authority aims to select several discharges from each area (i.e. Wales, Scotland, England) each year for feasibility and/or detailed design studies.

From the Coal Authority's point of view, it is desirable for treatment schemes to:

- most cost-effectively achieve a discharge water quality acceptable to the EA or SEPA
- require relatively little maintenance and low operating costs over the long-term.

On a technical level, the treatment method selected will be determined primarily by the chemical nature and rate of flow of the discharge, and the area and topography of the land available on which to site treatment works. Treatment schemes implemented to date fall into one of three categories:

- aeration (typically one aeration step is required per 50 mg/l iron - Younger 1999), followed by passive treatment, using wetlands
- aeration, followed by active treatment where (alkali,) coagulants and flocculants are added to aid precipitation of metal hydroxides (ochre) in settling basins (which should ideally be of sufficient capacity to provide > 24 hrs. retention time - Younger 1999). Note that sludge disposal is a critical step for the feasibility of active treatment. Traces of toxic metals can render disposal problematic, and addition of chemicals can increase sludge quantities.
- a combination of the above: most commonly, active treatment followed by a passive "polishing" wetland

As the discharges treated to date are predominantly net alkaline, it has not always been considered necessary to chemically neutralise the discharges to remove acidity. Of the above categories of treatment, passive treatment generally involves the lowest maintenance and operating costs, though requiring a greater area of land. Wetland area depends primarily on iron loadings (design removal rates of 10-20 $g/m^2/d$ are typically applied - Hedin et al. 1994), but hydraulic loading and residence time will also need to be considered. Time is also required for the reeds to become established, such that it may be necessary to incorporate a temporary active system to protect a watercourse from further pollution. This type of contingency scheme is being applied at Taff Merthyr (Wales) where a Millennium project to construct a water park downstream of the minewater discharges is due for completion in summer 1999, before a planned passive treatment scheme, utilising four settlement lagoons and sixteen reed-beds, will be complete. To avoid pollution of the man-made lakes, temporary treatment by aeration, settlement and filtration using filter cloths, will be implemented, utilising three temporary lagoons. These will ultimately be decommissioned and converted to additional reed beds, once the thirteen primary reed-beds are established.

Three of the treatment schemes shown in Table 2 treat minewaters which are being pumped from mine shafts. On-going pumping and treatment regimes are also being maintained in the Durham and Staffordshire coalfields. By maintaining pumping in this manner, regional levels of minewater are maintained below the level at which more widespread uncontrolled minewater discharges would occur.

Table 2. Minewaters for which treatment schemes are operational or near to completion

Discharge name	Date operational	Discharge type	Treatment scheme category
Woolley (NE)	June 1994	pumped shaft	**P:** aeration cascade, settling lagoons, wetland
Minto (S)	Summer 1998	gravity flow	**P:** wetland
Gwynfi (W)	Autumn 1998	gravity flow	**P:** wetland
Edmondsley (NE)	Summer 1999	pumped adit	**P:** wetland
Bullhouse (NE)	January 1999	gravity flow	**A:** aeration,(alkali dosing) & settlement
Monktonhall (S)	Summer 1999	pumped shaft	**C:** aeration, coagulant dosing, settlement, polishing wetland
Ynysarwed (W)	Under construction	gravity flow	**C:** short-term: neutralisation, settlement; long-term: wetlands.
Polkemmet (S)	Summer 1998	pumped shaft	**Semi-active** (contingency scheme): aeration, coagulant & flocculant dosing, settlement
Old Meadows (NW)	Spring 1999	pumped adit	**C:** aeration, alkali dosing settlement lagoons, polishing in wetland

Region: S = Scotland, W = Wales, NE = North East England, NW = North West England
Treatment: A = Active, P = Passive, C = combined

With the exception of the scheme at Woolley, all the completed treatment schemes are in the early phases of operation. The performance of these schemes will be continuously monitored and the accumulated information be made available to those designing future remediation measures.

REFERENCES

Banks D, Burke SP and Gray CG (1997a) The hydrogeochemistry of coal mine drainage and other ferruginous waters in North Derbyshire and South Yorkshire, UK. *Quarterly Journal of Engineering Geology* **30**: 257-280.

Banks D, Younger PL, Arnesen R-T, Iversen ER and Banks SB (1997b) Mine-water chemistry: the good, the bad and the ugly. *Environmental Geology* **32**: 157-174.

Bowen G, Dussek C and Hamilton RM (1998) Pollution resulting from the abandonment and subsequent flooding of Wheal Jane Mine in Cornwall, UK. *In:* Mather J, Banks D, Dumpleton S and Fermor M (eds.), "Groundwater contaminants and their migration". *Geological Society Special Publication* **128**: 93-99.

Butler D, Mills M, Davies G, Bourn G and Foolchand D (1994) *A survey of ferruginous minewater impacts in the Welsh Coalfields.* Welsh Office Contract No. WEP100/138/11. NRA Welsh Region Report, 41pp.

Environment Agency (1997) *Scoring and weighting system for River Water Quality Improvements.* Environment Agency Report, Version 4, 2.2.97, 38pp.

Firth S, Henry G, Cutter G and Moore L (1997) *Impact assessment of abandoned colliery minewaters in North East Region: final report.* Environment Agency, March 1997, 102pp.

Hedin RS, Narin RW and Kleinmann RLP (1994) Passive treatment of coal mine drainage. *U.S. Bureau of Mines Information Circular* **9389**, U.S. Dept. of Interior, 35 pp.

Scott Wilson CDM (1996) *Study of gravity drainage from abandoned mines.* Report in 4 volumes to the Coal Authority. Scott Wilson CDM, Chesterfield, Derbyshire.

Strömberg B and Banwart S (1994) Kinetic modelling of geochemical processes at the Aitik mining waste rock site in Northern Sweden. *Applied Geochemistry* **9**: 583-595.

Wood SC, Younger PL and Robins NS (1999) Long term changes in the quality of polluted minewater discharges from abandoned underground coal workings in Scotland. *Quarterly Journal of Engineering Geology* **32**: 69-79.

Younger PL (1995) Hydrogeochemistry of mine waters flowing from abandoned coal workings in County Durham, UK. *Quarterly Journal of Engineering Geology* **28**: S101-S113.

Younger PL (1999) *Minewater treatment technology: selection and implementation.* "Mine waste and minewater pollution", short course, The Mining Institute, Newcastle-upon-Tyne, March 22-24th 1999 (Univ. of Newcastle, Univ. of Sheffield and NUWATER Consulting Services Ltd.).

ACKNOWLEDGEMENTS

The authors are grateful to Keith Parker of the Coal Authority for his support of the presentation of this paper, and to Bernie Roome (Scott Wilson), Graham Tate, Simon Firth (EA) and Dave Holloway (SEPA) who have provided valuable information and discussion.

Protection areas of the São Pedro do Sul Spa – Portugal

L. M. FERREIRA GOMES, Department of Civil Engineering, University of Beira Interior, 6200 Covilhã, Portugal

F. J. AFONSO DE ALBUQUERQUE & H. FRESCO, S. Pedro do Sul Spa, Câmara Municipal de S. Pedro do Sul, 3660 S. Pedro do Sul, Portugal

ABSTRACT

This paper presents a case study of the protection areas of one of Portugal's most important spas. According to its chemical composition this thermal mineral water is rich in bicarbonate, sodium, carbonate, fluor and sulphate. The therapeutic indications are mainly illnesses related to rheumatism and to the respiratory system. As a complement the thermal mineral water is also used for geothermal purposes, as in the heating of greenhouses for the production of bananas and ananas.The study presents the geological and hydrogeological characterization of the area and its vulnerability to pollution and the physical and chemical characterization of the thermal mineral water. Finally, in accordance with Portuguese legislation, the protection areas are defined: immediate, intermediate and distant. The criteria and other aspects that led to their definition are also presented.

INTRODUCTION

Thermalism has a long tradition in Portugal, which owes mainly to the great diversity in chemism of its natural mineral waters, this as a consequence of the geological conditions of the territory. The spa of this case study is located in the village of S. Pedro do Sul, district of Viseu, in central Portugal.

The economic importance of the spa for the region can be inferred from the following numbers: in 1998 the number of spa visitors amounted to 20 012, with a direct revenue of 2,952,625 EURO and an indirect revenue of 15,009,000 EURO, considering that each visitor stays 15 days a year at the spa, spends an average of 50 EURO a day on food, lodging and other necessary goods. Consequently the protection areas are of vital importance so as to guarantee the quality of the mineral water, to prevent contamination problems which would lead to the closing of the spa and would be catastrophic since the spa provides for the way of living of many people of a most wonderful place in Portugal's inland.

The area of the study has two mineral water producing sectors that distance around 1.2 km from one another. The *Spa Sector* includes the bathhouse for medicinal treatments and the geothermal central (under construction) for the heating of non-mineral waters for domestic use (heating and water supply); it also includes a well (AC1) with a depth of 500m and an ancient spring (Traditional Spring, since the Roman empire) which together produce, by arthesianism, about 18 l/s at a temperature of approximately 67.5°C. The *Vau Sector* has a small mineral spring and two wells, the SDV1 with a depth of 216 m and SDV2 with 151m; the water flow of each, in pumping, can be of 10 l/s at a temperature of 67° C. At the moment, this sector is being used to heat greenhouses for the production of bananas and ananas.

PHYSICAL AND CHEMICAL CHARACTERISTICS OF THE MINERAL THERMAL WATER

The natural mineral thermal water of S. Pedro do Sul belongs to a group known as sulfurous waters; the results of complete studies are presented in Table 1 and indicate a high stability of the chemical composition. The most significative elements are: Bicarbonate (HCO_3^-), Sodium (Na^+), Fluoride (F^-) and Chloride (Cl^-).

Table 1. Physical and chemical results of the analysis of the natural mineral thermal water of the hydromineral and geothermal field of S. Pedro do Sul - Portugal

Parameter		Traditional mineral spring						AC1	SDV1	Average	Stand. desv.
		1903 (1)	1928 (1)	1985 (2)	1989 (2)	1994 (2)	1998 (2)	1998 (2)	1998 (2)		
Temperature - °C		68.7	67.5	67.5	67.0	66.5	67.3	67.3	67.4	67.4	0.58
pH		-	-	8.33	8.93	8.92	8.91	8.93	8.85	8.81	0.22
Total CO_2-mmol/l		-	-	2.08	1.98	1.97	1.95	1.98	2.00	1.99	0.04
Conductivity µS cm^{-1}		-	-	396	450	452	460	466	463	448	24
Alkalinity (HCl 0.1N) – ml/l		-	-	24.4	23.7	23.3	23.5	23.1	23.6	22.6	0.41
Total sulfuration (I_2 0,01N) - ml/l		36.8	23.1	15.5	23.1	26.4	17.1	18.1	24.6	23.1	6.3
Total silica -mg/l		66.8	70.4	75.9	70.4	72.9	75.8	75.5	78.8	73.3	3.7
Total solids-mg/l		-	-	363.3	341.1	352.1	363.2	359.9	368.3	358.0	9.0
Cations (mg/l)	Na^+	89.5	90.5	88.6	87.0	89.7	89.9	90.5	92.3	89.8	1.4
	Ca^{2+}	-	-	2.4	3.2	2.8	2.9	3.0	3.0	2.9	0.25
	K^+	-	-	3.59	3.5	3.3	3.5	3.4	3.4	3.4	0.09
	Mg^{2+}	-	-	0.73	0.02	0.03	<0.2	<0.1	<0.1	<0.2	-
	Li^+	-	-	0.64	0.55	0.56	0.58	0.58	0.59	0.58	0.03
	NH_4^+	0.53	0.02	0.36	0.25	0.34	0.30	0.31	0.31	0.30	0.13
Anions (mg/l)	HCO_3^-	-	-	120.2	103.7	114.7	112.9	114.7	115.9	113.7	5.0
	Cl^-	29.7	27.7	29.8	28.8	27.3	28.4	28.0	29.1	28.6	0.85
	SO_4^{2-}	17.7	25.0	8.9	8.6	9.0	9.3	9.3	9.8	12.1	5.28
	F^-	-	-	17.1	17.1	17.2	17.4	17.7	17.8	17.4	0.28
	CO_3^{2-}	-	-	6.4	8.6	5.7	5.7	6.3	6.0	6.5	1.0
	NO_3^-	-	-	<0.07	<0.07	<0.10	0.82	0.97	1.1	<0.5	-
	NO_2^-	-	-	<0.002	<0.002	<0.002	<0.02	<0.02	<0.02	<0.01	-
	HS^-	-	-	1.8	3.0	4.4	1.7	2.8	3.8	2.9	0.98
	H_3SiO^-	-	-	-	-	11.4	14.3	15.2	13.3	13.6	1.4
Secondary constituents (mg/l) (x10^{-3})	$H_2PO_4^-$	-	-	18	19	<40	-	-	-	-	-
	Al_3^+	-	-	337	640	52	18	38	20	-	-
	Mn^{2+}	-	-	6	2	4	2.1	1.1	2.6	-	-
	Br^{2-}	-	-	260	135	130	-	-	-	-	-
	B_3O_3	-	-	311	405	400	415	410	412	-	-
	Be	-	-	0.5	0.6	0.7	0.7	0.7	0.7	-	-
	Pb	-	-	4.7	11	6	-	-	-	-	-
	Cd	-	-	<0.3	2.1	<0.3	<1	<1	<1	-	-
	Nb	-	-	<0.4	<1	<0.5	-	-	-	-	-
	V	-	-	<0.5	<1	<3	<1	<1	<1	-	-
	Y	-	-	<0.3	<1	<0.1	<1	<1	<1	-	-
	Sn	-	-	<0.7	<1	<3	<5	<5	<5	-	-
	Cr	-	-	<0.1	8	<2	<4	<4	<4	-	-
	Fe^{2+}	-	-	26	60	<25	<7	<7	22	-	-
	Ba^{2+}	-	-	<2	<5	<2	4	3	4	-	-
	I^-	-	-	1	2	<1	-	-	-	-	-
	As_2O_3	-	-	5	<5	10	3	4	<3	-	-
	W	-	-	54	37	75	80	75	78	-	-
	Cu	-	-	0.4	<1	<1	<2	<2	<2	-	-
	Zn	-	-	4.5	8	11	2	<1	5	-	-
	Sb	-	-	<0.5	<5	7	<3	<3	<3	-	-
	Ni	-	-	<0.6	<1	<3	<5	<5	<5	-	-
	Co	-	-	<0.5	<1	<1	<3	<3	<3	-	-
	Mo	-	-	<1	<5	6	<5	<5	<5	-	-

(1) Results obtained by C. Lepierre (Machado, 1999). (2) Results obtained by IGM Laboratory.

GEOLOGICAL AND HYDROGEOLOGICAL ASPECTS

The geological conditions of the S. Pedro do Sul area have been studied by several researchers, but special attention is to be given to the works by Pereira and Ferreira (1985). It must be pointed out that the Spa and Vau sectors (Figure 1) are part of extensive granitoid massifs. On a regional scale the occurrence of the springs is favoured by the great active fault of Verin (Spain) – Régua – Penacova, which in the area of the study prolongs into the Ribamá fault (N0°-10°E) and can conduct flows from great distances and depths (Pereira and Ferreira,1985). On a local scale tectonic knots between N45°E (Spa fault) and N70°W condition the thermal springs.

From the elder to the more recent one, the region of the case study presents the following units:

> *Unit 1* - Precambrian / Cambrian;
>> constituted by schistous rocks, mainly schists and metagraywackes;
>
> *Unit 2* - Granitoids (Hercynian);
>> *2.1.* Vouzela granite: monzonitic granite medium to coarse grained, formed mainly by quartz, albite-oligoclase, potassium feldspar, biotite and mos-covite.
>>
>> *2.2.* S.Pedro do Sul granite: granite fine to medium grained, composed mainly by microcline, plagioclase, quartz, moscovite and biotite. Presents conside - rable radioactivity that may be related to the existence of uranium and thorium. This granite intrudes the Vouzela granite.
>>
>> *2.3.*Fataunços granite: granite close to granodiorite, fine to medium grained, with two micas.This granite intrudes the Vouzela granite.
>
> *Unit 3* - Quaternary;
>> *3.1.* Pleistocene: fluvial terraces, mainly with gravel.
>>
>> *3.2.* Holocene: alluvia, mainly with sand and gravel.

The schistous rocks and the granites present a fissural permeability, whereby the formers are practically impermeable. The mineral water discharge zones are located in the S. Pedro do Sul granites. According to A.CAVACO (1995) in these discharge zones the hydromineral aquifer functions as   confined; the obtained hydraulic parameters are: transmissivity (T) $\approx$109 m^2/day, storage coefficient (S) $\approx$ 4.3 x10^{-5}, hydraulic conductivity (k) $\approx$ 0.5 m/day.The presented values correspond to the conditions occurring at the tectonic knots to be found along the spa fault and were determined on the basis of pumping tests admitting a continuous model with a 200 m saturated thickness, after applying the Theis model.

The Quaternary deposits which reveal interstitial permeability do not condition directly the hydromineral circuit, whereby k = 370 m/day and T = 1850 m^2/day were the values determined for the Vouga river alluvia .

The hydrological data is based on charts between 1933 and 1960 at the S.Pedro do Sul rain measurement station: the annual rain fall average was of 1103 mm and the average annual air temperature was 13°C,; considering the results of the monthly sequential hydrologic balance one obtains an annual hydric surplus of 675 mm.

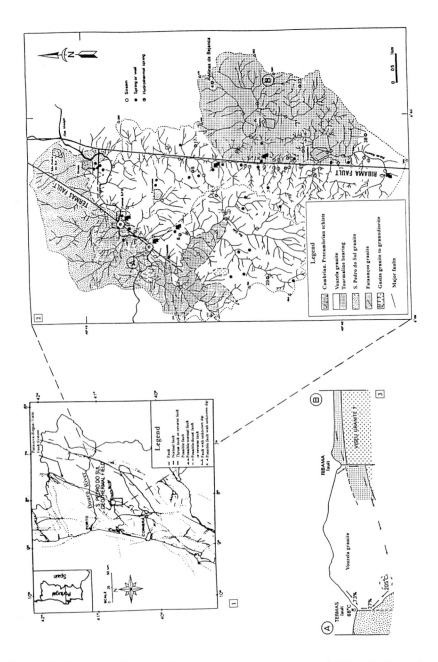

Figure 1.Hydrogeological setting of the hydromineral and geothermal field of S. Pedro do Sul: (1) Northern Portugal neotectonic map (modified from Cabral and Ribeiro,1989); (2) and (3) Hydrogeological outline (modified from Haven *et al*, 1985).

VULNERABILITY AND POLLUTION RISKS

In order to evaluate vulnerability, the DRASTIC index (Aller et al., 1987) has been used. Table 2 shows the obtained values and some qualitative aspects for the main hydrogeological units of the spa zone. Remind that the minimum and maximum vulnerability values of the DRASTIC index are respectively 65 and 223.

Table 2. DRASTIC Index values of vulnerability for the main hydrogeological units of the S.Pedro do Sul Spa zone.

Unit	Parameter (*)	Range	Rating	Weight	Partial index	DRASTIC total	Vulnerability
Alluvia	1	1.5 – 4.6	9	5	45		Generally high. The pollution can easily spread over great distances.
	2	101.6 – 177.8	6	4	24		
	3	sand and gravel	8	3	24		
	4	without	10	2	20	187	
	5	3%	10	1	10		
	6	sand and gravel	8	5	40		
	7	40 – 82	8	3	24		
Granite	1	4.6 – 9.2	7	5	35		Generally medium. In some fractures and rock alteration zones can sometimes be high
	2	50.8 – 101.6	3	4	12		
	3	granites	3	3	9		
	4	without	10	2	20	114	
	5	22%	5	1	5		
	6	igneous	6	5	30		
	7	<4	1	3	3		
Schists	1	4.6 – 9.2	7	5	35		Generally low. In zones with great rock alteration rates can locally present higher vulnerability.
	2	0 – 50.8	1	4	4		
	3	schists	3	3	9		
	4	without	10	2	20	96	
	5	6 - 12%	5	1	5		
	6	metamorphic	4	5	20		
	7	<4	1	3	3		

(*) Parameters:1- depth to the water table (m); 2-net recharge (mm); 3 - aquifer material; 4 - soil type; 5 - topography-slope; 6 - unsaturated zone; 7- hydraulic conductivity of the aquifer (m/dia).

As to pollution risks, the most significant ones are: i) *in the urban area* – pollution derived from human action such as septic tanks, outdated sewerage, Várzea cemetery, filling stations and small industrial activity; ii) *in the rural area* – the use of fertilizers and pesticides in small agriculture (although the forest is predominant),the greenhouses in the Vau sector as they must be considered a *contra natura* agriculture and, as such, as an additional risk.

PROTECTION AREAS

The protection areas are intended to reduce the risks of polluting the mineral water of the hidromineral and geothermal field of S.Pedro do Sul, or, in case of an accident, to prevent the pollution of reaching health endangering concentrations at the supply well or mineral spring. A.CAVACO (1995) made a study of the protection areas as they stand and its main criteria are here presented. Note that a possible use of the Vau sector for medicinal purposes in the future has not been regarded since at present only the greenhouses exist there. But if it were to be used for that end, the protection areas would have to be redefined.

For the definition of the protection areas two situations were considered:

i) *interference influences*, which result from the usage, in other places, of mineral or non-mineral water and that might interfere with the hydromineral circuit;

ii) *qualitative influences*, which are linked to pollution and contamination problems of the hydrogeological systems.

The following boundaries were established: Immediate zone, 1 day; Intermediate zone, 50-365 days; Distant zone, 25 years and / or limits of supply to the hydrographical basin. The evaluation of the interference influences was made on the ground of the classic equations of underground water flow for the collector wells in a continuous model. As to the qualitative influences, in which intervene complex phenomena of transport linked to processes of advection, dispersion, diffusion and others, an analysis of propagation time in a saturated zone was made. Additionally, the local estimation of vulnerability to pollution according to the DRASTIC index was taken into consideration.

Concerning eventual interference, calculations lead to believe that these can occur in fractured zones of the aquifer at 160, 1200 and 3000m after 1, 50 and 365 days respectively. For the estimation of propagation time in a saturated zone the method used was the US EPA (1987); note, for instance, that in granitic rocks and for transit times of 1, 50 and 365 days, distances need not be greater than 25, 200 and 500 m respectively.

So, taking into account all previously stated aspects, the boundaries of the protection areas (Figure 2) were set as follows and to fulfill these purposes:

Immediate zone - The most vulnerable protection area. Protects the discharge zone of the natural mineral water for medicinal purposes; covers 1.0 ha.

Intermediate zone -. Includes all areas surrounding the previously mentioned discharge zones in order to protect the mineral aquifer and even other non-mineral aquifers which might interfere with the mineral water circuit; covers a total area of 157 ha.

Distant zone - The area intended to protect the recharge zones, so that any eventual contamination does not reach the hydromineral circuit in dangerous concentrations; covers a total area of 979 ha.

For the different protection areas, Portuguese legislation sets the following restraints:

Immediate zone

1) It is forbidden to: a) erect any kind of building, b) perform drillings and underground works, c) make embankments, excavations or any kind of ground-levelling, d) use chemical or organic fertilizers, insecticides, pesticides or any other kind of chemical products, e) dump any kind of waste products and create garbage dumps, f) perform works for the conveying, treatment or collection of sewage.

2) Any of the ensuing acts must be previously authorized: the cutting of trees or shrubs, destruction of plantations, and the demolition of buildings of any kind.

3) The works referred in 1) a), b), c) and f) may be authorized by the competent authority of the Administration whenever necessary or advisable for the conservation and exploration of the natural mineral water.

Intermediate zone

1) Activities referred in points 1 and 2 of the Immediate zone are forbidden, except when duly authorized by the competent authority of the Administration and if by them provenly there is no interference in the natural mineral water or damage for the exploration.

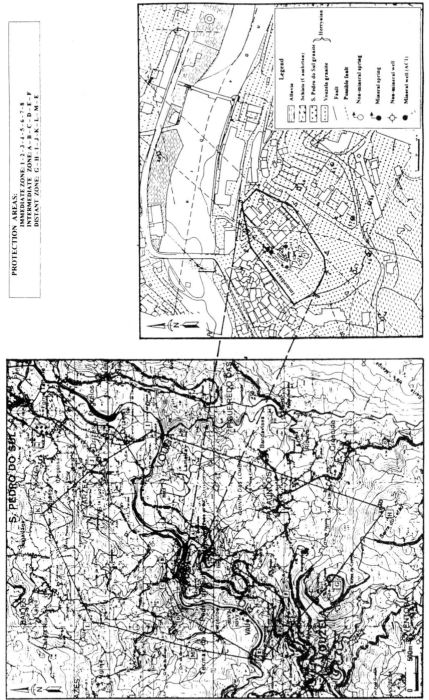

Figure 2. Outline of protection areas of the hydromineral and geothermal field of S.Pedro do Sul - Portugal

Distant zone

1) The activities mentioned in points 1) and 2) of the Immediate zone can be prohibited by order of the Minister of Industry and Energy when representing risks of interference or contamination of the natural mineral water.

REFERENCES

A. CAVACO(1995). Definição do perímetro de protecção ao Aquífero de S. Pedro do Sul. Relatório Final. Vol. I e II. Relatório Interno.Termas de S.Pedro do Sul. (in Portuguese).

Aller, et al. (1987). A standardized system for evaluation ground - water pollution potential using hiydrologic setting NWWA, NTIS, OIH, USA.

Cabral, J. and Ribeiro, A. (1989).Carta neotectónica de Portugal.Escala 1:1 000 000. Serviços Geológicos de Portugal. Lisboa. (in Portuguese)

DGGM (1990). Recursos geológicos. Legislação. D.L.90/90. Direcção Geral de Geologia e Minas. Lisboa. pp.11-33. (in Portuguese).

DGGM (1992). Termas e Águas Engarrafadas em Portugal. Direcção Geral de Geologia e Minas. Lisboa. (in Portuguese).

Haven *et al.* (1985) - Geochemical studies in the drainage basin of the Rio Vouga. Chemical Geology, 51; pp. 225-238.

Machado,M.J.C.; 1999. Estudo Físico – Quimico completo da água súlfurea das Termas de S.Pedro do Sul. IGM,Laboratório. 9p. (in Portuguese).

Pereira, E. e Ferreira, N. (1985). Geologia regional e controlo estrutural das nascentes termais de S. Pedro do Sul. Comun. Serv. Geol. Portugal. T. 71, fase 1, pp. 17-25. (in Portuguese).

US EPA (1987). Guidelines for delineation of wellhead proteccion areas. Office of groundwater protection, Wash,D.C. EPA 440/6-87-10.

Risk assessment of the former Houghton Main Coke Works

Dr. J. Oakeshott, BSc, MSc, PhD, FGS, CGeol. MCIWEM, Senior Hydrogeologist.,
Dr. J. H. Gettinby, BSc, PhD, Environmental Engineer and
Mr. J. Smithson, BSc, MSc, FGS, CGeol, Technical Director,
Allott & Lomax Consulting Engineers

1.0   INTRODUCTION

An area of the Houghton Main opencast coal site was utilised as a coking plant which operated between the late 1800's and the mid 1900's. It had originally been the intention that the contaminated material found in the location of the former coke works was to be placed within an engineered containment cell. However, discussions with the Environment Agency (EA) indicated that there was the possibility of retaining much of the material on site subject to the findings of a risk assessment. The EA were concerned that the Coal Measures Mexborough Rock aquifer which underlies the site may be adversely affected by contaminants leached from the made ground by infiltrating rainfall. The objective of the risk assessment was to develop site specific soil concentration thresholds for the area of the former coke works such that material below these thresholds could remain on site without adverse impact on the Mexborough Rock aquifer.

This paper reports on the risk assessment that was undertaken using the EA Research and Development Technical Report P13 (Ref. 1) and details the practicalities of obtaining soil trigger levels for a potentially contaminated site.

2.0   SITE INVESTIGATION

2.1 Background Water Quality
Previous water quality analyses from an existing monitoring borehole together with a sample obtained specifically for this assessment were reviewed (Ref. 2). The borehole is located some 500m north and also down hydraulic gradient of the former coke works area. Groundwater flow is currently towards the north east. This borehole sources the top of the Mexborough Rock aquifer at between 71.5 and 82m bgl. Analyses included, inter alia, a standard suite of metals (As, Cd, Cr, Pb, Hg, Cu, Ni, Zn) and organics (PAH, phenols) as well as ammoniacal nitrogen, cyanide and sulphate.

While the EA document (Ref.1) was published in 1998, it is apparent that it is not widely used, and even though it is an EA document, that many offices of the EA are not particularly familiar with it.. This report demonstrates the relatively straightforward application of the first two stages of the risk based approach, and how it can lead to significant cost savings on remediation while demonstrating no significant adverse impact on controlled waters.

2.2 Ground Investigation
A ground investigation specifically for this assessment was undertaken comprising the excavation of 8 trial pits to depths of about 3.80m, and supplemented the applicable information from two, more extensive, previous site investigations (Refs. 3, 4)

The aim of this investigation was two-fold:

i)   To obtain soil samples for corresponding soil and leachable chemical analyses in order to determine a site specific soil-water partition coefficient (K_d often also referred to as the sorption distribution coefficient or partition coefficient) for various determinands.

ii)   To obtain soil samples for geotechnical testing, in order to determine the relative proportions of water and air filled porosities, and the bulk density.

Geoenvironmental engineering, Thomas Telford, London, 1999, 491–498

There were two distinct types of made ground present - a re-worked mudstone deposit and an industrial fill deposit.

The re-worked mudstone was typically 2.20m in depth and generally consisted of a grey weathered mudstone with fine black ash, bricks, boulders and fragments of coal and shale. The industrial fill material was located in the north-eastern area of the former coke works. This material consisted of black, medium to fine ash, with fragments of coal, hardened tar, shale, cloth, brick, reinforced concrete, slag, wire, timber and weathered ironstone. The material had a slight organic odour and was generally found in depths not exceeding 3.80m.

Twelve soil samples were taken for analysis, of which 6 were selected for leachability testing. The samples for leachability analysis were selected on the basis of relatively elevated concentrations in the soil samples. All leachability tests were carried out using the EA Research and Technical Note 301 (NRA 1994) methodology.

The soil, leachability and water samples were all tested for the same suite of determinands tested for in the monitoring borehole. The determinands were selected on the basis of previous investigations where they were found to be elevated (Refs. 3, 4).

Representative samples of the made ground were tested for moisture content, bulk density, particle size and particle density/specific gravity following British Standard procedures (Ref. 5).

3.0   CALCULATION METHODOLOGY AND PARAMETERS

3.1 Target Concentrations (C_T)
The EA Technical Report (Ref. 1) refers to target concentrations (C_T) which represent the level of contamination that is acceptable at the controlled waters considered to be at risk. The C_T's were established with reference to appropriate environmental quality standards (EQS). After discussion with the EA, UK drinking water standards (Ref. 6) were used as the appropriate EQS. The exceptions were those determinands currently present above the relevant drinking water maximum permissible concentrations (mpc) in the Mexborough Rock aquifer. The mean background concentrations for ammoniacal nitrogen, sulphate and PAH are above the mpc, as shown in Table 1. These values were used as the EQS for these determinands.

Table 1		
Determinand	**Mean Background Concentration (mg/l)**	**mpc (mg/l)**
Ammoniacal nitrogen	1.3	0.5
Sulphate	269	250
PAH	0.04084	0.0002

3.2 Tier 2 Risk Assessment Methodology
The methodology of the EA Technical Report (Ref. 1) comprises a tiered approach. It was agreed with the EA that the risk assessment would be carried out to a Tier 2 level where soil threshold concentrations are derived on the basis of dilution by the receiving controlled water (the Mexborough Rock aquifer) of the leached contaminants arising from infiltration of the source (the former coke works made ground).

The Tier 2 methodology essentially consists of estimating on a site-specific basis a soil threshold concentration STC_2 which reflects the potential dilution of leached contaminants in the receiving water body by applying the dilution factor, DF, to the original, first estimate of on-site soil concentration, C_s or STC_1. This maximum allowable concentration, the Tier 2 soil threshold concentration STC_2, was calculated for each determinand using the formulae shown below.

$$STC_2 = C_S \times DF$$

Eq.3.1

which uses the Tier 1 on-site soil concentration, C_s :

$$C_s = \left\{ C_T \left(K_d + \frac{\theta_w + \theta_a H'}{\rho_b} \right) \right\}$$

Eq.3.2

where:
K_d	=	soil-water partition coefficient (l kg)
θ_w	=	water-filled soil porosity (l_{water}/l_{soil})
θ_a	=	air filled soil porosity (l_{air}/l_{soil})
H'	=	Henry's law constant (unitless)
ρ_b	=	dry soil bulk density (kg/l)

and also uses the dilution factor, DF:

$$DF \quad = \quad 1 + (Kid/IL)$$

Eq.3.3

where:
K	=	aquifer hydraulic conductivity (m/yr)
i	=	hydraulic gradient (m/m)
d	=	mixing zone depth (m)
I	=	infiltration rate (m/yr)
L	=	length of source parallel to flow (m)

3.3 Tier 1 On-Site Soil Concentration (C_s or STC_1)

The formula represents the derivation of the Tier 1 on site soil concentration (C_s) by application of a 'partitioning' or retardation factor ($K_d + \theta/\rho_b$) to the target concentration (C_T) value. The partitioning or retardation factor represents the ratio or relationship between the total and leachable fractions of the contaminants in the source. This reflects how soluble a specific contaminant is and how likely it is to leach or mobilise, and whether constituents within the soil, mainly organic carbon, are likely to retard the mobilisation of the contaminant. It also incorporates the preference of the contaminant for the vapour and aqueous phase. The various parameters in the formula are discussed below:

3.3.1 Target Concentration
These have been previously discussed in Section 3.1.

3.3.2 Soil-Water Partition Coefficient or Distribution Coefficient (K_d)
The partitioning of a contaminant is controlled by K_d which can be calculated using a number of different methods depending on the chemical nature of the contaminant. K_d varies dependent on soil type and contaminant. After consultation with the EA it was decided that an empirical, site specific K_d value was appropriate. A K_d value for each determinand was calculated from the ratio of the measured concentration of that particular determinand in the soil sample and the corresponding measured leachable concentration from the same sample. Thus a range of K_d values were calculated for each determinand as derived from the pairs of soil and leachable concentrations

Soil and leachable values were available from the two earlier site investigations (Refs. 3 & 4). However, the leachability tests were usually only undertaken for one or two specific determinands on any designated soil sample. The recent site investigation provided leachability testing for a full suite of determinands. These recent results were supplemented by soil and leachable testing from the earlier investigations to provide a larger sample set and thus more representative values where possible. In some cases, both soil and leachable concentrations were below detection limits and a K_d value was not calculated as the K_d value would purely be a function of the respective detection limits. Originally in cases where the soil concentrations were above detection limits but the leachable concentrations below, the detection limit was taken as equivalent to the leachable value for the K_d calculation and thus used to calculate a conservative STC_2 value. As discussed later, ultimately no below detection limit values were used.

3.3.3 Water Filled Porosity (θ_w) and Bulk Density (ρ_b)

Samples of made ground were also tested for a number of geotechnical parameters using sand replacement tests, nuclear density gauge tests and British Standard laboratory techniques. The parameters were determined from the testing. The mean values were calculated and are shown below:

Average moisture content (w) = 0.122 (12.2%)
Average bulk density (ρ_b)   = 1.562 Mg/m^3
Average particle density (ρ_s) = 2.295 Mg/m^3
   or specific gravity (G_s)

Further calculations were then undertaken using these parameters in order to estimate the degree of saturation S_r (Ref. 7) which represents the proportion of water filled porosity.

$$S_r = \frac{\rho_b w G_s}{\rho_w G_s(1 + w) - \rho_b} \times 100$$

$$\text{Eq.3.4}$$

where ρ_w, water density = 1.0 Mg/m^3

Thus S_r = 0.432

The degree of saturation, S_r, together with the specific gravity/average particle density and moisture content were used in the following formula (Ref. 7) to derive the total porosity, n:

$$n = \frac{e}{1 + e} = \frac{w G_s}{S_r + w G_s}$$

$$\text{Eq.3.5}$$

where e = void ratio (ratio)

Thus n = 0.39.

By applying the degree of saturation S_r to the porosity n, the water filled or saturated porosity was estimated as follows:

$\theta_w = n \times S_r = 0.17$

3.3.4 Air Filled Porosity (θ_a) and Henry's Law Constant or Coefficient (H')

A contaminant will partition between the water (dissolved) and vapour phases, representing a measure of the contaminants volatility. The distribution is dependent on the ratio of the contaminants solubility and vapour pressure (Henry's Law constant, H'). H' is really only significant where the contaminant is considered to be relatively volatile. Thus this factor was only used for the organic determinands, not the metal species. The appropriate H' values were taken from the EA Technical Report (Ref. 1). The H' is used in conjunction with the proportion of air filled porosity - the medium through which the vapour phase will migrate. The θ_a was estimated from the parameters used to calculate the water filled porosity θ_w and total porosity n as follows:

$\theta_a = n - \theta_w = 0.22$

3.4 Dilution Factor (DF)

The dilution factor is derived from the dilution of the volume of source leachate (ILx) by the receiving aquifer flux (Kidx) where x is an arbitrary unit width of contaminant source and aquifer:

DF = (Kidx + ILx)/ILx = 1 + (Kid/IL)

$$\text{Eq.3.6}$$

where: K   = aquifer hydraulic conductivity (m/yr)

    i   = hydraulic gradient (m/m)
    d   = mixing zone depth (m)
    I   = infiltration rate (m/yr)
    L   = length of source parallel to flow (m)

3.4.1 Hydraulic Conductivity (K)
Insitu permeability testing has not been undertaken at the site. An average hydraulic conductivity of 1×10^{-5} m/s (i.e. 0.864m/d, equivalent to 315.36m/yr) was used. This figure corresponds to a mid range value for Coal Measures aquifers given in Freeze and Cherry (Ref. 8). It represents a fairly conservative value for a sandstone and will contribute towards a lower dilution factor. This value had been previously used for the EA approved LandSim risk assessment undertaken for the proposed waste containment facility (Ref. 9).

$K = 315.36$m/yr

3.4.2 Hydraulic Gradient (i)
This was estimated from groundwater elevations recorded in 3 observation boreholes within the Coal Measures and monitoring the Mexborough Rock groundwater table. The elevations were recorded in October 1997 before significant dewatering from the opencast site commenced earlier this year. The dewatering is associated with the excavation of the Shafton coal seam which lies just above the Mexborough Rock.

$i = \dfrac{\text{difference in groundwater elevations along flow direction}}{\text{distance between respective elevations}} = \dfrac{31.31}{670}$

$i = 0.047$

3.4.3 Infiltration Rate (I)
This was taken as equivalent to the effective precipitation for a bare soil cover as estimated by the Meteorological Office for the nearest weather station to the site. The annual effective precipitation is 246mm/yr, i.e.

$I = 0.246$m/yr

3.4.4 Contaminant Source Length (L)
The value of L has been measured parallel to groundwater flow direction (south west to north east) and represents the maximum extent of the contaminated area, the former coke works, in this orientation.

$L = 200$m

3.4.5 Mixing Depth (d)
It is assumed, in the calculation of the dilution factor, that the leachate will not mix over the full saturated aquifer thickness. The depth over which the leachate will mix is estimated from the equation provided in the EA Technical Report (Ref. 1) shown below:

$$d = (0.0112L^2)^{0.5} + d_a \{1 - \exp.[(-LI)/(Kid_a)]\}$$

Eq.3.7

where d_a = saturated aquifer thickness (m).

The thickness of the Mexborough Rock aquifer (30m) was estimated from the logs of two of the Houghton Main shafts (Ref. 10).

Therefore using the above values, the zone depth d, was estimated to be 24.31m.

3.4.6 Mexborough Rock Aquifer Dilution Factor

On the basis of the preceding, evaluated parameters and applying Eq. 3.6, the Dilution Factor, DF was calculated to be 8.32. This factor has been applied to the various initial soil threshold concentrations (STC_1 or C_s). The STC_2 soil threshold concentrations have thus been calculated for each pair of soil and leachable concentrations.

4.0   DEVELOPMENT OF PROPOSED SOIL THRESHOLD CONCENTRATIONS

Soil threshold concentrations were developed from both the recent specific ground investigation chemical testing and also appropriate test data from the previous September 1996 and January/February 1997 investigations. The latter data enabled a larger sample set to be used. In addition, cadmium and cyanide concentrations in the recent soil and leachability testing were below detection limit and a reliable soil-water partition coefficient and thus retardation factor could not be estimated. While the absence of these determinands indicated that they were unlikely to present a contamination risk, the determinands were detected in previous investigations in the area. Chemical test data from the earlier investigations have therefore been used to 'fill' these gaps.

Table 2				
Determinand	Range of STC_2.values	Logarithmic Mean of STC_2 values	Minimum STC_2.value	EA agreed STC_2.value
	(where below detection limit leachable values used)		(excluding below detection limit leachable values)	
Phenol	117 - 657	288	117	120
PAH	1211 - 57,820	15,909	1,211	1210
Cyanide	8 - 11	9	8	
Ammoniacal Nitrogen	70 - 442	133	70	70
Sulphate	95431 - 661,812	202,032	95,431	
Arsenic	661 - 24,003	5,754	661	660
Cadmium	163		163	
Chromium	516 - 73,216	4,457	42,432	
Lead	269 - 6,947	1,220	6,957	
Mercury	66 - 275	147	100	100
Copper	56,662 - 1,810,601	343,685		
Nickel	832 - 28,742	2,062	832	830
Zinc	196,357 - 1,636,271	438,898	1,636,271	

Table 2 shows the range of STC_2s calculated from all the pairs of total soil / leachable concentrations including those where the leachable concentration only was below the detection limit. It can be seen that the ranges often cover several orders of magnitude for a specific determinand. Originally a logarithmic mean was calculated from the constituent values to provide a representative average value. However the EA requested a more conservative approach and that the minimum STC_2 for each determinand was selected. At this stage, the decision was taken to exclude any STC_2 calculated using a K_d value derived from below detection limit concentrations. It was considered that if a minimum value was to be employed, there should be confidence in its derivation. The previous use of below detection limit leachable values gave rise to a conservative K_d and thus STC_2 value. A mean value had originally been presented to the EA since there was some concern over the wide range of STC_2 values calculated for each determinand. However while a larger sample set was important for the estimation of a mean value, it became slightly less so for the selection of a minimum value. In that, if there is large range of values, then a large sample set provides a more statistically valid mean. However, if the minimum value is an order of magnitude or so less than the other results, the number of samples is not so critical. It can be seen from the table above that in some cases the minimum value derived from above detection limit concentrations only gives rise to a STC_2 above the lowest value previously calculated and sometimes above the previous average value (e.g. mercury, zinc, lead). This actually indicated that many of the leachable concentrations were below the detection limits, and thus unlikely to present a risk to the controlled waters. vcc

Some of the wide range of STC_2s is no doubt due to the fact that the investigations were undertaken at different times and analyses were conducted by different laboratories. For most of the determinands the mean values differed by at most 30% between site investigations, and the 'direction' of difference is not consistent i.e. for sulphate, the mean for the most recent investigation is lower than the mean for previous investigations while the converse is so for arsenic.

It is evident that some of the calculated mean STC_2s values are very high. These high values result from the use of the site specific empirical K_d value, and reflect the fact that these determinands are relatively insoluble. Indeed some of the average STC_2s values particularly those for metals are of economic resource levels. On the basis of the high STC_2s values and with lesser toxicity concerns, the EA excluded a number of the original determinands (sulphate, cyanide, cadmium, chromium, lead, copper, zinc). It should be stressed that the soil threshold concentrations are relevant to potential water pollution only, and do not provide protection levels with regards to the health risks to site workers during site operations associated with the movement and 'disposal' of material. Clearly, some of these levels would be unacceptable for retention on site without some form of cover as they potentially present a risk also to human health, flora and fauna.. It was recommended that these values were further reviewed and amended as necessary to take account of these health and environmental implications. The determinands and the ultimate agreed values are shown in far right column of Table 2. All the results from the original 1996 and 1997 investigations were reviewed against the final STC_2s. A significant, if relatively small, number of samples were above the STC_2s for PAH, phenol, arsenic and nickel. The EA requested that all trial pits where samples had either appeared or 'smelt' contaminated with coal tar compounds (regardless of test results) were also 'marked' as requiring remediation. A estimated volume of $16,000m^3$ out of an original total of $80,000m^3$ of former coking works waste was therefore deemed to require remediation. It is evident that the EA Technical Report P13 has been instrumental in reducing the cost of remediation (compared to the construction of a containment cell and excavating all the material) as well as safeguarding the underlying Coal Measures aquifer.

The nature of the defined contamination, predominantly organics, lends itself to bioremediation. The areas where metal contamination was found above the relevant STC_2. have been excavated and removed off site. The ex-situ bioremediation of the elevated PAH concentrations commenced in May 1999.

Acknowledgements

The authors would like to thank RJB (UK) Mining Ltd.

5.0   REFERENCES

1.   Methodology to determine the degree of soils clean-up required to protect water resources, Research and Development Technical Report P13, prepared by Dames & Moore, Environment Agency 1996.

2.   Groundwater Monitoring Boreholes, Houghton Main Site prepared by Allott and Lomax for RJB Mining (UK) Ltd., December 1997.

3.   Investigation of the Former Coke Works, Houghton Main Site prepared by Allott and Lomax for RJB Mining (UK) Ltd., October 1996.

4.   Contamination Appraisal of the Houghton Main Site, Barnsley prepared by Allott and Lomax for RJB Mining (UK) Ltd., March 1997.

5.   BS1377: British Standard methods of tests for soils for civil engineering purposes, British Standards Institution, 1990.

6.   Water Supply (Water Quality) Regulations 1989.

7.   Soil Mechanics, Principles and Practice. GE Barnes, 1995.

8.   Groundwater, RA Freeze and JA Cherry, 1979.

9.   Waste Management Licence Working Plan, Section A: Environmental Assessment, Houghton Main Site prepared by Allott and Lomax for RJB Mining (UK) Ltd., September 1997.

10. Sections of Strata of the Coal Measures of Yorkshire, Midland Institute of Mining, Civil and Mechanical Engineers, 1914.

GIS Technique Application to the Management of Sites Exposed to Ground Contamination from Herbicides

A. MARINO*, S. MARCHESE** & D. PERRET**
*ISPESL/DIPIA(Higher Institute for Occupational Safety and Health/Department of Productive Plants and Interactions with the Environment), Roma, Italy.
**University of Rome "La Sapienza", Dept. of Chemistry, Rome, Italy.

SUMMARY

This paper outlines the potential of GIS techniques as a tool for the management of areas exposed to ground contamination phenomena. In particular the possible patterns of surface and ground waters pollution from herbicides have been considered. Furthermore the interactions between agricultural activities and water resources have been analysed.

Remote sensing data have been processed in order to obtain land-use maps that have been inserted in a GIS database and integrated with further information like lithology, geological structure, hydrogeological features, DEM, etc. Different thematic layers (maps) related to the actual situation, and to the risk of ground contamination from herbicides, have been created and processed in the GIS.

As a random test, a sensitive and specific multiresidue method, based on reversed-phase liquid cromatography/mass spectrometry, with electrospray interphase (HPLC/ESI/MS), for determining 15 of the most representative compounds of a new class of POE herbicides in deep ground-water samples, collected in different times and sites, was applied. The results of the analysis have been interpolated and inserted, for the validation of the GIS in the database.

According to our experience the GIS methodology can be a reliable, sensible and easy to update support to the Competent Authorities in environmental management.

INTRODUCTION

The pollutant load of agricultural activities is often excessive, compromising the environment and in particular water resources. For this reason an efficient policy needs the new technologies to rapidly estimate the possible environmental hazard and better define the intervention.

In such a context, remote sensing and GIS methods offer good advantages in responding to the lack and need of rapid assessment of both spatial and temporal variability of parameters controlling the interactions between ground and surface waters, agricultural activities and urban settlements for that concerns the general dynamics of related phenomena.

It is a typical problem of natural hazards, in most case triggered and enhanced by the anthropogenic impact. Remote sensing plays different roles in the actions of forecast, prevention, monitoring and assessment. The steps leading to the definition of spatial vulnerability and risk necessarily pass through the combined merging of remote sensing data with those coming from other sources. This action is achieved by means of dedicated geographic information systems (GIS).

AREA OF STUDY

The area of study is the Valle del Fucino (Fucino Basin), a flat paleolacustrine basin located in the Abruzzi region. It extends NW-SE for about 17 km, having about 8 km in width. Its fluvial to lacustrine deposits (sand and clays) are surrounded by meso-cenozoic calcareous relieves. The perimeter of the basin is characterized by the presence of many alluvial cones.

Since a long time the area has been interested by a strong agricultural exploitation. Therefore the contamination of soils and water resources, raised in the last decades many conflicts and major issues.

METHODS AND RESULTS

GIS and Remote Sensing

Satellite images from Landsat TM, in different seasons, have been processed. Maximum likelihood algorithm (based on the probability that a pixel belongs to a particular class) and neural net techniques have been applied on enhanced images in order to obtain a supervised classification of land-use. The accuracy assessment gave a Kappa Coefficient equal to 0.82. The kappa coefficient expresses the proportionate reduction in error generated by a classification process compared with error of a completely random classification. The results of classification have been validated by direct survey.

The land-use map has been georeferenced assigning geographical coordinates to raster map data. The land-use map has been inserted in a GIS database and integrated with further information like lithology, geological structure, hydrology, hydrogeological characteristics, that were imported in the GIS as layers. A Digital Elevation Model (DEM) has been realized from interpolation of topographic data (IGM maps). Furthermore the same methodology allowed to reconstruct the spatial patterns of ground water from isopiestic lines.

The EASI-PACE (PCI) software allowed to consider and process information both in raster (like land-use map and DEM) and vectorial (like surface hydrology, isopiestic lines) format. The processing of data layers (Figure 1) was performed, using the same software through typical GIS operators like indexing, recoding, matrix analysis, proximity analysis. The matrix analysis assigns a different output value to each unique combination of input values. The proximity analysis creates an output layer showing successive zones of proximity (distances) to a specified entity or group of entities. The choice of recoding indexes, weights and matrices was based on references, personal experience and knowledge of the area. In particular the use of matrix analysis instead of simple weighted overlaying allowed a better control on parameters and variables. The recoding and indexing was applied in order to obtain a limited number of classes for each layer (map), for a better understanding of the information.

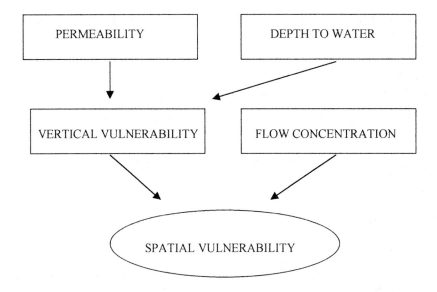

Figure1. Flow diagram of the Fucino Basin GIS

The permeability layer was based on a geological map as indexes were related to different lithologies (Tab. 1).

Table 1. Permeability indexes

Lithology	Index
Limestone	10
Dolomite and travertine	9
Sands and debris	4
Clay and peat	2

A depth to water layer was realized through the subtraction of groundwater altitude (DEM format) from topographical height (DEM format). Therefore the map values have been filtered and indexed (Tab. 2).

Table 2. Depth to water indexes

Groundwater depth	Index
0-1 m	10
2-3 m	9
4-9 m	7
10-15 m	5
16-23 m	3
24-30 m	2
$\geq$ 30 m	1

Vertical vulnerability layer was obtained combining permeability and depth to water indexes using a multiplicative matrix. The values in output have been recoded (Tab. 3)

Table 3. Vertical vulnerability indexes

Matrix output value	Index
≥45	10
40-44	9
35-39	8
30-34	7
25-29	6
20-24	5
15-19	4
10-14	3
5-9	2
0-4	1

Flow concentration layer was obtained from subtraction of groundwater altitude from a reference height (700m. a.s.l.). Output values have been recoded (Tab. 4).

Table 4. Flow concentration indexes

Difference between reference height (700m. a.s.l.) and groundwater altitude	Index
91-100 m	10
81-90 m	9
71-80 m	8
61-70 m	7
51-60 m	6
41-50 m	5
31-40 m	4
21-30 m	3
11-20 m	2
0-10 m·	1

Spatial vulnerability map was obtained from a matrix that combined, using multiplication, vertical vulnerability and flow concentration indexes. The obtained values have been recoded (Tab. 5).

Table 5. Spatial vulnerability indexes

Matrix output value	Index
91-100	10
81-90	9
71-80	8
61-70	7
51-60	6
41-50	5
31-40	4
21-30	3
11-20	2
0-10	1

Environmental Samples Analysis

Rapid famer acceptance of post-emergence (POE) herbicides can be ascribed to several factors including decreased use of chemicals, improved weed control efficiency, environmental and safety benefits, and soil conservation. In addition, recent advances in information systems for predicting crop damage on the basis of weed density, location, and soil type tend to favor the increasing use of such compounds.

A POE herbicides can be defined as a foliage applied product used to control weeds the have emerged in competition with the developing crops. However, many such herbicides still retain residual activity in the soil and thus can control late germination weeds. Their selectivity dictates the specific crop application for individual compounds.

Grab samples of deep ground water were collected every 15 days, from June to September 1997, in six different sites: two in the northern part of the basin (near Celano), two in the southwestern part (near Luco dei Marsi), and two in the western part (Cese). The water samples have been collected in brown glass bottles and kept at 4 °C in the dark until analyzed. In any case, environmental samples were analyzed after no more than a few days' storage.

A sensitive and specific multiresidue method, based on reversed phase liquid chromatography/mass spectrometry, with an electrospray interface (LC/ESI/MS), has been applied for determining 15 of most representative compounds of a new class of POE herbicides in water samples. The procedure used involved passing 2 L of groundwater through a 0.5 g graphitized carbon black (GCB) extraction cartridge. A conventional 4.6 mm i.d. reversed phase LC C-18 column, operating with a mobile phase flow rate of 1 mL/min, was used to chromatograph the analytes. A flow of 200 µL/min of the column effluent was diverted to the ESI source. The study demonstrated the sensitivity of the technique, with detection limit under 5 ng/L.

The results of the analysis have been interpolated and inserted, for validation of the GIS in the database. From this data processing can be outlined a temporal and areal variation in the use of the different herbicides related to the different crops and agricultural cycles. Furthermore, it is evident a pattern related to an increase, and therefore a possible concentration, of the pollutants moving from the relieves, through the alluvial cones, to the bottom of the basin. This evidence can be well related to the results of vulnerability evaluation performed using remote sensing and GIS techniques.

CONCLUSIONS
In a data rich environment it is important to find ways of transforming data into information. GIS layers, can be a very powerful tool to provide information from several different data. In the same time remote sensing fulfills the need of methods for monitoring and managing.

These emerging new technologies are operational now, but there is still a strong demand for further research and investigation to efficiently and fully utilize the potential of these new data sources.

Anyway according to out experience this methodology can be a reliable, sensible and easy to update support to competent Authorities in environmental management.

REFERENCES
Boni, C., P. Bono & G. Capelli 1988. *Carta idrogeologica del territorio della regione Lazio.* Roma, L. Salomone.

Brivio, P.A., G.M. Lechi & E. Zilioli 1990. *Il telerilevamento da aereo e da satellite.* Milano, Edizioni Delfino.

Brivio, P.A. & E. Zilioli 1995. *Il telerilevamento da satellite per lo studio del rischio ambientale.* Roma, Edizioni dell'Ulisse.

D'Ascenzo G, Gentili A, Marchese S, Marino A & Perret D 1998. *Multiresidue Method for Determination of Post-Emergence Herbicides in Water by HPLC/ESI/MS in Positive Ionization Mode.* Environ.Sci.Tecnol., vol.32, no.9, 1340-1347.

A.Poli, U., M. Ippoliti, A. Marino, I. Alberico & P.Bragatto 1996. Studio del rischio di inquinamento delle acque superficiali e del danno conseguente nell'area metropolitana di Roma fino alla foce del Fiume Tevere, mediante tecniche di telerilevamento e GIS. *Seminars of Metropolitan areas and rivers,*107-115. Roma: Acea.

System for the Rapid Assessment of Environmental Conditions to Aid the Urban Regeneration Process at London Docklands

DR ALBERT F. HOWLAND
A F Howland Associates, Norwich, UK

INTRODUCTION

The London Docklands Development Corporation (LDDC) was set up in 1981 to undertake a systematic regeneration of the run-down area of east London. This was dominated by the former docks complex that paralleled the River Thames east of the City of London to Woolwich and included land occupied by a number of associated industries. The area began to fall into decline during the 1950's and 60's as the continued operation of the docks became impractical and they went through a series of closures.

A similar decline was also taking place in other industrial areas in the UK leaving a sense of dereliction in many inner cities. A period of social unrest occurred in the 1970's that was largely driven by the lack of opportunity in these areas. In 1981 a substantial riot in Toxteth in Liverpool spurred Government to take positive action. Although many of the rundown areas had been subject to plans for improvement these had not been implemented. Government took the view that it was necessary to encourage development interest from the private sector in order to ensure that speedy action was taken to provide a successful solution.

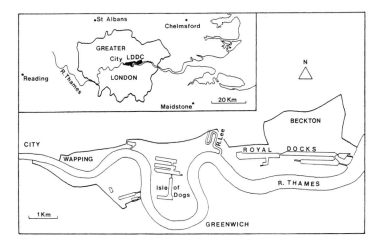

Figure 1   Plan and Location of the Docklands area of east London

Geoenvironmental engineering, Thomas Telford, London, 1999, 505–512

The extent of the degeneration of the docklands area of east London at the time of the formation of the LDDC was well reflected in land values. The general view was that ground conditions would warrant a disproportionate cost to the eventual resale value. Thus, the market anticipated high development costs that were not economic when balanced against the potential returns. To overcome the reluctance and to aid the regeneration process, the LDDC chose to market the area with sufficient information on the ground conditions and soil quality that discussions with potential developers could advance with realistic considerations being given to the ground costs. It therefore needed a means for collecting and presenting the necessary data.

DATA COLLECTION AND COLLATION

Shortly after its inception the LDDC began to collect existing site investigation and other relevant ground data from the area. This was to be made available to any interested party. Initially old site investigation reports were amassed from local authorities and other public bodies. However, it was found that the existing data was often not quantified and invariably did not cover the primary areas of interest. In order to overcome this, and provide detailed quantified information on the potential development areas, a systematic programme of investigation was implemented through term contracts. The first of these was let in October 1983 and these ran continuously to the closure of LDDC in 1998.

Two contracts were let, one for geotechnical site investigation and another for chemical and environmental site investigation. The provision of a separate contract for the chemical and environmental issues in civil engineering site investigation is believed to have been unique and recognised the importance of geoenvironmental issues in the regeneration process of the former industrial areas.

The use of the term contracts and the systematic collection of any other available information meant that a large amount of data was going to be collected in a relatively short space of time. It was expected that eventually some 10,000 borehole records would be processed. However, the LDDC had been set up to work with only a limited staff. In addition it was known that its offices would move to a number of different buildings during the course of its life. These practical constraints posed specific problems for the effective storage and subsequent retrieval of the data once collected (Howland, 1989, 1992). Computer based systems were evaluated as the best option.

As part of a wider advisory brief to the LDDC, A F Howland Associates produced four separate purpose written systems that stored and maintained the data on soil quality. These were developed progressively during a ten year period. This was partly led by an increasing appreciation of the geoenvironmental issues associated with the redevelopment of brownfield sites. Importantly, the systems were required to provide data for the entire docklands area. The basis was that they should store and allow access to primary factual data, but provide a means of filtering the data to provide a rapid response to a variety of needs within the regeneration process.

THE SYSTEMS

Four separate systems were developed. These were a geotechnical database, a geochemical database, a groundwater model and an historic land use register. The underlying principle with each of these was that systematic collection and collation of data meant that a regional understanding for the area could be developed. This allowed a rapid assessment of any site even before actual ground investigation had taken place.

The reliability of the assessment was much greater than could have been achieved from other published data because of the finer resolution of the information making up the systems. As additional data became available its reliability and accuracy could be assessed against the context of the regional understanding.

The regional approach has an advantage over a more usual site specific assessment. It takes account of the much greater amount of data making up the regional understanding and allows the specific data from any one site to be balanced against the expectation. This approach is particularly suited to risk assessments of geoenvironmental hazard, which require a source-pathway-target characterisation of contaminants and is therefore influenced by conditions that extend beyond arbitrary site boundaries.

Geotechnical Database

The Geotechnical Database was the first of the four systems. The primary source of the stored data was site investigation reports. These gave the variation of a number of discrete parameters at a point location, each recorded against depth. They were therefore of a form that is inherently suited to simple tabulation and storage by computer, while the use of a database offered a means of allowing more rigorous enquiries than simple listed retrievals.

The development software for the system was a high level relational database utility which was assessed as particularly suited to the style of the information to be stored and has been described fully elsewhere (Howland 1989). It enabled rapid and very selective retrieval of the data and allowed output in either printed or graphical format.

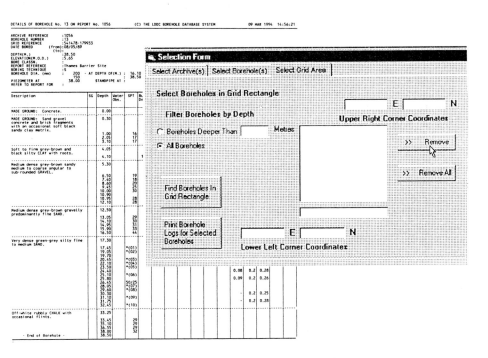

Figure 2   Typical Selection Screen and Output from the Geotechnical Database

Geochemical Database

Although the Chemical and Environmental Term Contract was in place from early on, the data from it was not included in the Geotechnical Database. The expectation was that the results would have a random character so that only limited regional interpretation would be possible.

This view was subsequently changed. Contaminants could be associated with specific industries and the possibility of contaminant migration meant that patterns in the distribution of the contaminants was likely to be present. A similar relational database approach was developed to allow the data to be stored. This was compatible with the Geotechnical Database in that the data was similarly referenced against depth and national grid reference and allowed for the storage of a wide array of possible determinands.

Where the boundaries of development sites changed during the planning procedures data for the new sites could be filtered from the database. The system also allowed the results to be compared to published guidelines to give a summary of hazard assessment.

Groundwater Model

Docklands is underlain by two principal aquifers, a lower aquifer that comprises the Chalk and the more sandy facies of the overlying solid geology and an upper aquifer that consists of the Thames Gravels a sandy gravel that blankets the entire area below a cover of made ground and alluvium. Over part of the area the two aquifers are in contact but become increasingly separated by a clay aquiclude towards the west. The natural hydrogeology of the area is consequently a complex system confused further by the increasing urbanisation.

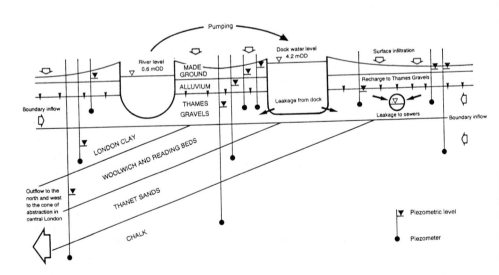

Figure 3   Schematic Representation of the Hydrogeological Controls in Docklands

The success of LDDC in regenerating the docklands area meant that the interaction between the development and the groundwater became increasingly important, particularly as a consequence of the scale of the associated engineering works. A full appreciation of the mechanics of the groundwater system was necessary both for effective engineering design and in order to ensure that movement controls to pollution were correctly established.

The multi-layer system in the docklands area meant that the model needed to be three dimensional. Although number of hydrogeological models existed. These were suited to a natural system in which the layers were horizontal and of uniform thickness. In the docklands area the geological strata dipped and so were of variable thickness and even wedged out and disappeared. This required a new modeling technique to be developed and was undertaken at the University of Birmingham by Professor Ken Rushton (Howland et al 1991).

The model gave the balance of the overall recharge and discharge to the area. Although this showed the situation at a point in time, time sensitive changes could be assessed by adjusting the various parameters. Thus, the consequences of natural variations, such as global warming, on flow patterns and groundwater levels could be determined. Similarly, the influence of new engineering works could be modelled. This could be a short term affect, perhaps during a dewatering phase for a new excavation, or a longer affect if the works were sufficient to disrupt an established flow pattern.

Changes to the groundwater regime can be used to assess the actual or potential for contaminant migration. When used in conjunction with the Geochemical Database, which provides information on the mobility of the contaminants present, the Groundwater Model gives the mechanism by which movement would take place and the speed at which it could occur. This provides a strong understanding of the source-pathway-target association for contaminant migration.

Historic Land Use Register
Once the mechanism of contaminant migration could be understood with the Groundwater Model, a means of providing for the rapid assessment of the potential for land to be contaminated was needed to supplement the actual data held in the Geochemical Database. This was carried out by a review of the past activities in the area. The concept had similarities to the philosophy of the register of contaminating use that was then proposed under Section 143 of the Environmental Protection Act. However, it took the principle further in that not only was the presence of a contaminating use established, but so was history of the site. Regardless, the exercise was not intended to define actual contamination, or assess the by-products that any industrial process may have produced, but merely to establish the potential for contamination by association with the activity or occupant.

The data was recorded in cartographic form using a series of transparent overlays to the published Ordnance Survey 1:2500 plan of the area. In order to provide a full picture data was retrieved from maps and plans at both larger and smaller scales, from books, directories and aerial photographs. The data was referenced to individual indexed plots shown on the overlay. Each plot represented the maximum extent of any industrial process or occupant, regardless of any periods of expansion or contraction.

Three overlays were used to represent heavy industry, light industry and artificial water. Heavy industry was separated from light industry both for ease of illustration and to identify those areas where chemical and associated industry, metallurgy, non-food animal products. oil, gas and electricity processing or generating took place. Light industry represents those activities such as wind and water power generation, food processing, engineering, textiles, leather, clothing and footwear, bricks, tiles, pottery, glass, paper, timber and large retail units. Artificial water includes docks, wharves, ponds, reservoirs and former river lines. The maximum extent only was recorded. Also included, were pumps, wells, and fountains since they can be associated with large underground cisterns.

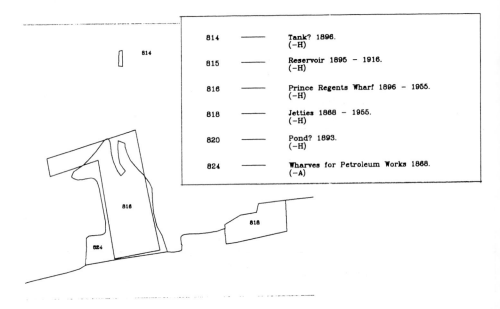

814	——	Tank? 1896. (–H)
815	——	Reservoir 1895 – 1916. (–H)
816	——	Prince Regents Wharf 1896 – 1955. (–H)
818	——	Jetties 1868 – 1955. (–H)
820	——	Pond? 1893. (–H)
824	——	Wharves for Petroleum Works 1868. (–A)

Figure 4   Typical Overlay Detail from the Historic Land Use Register

A subsequent development of the Historic Land Use Register was to convert it to a computer based system and free it from the constraints of a traditional paper cartographic system. A proprietary geographic information system (GIS) utility was used as the basis of the application. This used the digitised Ordnance Survey plans of docklands and allowed the digitised information from the map overlays of the Historic Land Use register to be referenced against this. An ability to represent the original indexed plots as polyline blocks on the GIS meant that they could be linked to an attribute list allowing the historic data for each plot to be instantly displayed to screen. The facility for moving around the background map and ability to move into an area of interest made retrieval of the data more intuitive for the casual user and quicker to implement.

The power of the GIS offered an opportunity to incorporate the data from the other systems to provide a single Geoenvironmental Register. This would provide a single system detailing not only the past history, but also the geotechnical and contamination data. It could also include details of any remediation undertaken in the area so that ongoing assessments of hazard could be made which allowed for the actual ground conditions. This final compilation was not fully completed before the LDDC was disbanded but showed that a simple means of interrogation of all geoenvironmental data was possible using a map based graphical interface.

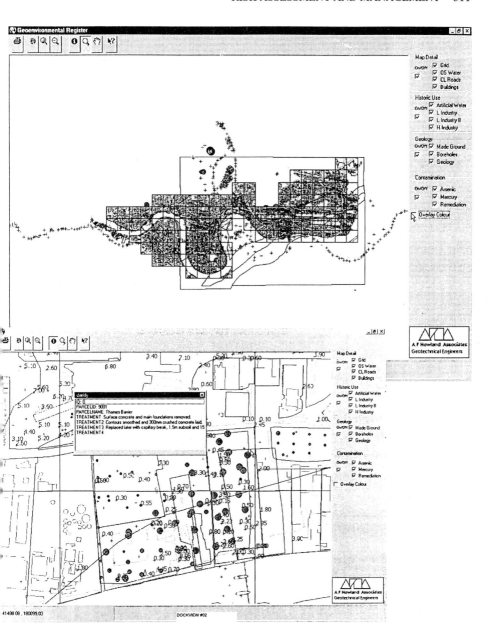

Figure 5   The Geoenvironmental Register for Docklands: Use of a Graphic Interface on a Geographic Information System to Zoom to Areas of Interest and Display a Variety of Data which can be Filtered to show only Pertinent Information

SUMMARY
A series of systems detailing geoenvironmental information collected from standard sources gave LDDC rapid access to information on soil quality. This also allowed the information to be retrieved. Output was in graphical or tabulated textual format. Some statistical processing of the data prior to output offered a useful summary of the data and enhanced its immediate value. In particular, chemical data was compared automatically to values of acceptability or other guideline criteria.

An understanding of the groundwater regime was provided by a dynamic computer model. This allowed rates and directions of groundwater movement to be determined under any set of conditions that may have developed through natural changes, or as a result of the development of the area.

A register of historic land use gave a cartographic index of the industrial activities in the area together with any water features and provided a measure of the potential for contamination to be present.

Each system operated independently, but when used together the combined data gave a total environmental statement, which benefited the regeneration process.

The final development of the systems was to combine the data held on each into a single geographic information system to provide a Geoenvironmental Register with the data searched and retrieved through a map based front end.

REFERENCES
Howland A F, 1989. Integration of the capture, storage and presentation of site investigation data by microcomputer by the London Docklands Development Corporation. Ground Engineering. 22, (3), 30-35.

Howland A F, 1992. Use of computers in the engineering geology of the urban renewal of London's Docklands. Q Jl Engng Geol. 25, 257-267

Howland A F, Rushton K R, Sutton S E and Tomlinson L M, 1993. A hydrogeological model for London Docklands; in Groundwater Problems in Urban Areas, Thomas Telford, London.

Sustainable brownfield redevelopment - risk management

P TEDD, J A CHARLES AND R DRISCOLL
Building Research Establishment, Watford, UK

INTRODUCTION

In opening the Land Reclamation Conference which was held in October 1976 at Grays, Essex, the Parliamentary Under Secretary of State emphasised that the Department of the Environment had the complementary aims of reclaiming the dereliction inherited from the industrial past and securing, by planning control and other means, that we do not create fresh areas which are left without effective treatment. He commented that although in the early 1970s an annual reclamation rate of 2200 ha was reached, this had not been wholly sustained (Barnett, 1976). In 1974 there were reckoned to be 43 000 ha of derelict land in England and by 1993 the area had decreased by 8% (Parliamentary Office of Science and Technology, 1998). However, the difficulties in developing an appropriate definition of derelict land were bound to reduce the value of reclamation statistics (Wallwork, 1974).

In more recent years the term *brownfield* has come into use. Needless to say, it is as difficult to obtain an acceptable and agreed definition of brownfield as it is to define dereliction. The United States Environmental Protection Agency defines brownfields as abandoned, idle or under-used industrial and commercial facilities where expansion or redevelopment is complicated by real or perceived contamination (McKenna, 1998). This would seem to link brownfield to contaminated sites which seems unduly restrictive. The Parliamentary Office of Science and Technology (1998) commented that because there is no agreed definition, many practitioners regard brownfield land as any land that has been previously developed. In a parliamentary answer, the Construction Minister is reported as saying that: *"There is, as yet, no specific definition of brownfield land, but it is usually taken to mean land previously used for urban uses ... these include residential, transport and utilities, industry and commerce, community services, previously developed vacant land and derelict land."* (Raynsford, 1998). In November 1996 a target was proposed that 60% of new housing should be built on brownfield land.

Although brownfield land is a world-wide phenomenon, the issues are particularly acute for Great Britain, a heavily populated island with a long industrial history. An example which illustrates the scale of the problem is that in 1996, a £1 billion plus package was announced for the regeneration of major coalfields. Some 910 ha of land were to be reclaimed for residential, commercial and retail uses (Sleep, 1996). Many of these sites will involve building on colliery spoil (Skinner et al, 1997).

The sustainability agenda requires the long-term productive re-use of brownfield land; the problem is that previous usage may have left a wide range of physical, chemical and biological hazards. Physical problems may include buried foundations and settlement of filled ground. The range of problems associated with chemical contamination is vast and chemical contamination can present an immediate or long term threat to human health, to plants, to amenity, to construction operations and to buildings and services. Biodegradation of organic matter may lead to the generation of gas.

SYSTEMS AT RISK

Three systems which may be at risk in brownfield developments can be identified: the human population, the natural environment and the built environment. In defining risks associated with brownfield development, there is often a lack of distinction between these three groupings. Linked to these three areas, there are risks to wealth and profit for developer, investigator, designer, builder, owner, occupier, insurer and finance institution. One unfortunate side effect of evaluating risk is that there is, of necessity, a concentration on problems and potentially negative effects. It needs to be made clear at the outset that the re-development of brownfield sites can have massive advantages and, while it is important to identify hazards and to evaluate risks, the benefits should not be overlooked.

Although the three systems listed above are now examined separately, they are, of course, interdependent and the issues need to be examined against a background of growing concern over degradation of the natural environment and the increasing prominence of the concept of sustainable construction involving the rate of use of renewable and non-renewable resources, and the emission of pollutants.

- *"Sustainable development meets the needs of the present without compromising the ability of future generations to meet their own needs"* (World Commission on Environment and Development [Brundtland Commission], 1987)
- *"Sustainable development also involves a changing conception of the relationship between the natural environment and mankind"* (Institution of Civil Engineers, 1996a)

Some of the issues involved in sustainability and acceptability in infrastructure development have been reviewed by the Institution of Civil Engineers (1996b).

Human population

For many years the redevelopment of derelict land and brownfield sites has been dominated by the hazards associated with *contamination* and the risks posed to human health. For example, in a paper on risk assessment and soil contamination it was affirmed that: *"Risk assessment is appraising the possibility and severity of potential adverse health events"* (Aldrich et al, 1998). No mention was made of potentially adverse impacts on the various aspects of the natural environment, let alone the built environment. However, in Part IIA of the Environmental Protection Act 1990 contaminated land receives specific definition: *"Contaminated land is identified on the basis of risk assessment. In accordance with the provisions of Part IIA and statutory guidance, land is only contaminated where it appears to the authority, by reason of substances in, on or under the land, that: (a) significant harm is being caused or there is a significant possibility of such harm being caused ; or (b) pollution of controlled waters is being, or is likely to be, caused."*

Harm is defined by reference to harm to health of living organisms or other interference with the ecological systems of which they form a part and, in the case of man, is stated to include harm to property. Where building development is proposed, risk assessment should consider a broad spectrum of potential targets which may be vulnerable to different hazards. Since brownfield sites are frequently contaminated, risk assessment should include health concerns and from the early stages of investigation through to the final use of the site, a range of people may be at risk. Where low-rise housing is built on the land, the occupiers are likely to be the people most at risk from many of the hazards. Society is increasingly risk averse and matters concerning health and safety rightly receive much attention and publicity (Bevan and Hind, 1999)..

Natural environment

Threats to the natural environment are often relatively narrowly conceived to concern soil and groundwater contamination. However, a much wider range of issues may need to be examined against a background of growing concern over degradation of the natural environment and the increasing prominence of environmental pressure groups promoting concepts such as biodiversity. It is not easy to see how an appropriate balance can be obtained between economic well-being and the natural environment. Furthermore, proposals to reclaim derelict land may now run into serious difficulties with opposition from those who consider that eco-systems established amidst the dereliction, and present as a direct result of it, should be preserved. The warning has been given that: *"If you take an abandoned quarry, though, you have to be quick if you want to develop it. Otherwise it will turn into a Site of Special Scientific Interest."* (Guthrie, 1999). This whole area is diffuse and difficult to evaluate as, unlike matters of human health and building damage, there are no widely agreed objectives or ground rules.

Built environment

The objective is to build safe, durable and economic structures. The site and the building development form an interactive system and it is important to evaluate the risk of adverse interactions during the lifetime of the development. Hazards to the built environment on a brownfield site can be physical, chemical or biological in character and concerns could include the following:

* poor load carrying properties of the ground,
* interaction between building materials and aggressive ground conditions,
* combustion,
* gas generation from biodegradation of organic matter and from other deleterious substances in the ground.

ASSESSING THE RISKS

It may be questioned whether current approaches to the sustainable redevelopment of brownfield land are suitable; in some cases solutions may be over-engineered (Wood and Griffiths, 1994). How do the risks compare with other types of risk? Do we, in fact, have an adequate understanding of brownfield site risks?

With an increasing proportion of building developments being located on brownfield sites with the various hazards that these may contain, risk assessment and risk management should have major roles to play. Quantitative risk assessment (QRA) can assist in determining the required level of remedial action to achieve a "suitable for use" situation. Although QRA has been widely used in the chemical and nuclear industries, it has received relatively little use in contaminated land; there has been a concentration on the source-pathway-receptor approach to assessing the risk to the human population.

Of course, matters of health and safety must be pre-eminent. Both individual and societal risk need to be evaluated. The latter is often expressed in the form of an F-N curve in which the number of fatalities (F) is plotted against the probability of occurrence (N). While there is no evidence to suggest that chemical contamination on brownfield sites has posed a major threat to human life, at least in the short term, long term health and quality of life could be affected and this is difficult to evaluate.

The financial risks need to be assessed. Hazards for the human population and for the natural and built environment can result in situations which result in a loss of wealth for clients, developers, investigators, designers, constructors, owners, occupiers, insurers and the public. However, an over-conservative approach to risk and the perceived need to minimise future potential liabilities can also lead to a loss of wealth to individuals and can lead to an unnecessarily large proportion of national resources being consumed.

The concept of sustainable development could mean that in future large scale redevelopment of brownfield sites, in addition to financial audits, there could be a requirement for audits concerning matters such as energy and entropy (Hudson, 1992) and ecology. The consumption of energy is fundamental to the economy; large centres of population require massive inputs of energy to remain viable. Entropy is a measure of disorder within a system and represents the energy no longer available to do useful work; human activities result in an increase in entropy and it has been claimed that our high entropy culture is not sustainable (Rifkin, 1980). An ecology audit is probably the most difficult as the value placed on various species of flora and fauna is highly subjective.

MANAGING THE RISKS
Risk management is closely linked to risk assessment and three particular aspects of risk management are now examined.

Poor physical behaviour
Where the load carrying characteristics of the ground are inadequate, there are several widely used methods of ground treatment including vibrated stone columns, dynamic compaction and pre-loading. BRE has carried out major studies of the effectiveness of these various treatment methods (Charles, 1993; Watts and Charles, 1997). The principal use of ground treatment in Britain is on filled ground. Little control may have been exercised over the placing of the waste material and the poor load carrying properties of many non-engineered fills have been associated with their heterogeneity and their loose, poorly compacted condition (Building Research Establishment, 1997/98). The major geotechnical problem is usually associated with settlement of the fill due to effects other than the weight of the building. Poorly compacted or excessively dry fill is likely to be vulnerable to a reduction in volume if the moisture content of the fill is increased. This phenomenon, which is usually termed collapse settlement or collapse compression, can occur without any increase in applied stress (Charles and Watts, 1996). The phenomenon can occur at depth within the fill either due to a rising ground water table or due to downward infiltration of surface water. As it is differential settlement rather than total settlement that damages buildings, local collapse compression from a surface source of water is of particular concern. Ground treatment should aim to overcome these problems

Contamination
While the diagnosis of the nature of the problems is an essential prerequisite for successful development of a brownfield site, in many cases some type of remediation work will be required prior to building development. As there are many different types of hazard, so there

are many different forms of solution. Decisions on how to manage physical and chemical hazards in brownfield sites depend critically on accepted criteria for performance "suitable for use". There is currently some uncertainty about these criteria. For contamination, a key issue is whether or not any action is required. If it is, which of the three following alternatives is most appropriate?

- Remove material ('dig-and-dump')?
- Contain material in the ground?
- Treat the ground to remove the contamination?

The vast majority of contaminated land in the UK is remediated directly by on-site containment or indirectly by excavating and transporting to contained landfill. Engineered cover layers and in-ground vertical barriers such as slurry trench cut-off walls have long been accepted by the various enforcement authorities as reliable means of containing contamination. Lack of long term validation data raises questions about the long term performance of containment systems. There may also be a preference for removal of the contamination off-site due to concerns over future liabilities, though this option may become increasingly financially unattractive as measures are taken to limit environmental impact (a) in transit and (b) at the other site.

Containment of contamination has a vital role in the utilisation of many brownfield sites. In this commonly used remediation method the objective is to block or control the pathway between a hazard and potential targets. Slurry trench cut-off walls using self-hardening cement-bentonite are the most common type of in-ground vertical barrier used in the UK to control the lateral migration of pollution and gas from contaminated land and landfill sites. It is estimated that more than 100 walls have been constructed during the last 20 years with a some several kilometres long. For a cut-off wall to function as a satisfactory barrier to leakage migration, it must have a low permeability and an adequate toe-in to an underlying aquiclude. Despite their extensive use in the UK for waste and pollution management, little is known about either the ground water contaminant conditions under which a slurry wall will remain intact without increase in permeability or the capacity of a slurry wall to "treat" contaminants as they migrate through the barrier. These uncertainties lead to difficulties in guaranteeing long-term retention of contaminants inside slurry-wall enclosed sites; BRE is undertaking a programme of field and laboratory research to reduce these uncertainties (Tedd et al. 1997a,b). Laboratory tests have shown that deterioration of slurry walls could occur in the long term in very chemically aggressive ground conditions (Tedd et al, 1998) but have also shown that certain contaminants are retained with the slurry wall.

Durability on construction materials in contaminated ground

The interaction between construction materials and chemically aggressive ground is being studied at BRE; materials at risk include concrete, mortars, metals, plastics and masonry. General guidance on the performance of building materials used in contaminated land has been given by Paul (1994). The effects of sulfates, acids and chlorides give rise to concern for both unreinforced and reinforced concrete. Magnesium, ammonia and phenol are also known to cause deterioration of concrete. High quality dense concrete is the primary prerequisite for durability and there is no substitute for quality of the materials used. The consequences of chemical attack on concrete are not easy to assess. Ultimately, chemical attack may cause the collapse of a structure, but such extreme examples are not often found. More commonly, localised deterioration and cracking will result in loss of strength. Chemical attack below ground is difficult to assess due to poor accessibility. Damage has generally been found on the tops of exposed piles or foundation walls. The form of concrete is important in determining the risks of attack. Slender sections are more at risk than massive foundations. Modern housing, which tends to be of lighter construction, should be carefully designed and specified. Slabs and floors on the ground are at risk especially where they can dry from the top which encourages the movement of contaminants into the concrete from the ground.

EXAMPLES OF DEVELOPMENTS ON BROWNFIELD SITES
Two examples of major building developments on brownfield sites illustrate some of the hazards and the solutions.

Snatchill, Corby
Ironstone was extracted from the late nineteenth century onwards in the Corby area. The development of mechanised opencast ironstone mining in the inter-war years created extensive areas of derelict land. The practice was to dump the overburden behind the advancing quarry face, thereby creating the distinctive "hill and dale" formation. Initially afforestation of hill and dale was considered an appropriate form of reclamation. From 1951, a policy of immediate reclamation was enforced; a level site being left by new workings and much of the existing hill and dale was levelled. By the mid-1970s, over 1000 ha of land affected by opencast mining in the Corby area had been restored to agricultural use.

The expansion of Corby involved housing and industrial developments on land previously worked for ironstone by opencast methods. The master plan published in 1965 envisaged that, by the early 1980s it would be necessary to develop over 100 ha of land in the Great Oakley area which had recently been restored. The principal hazard was considered to be that settlement of the 24m deep opencast backfill would damage the buildings. The solution was to adopt an appropriate form of ground treatment. At the Snatchill site in Corby, in collaboration with Corby Development Corporation, BRE carried out field investigations of the effectiveness of a number of ground treatment techniques (Charles et al, 1978). Experimental houses were built on the treated areas in 1975-76. Preloading the 24m deep fill with a 9m high surcharge of fill was demonstrated to be a particularly effective form of treatment. In the 1990s, a number of housing developments have been carried out in the vicinity following preloading with a 7m high surcharge.

Hampton, Peterborough
For over one hundred years, Lower Oxford Clay was excavated at Fletton for brick making and much of the site was derelict for decades. Between 1965 and 1991, large areas of land were restored to agriculture through a land reclamation project (Brown and Snell, 1976). This involved disposing of pulverised fuel ash (pfa) in the old clay pits which were between 10 and 20m deep; the pfa was pumped as a slurry into the pits. When a pit was filled, it was covered with a layer of topsoil. When it was decided to redevelop the area as a township, there were several different landforms including pfa filled pits, pits not filled with pfa, and ridge and furrow formation.

A wide variety of possible hazards were identified, including several geotechnical problems. A key factor in the decision to develop a township on the site was the feasibility of development on the lagoon pfa areas and a major programme of field and laboratory testing was undertaken (Humpheson et al, 1991). The ridge and furrow area required a major earthmoving operation (Patel, 1995) and much of the clay was considerably wet of optimum moisture content with a very low undrained shear strength.

It has been claimed that the new Hampton township being developed on this 1000 ha site at Peterborough represents a unique example of an entire community being constructed in harmony with the environment on a site which was once a derelict landscape of exhausted clay workings and water filled pits. Over 5000 houses will be built to accommodate 13000 people. One constraint on development, the significance of which might not have been immediately recognised by a civil engineer, has been the presence of several thousand great crested newts. These had colonised parts of the old clay pits (Prentice, 1997). The cost of the large newt reserve adjacent to the township and the relocation of the newts has been substantial with a figure of £15 million quoted in newspaper reports (Brown, 1996; Clover, 1996).

CONCLUSIONS

As there can be a multiplicity of possible hazards on brownfield sites, it is vital to identify the most significant problems and to evaluate the risks that they pose. It is also necessary to define what is the acceptable level of risk. On housing developments, risks to human health from contamination may be a significant issue, but this should not distract attention from the hazards to the built environment. Although there is a need for improved techniques of risk assessment and management, it should be emphasised that the redevelopment of brownfield sites can have massive advantages and that greenfield sites are not necessarily problem free. The safeguarding of the natural environment is a particularly contentious subject which is likely to be increasingly prominent.

ACKNOWLEDGEMENTS

Research on building on brownfield land is carried out for DETR under the Building Regulations Framework Agreement and the Safety and Health Business Plan.

REFERENCES

Aldrich T E, Torres C and Lilquist D (1998). Risk assessment and soil contamination. Land Contamination and Reclamation, vol. 6, no 4, pp 207-213.

Barnett G (1976). Opening address by Parliamentary Under Secretary of State. Proceedings on Land Reclamation Conference, Grays, Essex, October, pp 2-4.

Bevan S and Hind G (1999). Birth defects cluster found near toxic dump. The Sunday Times, 11th April.

Brown J and Snell P A (1976). The use of pfa in land reclamation. Proceedings on Land Reclamation Conference, Grays, Essex, October, pp 135-141.

Brown P (1996). Newts undermine Hanson project. The Guardian, 11th June.

Building Research Establishment (1997/98). Low-rise buildings on fill. Digest 427 (3 parts).

Charles J A (1993). Building on fill: geotechnical aspects. BRE Report BR230.

Charles J A Earle E W and Burford D (1978). Treatment and subsequent performance of cohesive fill left by opencast ironstone mining at Snatchill experimental housing site. Corby. Clay Fills. Proceedings of conference held at Institution of Civil Engineers, London, November 1978, pp 63-72. ICE, London, 1979.

Charles J A and Watts K S (1996). The assessment of the collapse potential of fills and its significance for building on fill. Proceedings of Institution of Civil Engineers, Geotechnical Engineering, vol. 119, January, pp 15-28.

Clover C (1996). Great newt hunt to save 5000 homes. The Daily Telegraph, 2nd September.

Guthrie P (1999). Quoted by A Mylius in "Brown paradise". New Civil Engineer, 28th January, pp 20-22.

Hudson J A (1992). Rock engineering systems – theory and practice. Ellis Horwood, Chichester. 185pp.

Humpheson C, Simpson B and Charles J A (1991). Investigation of hydraulically placed pfa as a foundation for buildings. Ground Movements and Structures. Proceedings of 4th International Conference held in Cardiff, July 1991, pp 68-88. Pentech Press, London.

Institution of Civil Engineers (1996a). Whither civil engineering? ICE, London. 59pp.

Institution of Civil Engineers (1996b). Sustainability and acceptability in infrastructure development. ICE, London. 96pp.

McKenna G F (1998). Innovation in brownfields site assessment. Risk-based corrective action and brownfield restorations. Geotechnical Special Publication no 82, pp 16-29. ASCE, Reston, Virginia.

Parliamentary Office of Science and Technology (1998). A brown and pleasant land – household growth and brownfield sites. Report 117, July. House of Commons, London.

Patel D (1995). Opening speaker at British Geotechnical Meeting on Building on fill held at Institution of Civil Engineers on 12th April 1995. Meeting report by K S Watts. Ground Engineering, July/August, pp 32-35.

Paul V (1994). The performance of building materials in contaminated land. Building Research Establishment Report BR255, Construction Research Communications.

Prentice E-A (1997). A town rises from the pits. The Times, Wednesday 23rd April.

Raynsford N (1998). Quoted by S Aker in "Talking Point", Ground Engineering, vol. 31, no 11, November, p3.

Rifkin J (1980). Entropy – a new world view. © Foundation on economic trends, 1980. Published by Paladin Books, London, 1985. 335pp.

Skinner H D, Watts K S and Charles J A (1997). Building on colliery spoil: some geotechnical considerations. Ground Engineering, vol. 30, no 5, June, pp 35-40.

Sleep K (1996). Coal deal will yield £1 billion in work. Construction News, 14th November, p6.

Tedd P, Butcher A P and Powell J J M (1997a). Assessment of the piezocone to measure the in-situ properties of cement-bentonite slurry trench cut-off walls. Geoenvironmental engineering – contaminated ground: fate of pollutants and remediation. Proceedings of British Geotechnical Society Conference, Cardiff, September, pp 48-55. Thomas Telford, London.

Tedd P, Holton I R, Butcher A P, Wallace S and Daly P J (1997b). Investigation of the performance of cement-bentonite cut-off walls in aggressive ground at a disused gasworks site. Proceedings of 1st International Containment Technology Conference, St Petersburg, Florida, pp 125-132. Also in Land Contamination & Reclamation, vol. 15, no 3, pp 217-223.

Tedd P, Holton I R, Wallace S, Daly P J (1998). Slurry trench cut-off walls: two years of results from the BG-BRE test site. ConSoil98. Proceedings of the 6th International Conference on contaminated land, vol. 2, pp 1065-1066. Thomas Telford, London.

Wallwork K (1974). Derelict land – origins and prospects of a land-use problem. David and Charles, Newton Abbot. 333pp.

Watts K S and Charles J A (1997). Treatment of compressible foundation soils. Ground improvement geosystems: densification and reinforcement. Proceedings of 3rd international conference, London, June, pp 109-119.

Wood A A and Griffiths C M (1994). Debate: contaminated sites are being over-engineered. Proceedings of Institution of Civil Engineers, Civil Engineering, vol. 102, August, pp 97-105.

World Commission on Environment and Development (1987). Quoted in Institution of Civil Engineers (1996b).

The Development of an Environmental Management System

DR. SALLY UREN, MARTIN BROCK, MATT DORMER, PROFESSOR QUENTIN LEIPER*
Stanger Science & Environment, The Lansdowne Building, Lansdowne Road, Croydon CR0 2BX.
*Company Chief Engineer, Carillion plc

INTRODUCTION

Carillion is an organization employing around 15,000 people on approximately 600 sites and across 30 countries worldwide. It is involved in many aspects of the built environment from civil engineering, design, building and refurbishment through to rail track construction, maintenance, facilities management services and the delivery of projects funded under the Private Finance Initiative (PFI). Carillion's projects are therefore as diverse as the construction of Manchester Airport's second runway, the project management of the refurbishment and rebuilding of the Royal Opera House and the provision of a new PFI funded hospital in Swindon.

Carillion recognises its responsibilities to protect and enhance the natural environment in and around the sites at which it operates across the globe. Furthermore, due to the nature of its activities, Carillion Business Groups operate on the environmental front line, often involved in projects where public interest is high. In response to this public interest Carillion (formerly Tarmac Construction Services) convened an Independent Environmental Advisory Panel in 1994 to provide an independent and experienced resource to help it formulate an effective environmental position. The Panel advises on direction and goals to help develop strategy and reviews the Company's performance against these targets on an annual basis. In 1997 the Panel's challenge centered on urging the Group to put in place an Environmental Management System (EMS) at every site.

Given the scale, nature and diversity of Carillion's activities, the implementation of a company wide EMS was a challenging task and could not be achieved over a short timescale. It was necessary for the EMS to have a structure which was applicable to every part of the business, but was also sufficiently flexible to accommodate differences in the range of activities carried out as well as differences in location. To meet these requirements, Carillion approached EMS development in a systematic manner, with the formation of sequential annual targets.

This paper describes the development of Environmental Management Systems within Carillion, the progress that has been made so far, and the further challenges that will need to be met by Carillion and the wider industry in the drive towards sustainability.

Geoenvironmental engineering, Thomas Telford, London, 1999, 521–528

WHY AN ENVIRONMENTAL MANAGEMENT SYSTEM?

An EMS is a means of identifying, measuring, monitoring, controlling and reducing the impacts of a company's activities on the environment, resulting in continual improvement and legislative compliance.

There are two voluntary EMS specifications that organisations can use to develop an EMS for their business. The most widely adopted is ISO 14001, produced by the International Organization for Standardisation and adopted in the UK as BS EN ISO 14001:1996. Certification to ISO 14001 has been available since the standard was in draft form and the number of certificates worldwide is growing rapidly, having recently passed the 10,000 mark. The other EMS specification is the Eco Management & Audit Scheme (EMAS), which is applicable within Europe. This is a scheme rather than a standard which equivalent to ISO 14001 with the additional requirement for a published and validated Environmental Statement.

There are four stages to an EMS based upon ISO 14001: planning the way forward; implementing actions throughout the organisation; checking that the system is working as planned, and using the feedback to fine tune, where necessary, to improve the system.

The planning stage requires the identification of significant environmental aspects which are those activities which can have an impact on the environment, which in turn can be either positive or negative. The planning stage also involves identification of environmental legislation relevant to a company's activities and the setting of objectives and targets by the organisation. This stage includes production of an environmental policy.

In the implementation phase, the policy statements are converted into actions. Environmental responsibilities throughout the organisation are set and awareness raised. Competence is ensured where activities are associated with significant environmental aspects and operational controls are applied to manage, control and monitor the significant aspects/impacts. Training is clearly an essential part of this stage.

The third element of the management system includes monitoring of environmental performance, environmental compliance and progress against objectives & targets. It is accompanied by internal auditing of the EMS to investigate the degree of compliance with the company's environmental policy and to investigate how successfully procedures are being implemented.

The final element, which " completes the loop", is to use the results of audits and management review to implement changes to the system to ensure that it continues to be appropriate and that continual performance improvement is being achieved.

These four stages have been translated by Carillion to produce the EMS Cycle (Figure 1).

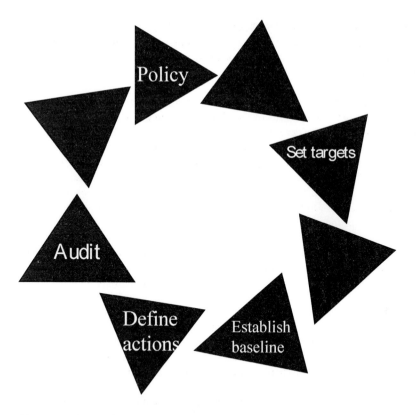

Figure 1. The Carillion EMS Cycle

BENEFITS OF AN EMS
Implementation of an EMS has enabled Carillion to;

- identify its environmental impacts;

- in many cases, purchase less raw materials and produce less waste, resulting in financial benefits which are passed directly to the bottom line

- gain a better understanding and management of environmental legislation to minimise the risk of prosecution;

- meet and exceed client requirements, in some instances generating a competitive advantage;

- improve relationships with the public and finally;

- harness staff commitment.

CARILLION'S APPRAOCH TO EMS DEVELOPMENT

Carillion approached EMS development by defining annual business-specific targets, which recognised the varying levels of complexity, performance and existing management tools available within each of the businesses.

The Panel's challenge in 1997 was to implement an EMS at all sites by 2002, so a means had to be found of translating the principles of environmental management into a practical site management tool. The aim was to achieve this by integrating environmental management into existing quality and safety systems; principally the existing Project Management Plan (PMP) which was already in place at a large number of sites. As with any EMS, a key feature was and continues to be, training and communication, enabling the whole process to be kept 'live'.

The working definition of the integrated management system was taken to be: 'the procedures for the planning, monitoring and control of all aspects of a construction project that are necessary to ensure that all those involved are aware of their responsibilities and perform them safely, to the required quality and with due regard to the environment'

The PMPs are intended to deliver a process where all three issues (safety, quality and environment) can be focused according to risk, referenced to existing best practice. Beneath the PMP are specific activity plans that again take account of the quality, safety and environmental factors of the specific activity (Figure 2). The PMP is intended to be formulated in outline at project inception (tender) stage, and further developed as the project continues, tracking the project through to completion, where it provides a record of problems, solutions and evidence for eventual handover to the client. It therefore provides an approach to quality and environmental issues which is comparable to the CDM Regulations for safety.

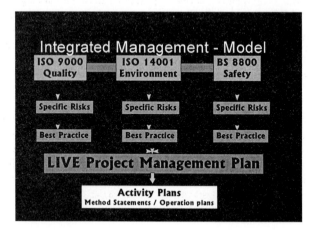

Figure 2. Translation of an Integrated Management System for Application on Site

Within the PMP the environmental component of the integrated management system is often detailed in an Environmental Plan, comprising;.

- Environmental Policy Statements (Carillion and client, if appropriate)

- Identification of project environmental issues;

- Identification of significant environmental aspects and impacts;

- Environmental Management Structure.   A management structure for the control of environmental aspects would be provided in consultation with the Project Team and the management procedures and controls described.

- Details of training to be given to all site staff, including site management and sub-contractors;

- Schedule of Consents, Licenses and Authorisations. All relevant consents, licenses and authorizations would be listed and updated as the precise regulatory requirements for the works are more clearly identified;

- Method Statements for specific activities such as the monitoring of water quality, noise and vibration or dust, would be referenced from the Environmental Plan;

- Environmental Emergency Plan.

In developing a PMP the first key step is conducting a risk assessment to identify environmental, safety and quality hazards, to focus on those that are significant and where effort is most needed. This process is used as the basis for planning and controlling site activities and operations and for identifying 'minimising actions' whereby the use of raw materials, the production of waste, the use of energy and safety risks can potentially be minimised. Following the EMS model, auditing and checking programmes are also linked to key risks and prioritised accordingly.

In order to ensure that management of environmental issues is a live process, Carillion committed to embedding the emerging EMS with a structured programme of environmental training based upon the output of a Training Needs Analysis undertaken in 1997 in each Business Group. Training courses are run through the Company's in house training centre by Carillion's specialist environmental consultancy, Stanger Science and Environment. Stanger have now developed six bespoke environmental awareness modules, including courses for project managers, site managers, designers and office managers.

PROGRESS SO FAR

During 1998, Carillion successfully implemented many of the steps towards a formal EMS, including the development of a register of legislation, the evaluation of significant environmental aspects and the development of objectives and targets. An additional environmental target to audit the progress made in 1997 was implemented to ensure that the EMS is being effectively maintained and improved.

The first Carillion Business Group to achieve certification of its EMS to ISO 14001 was Crown House Engineering, whose manufacturing centre achieved certification one year ahead of schedule in November 1998. Carillion Construction and Engineering has also now developed and implemented an EMS based on ISO 14001 and achieved certification of the system in July 1999. Carillion Building is well advanced in its progress towards implementing an Integrated Management System and also aims to achieve certification to ISO 14001 in 1999. Schal Construction Management and Stanger Science and Environment are other Carillion Business Units aiming for certification in 1999.

Since 1998, the Company has paid close attention to how its environmental performance compares to that of other organisations both within and outside the sectors in which it operates. The annual survey of corporate environmental engagement within the Major Contractor's Group was a useful benchmarking exercise, with the 1998 survey continuing to show Carillion as one of the leaders in the development of environmental improvement tools for the construction industry. The survey also showed an improvement in the overall performance of the industry.

EMBRACING SUSTAINABILITY

It is not intended that implementation of an EMS throughout Carillion will be the culmination of the process that was set in motion by establishing the Panel in 1994. There are already efforts to move beyond 'greening' to tackle the more fundamental issue of sustainability.

Sustainable development was defined by Brundtland in 1990 as "development that meets the needs of the present without compromising the ability of future generations to meet their own needs". Recently the understanding has been refined to embrace the so-called 'Triple Bottom Line' which equates to environmental, social and economic accountability. In simple terms, sustainability can be interpreted as delivering credible financial performance (economic accountability) whilst successfully meeting social and environmental responsibilities.

Sustainable development is central to Government policy with the application of the themes of sustainable development for the construction industry explored in a consultation paper covering sustainable construction. Sustainable construction has many different definitions, one interpretation is the adoption of a more strategic approach to design, construction and maintenance in order to optimise whole life performance. At a more practical level interpretations include innovative ways to re-use and recycle materials (addressing the environmental bottom line), effective dialogue with stakeholders (the social bottom line) and innovative designs to reduce raw material consumption (the financial and environmental bottom line).

There are obvious constraints to making construction industry more sustainable. How for example do you measure sustainability? Whereas systems have been developed to measure environmental performance, there are not yet the equivalent systems to measure the financial and social benefits of environmental initiatives. One approach to addressing the financial performance is 'green accounting' - integrating environmental reporting with financial reporting. However, green accounting systems are still in their infancy. Another constraint to the adoption of sustainable construction practices is the culture of the industry. Although improving, environmental awareness is still not high in the construction industry, the understanding of sustainability is even lower. It is also a low priority on management agendas in an industry that is faced with severe margins, poor returns and clients who generally have not yet seen the environment as a real business driver.

In the last 12 months Carillion has begun to consider ways of delivering a practical translation of sustainable construction. Through the use of a tool called The Natural Step, sustainability action plans have now been developed for two projects, including the PFI hospital at Swindon. This action plan includes such features as green transport plans and is looking at ways of offsetting carbon dioxide emissions in construction through tree planting for example. Through consideration of sustainability, new business opportunities have been identified for example Carillion Contract Housing is looking at ways of differentiating itself in the social housing market by offering sustainable housing incorporating such features as the use of timber from sustainable sources. The consideration of sustainable construction has also allowed Carillion to enter into new partnerships, for example Carillion is a partner company with CIRIA in a new project which will be trailing the use of sustainable indicators in construction projects.

CONCLUSIONS
The implementation of an EMS in all Business Groups and Units has allowed Carillion to identify, measure and improve its environmental performance. The success of implementation has been facilitated by integrating new procedures with existing quality and safety management systems, the development of flexible tools such as the Project Management Plan and by a strong commitment to training. The next challenge for Carillion is to use the EMS to deliver sustainability for the simple reason that a construction business that has implemented systems to protect the environment effectively and uses natural resources prudently, while recognising the needs of everyone affected by its activities, will be well placed to succeed in the coming millennium.

REFERENCES
BS EN ISO 14001: 1996.   Environmental Management Systems - Specification with Guidance for use. British Standards Institution.

Eco-Management and Audit Scheme. European Commission DGXI.E.1

DETR (1998).   Opportunities for Change - Consultation Paper on a UK Strategy for Sustainable Development.

DETR (1999). A Better Quality of Life - A Strategy for Sustainable Development for the UK.

Tarmac plc (1999).   The Fourth Report of the Independent Environmental Advisory Panel - Tarmac in the Environment, 1998/99.

Recycling and Reuse of Waste Materials

Re-use of Coal Fired Power Station Sites

Hugh Brocklebank, Mott MacDonald, Birmingham, United Kingdom and EurIng **Peter Smith,** Mott MacDonald, Croydon, United Kingdom.

INTRODUCTION

Redevelopment and re-use of old coal fired power station sites involves overcoming a spectrum of environmental and engineering problems to ensure that land is brought back into use, in a safe and economic manner, taking full account of environmental and planning considerations. Burning of coal for generation of electric power results in the production of substantial quantities of pulverised fuel ash (PFA), which are left as unwanted waste by-products, and are handled and stored on the land either in the form of slurry in lagoons or tipped in stockpiles. Other potential hazards from redundant power stations include made ground, buried foundations, asbestos, coal storage areas, railway sidings and workshops. Redevelopment of coal fired power station sites requires a clear understanding of the nature of PFA materials and knowledge of their deposition across the site to enable the waste by-product to be brought back into use.

This paper reviews some previous studies on the geotechnical and environmental impacts of PFA at former coal fired power station sites. It presents site specific data for the former power station site at Hams Hall, located near Birmingham, United Kingdom. Redevelopment of the three old coal fired power stations at Hams Hall, which formed the overall complex, involved construction of infrastructure for new distribution and manufacturing industries, with the provision of rail access and creation of a separate freight terminal.

REVIEW OF ENGINEERING PROPERTIES

Previous studies on PFA have been carried out to assess engineering properties, assessing the use of PFA to act as an engineered fill for infrastructure improvements.

Results at Hams Hall

A review of engineering properties of 'lagoon PFA' based on laboratory tests on disturbed samples taken from the Hams Hall power station site was undertaken by Powrie(1995). Laboratory testing included particle specific density, particle size distributions, coefficient of permeability, one dimensional consolidation tests and consolidated undrained triaxial shear tests. The results were found to lie within the ranges encountered at other PFA sources, despite the fact that all the PFA samples arose from a variety of sources of coal deposits. Of particular interest at Hams Hall is the established variability in compaction characteristics as follows:

Maximum dry density:	1.85 to 2.18Mg/m^3
Optimum moisture content:	25% to 35%
Particle specific gravity:	1.85 to 2.18

Geoenvironmental engineering, Thomas Telford, London, 1999, 531–538

Earthworks Classification Tests at Other Sites
Test results on PFA at North Earley Power Station, Reading, (Atherton 1987) indicated that for 80% of the samples that the optimum moisture content test results lie between 32% and 42% with an average of 37%. Compacted fill generally had CBR at 5% or better with settlement under a uniformly loaded plate at $200kN/m^2$ being between 5.1mm and 6.7mm. In-situ density in these circumstances was reported to vary between 91% and 96% of the maximum dry density. With moisture content recorded as increasing in PFA by 10% during light drizzle, the results demonstrated the highly sensitive nature of PFA; moreover the material quickly becomes unsuitable for compaction until it is allowed to dry out in suitable weather conditions.

In-situ Field Trials
Hydraulically placed PFA within old clay pits near Peterborough (Humpheson 1992) extending to depths of between 10m and 20m with a high long term water table was investigated to establish the feasibility of building developments on PFA filled areas. Investigations included field trials using full size house raft foundations, pad foundations and strip foundations. Back analysis of the settlement measurements over a 5 month period found that the calculated stiffness values, expressed in terms of drained Young's Modulus E' ranged between 6 and 45 MPa, with the majority of values between 18 and 38MPa. These results demonstrated that ground movements are expected to be less than 12 mm at net bearing pressures between 42kPa and 52kPa, typical for low-rise development.

PREVIOUS GROUNDWATER POLLUTION STUDIES
In relation to environmental impacts previous studies have focussed on the potential implications for groundwater pollution from leached PFA materials. The chemical composition of PFA is affected by the mineral content of the coal that is used in the combustion process. On average trace elements within coal are equivalent to the average concentration in the earth's crust. After the combustion process, which does not destroy these elements, the ash contains the same type of elements as those present in the source coal, but consequently at higher concentration.

Types of Previous Studies
Research on the impact of ash on groundwater systems has resulted in two main types of studies:
a) field studies to establish changes in the depth profile of in-situ ash due to the influence of the weathering process and
b) laboratory experiments to assess the influence of infiltration of water through the ash and consequent leachate generation.

In-situ Depth Profile Testing
A field study carried out by Lee(1995) tested samples of PFA from a 20 year old mound adjacent to Drax Power Station, East Yorkshire to determine changes in the chemical composition in porewater as a function of depth. Most of the anions and cations in the porewater showed increases in concentration with increasing depth with the pH being near neutral. This confirms that PFA is reactive with porewater and leachate forms on

contact with infiltrating water. The dominant anion in porewater was found to be sulphate (SO_4^{2-}) and the most concentrated trace element was boron (B). Chemical concentrations of boron (B), chromium (Cr), Nickel (Ni), arsenic (As) and lead (Pb) increased with depth indicating that infiltration of rainwater and natural weathering processes contribute to the formation of leachate within the ash. This study provides field evidence of the generation of leachate in PFA, and the potential for the leachate to enter surface water or percolate into groundwater if plausible pathways exist. It should be noted that dilution is always likely to occur as the leachate emerges into surface water or groundwater.

Laboratory Leaching Studies

Laboratory studies have investigated the effect of changing pH of the leaching medium. For example Karuppiah(1997) investigated the concentration of metals leached from coal combustion ash. Metal concentrations were highest from ash leached with hydrochloric acid (pH=4) and significantly lower when leached with acetic acid (pH=7). Lee(1997) tested leachate water from an ash mound and measured high concentrations of calcium (Ca), sodium (Na) and sulphate (SO_4^{2-}), reflecting the chemical characteristics of the PFA. Boron, lithium and arsenic were the important trace elements, recording the highest concentration of 22.3, 24.6 and 16.9 mg/l respectively. These studies show that there is potential for leachate formation as porewater passes through the ash. The degree of acidity of the infiltrating water has been found to be important in controlling the concentration of elements in the leachate.

POWER STATION DEMOLITION

Experience gained by others during the demolition of redundant power station sites shows that careful control is required to maximise recovery of materials and to prevent pollution and nuisance to third parties. Clearly there is potential for recovery and recycling of structural metal material and to crush concrete and brick rubble for re-use as granular fill in civil engineering and infrastructure works such as occurred at Connah's Quay coal fired power station, North Wales (Altoft 1996). Unfortunately, if demolition is carried out in an unplanned manner, there is the potential for mixing of normally recoverable material with hazardous material, leading to unnecessary costs to recover re-useable materials or the disposal of more contaminated material. Hazardous materials that can be encountered during demolition include asbestos from station boilers, pipe lagging and buildings; oils with polychlorinated biphenyls (PCB) content when used as dielectric fluid; arsenic-salt or creosote impregnated timbers from the cooling towers; coal stockpiles and other solvents and fuels in storage and workshop areas.

DEVELOPMENT AT HAMS HALL

The Powergen site at Hams Hall comprised three coal fired power stations that had generated electricity over a period of some 65 years with rail connections from the Birmingham/Nuneaton railway line. A decision was taken to develop approximately 350 acres of the land for distribution and manufacturing purposes with the provision of rail access to the commercial sites and a separate freight terminal to provide a major new national distribution part in the West Midlands. This involved providing new infrastructure and improved access to the development and gaining planning consent to enable individual plots of land to be released for general and light industrial/commercial development.

DESIGN PROCESS
Development proceeded in a staged manner to enable infrastructure to be developed gradually in concert with a planned release of sites for development.

Desk Study and Preliminary Investigation
At an early stage in the project it was essential to collect and collate available information on the previous use and condition of the land. A comprehensive geo-environmental desk study was undertaken with an initial walkover survey to provide essential up to date information for planning application purposes. This study was supported with a preliminary site investigation to focus on environmental factors pertaining to the site.

Development of Suitable Remediation Strategies
Chemical laboratory test results were interpreted using qualitative risk assessment against recognised UK guidance criteria to enable remediation strategies to be developed and to overcome specific human health and environmental hazards. Depending on the client's redevelopment requirements and the nature of the ground, plots were either sold or completely remediated before sale. Extensive consultation was undertaken with Local and Regulatory Authorities to ensure a clear and consistent redevelopment throughout the investigation and design process.

Detailed Design
Remediation strategies were incorporated into Contract Documents for each phase of reclamation with programmes of monitoring and validation incorporated as necessary. The staged remediation and reclamation was cost effective and ensured that full consideration was taken of environmental and planning matters.

GEO-ENVIRONMENTAL FINDINGS
All of the problems typically encountered during the remediation of other coal fired power stations were encountered at the Hams Hall site. Because the site had been previously decommissioned and largely demolished to ground in a somewhat haphazard manner, it was clear that there was a risk that some hazardous material had not been fully addressed and that old foundations and underground service corridors were still in existence. In addition, historical research showed that there were some specific problematical areas on site, including an old licensed landfill site, an asbestos tip, coal stock-yards and a fire practice area.

FINDINGS OF CHEMCIAL TESTING ON PFA
Comprehensive chemical laboratory testing was undertaken on samples of PFA and other materials present on the Hams Hall site as part of the detailed validation of the remediation process. Results of the tests on 54 soil samples to determine leachate properties using the National Rivers Authority test method are detailed in Table 1. Chemical concentrations of most results are within accepted leachability trigger concentrations (NRA proposed, pre-1995), with sulphate and arsenic concentrations

being significantly above. As was expected from similar previous studies these results show that there is potential for leachate to form under the acidic test conditions At Hams Hall the in-situ soil pH of PFA, which averaged 9.35 with 95% of the results falling in the range 8 to 11, would tend to reduce metallic dissolution into the leachate compared to the laboratory pH conditions. Potential environmental risks, however were mitigated by careful evaluation of site hydrogeological conditions and by positioning compacted PFA wherever possible in hydraulic isolation from underlying aquifers and surface water channels to eliminate significant pathways. With industrial developments, provision of buildings and hard surfaces inevitably reduce infiltration of rainwater through PFA, substantially reducing potential for leachate generation. A programme of monitoring was initiated to validate the risk assessment procedures and to demonstrate that the adopted construction poses no future significant environmental risk.

FINDINGS OF ADDITIONAL GEOTECHNICAL TESTING ON PFA
Moisture Content
Typically PFA comprised a fine sandy silt and is 6m to 8m in thickness in the lagoon areas. Moisture content ranges from 30% to 40% near the upper surface and increases with depth to between 50% and 70% near the base of this material. Being a hygroscopic material, PFA will only release moisture slowly in a lagoon environment, whereas PFA excavated and stockpiled in drying weather conditions will release moisture readily during processing.

In-situ Strength
From standard penetration tests carried out in uncompacted PFA, N values were found to be predominatly in the range of 1 to 5, indicating a relative density varying from very loose to loose, with occasional dense and cemented zones also present. This type of test is not entirely reliable in silt sized deposits because pore pressures could result in the strength being underestimated as a result of dynamic disturbance during the test. A more reliable estimate of undrained strength is obtained using static cone penetration tests, which measures end resistance and frictional resistance whilst a cone is advanced at a continuous and slow rate. Tests carried out in lagoon ash placed in 1992 gave a typical range in cone end resistance of between 0.8MPa and 3.2MPa. Lower values of cone end resistance occur locally, at the western end of the lagoon, which is nearer to the source of the delivery of the PFA. Occasional thin layers with higher cone end resistance, between 4MPa and 19MPa could reflect the presence of desiccated zones or cemented zones within the PFA sequence.

Compaction Properties
Compaction characteristics of PFA vary considerably within the single source of material tested, due to variation in particle size distribution, moisture content and specific density. Experienced gained at Hams Hall for compaction of these materials has demonstrated that when PFA is handled carefully and in the appropriate weather conditions, it can be compacted to achieve an end product specification of 95% of the maximum dry density. Under these circumstances it is possible to achieve California Bearing Values (CBR) in the range of 5% to 12%.

Site Control of End Product Specification
Sometimes there is difficulty in interpreting the results of in-situ dry density tests compared with maximum dry density values. Results of density tests can fall outside the acceptance criteria, even though visual observations and other tests such as in-situ CBR and plate load tests indicate sound structural fill has been placed.   The key to understanding this scenario is acceptance that values of maximum dry density vary with optimum moisture content and specific gravity.

GEO-ENVRIORNMENTAL RISK REDUCTION MEASURES
Table 3 illustrates the variety of remedial techniques adopted to reduce risks to levels acceptable to both Regulators and future investors. The hazards identified essentially fall into categories including PFA re-use in earthworks, re-use of material excavated from buried concrete foundations and the management of hazardous materials such as asbestos, landfill and hydrocarbon spills.   Techniques adopted to address engineering and environmental issues relied upon integrating waste minimisation into the design process. This approach, developed at an early stage in the project, ensured the economic re-use of available materials on site, considered with minimal negative environmental impacts from the importing and exporting of large quantities of materials to the site. This strategy required classifying waste sources, adopting appropriate segregation processes and splitting the waste stream into those for recycling, or off site disposal, or on site containment and on site treatment components as illustrated in Table 4.

VALIDATION TESTING ON COMPACTED PFA
After completion of general earthworks and associated earthworks control tests, a further programme of specific validation testing was undertaken to demonstrate the adequacy of engineering properties of the compacted PFA. Results summarised in Table 2 show that there was considerable improvement in strength characteristics compared to the uncompacted material. Standard penetration tests have increased from a range of 1 to 5 for uncompacted PFA to an average of 21.73. Similarly static cone end resistance has increased form a range of 0.8MPa to 3.2MPa for uncompacted PFA to between 5MPa and 13.4MPa within compacted fill.   These results demonstrate that PFA can be effectively recycled as structural fill with effective environmental controls, which is particularly relevant in overseas countries reliant on coal as a vital source of energy.

REFERENCES
ALTOFT, R. G. 1996. Demolition of the Connah's Quay coal-fired power station. *Proceedings of the Institution Civil Engineers,* **114**, May, 63-72.
ATHERTON, D. 1987. Reclaimed PFA - field trials to determine compaction characteristics and a performance specification. In: *Compaction Technology*, Thomas Telford, Paper 11, 139-162.
HUMPHESON, C., SIMPSON, B. and ChARLES, J.A. 1992. Investigation of Hydraulically placed PFA as a foundation for buildings. *Ground movements and structures: Fourth International Conference, University of Wales College of Cardiff*, 68-88.
KARUPPIAH, M. and GUPTA, G (1997). Toxicity of and metal in coal combustion ash leachate. *Journal of Hazardous Materials*, **56**, 53-58.

LEE, S and HAHN, J. (1997). Geochemistry of leachate from fly ash disposal mound. *Journal of Environmental Science and Health, Part A: Environmental Science and Engineering & Toxic Hazardous Substance Control*, **32**, 649-669.

LEE, S and SPEARS, D.A. (1995). The long term weathering of PFA and implications for groundwater pollution. *Quarterly Journal of Engineering Geology*, **28**, S1-S15.

POWRIE, W., GOURVENEC, S. M. and MUNDY, N.C. 1995. Geotechnical properties of pulverised fuel ash. *Proc. 11th European Conference on Soil Mechanics and Foundation Engineering, Copenhagen*, Danish Geotechnical Society Bulletin 11(2), 2.113-2.120.

Table 1 Summary of Leachate Test Results on PFA

Parameter	Minimum concentration (μg/l)	Maximum concentration (μg/l)	Upper Tame Catchment Trigger concentration (μg/l)	Samples exceeding trigger value (%)
Arsenic	Less than 50	380.00	50	20.4
Cadmium	Less than 5	18.60	5	1.9
Chromium	Less than 250	350.00	250	3.7
Copper	Less than 50	130.00	100	1.9
Mercury	Less than 0.1	0.30	2	0.0
Nickel	Less than 50	240.00	200	1.9
Lead	Less than 5	60.00	250	0.0
Zinc	Less than 500	360.00	500	0.0
Boron	Less than 5	4100.00	Not provided	Not assessed
Sulphate	Less than 50	421,000.00	250,000	16.7

Table 2 In-situ Geotechnical Tests on Compacted PFA

Test Type	Test Result
CBR	CBR greater than 5%
Standard penetration test	N value 7 to 50, mean 21.73
Static cone penetration test results	Cone end resistance generally between 5MPa and 13.4MPa
Plate load tests with a 600mm diameter plate	Under an applied load of 120 kN/m^2, plate settlement ranged between 1.55mm and 3.66mm

Table 3 Geo-Environmental Risk Reduction Measures

Hazard	Risk Reduction Measures	Monitoring and Validation
PFA lagoon	Sealing and infilling of empty lagoon with compacted fill Compaction of PFA and general fill Capping with landscape cover	Groundwater level and quality monitoring Ground movement monitoring with magnetic extensometers In-situ CBR, plate load validation tests
PFA stockpiles	Screening and drying Capping with landscape cover	In-situ CBR, plate load, SPT and static cone penetration tests
Under ground structures	Excavation, crushing and screening of concrete for re-use as granular fill	Particle size and 10% fines validation tests
Asbestos Tip	Licensed on-site containment with 2m cover and passive gas venting	Gas monitoring
Fly-tipped Asbestos	Excavation and off-site disposal in licensed landfill facility	Trial pits to validate removal of fly-tipped asbestos
Railway sidings	Containment of slightly contaminated material within landscape bunds	Ex-situ treatment and on-site testing to select material for landscape bunds
Railway workshop	Excavation and disposal of hydrocarbon contamination in a licensed landfill facility	Phased investigation and validation of removal of hydrocarbon contaminated soil
Landfill sites/fire practice area	Excavation and disposal of contamination in a licensed landfill facility	Chemical testing to validate removal of contaminated soil Gas monitoring

Table 4 Integration of Waste Minimisation into the Design Process

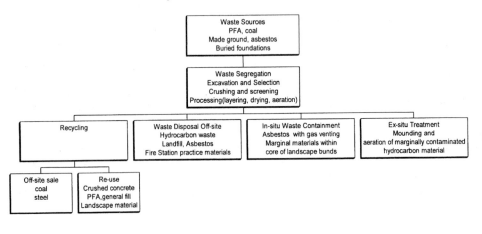

The restoration of a former coal fired power station into a coastal parkland

S.D.Lawrence, BSc (Hons), MSc, MCIWEM and I.W.Flower, BEng, MICE, MHIT, Mott MacDonald, Cardiff, UK.

ABSTRACT

During December 1996 Mott MacDonald were commissioned by Carmarthenshire County Council's Millennium Coastal Park Project Team to design a scheme which would rejuvenate the site of the former coal fired Carmarthen Bay Power Station. The former station is situated adjacent to the harbour town of Burry Port, Carmarthenshire, Wales, UK and the Design Brief was to restore this former industrial area into a woodland section of a 13.2 miles (22km) Coastal Park which is presently being created to celebrate the coming of the new Millennium. The Park is being designed to encourage increased public patronage of the area through the creation of water, landscape and ecological features. The underlying core strategy throughout the entire design, construction and maintenance of the Park is one of sustainability.

This paper discusses the sustainable re-use of contaminated materials found at the Site, these include pulverised fuel ash (pfa) and arisings from the demolition to the ground level of the former power station turbine hall. Further contaminated materials included asbestos, from pipe lagging and general insulation, and heavy metals from locally derived slags and coals. The work involved the investigation of the geotechnical and geochemical properties of the on-site materials. In designing suitable on-site engineering solutions the use of a Dutch derived risk assessment model *Risc Human* (1995)[1] was used.

INTRODUCTION

During December 1996 a team lead by consultants Mott MacDonald were commissioned by Carmarthenshire County Council to take on the role of Design Engineer in order to rejuvenate a coastal industrial area in South Wales, UK. The Burry Port Site lies approximately seven miles to the west of Llanelli, Carmarthenshire at Ordnance Survey Grid Reference SN 4601. The Site is set overlooking the internationally recognised Loughor Estuary (SSSI, Ramsar, SPA, cSAC), covering approximately 250ha, lying on a flat coastal strip with a 3km shoreline. The boundaries of the Site are defined by the Loughor Estuary to the south and by the town of Burry Port and a main road (A484) to the north. To the west the Site includes the former Pembrey Harbour, and it extends eastwards to the village recreational grounds at Pwll. Areas of residential housing are excluded from the Site, as is the main Cardiff-Fishguard rail line that bisects the Site.

The main features of the Site are the two former industrial harbours built during the last Century, Burry Port and Pembrey, which have local historical interest for the exportation of coal and metal refining processes located around the harbour. Pembrey Harbour has become entirely silt filled and is no longer able to function as a harbour. Instead, it has become a haven for wildfowl and an attraction for holidaying members of the public. Burry Port Harbour operated as a commercial port until 1951, but is also susceptible to siltation which requires periodic dredging to allow the harbour to continue to function as a haven for leisure boats and small fishing vessels. The former Carmarthen Bay Power Station site dominates the majority of the Site. The turbine hall was demolished during 1989 creating a pond approximately 200m in length and 30m wide; the Station's remaining stockpiles of pulverised fuel ash (pfa) were grassed, and the remaining overhead power lines were removed, or re-routed where necessary, under a separate contract during 1997.

When completed the Burry Port Site will form part of Carmarthenshire County Council's Millennium Coastal Park extending some 22km from the Loughor Bridge in the east to Pembrey Country Park in the west. With a value of £29m, the works are being undertaken with the support of the Welsh Development Agency (WDA) and involve the transformation and creation of lakes, gardens, woodlands and other landmark features. Furthermore, the project has received 50 per cent funding from the Millennium Commission.

Burry Port Site Proposals

In 1995, an application to the Millennium Commission for the Park as a whole included a proposal for the harbour area to become a key entrance to the Park, with leisure and recreational facilities. The former turbine hall area is to become a wildlife water habitat, and the pfa stockpile areas landscaped for walkers, cyclists and equestrians which is to include an earth sculpture some 150m in length and 15m in height moulded in pfa.

DESK STUDY

Initially, the design team published a Desk Study. The Study investigated the following key aspects:

site history;
geology;
mining;
hydrology;
previous site investigations;
landscape appraisal;
ecological appraisal;
opportunities;
constraints;
a planning appraisal;
a market appraisal; and,
a contract procurement strategy.

The Desk Study identified the possibility of using site derived pfa and harbour silt as a medium for vegetative growth. Therefore, amongst other actions, the Desk Study recommended that a geotechnical and geochemical invasive site investigation be undertaken, and that growth trials upon the pfa and silt should be undertaken.

The Desk Study reviewed previous work and identified that an earlier investigation by the Laboratory of the Government Chemist (1997)[2] discussed the presence of asbestos fibres in an area of the Site to the north of the turbine hall pond; asbestos cement board within the pond; and a licenced asbestos tip. Rumours of asbestos buried on the site abounded amongst the community.

SITE INVESTIGATION
The recommended invasive site investigation was undertaken during the months April to June 1997. The scope of the work comprised the drilling of cable percussion and rotary cored boreholes; the installation of water and gas monitoring standpipes and piezometers; trial pitting; static cone penetration testing (cpt); hand-dug pits; and the laboratory analysis of indicative samples.   This range of techniques was used to investigate the following conditions:

geotechnical ground conditions;
hydrology;
hydrogeology;
soil chemistry;
surface water chemistry;
groundwater chemistry; and,
gaseous chemistry of the soil.

The strategy of the site investigation was based upon UK Best Practice utilising such guidance documents as *Sampling Strategies for Contaminated Land* (DoE, 1994)[3], and *The Code of Practice for the Identification of Potentially Contaminated Land and its Investigation: DD175* (BSI, 1988)[4].   Proposed end use of individual areas of the site was intrinsic to the sampling strategy adopted.   Therefore, to aid in the designing of the sampling strategy the Site was sub-divided into Zones.   These Zones delineate the former industrial uses of the Site as identified within the Desk Study.

Site Investigation Findings

Geological
The site investigation identified the following strata types (with approximate depths below ground level) on Site:

0-10m	Made Ground comprising	a) granular material including brick, ash and slag
		b) pulverised fuel ash (pfa)

The Desk Study identified the presence of a former Control of Pollution Act (CoPA, 1974)[5] licensed asbestos disposal facility. The facility showed signs of degradation.   The original CoPA (1974)[5] licence stipulated that the disposal area was to be covered with a concrete slab engraved with "ASBESTOS BURIED HERE." The requirement was later modified to 1m of soil cover.   The area was investigated with a cpt rig, a concrete cap was not located.

The materials encountered were tested for their engineering properties. Laboratory tests were performed, in addition to field tests and observations, including moisture content determination, one-dimensional consolidation, undrained shear strength and compaction. Particular attention was afforded to the possibility of using on-site pfa and granular Made Ground as a resource.

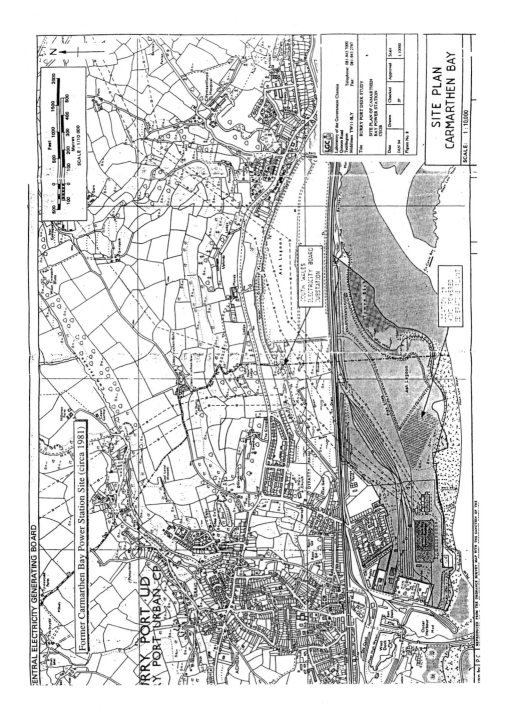

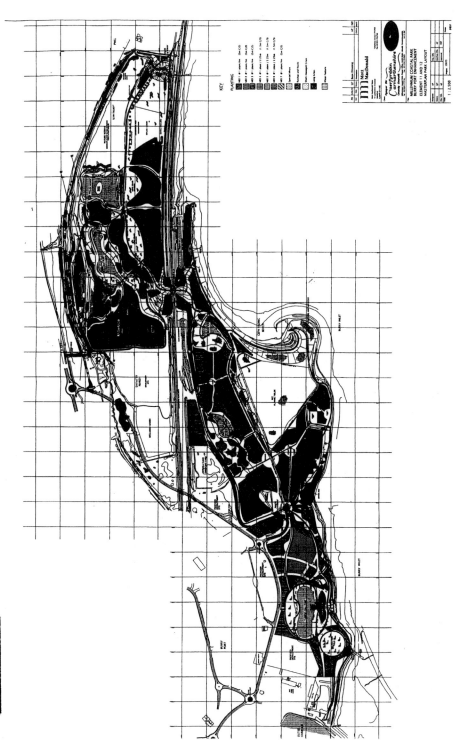

Proposed Masterplan

The diagnosed parameters for pfa are summarised in the following table:

Test/Method	Parameter Tested	Units	Results		Adjectival/Comments
			Range	Mean	
Standard Penetration Test	Relative density	'N' values	1-12	3	Very loose to medium dense
Particle Size Analysis	Particle size	n/a			Silty sand to sandy silt
Compaction Test	Optimum water content for compaction	%	7.9-33.0		
	Maximum dry density	Kgm^{-3}	850-1230		
Moisture Content Test	Moisture content	%	18-85	41	Above optimum for compaction
Atterberg Limit	Plasticity	n/a			Non-plastic
Particle Density Test	Specific gravity	Mgm^{-3}	1.68-2.20		Low by comparison with previously published figures

Previous experience with pfa indicated that the effective angle of friction was between 30-36°.

The lithology distinguished as coarse granular Made Ground varied somewhat in its physical properties:

Test/Method	Parameter Tested	Units	Results		Adjectival/Comments
			Range	Mean	
Standard Penetration Test	Relative density	'N' values	4-50	18	Very loose to very dense
Particle Size Analysis	Particle size	n/a			Mixed fine to medium sand and coarse gravel with cobbles and boulders

Geochemical

Following the execution of a sampling strategy and laboratory analysis data sets for each Zone were assessed against the following U.K. and global guidelines:

ICRCL Guidance Note 59/83 (1987)[6];
Revised Dutch Guidelines (1996)[7];
BRE Digest 363 (1996)[8];
Welsh Office Circular 16/89 (1989)[9]; and,
Institute of Occupational Medicine (1988)[10].

To evaluate each data set the mean and standard deviation of each contaminant's concentration was calculated. The data was assumed to be normally distributed, therefore at a calculated concentration value of two standard deviations above the mean it was considered that 95% of the site material was below this value:

$$\bar{x} + 2 \cdot \sqrt{\frac{n \sum x^2 - (\sum x)^2}{n(n-1)}}$$

Where

x : contaminant concentration

n : size of data set

Land associated with pfa was found to be above guidance concentrations with the following contaminants:

arsenic (two standard deviations above the mean, 207.5mg.kg^{-1});
nickel (two standard deviations above the mean, 341.1mg.kg^{-1}); and,
zinc (two standard deviations above the mean, 782.7mg.kg^{-1}).
The pH of the pfa was found to be alkaline, with a mean of pH 8.4.

The following contaminants within the granular Made Ground were also found to be above guidance concentrations:

arsenic (two standard deviations above the mean, 1172.6mg.kg^{-1});
cadmium (two standard deviations above the mean, 19.2mg.kg^{-1});
lead (two standard deviations above the mean, 3650.7mg.kg^{-1});
copper (two standard deviations above the mean, 2210.0mg.kg^{-1});
nickel (two standard deviations above the mean, 792.5mg.kg^{-1}); and,
zinc (two standard deviations above the mean, 49468.6mg.kg^{-1}).
The granular Made Ground was found to be alkaline, due to the presence of concrete fragments, with a mean of pH 8.3. Also, asbestos fragments were found to be present.

Gaseous and groundwater data from both the pfa and granular Made Ground were not considered to be contaminated.

Growth Trials

Vegetative growth trails undertaken by ERA, (1998)[11] where undertaken upon the pfa and harbour silt. Pot samples under glass of the pfa and silt demonstrated good germination of Perennial Ryegrass (*Lolium perenne*). Increased growth was noted with the application of water and an NPK fertiliser (15:15:15). Further external site trials on harbour silt confirmed these observations.

INTERPRETATION OF THE DATA

The intrinsic factor in the interpretation of results was the proposed end-use as public amenity land which included open spaces, pathways, cycle ways and a camp site.

Geotechnical

For pfa, the greatest discrepancy between the geotechnical analytical results and the requirements for a suitable engineering material existed in the moisture content. The optimum moisture content for compaction (7.9-33%), which was found to be considerably lower than the in-situ moisture content (18-85%). Additionally, a characteristic property of pfa is its low density. These properties are useful in situations where construction is on a low load-bearing capacity material, for example as a landscaping material. Unfortunately, these characteristic meant that the pfa has a low trafficability reducing its suitability as a footpath and cycleway material. The granular Made Ground material was considered to be good engineering fill material.

Geochemical

The Dutch derived environmental risk assessment model *Risc Human* (1995)[1] formed the cornerstone of the contaminative risk assessment process. The computer based model allowed the simulation of the site investigation data with regard to human risk via the following pathways:

the ingestion of soil;
the inhalation of soil;
dermal contact of soil;
the inhalation of indoor air;
the ingestion of drinking water; and,
the inhalation of vapours.

The model calculated the Lifelong Average Human Daily Uptake (LAHDU) associated with each chemical. The LAHDU is the sum of the uptake of the chemical by the applicable pathways. The model allows the comparison of the LAHDU with the Maximum Tolerable Risk (MTR), a value which is provided by the model and is indicative of the concentration of a chemical, based upon body weight for oral exposure, which can be daily ingested by a human during an entire lifetime without the prospect of adverse effects on health.

*pfa (**all pathways permitted***)

Chemical	Concentration $(mg.kg^{-1}dry\ soil)$	LAHDU $(mg.kg^{-1}d^{-1})$	MTR $(mg.kg^{-1}d^{-1})$
Arsenic	2.08×10^2	3.15×10^{-4}	2.10×10^{-3}
Nickel	3.41×10^2	5.18×10^{-4}	5.00×10^{-2}
Zinc	7.83×10^2	1.19×10^{-3}	1

Following the application of the model to the site derived pfa data it was concluded that the material did not exhibit a chemical risk with regard to the proposed site end use.

Granular Made Ground (**all pathways permitted**)

Chemical	Concentration (mg.kg^{-1}dry soil)	LAHDU (mg.kg^{-1}d^{-1})	MTR (mg.kg^{-1}d^{-1})
Arsenic	1.17×10^3	1.78×10^{-3}	2.10×10^{-3}
Nickel	7.93×10^2	1.20×10^{-3}	5.00×10^{-2}
Zinc	4.95×10^4	7.52×10^{-3}	1
Cadmium	1.92×10^1	2.92×10^{-5}	1.00×10^{-3}
Lead	3.65×10^3	5.55×10^{-3}	3.60×10^{-3}
Copper	2.21×10^3	3.36×10^{-3}	1.40×10^{-1}

The model demonstrated that the granular Made Ground, with particular regard to lead and arsenic concentrations, should be considered as a risk to humans.

Granular Made Ground (**with the following pathways removed: the ingestion of soil and dermal contact with soil**)

Chemical	Concentration (mg.kg^{-1}dry soil)	LAHDU (mg.kg^{-1}d^{-1})	MTR (mg.kg^{-1}d^{-1})
Arsenic	1.17×10^3	1.11×10^{-5}	2.10×10^{-3}
Nickel	7.93×10^2	7.52×10^{-6}	5.00×10^{-2}
Zinc	4.95×10^4	4.69×10^{-4}	1
Cadmium	1.92×10^1	1.82×10^{-7}	1.00×10^{-3}
Lead	3.65×10^3	3.46×10^{-5}	3.60×10^{-3}
Copper	2.21×10^3	2.10×10^{-5}	1.40×10^{-1}

Modelling the granular Made Ground with the removal of the pathways of soil ingestion and dermal contact it was demonstrated that this material could be considered as a low risk to human health.

SITE INVESTIGATION DISCUSSION
The pfa was identified as being suitable for re-use within landscaped features, including the Earth Sculpture, and as a cover material. The pfa and harbour silt were considered suitable growing medium following the application of water and NPK fertiliser. The main concerns were that the pfa had a high moisture content and a low density which would prohibit a high trafficability; and its friable nature which when dry would create dust problems.

The granular Made Ground was considered to be a good engineering fill material, but would require a capping layer to prevent the human ingestion and dermal contact due to the lead and arsenic concentrations identified, and the presence of asbestos.

SITE INVESTIGATION CONCLUSIONS

Adopting a sustainable approach in utilising site derived materials was concluded that the pfa is suitable for landscaping. Beneath the pfa formed Earth Sculpture granular Made Ground should used as a drainage material: the pfa in this application would afford the necessary cover. Prior to movement of the granular Made Ground site testing and visual inspections for asbestos must be made. Additionally, during movement the material must be kept damp to suppress the risk of asbestos fibres becoming airborne.

The former asbestos disposal site could be capped with pfa underlain by a drainage blanket composed of granular Made Ground. It was concluded that the model demonstrated that pfa could be utilised as a growing medium, and that the growth trials (ERA, 1998)[11] also indicated that the harbour silt would support plant growth. Due to the contaminants present at the site the vegetation grown should not be animal grazed or suitable for human consumption. It was concluded that groundwaters percolating through the pfa and silt would dissolve metal salts from the upper layers which would increase the material's capacity to support vegetation. To allow this salt movement to occur a period of rainfall would be required prior to planting. The application of fertiliser could be undertaken in the form of sewage cake.

CONSTRUCTION

The anecdotal evidence of asbestos cement board in the pond meant that removal of material from the pond was likely to result in the need to dispose of asbestos contaminated material off site. To avoid this, a pond was designed by TACP, landscape architects, which avoided material removal, but reprofiled the pond base to provide a combination of steep and shallow pond sides to a depth of 2m to ensure that the freshwater did not stagnate. The base of the pond was found to be concrete, this was blinded with locally derived sand before it refilled from the adjacent catchment area.

Granular Made Ground

The area to the north of the pond, where an earlier investigation undertaken by the Laboratory of the Government Chemist (1997)[2] had identified asbestos, was marked out and was not excavated. This area was capped to a depth of 1m with site derived inert fill. In other areas around the pond further tests were undertaken using a site based scientist to verify the absence of asbestos. These areas were excavated to a depth of up to 2m to recover mass concrete. This was then crushed to provide graded course material for elsewhere on the site.

pfa

The pfa was excavated over large areas of the site on a cut and fill basis to provide extension ponds for wildlife, and to create a topography designed by the landscape architect. Approximately 500,000m^3 was excavated, and used in part for the construction of an earth sculpture designed by Richard Harris, a local sculptor. Material was also exported to a neighbouring low lying area, adjacent to the Park, which is designated as a site for an office development in the future. Crushed arisings from the power station were used beneath both the earth sculpture and development site to act as a drainage layer to assist in the control of settlement. Additionally, 180,000m^3 of pfa was exported to an adjacent contract to act as a cover to a landbridge over the railway which bisects the site.

Inert growing material was in short supply. The importation of 18,000m^3 of subsoil from a site 10 miles away has provided cover for 16ha of the parkland that is destined to be grassland and meadow. Cell grown trees have been planted in litter pits, filled with mixture of arisings, mushroom compost and sewage cake from a local sewage works; or bare rooted in pfa which has been ripped in two directions and dressed with sewage cake. Finally, in the vicinity of the former power station, material from the harbour has been dredged and spread in layers of approximately 600mm thickness prior to planting.

ACKNOWLEDGEMENTS
The authors wish to acknowledge Carmarthenshire County Council's Millennium Coastal Park Team's support in the production of this paper. The comments and conclusions made are entirely those of the authors.

REFERENCES
[1]   M. de Boer et al., (1995). Van Hall Institute. *Risc Human Version 2.01.*

[2]   Laboratory of the Government Chemist, (1997). *Chemical Contamination Site Investigation Report. Carmarthen Bay Power Station Site. Burry Port, Carmarthenshire. May 1996. For Carmarthenshire County Council.*

[3]   DoE (Department of the Environment). Contaminated Land Research Report No.4, (1994). *Sampling Strategies for Contaminated Land.*

[4]   BSI (British Standards Institution). DD175, (1988). *Code of Practice for the Identification of Potentially Contaminated Land and its Investigation.*

[5]   HMSO (Her Majesty's Stationary Office), (1974). *Control of Pollution Act.*

[6]   ICRCL (Interdepartmental Committee on the Redevelopment of Contaminated Land). Guidance Note 59/83, (1987). *Guidance on the Assessment and Redevelopment of Contaminated Land.*

[7]   Dutch Ministry of Housing, Spatial Planning and the Environment, (1994). *Intervention and Target Values - Soil Quality Standards.*

[8]   BRE (Building Research Establishment). BRE Digest 363, (1996). *Sulfate and Acid Resistance of Concrete in the Ground.*

[9]   Welsh Office. Circular 16/89, (1989). *Water and the Environment.*

[10] Institute of Occupational Medicine, (1988). *The Release of Dispersed Asbestos Fibres from Soils.*

[11] Environmental Research and Advisory Partnership, (1998). *Millennium Coastal Park, Burry Port. Final Report - Plant Growth Experimental Trials on Silt.*

Drying of sewage sludge by using wasted edible oil under heating and decompression

S. Nakazono*, E. Nakazono*, N. Tokutome*, K. Nakaji**, K. Okano**, S. Yamanaka**, J. Chikushi**, and M. Ohtsubo**
* Prorex, Inc., Tenjin 1-13-21, Fukuoka, Japan 810-0001
**Faculty of Agriculture, Kyushu University, Fukuoka, Japan 810-8581

ABSTRACT

This paper presents the picture of a newly developed system for sewage sludge treatment, which makes it possible to reduce the volume and weight of the sludge. In this system the sludge is dehydrated using wasted edible oil as a heating medium under a reduced pressure, where the oil serves to remove water in the sludge uniformly and quickly, leading to the reduction in volume and weight of the sludge. The system consists of the following major processes: (1) pretreatment of the sludge cake, (2) dehydration of the cake in the heated oil medium (100C) under a reduced pressure (-370 mm Hg), (3) squeezing and refinement of the oil (which is reused), (4) condensation of the removed moisture, (5) deodorization of the treated sludge. The pilot plant of the system achieved the following results. Deodorizing process was attained because the sludge was treated in a closed chamber under a reduced pressure. The initial water content of the sludge cake of 400% was reduced to <6% in an hour, turning the cake into a solid material that is easy to handle. The solid products of sludge can be utilized as a useful material for agriculture since the sludge is sterilized and the organic nutrients in the sludge are retained after treatment. Wasted edible oil, posing heavy load to the aquatic environment, can be used as a heating medium of the system and as the source of heat to operate the system.

INTRODUCTION

Treatment of sewage sludge is becoming a concerned issue in Japan from the environmental point of view associated with the increased sludge production resulting from growing sewage projects. Several methods to manage sewage sludge, such as landfill, dumping to the ocean, incineration, and sintering have been employed, which however gave rise to serious problems like production of dioxin, marine pollution, and the shortage of dumping sites, etc.

This paper presents the picture of a newly developed system for sewage sludge treatment that enables one to reduce the volume and weight of the sludge, leading to utilization as soil improvement material, fertilizer, and fuel. This system is based on the drying technique of animal and seafood residue by using wasted edible oil under a reduced pressure (Nakazono, 1998). This technique has been systematized in an industrial scale, and some of processed

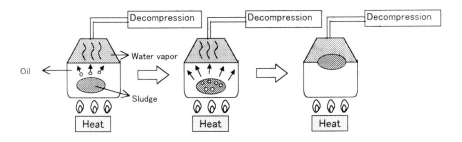

Fig.1. Principle of dehydration of sludge.

product of animal and seafood reside are commercially available for feed of the livestock and fish culture, and fertilizer.

PRINCIPLE AND FEATURES OF THE SYSTEM

The diagram of the system is illustrated in Fig.1. Sludge is dehydrated by mixing the sludge with the edible oil heated at about 100C under a reduced pressure. In the first stage, the sludge containing water, because of its smaller specific gravity than oil, settles at the bottom of the oil phase. Upon heating, moisture evaporation from the surface of sludge particles is initiated, leading to the gradual floating of the sludge. In the second stage, evaporation of the moisture from the core of sludge particles is accelerated due to a reduced pressure. In the third stage, the sludge is almost dried, floating onto the surface of the oil phase.

Thus the moisture in the sludge is uniformly dehydrated in the drying process, since the sludge exhibits fluid behavior and the heat is effectively transmitted through the sludge in the boiling process. The plant can be promptly re-operated even when an operation suddenly comes to a halt since a batch system is employed for the plant operation.

PROCESS FLOW

Pre-treatment process

The flow diagram of the sludge treatment is illustrated in Fig.2. The sludge with 400% water content (the mass of water per unit mass of dry sludge) is transported from the sludge acceptance hopper to the preheating tank by the conveyer, being mixed with oil in the tank and heated up to about 85C. The sludge thus mixed with oil exhibits the low viscosity enough to ensure fluid behavior of the sludge.

Dehydration process

Drying of sewage is performed in the dehydration unit that is the core part of this system (Fig.2). The sludge and oil are drawn from the pre-treatment tank into the dehydration unit by maintaining the unit at a reduced pressure. The composite of sludge and oil is heated by introducing the steam with a pressure of 7 kg cm^{-2} and a temperature of 169C into the mixing and heating rod, and the jacket (Fig.3). The time of 55 to 75 minutes is needed to complete one batch dehydration. Then the temperature of the composite is maintained at 85-90C under a pressure of about -370 mm Hg. The evaporated moisture is led to the cyclone where fine particulate solid is removed, being condensed by the condenser and sent to the water treatment

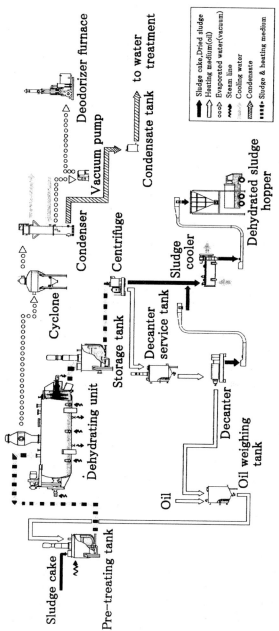

Fig.2. Flow diagram of the sludge drying system.

line. The composite of the dried sludge and the oil with a water content of about 6% is led to the storage tank.

De-oiling process
The composite of sludge and oil in the storage tank is transferred to the centrifuge, where oil is separated from the dried sludge. The separated oil is collected in the decanter service tank as the oil contains fine particulate sludge, being transported to the decanter and refined to serve for reuse. In one treatment process of the sludge, 70 kg edible oil is consumed to treat 1 ton sludge and the oil equivalent to this amount is supplied for the subsequent treatment.

Merchandizing process of dried sludge
The dried sludge is transferred to the sludge cooler, being cooled below 40C and then stored in the dehydrated sludge hopper. The water content of the dried sludge was reduced to less than 6% and the volume by 80% of the pre-drying.

Deodorization process
The diffusion of the odor in the drying process can be prevented as the dehydration unit is airtight under decompression. Non-condensable gas produced in other processes is led to the deodorizer furnace, being decomposed by burning at a temperature of about 800C.

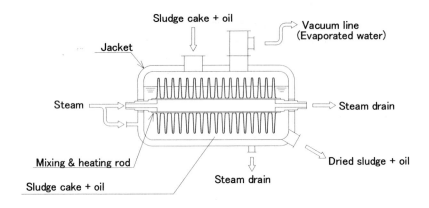

Fig.3. Diagram of the dehydration unit.

QUALITY OF CONDENSED WATER
Water quality of the condensed water collected from the vapor in the condenser (Fig.2) is shown in Table 1. The measured values were extremely high for BOD, T-N and normal hexane. However, the effect of such pollutants in the vapor on the sewage treatment line would be small since the load of these pollutants is not great compared to those of waste water flowing into the sewage treatment plant.

Table 1. Analytical results of condensed water

SS	4.9 - 28.0 (mg l^{-1})
BOD	1600 - 4200
COD	240 - 900
T-N	530 - 1900
T-P	<0.2
n-C_6H_{14}	10 - 170
Cr	N.D.
Zn	N.D.
Cd	N.D.
Pb	0.002 - 0.032
Cu	N.D.
Hg	0.001 - 0.0015
As	N.D.
pH	9.0 - 9.7

The concentrations of Pb and Hg were extremely low, and those of Cr, Zn, Cd, Cu, and As were below the limits to be detected.

OIL TYPE AND ITS DETERIORATION
The edible oil wasted in 1988 amounted to 410 k ton and only its 46% was reused for feed additives or for fat decomposer. The edible oil disposed from households, hotels, and food service industries is used as a heat medium to dehydrate the sludge in this system. Table 2 shows the properties of oil before and after drying treatment of the sludge. The acid value as an indicator of the oxidation of oil is below 20 even after treatment. This oxidation degree is below the tolerable limit of the oil for feed of livestock.

CHARACTERISTICS OF THE DRIED SLUDGE
The contents of hazardous substances including eight polycyclic aromatic hydrocarbons (PAH) and three plastic additives for pre- and post-processing are shown in Table 3. These substances contained initially in the sludge were reduced to less than 2% in the drying process (Matsui et al., 1996). This reduction would have been brought about by their transfer from the sludge to oil and gaseous phases, and by the degradation of these substances during processing. The sterilization of the sludge was also confirmed.

The sludge produced in the drying process is shown in Photo 1. The dried sludge is a granular material with the diameter of 0.4 to 2 mm, exhibiting dark-brown color. The water and oil

Table 2. Properties of wasted edible oil

	Solid fraction (%)	Acid value (mg-KOH·g-fat^{-1})	Viscocity(80 °C) (mPa·s)
Pre- treatment	0 - 0.5	4.4 - 7.6	33 - 44
Post- treatment	2.8 - 13	7.1 - 14	39 - 89

Table1 3. Comparison of hazadous substances in wasted edible oil for pre- and post-treatment

Polycyclic aromatic hydrocarbons	Pre-treatment (%)			Post-treatment (%)			
	Sludge	Oil	Total	Sludge	Oil	Drainage	Total*
Naphthalene	88.1	11.9	100	0.7	18.5	0.9	20.1
2-Methylnaphthalene	90.5	9.5	100	1	22.1	0.7	23.8
Biphenyl	95.3	4.7	100	1.2	23.3	0.7	25.2
Dibenzothiophene	78.7	21.3	100	1.2	61.8	0	62.9
Phenanthrene	60.7	39.3	100	1.3	70.8	0.1	72.2
Anthracene	88.5	11.5	100	0.9	54.5	N.D.	55.4
Fluoranthene	68.9	31.1	100	0.7	72.4	N.D.	73.1
Pyrene	59.2	40.8	100	0.7	75.3	N.D.	76.1
Plastic additives							
Butylated Hydroxytoluene	52.1	47.9	100	0.5	49.2	0.4	50.2
Dibutyl Phthalate	12.1	87.9	100	2	82.8	0	84.9
Bis Phthalate	7.3	92.7	100	0.5	117.2	0	117.7

*Percentage against the total in the pre-treatment column

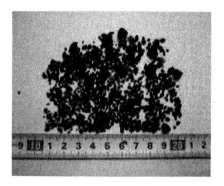

Photo 1. View of the dried sludge.

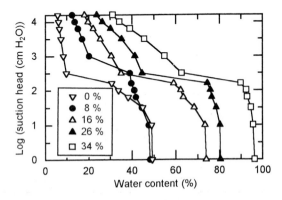

Fig.4. Suction versus water content curves for sludge added soil.

contents of the dried sludge are 6 and 30%, respectively. There is not odor peculiar to raw sludge although odor of oil remains. Because of low water content, the dried sludge is easy to handle and can be preserved for a long period under a normal temperature.

The chemistry data of the dried sludge is presented in Table 4. Most of the component of fertilizer in the sludge satisfies the recommended limits in the standard for sludge manure. The concentration of hazardous substances in the leachates obtained from elution tests and the mass concentration of heavy metals were less than the tolerable limits in the regulated standard for manure. The dried sludge therefore could be used as a soil improvement material and fertilizer.

Table 4. Chemical analysis of dried sludge

Determination	Mean value	Regulation for manure*1	Recommendation for sludge manure (Compost)*2
Component of fertilizer (%)			
C	49.2	-	-
N	4.4	-	<2(1.5)
C/N	11.1	-	>10(20)
Alkali	2.17	-	>25
Water Content	1.1	-	>100
Organic Matter	79.9	-	<35
P_2O_5	3.83	-	<2
K	0.16	-	-
pH	6	-	>(8.5)
Elution test (mg l^{-1})			
As	0.082	0.3	-
Cd	<0.01	0.3	-
Hg	<0.0005	0.005	-
Pb	<0.01	0.3	-
Organic Phosphous	<0.1	1	-
Cr^{6+}	<0.05	1.5	-
T-CN	0.038	1	-
HgR	<0.0005	N.D.	-
PCB	<0.0005	0.003	-
Content in the solid (mg kg^{-1})			
As	5.25	50	-
Cd	1.97	5	-
Hg	0.6	2	-
Cu	290	-	>600
Zn	480	-	>1800

* 1 Heavy metal standard for special manure
* 2 Recommended standard for organic manure quality
      (Notification by Ministry of Agriculture, Forestry and Fisheries, Japan)

Application of sewage sludge to the field benefits the soil not only as a nutrient but as a soil improvement material. Several studies on the effect of the sludge on the physical properties of soils have been conducted (Clapp et. al., 1983; Ekwue and Stone, 1997; Lindsay and Logan, 1998). In the present study the water holding capability of sludge-added soil was examined. Figure 4 shows the suction versus water content relationships for the soil sample incorporated with the sludge at five levels. The soil sample used for the test is a sandy soil derived from granite, with the clay, silt and sand fraction of 4, 9 and 87%, respectively. By adding the sludge to the soil sample, the suction versus water content curve shifted toward higher water contents, indicating that incorporation of the sludge to the sandy soil enhance the water holding capability.

CONCLUSION REMARKS

The processing system presented in this study is capable of reducing the water content of the sludge from initial 400% to less than 6% and its volume by 80%. The hazardous substances such as polycyclic aromatic hydrocarbons in the sludge were removed to a great extent in the drying process. The amount of heavy metals in the processed sludge was below the tolerable limits in the standard, which allows the sludge to use as fertilizer to the farmland. The research project on reuse of the sludge processed by this system started in 1980, and several patents have been acquired during this period. At present the system with sludge treatment capability of 90 ton per day is under construction in Fukuoka Prefecture, Japan. This system would contribute to reduce the space of sludge processing plant, reuse the sludge as a resource and reduce the load of wastes to aquatic environment.

REFERENCES

Clapp, C.E., S.A. Stark D.E. Clay, and W.E. Larson. 1986. Sewage sludge organic matter and soil properties. P.209-253. In Y. Chen and Y. Avnimelech (ed.) The role of organic matter in modern agriculture. Developments in plant and soil sciences. Martinus Nijhoff Publ., Dordrecht, the Netherlands.

Ekwue, E.I., and R.J. Stone. 1997. Density-moisture relations of some Trinidanian soils incorporated with sewage sludge. Transaction of the ASAE 40: 317-323.

Lindsay, B.J., and T.J. Logan. 1998. Field response of soil physical properties to sewage sludge. J. Environ. Qual. 27: 534-542.

Matsui, S., H. Yabushita, M. Shima, and H. Yamada. 1996. A study on the extraction and removal of trace hazardous substances in the sewage sludge processed with edible oil. Gesuido Kyokaishi (J. of Japan Sewage Works Association. 34; Research Journal. 17): 53-62.

Nakazono, S. 1998. Technology development for recycle of municipal waste processed using dehydration system with heated oil under decomposition. Proceedings of the 9th Annual Conference of the Japan Society of Waste Management Experts. P.366-369.

Swelling behaviour of stabilized soils

SUDHAKAR M. RAO[1], B.V.VENKATARAMA REDDY[2], M. MUTTHARAM[3]
[1]Assistant Professor, [2]Principal Research Scientist, [3]Research Student
Department of Civil Engineering, Indian Institute of Science, Bangalore 560012, India

ABSTRACT
Soil deposits stabilized with waste materials - wood-ash and organic matter (leaves, grass, etc.) exist in black cotton soil areas in North Karnataka. The wood-ash modified BC soils are apparently stabilized by hydrated lime produced by biochemical, dissolution and hydration reactions. The influence of cyclic wetting and drying on the swelling behaviour of wood-ash modified black cotton soil (BCS) and a laboratory stabilized black cotton (BC) soil are examined in this study. The laboratory stabilized BC soils were stabilized using 2 to 7% (on weight basis) hydrated lime. The study was required to assess the long-term behaviour of earth structures and foundations on stabilized soils. Experimental results showed that chemical stabilization controlled the wetting induced volume changes of the ash-modified soil specimens and laboratory stabilized BCS specimens in the compacted state. Cyclic wetting and drying however caused the ash-modified soil specimens to collapse significantly at the experimental pressures. Experimental results with laboratory stabilized BCS specimens indicated that if a soil is stabilized with inadequate lime content, then the beneficiary effect of lime stabilization is lost on cyclic wetting and drying.

INTRODUCTION
Expansive black cotton soils occur in climatic zones characterized by alternate wet and dry seasons. The expansive soils experience periodic swelling and shrinkage during the alternate wet and dry seasons. Such cyclic swell-shrink movements of the ground cause considerable damage to the structures founded on them. The influence of cyclic wetting and drying on the swelling behaviour of natural expansive soils is well-documented (Popescu 1980, Chen and Ma 1987, Subba Rao and Satyadas 1987, Dif and Bluemel 1991, Day 1994). However, the impact of cyclic wetting and drying on the swelling behaviour of chemically stabilized expansive soils has not been examined. Such a study is required to assess the long-term behaviour of earth structures and foundations on chemically stabilized soils.

During a recent survey of black cotton soil zones in Karnataka State in India, indigenously stabilized black cotton soil deposits were encountered. These modified BCS deposits ranged in area from < 0.5 to 8 hectares and varied in thickness from 0.5 to 3m and are in existence for more than 400 years. These soils are stabilized with waste materials - wood-ash and organic matter (leaves, grass, etc.). It is speculated that organic acids produced by biochemical degradation of organic matter dissolved calcite in the wood-ash and released calcium ions. Hydrolysis of calcium ions produced hydrated lime that stabilized the soil. Hydrated lime is commonly used to stabilize expansive soils (Bell, 1993). The effect of cyclic wetting and drying on the swelling behaviour of chemically stabilized soils has however not been

Geoenvironmental engineering, Thomas Telford, London, 1999, 559–564

examined. The influence of cyclic wetting and drying on the swelling behaviour of ash-modified soils and lime stabilized BCS specimens are hence examined in this study.

MATERIALS AND METHODS
Soils
Representative samples of wood-ash-modified BCS were collected at 0.5m depth from two deposits in Belgaum District in Karnataka State and are designated as ash-modified soil 1 and ash-modified soil 2 respectively. Representative samples of BCS were also collected from 0.5m depth at locations close to ash-modified soil deposits 1 and 2 respectively.

Property Evaluation
The index properties and Modified Proctor Compaction characteristics of the ash-modified and natural BCS specimens were determined for soil fractions passing 425μm sieve and the values are given in Table 1. Modified Proctor Compaction test was preferred to Standard Proctor Compaction test as the swell potentials of compacted specimens are more pronounced at higher dry density and lower moisture content (Holtz and Gibbs, 1956). The swell/collapse potentials of natural and ash-modified BCS specimens were also performed on 425μm passing soil fractions. The specimens were statically compacted at their respective Modified Proctor OMC (optimum moisture content) and MDD (maximum dry density) values in standard consolidation rings (75mm diameter, 25mm height) and inundated at pressures of 6.25 kPa, 100 kPa and 200 kPa respectively. The resulting changes in the dial gauge readings of the compacted specimens were noted until they became nearly constant. The percent swell/collapse of a soil specimen was calculated as $\Delta h / h_i$, where Δh is the change in height of the specimen on wetting at a given pressure and h_i is the initial height of the specimen.

Table 1: Index properties and compaction characteristics of soils

Soil property	BCS-1	BCS-2	Ash-modified soil-1	Ash-modified soil-2	2% lime treated BCS-1	4% lime treated BCS-1	7% lime treated BCS-1
Liquid limit %	65	55	61	53	61	60	58
Plastic limit %	19	25	42	37	38	42	43
Sand %	28	20	32	29	16	18	21
Silt %	32	29	49	57	61	65	79
Clay %	40	51	19	14	23	17	-
*MDD Mg/m³	1.73	1.69	1.28	1.41	1.60	1.34	1.30
*OMC %	20	21	37	30	23	32	33

* Modified Proctor values

Lime Stabilized Specimens
As a part of the experimental programme, hydrated lime treated BCS-1 specimens were prepared. . The initial consumption of lime (ICL) test as per British Standard 1924: Part 2: 1990 (Methods of test for stabilized soils) indicated a value of 3% for the BCS-1 specimen. The selected hydrated lime additions ranged on both sides of this ICL value and corresponded to 2, 4 and 7% of soil by weight. Appropriate weights of hydrated lime (Laboratory reagent grade) and BCS-1 specimens were mixed in the dry state. Water was added to the lime-soil mixtures until water content of each mixture approximated the Modified Proctor OMC of the ash-modified soil-1 specimen (37%). Further, the moist soil-lime mixtures were statically compacted in consolidation rings at the Modified Proctor MDD of the ash-modified soil-1 specimen (1.28Mg/m³). The lime treated BCS-1 specimens and

ash-modified soil-1 specimen were compacted at same dry density and water content to facilitate comparisons of their swell/collapse potentials. The lime added specimens were cured in a desiccator for 10 days. Duration of curing period was not considered in this study. The 10 day cured specimens were used in the testing programme The index properties of the lime treated soils are included in Table 1.

Wetting-Drying Procedure

The BCS specimens, the ash-modified specimens and the lime-cured specimens contained in consolidation rings were subjected to 4 cycles of alternate wetting and drying in the laboratory by an earlier described procedure of Rao et al. (1995). The natural and wood-ash-modified specimens were compacted at their respective Modified Proctor MDD and OMC values. The lime treated BCS-1 specimens were compacted to the same dry density and water content as the ash-modified soil-1 specimen. The drying of the laboratory specimens was performed at 40^0C aided by a warm-air circulator. The dried specimens were wetted by allowing them to absorb water due to capillary from a wetted sand bath. A degree of saturation of 90 to 93% was achieved by this saturation technique.

Table 2: Index properties of compacted and desiccated soils

Soil property	Soil state	BCS-1	BCS-2	Ash-modified soil-1	Ash-modified soil-2	2% lime treated BCS-1	4% lime treated BCS-1	7% lime treated BCS-1
ρ_d Mg/m^3	*Compt.	1.73	1.69	1.28	1.41	1.28	1.28	1.28
	**Desicct	1.49	1.40	0.91	1.09	1.25	1.30	1.41
w%	Compt.	20	21	37	30	37	37	37
	Desicct.	6	3	4	2	4	4	4
e	Compt.	0.60	0.63	1.06	0.83	1.11	1.13	1.14
	Desicct.	0.86	0.96	1.90	1.37	1.17	1.09	0.94
S_r %	Compt.	91	92	92	93	90	89	89
	Desicct.	19	9	6	4	9	10	12

*Compt. = Compacted, **Desicct. = Desiccated

Completion of four cycles of wetting and drying required about 10 to 15 days. As the wetting induced deformations of compacted expansive soils become nearly constant after three to four cycles of wetting and drying (Gangadhara, 1998), the laboratory wetting-drying procedure was terminated after four cycles. Specimens in the shrunken state with a history of four cycles of wetting and drying are referred to as desiccated specimens in this study. The void ratio (e), water content (w), degree of saturation (S_r), and dry density (ρ_d) of the various specimens in the compacted and desiccated states are given in Table 2.

The swell/collapse potentials of the desiccated specimens were determined at vertical pressures of 6.25 kPa, 100 kPa and 200 kPa respectively. The swell/collapse potentials of the desiccated specimens were calculated as $\Delta h/h_{shrunken}$, where Δh is the wetting induced change in height of the desiccated specimen at a given pressure and $h_{shrunken}$ is the height of the desiccated specimen prior to wetting.

RESULTS AND DISCUSSION
Ash-Modified Soil Specimens and BCS Specimens
Compacted State

Table 3 compares the swell potentials of the compacted ash-modified soil specimens and BC soil specimens. All specimens were compacted at their respective Modified Proctor MDD and OMC values. Both the ash-modified soils exhibit low swell potentials of less than three percent at 6.25 kPa. Comparatively, the BCS specimens exhibit higher swell potentials of 6 to 10% at 6.25 kPa. The low swell potentials of the ash-modified soil specimens are apparently due to their lower dry densities and higher water contents (Holtz and Gibbs, 1956). The low MDD's and high OMC's of the ash-modified soils (Table 2) are in turn a consequence of their chemical stabilization (Rao et al., 1999). Interestingly, the ash-modified soil specimens experience slight collapse on wetting at 100 and 200 kPa respectively, while, the swell potentials of the compacted BCS specimens decrease progressively at these pressures. Results in Table 3 demonstrate the effectiveness of wood-ash modification in controlling the swelling of the BC soil. To estimate their long term swelling behaviour, the wood-ash modified soil specimens were subjected to a few cycles of wetting and drying and the swell potentials of the desiccated specimens were examined.

Table 3: Swell/collapse potentials of soils

Soil state	Pressure kPa	Swell potential / collapse potential, %			
		BCS-1 specimen	BCS-2 specimen	Ash-modified soil-1 specimen	Ash-modified soil-2 specimen
	6.25 kPa	10.14	5.89	2.87	2.82
Compacted state	100 kPa	4.76	2.30	-0.10	-0.20
	200 kPa	2.70	1.60	-0.33	-0.42
Desiccated state	6.25 kPa	12.77	5.17	-1.89	-1.64
	100 kPa	-2.65	-5.77	-11.29	-9.21
	200 kPa	-5.34	-7.82	-14.43	-11.03

Desiccated State

Table 3 examines the wetting induced volume change (at various pressures) of ash-modified soil specimens and BCS specimens subjected to four cycles wetting and drying. The collapse tendency of the ash-modified soil specimens becomes significantly pronounced after four cycles of wetting and drying and the specimens exhibit collapse potentials as high as -11 to -14% percent on wetting at 200 kPa. Interestingly, the swell potentials of the desiccated BCS specimens showed a small increase (BCS-1 specimen) or were unaffected (BCS-2 specimen) on wetting at 6.25 kPa. On wetting at 100 and 200 kPa, both the cyclically wetted and dried specimens experienced collapse strains that were however much lower than that experienced by the desiccated ash-modified soil specimens. The transformation of the ash-modified soil specimens from a low swelling to a highly collapsing soil may be attributed to loss of chemical bonds (discussed later) and reduction in dry density and water contents after four cycles of wetting and drying (Table 2). The marked reduction in swell potentials and manifestation of collapse tendency of the desiccated BCS specimens can be attributed to the reduction in dry density and water content of the specimens (Table 2).

Ash-Modified Specimens and Lime-Treated BCS Specimens

Results in the previous section had shown that the beneficiary effects of ash-modification in controlling the swelling behaviour of BC soils is lost on subjecting them to cycles of wetting and drying. Prompted by the results obtained for ash-modified soils, the influence of cyclic wetting and drying on the swelling behaviour of laboratory stabilized soils is examined.

Compacted State

Table 4 compares the swell/collapse potentials of ash-modified soil-1 specimen with that of lime stabilized BCS-1 specimens at various pressures. The lime treated specimens and the ash-modified soil-1 specimen were compacted at the same dry density ($1.28Mg/m^3$) and water content (37%). Despite their similar dry density and water content, the ash- modified soil specimen and lime-treated BCS-1 specimens exhibit different wetting induced volume change behaviour at 6.25 kPa. While, the compacted ash-modified soil specimen swelled by 2.9%, the compacted lime-treated specimens collapsed by -0.02 to -0.1%. However, on wetting at 100 and 200 kPa, both soil types collapsed by -0.07 to -0.33%.

Table 4: Swell/collapse potentials of soils

Soil state	Pressure kPa	Swell potential / Collapse potential, %			
		Ash-modified soil-1 specimen	2% lime treated BCS-1 specimen	4% lime treated BCS-1 specimen	7% lime treated BCS-1 specimen
Compacted state	6.25	2.87	-0.02	-0.04	-0.10
	100	-0.10	-0.07	-0.08	-0.14
	200	-0.33	-0.10	-0.13	-0.18
Desiccated state	6.25	-1.89	4.16	1.88	1.16
	100	-11.29	-4.88	-1.27	0.48
	200	-14.43	-8.19	-2.12	-0.07

Desiccated State

Table 4 also compares the swell/collapse behaviour of desiccated ash-modified soil-1 specimen and the desiccated lime treated BCS-1 specimens at various pressures. Data in Table 4 shows that the beneficiary effect of lime stabilization in controlling the wetting induced volume change of the laboratory stabilized specimens is affected by the cyclic wetting-drying process. The desiccated, ash-modified soil specimen and desiccated, lime treated specimens respond similarly to wetting at 6.25 kPa and exhibit swell potentials ranging from 1.2 to 4.2%. The desiccated lime treated specimen with a lime content of 2% was most affected by the cyclic wetting and drying. The swell potential of the desiccated ash-modified specimen ranged between that of the desiccated lime treated specimens. On wetting at 100 and 200 kPa, the desiccated 2% lime treated specimen exhibited notable collapse potentials of -5 to -8%, that were however lesser than the collapse potentials experienced by the desiccated ash-modified soil-1 specimen. The desiccated 4% lime treated BCS-1 specimen exhibited much lower collapse potentials of -1 to -2% on wetting at 100 and 200 kPa. Comparatively, the desiccated 7% lime treated specimen exhibited a negligible swell potential of 0.5% and collapse potential of -0.1% on wetting at 100 and 200 kPa respectively. The variations in the wetting induced deformations of the desiccated lime treated BCS specimens concur with their dry densities (Table 2). The results suggest that if a soil is not stabilized using adequate lime content, the beneficiary effect of lime stabilization will be eventually lost due to periodic climatic changes. Apparently, chemical stabilization of the ash-modified soil-1 specimen was not adequate. Therefore, the beneficiary effect of chemical stabilization of the wood-ash-modified specimen was lost on cyclic wetting and drying.

CONCLUSIONS

Lime stabilization controlled the wetting induced volume changes of the ash-modified soil specimens and laboratory stabilized BCS specimens in the compacted state. Cyclic wetting and drying of the ash-modified soil specimens however caused them to experience significant collapse on wetting at the experimental pressures. The transformation of the ash-modified soil specimens from a low swelling to a highly collapsing soil is attributed to loss of chemical bonds and reduction in dry density and water content after four cycles of wetting and drying. Experimental results with laboratory stabilized BCS specimens indicated that if a soil is not stabilized using adequate lime content, then the beneficiary effect of lime stabilization is lost on cyclic wetting and drying.

REFERENCES

Bell, F. G., 1993, Engineering treatment of soils, E & F N Spon, London.

Chen, F. H., and Ma, G. S. 1987, Swelling and shrinkage behaviour of expansive clays, Proceedings 6th International Conference on Expansive Soils, Vol. 1, New Delhi, India, pp. 127-129.

Day, R. W., 1994, Swell-shrink behaviour of compacted clay, Journal of Geotechnical Engineering, ASCE, Vol. 120, pp. 618-623.

Dif, A. E., and Bluemel, W. F., 1991, Expansive soils under cyclic drying and wetting, ASTM Geotechniical Testing Journal, Vol. 14, pp.96-102.

Gangadhara, S, 1998, Cyclic swell-shrink behaviour of laboratory compacted expansive soils, Ph.D. Dissertation, Indian Institute of Science, Bangalore, India.

Holtz, W. G., and Gibbs, H. J., 1956, Engineering properties of expansive soils, Transactions ASCE, Vol. 121, pp. 641-677.

Popesco, M., 1980, Behaviour of expansive soils with crumb structure, Proceedings 4th International Conference on Expansive Soils, Vol. 1, New York, pp. 158-171.

Rao, S. M., Sridharan, A., and Ramanath, K. P., 1995, Collapse behaviour of an artificially cemented clay, ASTM Geotechnical Testing Journal, Vol. 18, pp.334-341.

Rao, S. M., Venkatarama Reddy, B.V., and Muttharam, M., 1999, Engineering behaviour of wood-ash modified soils, Ground Improvement (Institution of Civil Engineers, London), Provisionally Accepted.

Subba Rao, K. S., and Satyadas, C. G., 1987, Swelling potential with cycles of swelling and partial shrinkage, Proceedings 6th International Conference on Expansive Soils, Vol. 1, New Delhi, India, pp. 137-142.

The Use of Spent Oil Shale in Earthwork Construction

MG WINTER

Transport Research Laboratory, Edinburgh, UK

ABSTRACT
Spent oil shale (or blaes) is a potentially valuable engineering material and is present in large quantities in the West Lothian area of Scotland. It can be used successfully as general fill or capping layer. However, due to its high quality it may be more suited to use as selected granular fill or sub-base. In particular, cement stabilisation will reduce frost susceptibility and may be a particularly appropriate outlet for spent oil shale for use as sub-base. However, an increase in control and testing may be required, having an effect on the cost of using such materials. Conditions under which spent oil shale should not be used are also identified.

INTRODUCTION
The oil shale industry in the West Lothian area of Central Scotland was founded on James Young's invention of a process for obtaining oil from bituminous coal. The industry began in 1851 and reached its peak in 1913, before closure of the last oil shale mine in 1962. The legacy of this industry is an estimated 100 million tonnes of spent oil shale stored in around 30 tips or bings. Some of the bings have been landscaped and integrated into the surrounding environment as distinctive visual features and, in one case, a nature reserve has been created. Nonetheless a great deal of spent oil shale remains available for reuse in construction works to the benefit of the areas surrounding the bings.

The potential uses of spent oil shale (or blaes) in road construction include general fill, selected granular fill, capping and sub-base. The specifications for these materials are given in the Manual of Contract Documents for Highway Works (MCHW 1). The chemical and physical properties of spent oil shale are described and each potential application is examined in this context and that of the specification. Conditions under which spent oil shale should not be used are also identified. Broad conclusions are drawn within the environmental and economic contexts.

THE OIL SHALE INDUSTRY
The oil shale industry in Scotland was founded on James 'Paraffin' Young's invention of a process for obtaining oil from bituminous coal by low temperature distillation and refining the oil by further distillation and chemical treatment (British Patent No. 13292). The industry began in 1851 and reached its peak output in 1913 when 3.3 million tonnes of oil shale was extracted, producing an estimated 73 million gallons of crude oil and naptha. The industry declined rapidly after the First World War and in 1938 less than half the quantity of oil shale was mined compared to the peak. Despite a brief recovery, during the Second

World War, the last Scottish oil shale was mined at Bathgate in May 1962. Kerr (1994) provides an excellent reference on this subject.

The oil shale exists in a layer around 900m thick and forms part of the Carboniferous sandstone series. It lies below the coal and limestone beds that were worked in Lothian and Lanarkshire but above the volcanics, which form, for example, Arthur's Seat in Edinburgh. The shale measures are found throughout Scotland but accessible shales with recoverable amounts of oil were found in West Lothian, parts of Midlothian, Lanarkshire and Fife.

The unprocessed shale consisted of tough, fine-grained, thinly laminated material and was generally brown or black in colour. It had a rubber-like consistency and miners identified them by their ability to be paired with a knife and to leave a brown streak when rubbed.

The processed (or spent) oil shale, together with materials considered unsuitable for processing, was deposited in spoil heaps, or bings, on land adjacent to the mines and refineries. The bings are mainly located in West Lothian with a small number in the surrounding areas. The Five Sisters bing, near West Calder, is pictured in Figure 1. In 1962 it was estimated that between 100 and 150 million tonnes of such material was stored in 30 bings (Burns, 1978). More recent estimates indicate that around 100 million tonnes of spent oil shale occupy an area of 395 hectares of land around Livingston and Bathgate, about 20km to the west of Edinburgh (Whitbread *et al.*, 1991; Sherwood, 1994).

CHEMICAL AND PHYSICAL PROPERTIES
Chemical Properties

Studies indicate that spent oil shale can vary in colour from pink, red and yellow to dark blue (Lake *et al.*, 1966; Burns, 1978).

The typical chemical composition of spent oil shale is given in Table 1 (data for burnt colliery spoil is given for comparison). The range of total sulphate contents for spent oil shale has been reported as between 2.2%(SO_3) and 2.8%(SO_3), with one value of 0.7% (Burns, 1978), and is similar to the range for burnt colliery spoil (Sherwood, 1994).

Table 1 – Typical chemical compositions of spent oil shale (data for burnt colliery shale is shown for comparison).

Chemical Component	Composition (%)	
	Spent Oil Shale (Burns, 1978)	Burnt Colliery Spoil (Sherwood and Ryley, 1970)
SiO_2	48.5	45-60
Al_2O_3	25.2	21-31
FE_2O_3	12.1	4-13
CaO	5.3	0.5-6
MgO	2.2	1-3
Na_2O	Not Recorded	0.2-0.6
K_2O	Not Recorded	2-3.5
SO_3	3.2	0.1-5
Loss on Ignition	0.03	2%-6%

There appears to be some ambiguity in the standard (Anon, 1975) used to determine the values of sulphate content. After some study and debate it appears most likely that the values quoted by Burns (1978) should be reported in units of g(SO_3)/L. Consequently, sulphates may cause problems by migrating from the shale and reacting with lime, concrete, cement bound or other cementitious materials (MCHW 1). Problems are also likely to arise if spent oil shale is placed in contact with metallic objects (Winter and Butler, 1999). Because the shale was heated to extract the oil, spontaneous combustion is clearly not a problem. Neither is the presence of sulphides as these will have been driven off or converted to sulphates during the extraction process (Sherwood, 1994).

Physical Properties

Considerable variations are commonly found in the physical properties of spent oil shales, particularly the grading. However, Burns (1978) noted that the particle size distributions for such materials were generally within the grading limits for Type 2 (Figure 2) granular sub-base materials (Specification for Road and Bridge Works, 1976) and were very close to the requirements for Type 1 granular sub-base materials. (These limits are essentially the same as those in MCHW 1.)

Burns (1978) reported the results of standard 2.5kg rammer compaction tests. These indicated maximum dry densities in the range 1.30Mg/m³ to 1.44Mg/m³ and optimum moisture contents in the range 22% to 34%. The relatively high optimum moisture contents and correspondingly low maximum dry densities were probably due to the porous nature of the material. This is supported by the high values of water absorption (10% to 21%) and the low particle densities for the size range 2.4mm to 37.5mm (2.06Mg/m³ to 2.42Mg/m³). Particle densities for sizes smaller than 2.4mm were between 2.58Mg/m³ and 2.76Mg/m³.

Figure 1 – Five Sisters spent oil shale bing, near West Calder, West Lothian.

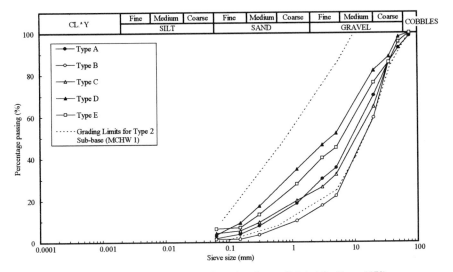

Figure 2 - Particle size distributions of samples of spent oil shale (after Burns, 1978).

The degree of crushing, or fragmentation, that will occur during the compaction of a granular material generally depends upon the original particle size distribution, the crushing strength of the grains, and the stress level applied. Essentially, if a brittle particle is subjected to a stress in excess of its ultimate strength it will break. The stress may place the particle in tension, compression, flexure, shear or torsion. In addition, a particle may be worn away by another particle: that is, subjected to attrition.

Burns (1978) observed that spent oil shale was subject to crushing as a result of the compaction process (Figure 3). While fragmentation during compaction is often viewed as a problem, crushing may confer benefits to the compacted mass. Small particles resulting from crushing may fill voids and thus produce higher densities and lower air voids than would otherwise be the case. This will, in turn, tend to produce a more stable compacted mass. The fragmentation of earthwork materials is discussed in more detail by Winter (1998).

Most spent oil shales are frost susceptible and therefore should not be placed within 450mm of the road surface. However, the addition of cement reduces the frost susceptibility. Burns (1978) observed that the addition of 5% cement reduced the voids content and, thus, the frost induced heave to acceptable levels for the materials tested. However, he did note that larger proportions of cement might be required for materials from some spent oil shale sources to reduce heave to acceptable levels.

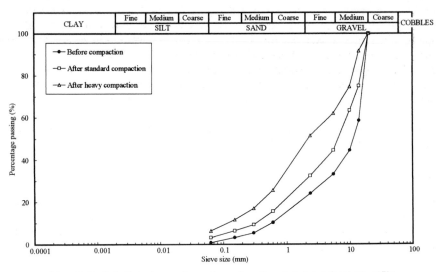

Figure 3 - Typical effect of compaction on the grading of spent oil shale (after Burns, 1978).

USE OF SPENT OIL SHALE
Historical Use
Spent oil shale bings may contain both spent oil shale and material excavated from non-oil bearing areas. The heating of oil shale produces a stable granular material that can be used in place of natural aggregates. The optimum use of spent oil shale is as a selected granular fill or as an unbound granular sub-base although it could be used as general fill.

Nevertheless, Burns (1978) estimated that some 20 to 25 million tonnes of spent oil shale had been used in Central Scotland as general fill since the early 1960s, principally on the M8, M9 and M90 motorways. It has also been used in other bulk fill applications such as the Edinburgh Airport runway, where some 200,000 tonnes were placed. More recently some 15,000m^3 were place as general fill during the construction of the underpass at the M8/M9 Newbridge interchange. Small quantities of spent oil shale have also been used as general fill on other recent road construction projects in Central Scotland. Stabilising spent oil shale with cement on a pre-mixed basis has proved commercially viable and Burns estimated that some 0.5 to 1.0 million tonnes had been so used, including as sub-base on the M9 motorway.

General Fill
The assessment of acceptability for general fill should be by conventional means. For a granular material such as spent oil shale this is generally by control of the moisture content within set limits of the optimum moisture content (see MCHW 1, Table 6/1). However, given that spent oil shale is generally acceptable for use as selected fill it may be uneconomic to use these materials as general fill.

Selected Granular Fill
Spent oil shale is susceptible to crushing during compaction (Figure 3). However, crushing may yield higher densities and lower air voids than would otherwise be the case, thus tending to produce a more stable compacted mass. Indeed, this indicates one barrier to the use of spent oil shale (and burnt colliery spoil) as selected granular fill. Previous specifications have relaxed the 10% Fines Value required for well burnt non-plastic shale sub-bases (Specification for Road and Bridge Works, 1976; Burns, 1978) but the current Specification (MCHW 1) retains a requirement for a minimum 10% Fines Value of 50kN. Such a requirement is unlikely to be met by many spent oil shales and Dawson (1989) described the 10% Fines Value as too severe a test of the ability of well-graded aggregates to withstand compaction stresses. There is a strong argument for reducing the 10% fines value requirement for spent oil shale, provided that the compacted material is within the particle size distribution limits defined in the specification.

Capping Layer
The Specification (MCHW 1) allows spent oil shale to be used as granular capping, provided that the grading and other specification requirements are met. However, as for selected granular fills, the 10% Fines Value requirement (30kN in this case) is unlikely to be met. Similar arguments may be applied to relaxing the 10% fines value as are made above for selected granular fill.

While spent oil shale may be stabilised to form a capping layer there would generally be little reason for doing so as it would be suitable for use in an unbound form (Sherwood, 1994). The only exception to this would be if the capping layer were within 450mm of the completed road surface as many spent oil shales are highly frost susceptible and thus cannot be in such a location. The addition of 5% or more cement has been shown to reduce the frost susceptibility to acceptable levels (Burns, 1978). Frost susceptibility is related to voids content and the use of cement reduces the voids content while having the added benefit of increasing the inter-particle strength. For this reason, many sub-bases constructed from spent oil shale have been stabilised with cement. Such a treatment will be suitable only for materials with low sulphate contents (HA74 - DMRB 4.1.6) when ordinary portland cements are used. However, the use of sulphate resisting cements may prove effective with higher sulphate materials. To ensure that the material is not susceptible to heave as a result of the reactions between sulphates and the cement it would be prudent to use the durability test given in the Specification (MCHW 1, Clause 1036) for CBM1. However, as the frequency of testing may have to be increased, such a process may become uneconomical. The use of sulphate resisting cements may prove effective with higher sulphate materials. Comprehensive warnings have been given on the damage caused by mixing high sulphate materials with cement and lime (Perry *et al.*, 1996a; 1996b).

Sub-Base
Figure 2 shows spent oil shale can be obtained to satisfy the grading requirements of granular sub-base materials and can therefore satisfy the more relaxed criteria for selected granular materials. Well-burnt colliery spoil is mentioned by name in the Specification (MCHW 1) for many selected granular fill applications. However, Clauses 803 and 804 of the Specification (MCHW 1), which refer to granular sub-base, allow "well burnt non-plastic shale" which embraces both burnt colliery spoil and spent oil shale. The clauses for selected granular fill could usefully be phrased in the same manner rather than specifically requiring burnt colliery spoil (Sherwood, 1994), effectively excluding spent oil shale. Similarly, where burnt colliery spoil is excluded then it would be prudent to also exclude spent oil shale by the use of the phrase "well burnt non-plastic shale".

Burns (1978) noted that in the Specification for Road and Bridge Works (1976) sub-base materials, with the exception of well burnt non-plastic shales, were required to have a minimum 10% Fines Value of 50kN. This exception specifically acknowledged that crushing could occur with this material with no apparent structural disbenefit to the completed sub-base. Under the current Specification (MCHW 1) well-burnt non-plastic shales are specifically included in the allowable materials but the exception to the minimum 10% Fines Value of 50kN has been removed. Sherwood (1995a; 1995b) notes that although many burnt colliery spoils (and, by implication, spent oil shales) would be suitable for such a use it is unlikely that many would achieve a 10% Fines Value of 50kN. Further, it is quite possible that a more stable material may result as voids are filled with the smaller particles created by crushing.

Similar arguments may be applied to relaxing the 10% fines value as are made above for selected granular fill and capping layer.

Other Considerations
Visual inspection of spent oil shale at the tip is required to ensure that the material delivered to site does not vary too frequently. It is important that all design and specification tests should be carried out on the material in the crushed, post-compaction state, not on the excavated or delivered material.

If spent oil shale is to be placed within 500mm of metallic items, lime, concrete, cement bound or other cementitious materials then the limits in the Specification (MCHW 1, Clauses 601.13 and 601.14) apply. It seems unlikely that it could be used in close proximity to such materials given the high sulphate content of many spent oil shales. Care is also required if spent oil shale is to be stabilised using cement. For lower sulphate materials it may be possible to use ordinary Portland cement. However, for higher sulphate materials sulphate resisting cements may be required. Doubts also remain as to the potential reactions between spent oil shale and polymeric reinforcing materials.

Notwithstanding the above comments on frost susceptibility and the interaction with cementitious products, the durability of spent oil shales is unlikely to pose a problem. However, it is nonetheless important to ensure that compaction is sufficient to minimise the air voids of the earthworks and that the requirements of MCHW 1 (Clause 602.15) are followed to ensure that water ingress does not cause a long-term durability problem.

DISCUSSION

The transport of waste materials, such as spent oil shale, to site incurs a cost for haulage. If a waste material is not available within an economic haulage distance of the site then it may be rejected on economic grounds. It is generally accepted that the maximum economic haulage distance for the import of waste materials will depend on a number of factors, including the following:
- the location and nature of the site;
- the location and nature of the waste material source;
- the method of transport;
- the relative costs of the waste material and its alternative.

However, if the waste material is located within economic haulage of the site then its use should be considered and the factors to be taken into account are summarised by Anon (1985) and Sherwood (1994).

The advantages of using waste materials are:
- removal of waste tips;
- avoidance of borrow pits and consequent savings in the finite sources of natural aggregates;
- avoidance of liability for landfill tax (where appropriate).

It should, however, be noted that it is seldom possible to remove all of a tip due to the mix of materials contained therein. Indeed, planning permission is generally required to remove material from an old tip due to the possibility of aggravating the existing dereliction of an area. In some cases, landscaping a tip to sympathetically blend into the local area may be a more environmentally beneficial solution than seeking engineering uses for the materials contained therein (Sherwood, 1994).

The disadvantages of using waste materials are:
- increased haulage costs;
- disturbance caused by haulage;
- greater variability of waste materials.

Increased haulage costs are almost certain to be a factor in the substitution of waste materials for site-won bulk fill and for some materials, haulage costs may increase if waste materials are used. This will however depend upon the relative locations of the waste materials and borrow-pits or quarries from which the selected fills would otherwise be obtained.

If haulage costs are increased by the use of waste fills then the disturbance caused by haulage will also increase and environmental disbenefits will accrue if public roads are used. These disbenefits will include congestion, noise, pollution, dust and deposition of the material along the haulage route. Such disbenefits are, with the exception of congestion, difficult to price and are frequently ignored by engineers.

Inspection of waste materials, such as spent oil shale, at the tip is one means of managing their potentially greater variability compared to conventional materials.

Clearly a blend between economic and environmental benefits must be achieved if waste materials are to be successfully used in road construction. Without an economic benefit contractors will not be encouraged to use waste materials. If some economic benefit exists then it is important that environmental benefits can be demonstrated by their use. Such benefits might include savings in the use of finite natural aggregate resources. Designers should identify sources of waste materials and allow contractors to select on an economic basis, within the prevailing environmental legislation (HA44 - DMRB 4.1.1).

SUMMARY
The crushing of particles during compaction is often viewed as a problem. However, crushing may confer benefits to the compacted mass. Small particles resulting from crushing may fill voids and thus produce higher densities and lower air voids than would otherwise be the case. This will, in turn, tend to produce a more stable compacted mass.

Spent oil shale can be used successfully as general fill. However, due to its high quality as aggregates it is more suited to use as selected granular fill. In particular, given its frost susceptibility, stabilisation as sub-base seems a particularly appropriate outlet for this material. However, it would be prudent to use the chemical durability test given in the Specification (MCHW 1, Clause 1036) for CBM1 to ensure that the material is not susceptible to heave as a result of the reactions between sulphates and the cement.

Spent oil shale can be obtained to satisfy the grading requirements of granular sub-base materials and can therefore satisfy the more relaxed criteria for selected granular materials. The relevant clause of the Specification for granular sub-base refers to "well burnt non-plastic shale" which embraces both spent oil shale and burnt colliery spoil. The clauses for selected granular fill refer specifically to well-burnt colliery spoil whether including or excluding such materials from use. In either case spent oil shale is not covered, effectively disallowing its use where it could be used and allowing its use where it should not be used.

The minimum 10% Fines Value requirement of 50kN for granular sub-base materials is severe. In previous specifications this requirement was waived for burnt non-plastic shales, such as spent oil shale. Although limited crushing during compaction will tend to produce a more stable compacted mass many such materials will be deemed unsuitable on the basis of the 10% fines value. The minimum 10% Fines Value of 30kN required for some classes of capping layer also seems severe.

Spent oil shales generally have high sulphate contents. Consequently, these materials are unlikely to be suitable for use close to concrete structures or metallic items. There also remain doubts as to the potential reactions between spent oil shales with high sulphate contents and polymeric reinforcing materials.

Clearly a blend between economic and environmental benefits must be achieved if waste materials are to be successfully used in road construction. Without an economic benefit contractors will not be encouraged to use waste materials. Designers should identify sources of waste materials and allow contractors to select on an economic basis, within the prevailing environmental legislation. If some economic benefit exists then it is important that environmental benefits can be demonstrated by their use. Such benefits might include savings in the use of finite natural aggregate resources.

ACKNOWLEDGEMENTS
The author is grateful to Mr RA Snowdon (TRL) for helpful discussions and suggestions. Dr J Perry, Dr GD Matheson and P McMillan variously performed the technical and quality audits for this report; the author is grateful for their helpful and constructive suggestions.

The work described in this paper forms part of a Scottish Office Development Department research programme (Programme Director, P McMillan), and the paper is published by permission of the Scottish Development Department and the Chief Executive of TRL Ltd.

REFERENCES
Anon (1975). *Methods of test for soils for civil engineering purposes,* BS 1377. London: British Standards Institution.

Anon (1985). *Guide to the use of industrial by-products and waste materials in building and civil engineering,* BS 6543. London: British Standards Institution.

Burns, J (1978). The use of waste and low-grade materials in road construction: 6. Spent oil shale. *TRRL Laboratory Report LR 818.* Crowthorne: Transport Research Laboratory.

Dawson, AR (1989). The degradation of furnace bottom ash under compaction. *Unbound Aggregates in Roads: UNBAR 3,* pp. 169-174. London: Butterworths.

Design Manual for Roads and Bridges. London: The Stationery Office.
    HA 44 - Earthworks: Design and Preparation of Contract Documents (DMRB 4.1.1).
    HA 74 - Design and Construction of Lime Stabilised Capping Layers (DMRB 4.1.6).

Kerr, D (1994). *Shale oil Scotland: The World's pioneering oil industry*. Edinburgh: Scotch Publishing.

Lake, J, CK Fraser and J Burns (1966). Investigation of the physical and chemical properties of spent oil shale. *Roads and Road Construction*, **44**, 522, pp. 155-159.

Manual of Contract Documents for Highway Works. London: The Stationery Office.
  Volume 1: Specification for Highway Works (December 1991, reprinted August 1993 and August 1994 with amendments) (MCHW 1).

Perry, J, DJ MacNeil and PE Wilson (1996a). The uses of lime in ground investigation: a review of work undertaken at the Transport Research Laboratory. *Lime Stabilisation*. London: Thomas Telford.

Perry, J, RA Snowdon and PE Wilson (1996b). Site investigation for lime stabilisation of highway works. *Advances in Site Investigation Practice*, pp. 85-96. London: Thomas Telford.

Sherwood, PT (1994). The use of waste materials in fill and capping layers. *TRL Contractor Report CR 353*. Crowthorne: Transport Research Laboratory.

Sherwood, PT (1995a). The use of waste and recycled materials in roads. *Proceedings, Institution of Civil Engineers*, **111**, 2, pp. 116- 124. London: Thomas Telford.

Sherwood, PT (1995b). *Alternative materials in road construction: a guide to the use of waste, recycled materials and by-products*. London: Thomas Telford.

Sherwood, PT and MD Ryley (1970). The effect of sulphates in colliery shale on its use for roadmaking. TRRL Laboratory Report LR 324. Crowthorne: Transport Research Laboratory.

Specification for Road and Bridge Works (1976). Fifth Edition. Department of Transport, Scottish Development Department, Welsh Office, Department of the Environment for Northern Ireland. London: The Stationery Office.

Whitbread, M, A Marsey and C Tannel (1991). *Occurrence and the utilisation of mineral and construction wastes*. Department of the Environment. London: The Stationery Office.

Winter, MG (1998). The determination of the acceptability of selected fragmenting materials for earthworks compaction. *TRL Report 308*. Crowthorne: Transport Research Laboratory.

Winter, MG and AM Butler (1999). Gone to the wall. *Surveyor*, 6 May, 12-13.

Contaminated Water

Groundwater Monitoring System in Road Belts

GORAN RASULA, B.Sc in Hydrogeol.
Jaroslav Cerni Institute for the Development of Water Resources, P.O. Box 33-54, Belgrade, Yugoslavia, E-mail: rasula@beotel.yu Web page: www.geocities.com/CapeCanaveral/Hall/8183

ABSTRACT

Protection of groundwater already represents, and in the future will also be, particularly important environmental, socio-economic, social welfare and political task. The fact that expenses necessary for preventive measures are 10 to 20 times lower than funds needed to clean up and revitalize polluted aquifers, and the fact that the area occupied by future traffic and other infrastructure facilities will aggravate and become a burden to protection, both call for an urgent introduction of modern methods of monitoring and preserving all existing and potential groundwater resources. The basic principle of groundwater protection against pollution implies constant control and preventive protection measures, namely prevention of incidental and other pollution. Firstly, the very construction of communications and all accompanying facilities brings about changes and deteriorates the existent natural-geological environment, and later on, during exploitation, they become structures in whose immediate surrounds surface and ground waters may be polluted by toxic and waste substances transported by such communications. The experience gained through activities undertaken so far: the realization of the first "pilot model" of groundwater monitoring in the zone of Makis, the investigations and monitoring of critical sites polluted with xylene (CIP, Belgrade, 1994/95), and information on up-to-date achievements in this field (obtained while working on the Study of methodology of hydrogeological investigations needed for groundwater quality monitoring system - Jaroslav Cerni Institute for the Development of Water Resources, Belgrade, 1997), stress a necessity for obligatory designing and constructing of operating systems for groundwater quality monitoring, particularly in the zones of infrastructure facilities (roads, railway, gas pipelines, oil pipelines), which can be considered as potential linear pollutants.

In the light of that, this paper contributes to the idea of developing hydrogeological investigation methodology for the need of designing and operating a monitoring system in a road belt, as well as of its legal and practical inclusion in the engineering documents of each project. Beside that an emphasis is put on multidisciplinary approach to system implementation, for the purpose of its full automation, rational management, and designing of effective and optimal measures for groundwater protection.

INTRODUCTION

Since 1984, when an incidental spillage of a great quantity of xylene happened in "Makis - Belgrade" marshalling yard, the Department of Geotechnics, CIP Institute of Transportation, Belgrade and 'Jaroslav Cerni' Institute for the Development of Water Resources, Belgrade have been continually performing certain research/study hydrogeological investigations, synchronized with extensive remedial measures, in order to define groundwater and surface water pollution in the wide zone of the Makis marshalling yard and highway Beograd-Obrenovac (Fig. 1), transform pollutants and physically migrate them. In the last Phase V investigations (1996) a pilot monitoring system was established to periodically monitor the groundwater and surface water regime and quality, and after this (1998), an active groundwater quality monitoring stations were proposed on the basis of a Work Resumption Programme.

This paper should help to identify possibilities for a quality hydrogeological approach to design modern groundwater protection system, which is also important beside the belts of transport infrastructure, in the zones of dump areas, landfills with adequate sanitary measures and disposal areas for industrial wastes, etc. and particularly in the areas communicating with, or adjacent to existing, or potential, groundwater source zones.

THE CONCEPT AND SCOPE OF PERFORMED HYDROGEOLOGICAL INVESTIGATIONS FOR THE PILOT MONITORING SYSTEM

The basis for groundwater protection is generally a set of prevention and treatment procedures and measures which should primarily prevent pollution from occurring, particularly in water source zones and on the grounds containing considerable groundwater reserve. Complex and comprehensive methods for solving concrete problems directly depend upon the geological structure, i.e. the hydrogeological properties of rock; namely, it is necessary to investigate the terrain thoroughly. This calls for development of appropriate and concrete methodology of hydrogeological investigations in full accordance with the local conditions. The investigations themselves were so designed as to reliably and in a modern way enable determination of the main hydrogeological parameters for the investigated area, definition of the nature and optimum number and technical characteristics of the groundwater monitoring system and structures, finding the way how to achieve the rational operation by testing of the first pilot model of monitoring system and give a quality forecast of the system operation in the future.

The investigation programme comprised the following:
* A detailed reconnaissance of the site in a wider area of investigation, hydrogeological mapping and registration of all the existing water structures, their output and other parameters;
* Boring of ten new piezometric holes (Fig.1.) down to the impervious strata (approximately 25 m per one structure) in order to determine all local hydrogeological characteristics of the ground and the key hydrogeological parameters of the analyzed aquifer;
* Recording and land registering of all active and potential pollutants, which are in indirect or direct connection with the water source zone, especially the present facilities in marshalling yard and surrounds, frequency and quantity of toxic materials transport on the highway and the like, and to understand the way in which the structures and people function and might, for some reason, cause any other groundwater quality degradation.
* Determination of detailed hydrogeological characteristics in the zone of extensive and immediate protection :
  - the aquifer type and its potential,
  - the infiltration characteristics and geometry of the aquifer complex, and the capacity of the present water intake structures and the water source in the zone of Makis,
  - conditions of groundwater recharge, migration and outflow,
  - detailed infiltration characteristics and position of the roof/surface complex, in the zone of impact of the water source, particularly at the highway belt and at the contact of marshalling yard and base alluvial complex,

- the groundwater table regime, and hydrodynamic relations between surface and ground waters,
- in the field and laboratory, the transporting velocity of several expected pollutants through the protective, roof humus-clay soil complex and aquifer complex, were analyzed and also the sorption, dispersion and other parameters, and pollutant transformation,
- "Zero state" of hydrochemical groundwater parameters were analyzed to determine four parameters which should be continually monitored over one year period.

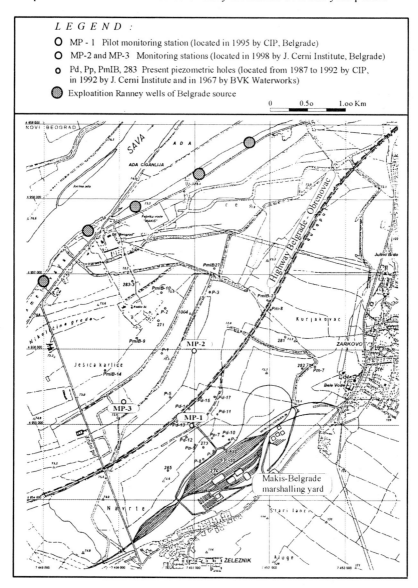

Figure 1.   Situation map with all analyzed structures in the wide investigated area

The results of the investigations, designed and performed as described above offered a basis for scheduling the best possible conditions for placing the 'pilot model' of groundwater monitoring station. In the first step, the results of previous field investigations and laboratory tests were systematized. Then hydrogeological, hydrochemical, hydrodynamical and other basic data were interpreted as well as the results of earlier hydrogeological investigations preceding the opening of the Belgrade source in the zone of Makis. Finally, a complex analysis and synthesis of the collected data was done and the best technical solutions for groundwater monitoring were defined.

GENERAL HYDROGEOLOGY OF THE TERRAIN
The alluvial region on the right hand bank of the Sava river, in the south part of Belgrade (Fig. 1), covers an area of about 10 km^2. With regard to geomorphology the said region is in the form of a gentle plateau, average altitude 73 above sea level. In some parts the terrain is marshy or contains traces of the old meliorate channels or meanders of the Zeleznicka river which flows through the central part of the terrain. From the Sava river basin to the east, the terrain gently rises towards the marshalling yard 78 m above sea level and further eastwards passes into steep hilly background of Zarkovo and Zeleznik with their highest crests rising to altitude 130-171 above sea level. A distinctive characteristic of the region is its moderate continental climate. Mean precipitation according to the records of the hydrometeorologic station in Belgrade for the period 1965-1994 amounted to 781 mm. The hydrographic network consists of the Sava river stream and several minor permanent surface streams of which the most significant ones are Zeleznicka and Topciderska rivers.

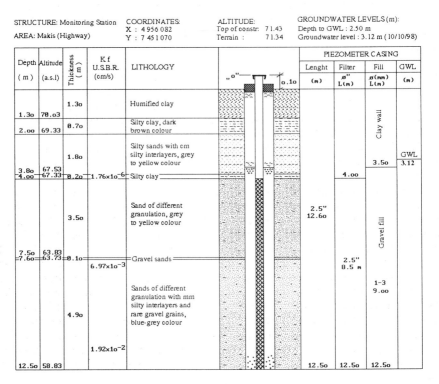

Figure 2. Hydrogeological profile of MP-1 "pilot" monitoring station

Alluvial sediments occur everywhere in the region varying in depth between 10 and 25 meters. They are represented by sandy gravel and sands in which the main water-bearing environment is formed (Fig 2), while the roof ground complex, conditionally of low permeability is made of humified sands - clays and silts of different thickness. The confined aquifer is replenished with atmospheric and surface water percolating through the diluvial and proluvial sediments on the eastern flanks of Zarkovo and from the Sava river in the southern part of terrain.

The natural direction of groundwater flowout is towards the southwest sections of the Sava river, with variation that depends on the exploitation in Ranney wells of Belgrade source. The confined aquifer regime in the sand-gravel complex on one part and the small depth (less than 3 m) and filtering properties of the roof protective complex on the other part may have direct effect upon the quality of groundwater.

"PILOT" MODEL OF GROUNDWATER MONITORING SYSTEM
Any complex and comprehensive approach to concrete problems directly depends on the geological structure and hydrogeological characteristics, that is, on the availability of good results of soil investigations. Initial steps in the hydrogeological investigations of the Sava river alluvium, in the zone of Belgrade-Obrenovac highway and Makis marshalling yard, were to compose a register of all the existing water intake structures (wells, piezometers) and a register of active and potential contaminants from the highway transport and within the railway facilities that may directly or indirectly cause degradation of groundwater quality in the future, and to have the above actions systematically monitored. Subsequent hydrogeological activities were focused on the identification of general hydrogeological and hydrochemical parameters, geometry and percolating properties of the roof complex and groundwater quality in the aquifer.

When all the piezometer boreholes from the earlier investigation phases (1982-1994) were prospected and their functional properties were determined,   work started to make ten new piezometers and form a network of a total of 40 observation stations. Afterwards the hydrochemical zero quality of groundwater was to be determined. Several physical chemical parameters were selected and successively monitored over time. These were temperature, conductivity, pH, xylene concentrations, Fe ions, Mn ions, phenol and ammonia. Some of these parameters were monitored by digital measurements at each 0.50 m of the water column in the network of observation stations - piezometers while the others were monitored by taking water samples and performing dedicated chemical analysis. To monitor xylene concentrations, beside laboratory analysis a special probe was constructed to measure variations in specific electrical resistivity of groundwater. There were ten series of measurements in the course of one year. Results were selected for this presentation from one serie of conductivity, resistivity and temperature monitoring on a pilot monitoring station MP-1 shown at Fig. 3&4. By continual monitoring of groundwater temperature regimes in plan and profile, it was stated that there were zones and localities where surface water from the existing meliorate channel network directly infiltrated into the underground. All the results obtained indicate that in the process of a pilot monitor station installation it was justified to carry out detailed hydrogeological investigations that   enabled rational selection of appropriate modern measuring equipment and technology for continuous monitoring of the selected parameters of groundwater quality. To this end, a specific stations were designed and put into operation enabling start-up of observations in 1998 (MP-2 and MP-3 shown on Fig. 1).

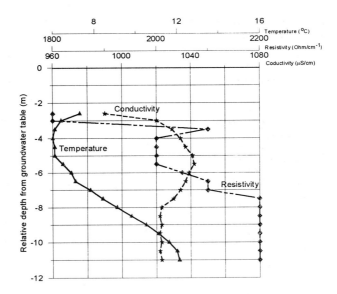

Figure 3. Diagrams of groundwater conductivity, specific electric resistivity and temperature at the MP-1 monitoring station on December 17, 1996

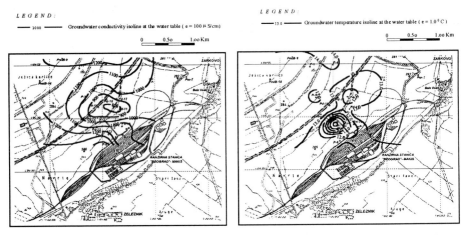

Figure 4. Maps of conductivity and temperature isolines at the groundwater table on December 17, 1996

MONITORING SYSTEM - BASIC FEATURES AND FUNCTIONS

The essence of a groundwater quality monitoring system in road belts is a modern, rational, effective and permanent water quality control, a way to direct all activities in the domain of preventive protection against degradation and pollution, to find, in incidental cases, effective and adequate remedial measures, and to monitor and control the aftermath of incidents. With regard to the zones of large important water supply or perspective systems and regional water sources, which are tangent by highways, the Monitoring system would unite functions of a large number of services and institutions that are responsible for the monitoring and guidance of the system, while with regard to minor water sources or prospective regions only the basic segment of a hydro-ecological station and control module would suffice. (Fig. 5).

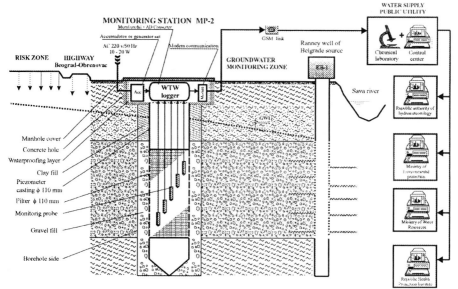

Figure 5. Functioning of the local MP-2 gauging station and groundwater quality monitoring system in highway belt Belgrade-Obrenovac

The basic parts of the Monitoring system are:
1) Hydro-ecological gauging stations (MP-2 and MP-3), hydrogeological investigation Team, authorized control chemical laboratory and a technical service for work coordination;
2) Module to control and manage the Monitoring system;
3) Authorities and a network of interrelated users and participants in the system operation.

A hydro-ecological monitoring station or a cluster of them depending on the area to be protected could be located between the risk zone of contamination and the water structures in use (wells, galleries, intake structures, terrain to protect, etc). That is, in fact, a standard investigation piezometer ϕ 146/110 mm borehole. The equipment and type of piezometer depend primarily on the lithology and type of water intake structures in the zone of water sources. After placing the piezometer construction, special waterproofing and heat insulating material are inserted in the manhole at the top. A digital instrument or logger for measuring the temperature, pH values, O_2 concentration, turbidity, conductivity, specific electrical resistivity, and a device for automatic continual groundwater table gauging are inserted, and lowered to the piezometer level through special measuring tips (Fig. 5). The gauging device are fed with power via a 220 V/10-20W conductor or from a generator set suitable for continuous operation of minimum six months. An A/D converter (analog-digital) for signal conversion, and an independent modem device for direct, wireless connection to standard phone or GSM link are integral parts of this device. Authorized chemical laboratory for monitoring the hydro-ecological station in operation serves for permanent surveillance and identification of irregularities, if any, then for groundwater sampling, detailed laboratory analysis and indications to the control module of the type of measures to be undertaken with regard to quality.

A PC server with a wide network of computer terminals will be incorporated in the control module to serve for a complete control of the Monitoring system. They will have constant modem connection with a network of interrelated users. For the operation of the system it will be necessary to prepare control-information software containing an active analytical and graphic database on all the parameters of the system functioning, and a processing control module for controlling, managing the system and automatic signaling of even minor incidents or changes in the groundwater quality.

An additional part of the system will be a network of interrelated users, that will be able to have at any moment precise information about the groundwater quality, about any adequate protection measures undertaken, and in incidental cases about other effective measures undertaken to solve the problem. This network will be available to all educational and scientific institutions, professional and other teams, allowing them to further their developments in the environmental field, and use their experience wherever needed. This system will also be open and available to the media, that will accurately and timely inform the public, the system getting preventive and educational character, then.

Instead of a conclusion

Specific hydrogeological investigations in the process of groundwater protection should be essential for the establishment of an active groundwater quality monitoring system in line infrastructure belts and their facilities, which will be a base for defining conditions for prospective expansion of the existing water sources and opening of new ones. In the present water supply systems this should become the framework, for global environmental protection in the future. If opposite, with further traffic development, and new economic trends a situation may arise, that, after 2000, many important and potential groundwater sources may not be adequately exploited in spite of their considerable potency.

REFERENCES
Rasula G., 1996: *"Importance of systematic groundwater quality measurement by establishing an active monitoring system"*, Proceedings of 17th Yugoslav Symposium 'Waterworks and Canalization', Edited by Association of Yugoslav engineers and technicians, Pages 111-115, Sabac, Yugoslavia

Rasula G., Rasula M., 1997: *"Information system software as a base for groundwater monitoring system functioning"*, Proceedings of Yugoslav Symposium 'Water protection 97', Edited by Association of Yugoslav engineers and technicians, Pages 415-419, Sombor, Yugoslavia

Rasula G., Rasula M., 1998: *"Groundwater monitoring in the road belts"*, Proceedings of Yugoslav Experts Symposium 'Road and environment', Federal Ministry of environmental protection, Pages 239-245, Zabljak, Montenegro

Rasula M., Rasula G., 1998: *"Groundwater quality monitoring as a base parameter in existence of the alluvial aquifers"*, Proceedings of the International Conference on the World Water Resources at the beginning of the 21st Century :"Water- a looming crisis?", Edited by H. Zebidi, Pages 205-211, IHP-V, No.18, UNESCO, Paris, France

Arsenic contamination of pond water and water purification system using pond water in Bangladesh

H. YOKOTA, M. SEZAKI, K. TANABE, Y. AKIYOSHI
Miyazaki University, Miyazaki, Japan
T. MIYATA, K. KAWAHARA, S. TSUSHIMA, H. HIRONAKA, H. TAKAFUJI
Asia Arsenic Network, Hyuga, Japan
M. RAHMAN
Bogra Technical Training Centre, Bogra, Bangladesh
SK. AKHTAR AHMAD, M. H. SALIM ULLAH SAYED, M. H. FARUQUEE
National Institute of Preventative & Social Medicine (NIPSOM), Dhaka, Bangladesh

1. INTRODUCTION

Arsenic-contaminated groundwater has been found in 59 districts (as of January 1999) out of a total of 64 districts in Bangladesh where almost all drinking water is supplied from groundwater. The cause of arsenic contamination of groundwater is not clear yet, and it is estimated that about 40 million people[1] are at risk of arsenic poisoning accordingly. The arsenic-free water supply system is, therefore, urgently needed in Bangladesh. Since March 1997 we have been investigating the causes of the arsenic contamination in groundwater and developing arsenic-free water supply systems in Samta village of Jessore district which we have chosen as a model village for our research. One of the water supply systems we have tried is to use the water in ponds; namely, a Pond Sand Filter system ("PSF" hereafter). Pond water is supposed to be not contaminated by arsenic, and ponds are available in abundance all over the country and currently used for washing, bathing, etc.

Prior to the construction of PSF the water quality of ponds was examined, and arsenic, though at low levels, was found in the half of the ponds surveyed. Previously, it was believed that surface water such as in ponds and rivers is not polluted by arsenic in Bangladesh. We investigated the causes of arsenic contamination of pond water. The water treatment system using pond water was completed at the end of January 1999 in Samta village, and the treated water is now being supplied. This paper shows the results of our investigations on arsenic pollution of pond water together with that of groundwater and the quality of the treated water by PSF.

Geoenvironmental engineering, Thomas Telford, London, 1999, 585–592

2. ARSENIC POLLUTION OF GROUNDWATER IN SAMTA

2.1 Arsenic Distribution in Dry Season

In dry season, March 1997, we analyzed the arsenic concentrations[2] in water of all the tubewells (282) used for drinking and cooking purposes in Samta village where about 3,600 people live. The arsenic levels here were measured by a field kit[3] based on the Gutzeit colorimetric analysis. The arsenic level of each tubewell was plotted on the map of Samta village as shown in Fig.1. The result was that 95% of the tubewells had arsenic concentrations above 0.01mg/l (the WHO guideline) and 90% above 0.05mg/l (the Bangladesh standard). High arsenic concentrations over 0.50mg/l were distributed in the south with a belt-like shape from west to east, and arsenic levels fell toward the north.

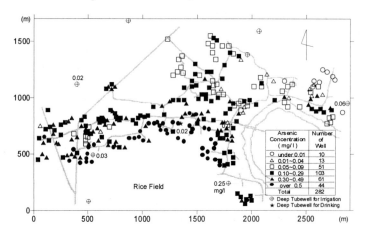

Figure 1. Distribution of arsenic contamination in groundwater (March 1997)

2.2 Flow of Groundwater and Arsenic Concentration

The contour lines of groundwater levels in the dry season (May 1998) are shown in Fig.2.

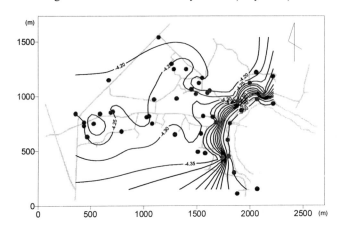

Figure 2. Contour lines of groundwater levels in dry season(May 1998)

The lines were obtained by measuring the water level of 38 sample tubewells. From the contour lines in Fig.2 it can be seen that groundwater flows from the north to the south, drifts at the low water table of the south, and flows into the river on the east. If the distribution of arsenic concentrations of Fig.1 was compared with Fig.2, it could be seen that the high arsenic zone (As $\geqq$ 0.5mg/l) almost agrees with the drifting zone of groundwater. When groundwater is under the condition of drifting, the aquifer might be in the state of reduction under which arsenic precipitated with iron could be released into groundwater. This is now being examined by using core samples obtained during the drilling test performed in Samta.

2.3 Arsenic Concentration in the Rainy Season
The arsenic levels in the rainy season were overall higher than those in the dry season, though the distribution of arsenic concentrations was similar to that of the dry season. In Fig.3 arsenic levels of the 38 sample tubewells are compared between the dry and rainy seasons. The contour map depicts the groundwater levels in the rainy season. The marks ●, ▲ and ■ in Fig.3 show the location of the tubewells surveyed with arsenic levels that were "clearly higher" (●), "almost same" (▲) and "less" (■) in the rainy season compared with those in the dry season. The flow of groundwater in the rainy season is fairly different from that in the dry season. In the rainy season, as some of tubewell water shows quick response to rainfall, there might be local flow of groundwater. If that is the case, oxygen may go into the aquifer where minerals, such as pyrite, might be oxidized to dissolve arsenic in groundwater.   Trace element analysis and elution test using core samples of the drilling test are now carried out to examine the above.

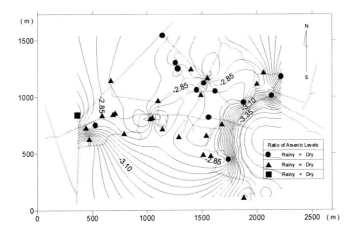

Figure 3. Ratio of arsenic concentration in rainy season to those in dry season

3. ARSENIC CONTAMINATION OF POND WATER
3.1 Outline of PSF
Prior to the construction of a PSF, 14 ponds were surveyed in May 1998 with a view to

obtaining the causes of arsenic pollution of ponds. The locations of the ponds and tubewells , for example P6 and 160 respectively, are shown in Fig.4. We measured the arsenic concentration and the water levels of both ponds and shallow tubewells nearby to check the infiltration of groundwater into the ponds. The waste water from tubewells mainly flows into the ponds since there is no sewerage in Samta. We, therefore, examined the water quality such as electrical conductivity (EC), redox potential (ORP), pH, temperature and turbidity of pond water. Table 1 shows the results of the investigation.

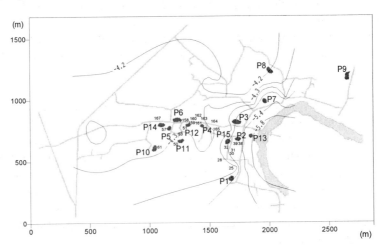

Figure 4. Locations of ponds and shallow tubewells

3.2 Investigation about causes of arsenic contamination of ponds

Arsenic, though at low levels, was found in 7 ponds as shown in the second row of Table 1. The level of water surface of 9 ponds out of 14 ponds was measured and is shown in the third row. The level was standardized by a temporary bench mark in the centre of Samta

Table 1. Survey results of ponds

Pond	As	Water level(m)		Turbidity	EC	pH	ORP	Temp
	(mg/l)	Pond	Well		(μ S/cm)		(mV)	(°C)
P1	0.02			Turbid	525	8.7	145	35.1
P2	0.00			Turbid	488	9.3	124	33.3
P3	0.02	−5.11	−4.31	Turbid	545	8.7	165	33.9
P4	0.04	−4.87	−4.27	Turbid	711	8.6	146	33.8
P5	0.00	−3.24	−4.26	Turbid	523	8.4	166	35.7
P6	0.00	−3.86	−4.27	Clean	786	7.7	119	36.6
P7	0.00			Clean	408	8.8	156	34.3
P8	0.02			Turbid	555	8.0	165	35.1
P9	0.03			Turbid				
P10	0.04	−4.02	−4.17		417	9.0	138	36.0
P11	0.03	−3.95	−4.35	Turbid				
P12	0.00	−2.78	−4.27	Turbid				
P13	0.00	−3.92	−4.14	Clean				
P14	0.00	−4.67	−3.42	Clean	359	8.4	140	

village. The level of pond water was compared with the highest water levels of shallow tubewells near the ponds as shown in the fourth row of Table 1, from which it is observed that the former is higher than the latter. Since pond water is not connected to groundwater, it is estimated that arsenic, which is generally contained in the groundwater, does not go to the ponds through infiltration.

Waste water, especially that from shallow tubewells, often flows into ponds. The waste water of a tubewell (As=0.60mg/l) near the pond 11 firstly stayed in a puddle around the well and then flowed into a pond. The arsenic concentration of the puddle and the pond were 0.40mg/l and 0.03mg/l respectively. The same relation between wasted tubewell water and arsenic contamination of pond water was also recognized with the pond 4 which was selected for the PSF.

On the other hand, there was a shallow tubewell, one of the most highly arsenic-contaminated tubewells in Samta (As=1.16mg/l), near pond 12, into which the drainage from the tubewell had flowed for about ten years. After the tubewell was sealed about two years ago, a new deep tubewell was installed in the area. The arsenic concentration of the deep tubewell has ordinarily varied in the range of 0.03-0.07mg/l, showing rarely higher concentration. We could not detect any arsenic in the water of the pond 12 in spite of the history of arsenic contamination from the old tubewell mentioned above. It is considered that the arsenic had precipitated to the bottom of the pond during the past two years after the seal of the old shallow tubewell.

The correlation between the arsenic pollution of ponds and the drainage from tubewells was examined with the above-mentioned 3 ponds only. From the examination it was observed that the drainage from an arsenic-polluted tubewell is now causing another arsenic pollution of pond water. Also, there was a case of no arsenic contamination of pond water, although high arsenic-contaminated tubewell water had previously flowed into the pond for about ten years but its inflow stopped two years ago. With regard to the remaining ponds we did not survey the relation between drainage from tubewell and arsenic contamination of pond water. It was because we confirmed from the above examination of 3 ponds that the arsenic of pond water would be removed by re-excavating the pond when it was dried up in the dry season and by stopping the inflow of drainage from tubewells nearby.

The turbidity colour of pond water is shown in the fifth row of Table 1. The "turbid" in the row means the colour was blue-green or red-brown, which was caused by eutrophication. Such colours were seen in 6 ponds out of 7 ponds that were contaminated by arsenic. No turbidity colour was observed in more than half in the case of not arsenic-contaminated ponds. We examined the soil at the bottoms of both arsenic and non-arsenic ponds. It was found that the bottom of arsenic-contaminated ponds was very slimy with thick mud layer and non-arsenic ponds had some lives such as pond snail and conch at the bottom. The anaerobic condition at the bottom of ponds, where the arsenic might precipitate from the waste water of tubewells or dissolve from clay ground surrounding the pond, is considered to play a major role in the arsenic contamination in case of ponds. This will be further investigated when the pond is re-excavated in May 1999.

4. POND SAND FILTER (PSF)

4.1 Outline of PSF

The PSF is comprised of two systems as shown in Fig.5. One is a Horizontal Roughing Filter[4] ("HRF" hereafter) developed by the All Indian Institute of Hygiene and Public Health (AIIH&PH) and the other is a Slow Sand Filter ("SSF" hereafter). The HRF is installed as an alternative pre-treatment for SSF. As shown in Fig.5 (a), the HRF is divided into three parts; namely, the inlet structure, the filter bed composed of three compartments, and the outlet structure. The cross section of HRF is 1.0m (W) × 1.4m (D). The length of both inlet & outlet structures is 0.8 m, and the first, second and third compartments of the filter bed have their length of 1.0, 2.0 and 2.0 metres respectively. The filter bed is packed with gravel of diameters of 15mm in the first compartment, 10mm in the second and 5mm in the third.

Walls separating each compartment have many small holes and the raw water in inlet, which is pumped up from the pond, is to flow horizontally through the void of the packed gravel in each compartment of the filter bed. Since the surface of gravel provides a large bed for the settlement of suspended solid articles in the flowing water, HRF is able to decrease suspended solids concentration of raw water. HRF could be cleaned by draining off the accumulated solid articles through pre-laid pipes at the bottom of the filter bed (See Fig.5 (b)).

The SSF works as a biological filter to improve the water quality further after HRF. The thickness of sand media and gravel media are 600mm and 400 mm, respectively. The water treatment capacity of the PSF is 1000 litre/hour. It means about one hundred households can use the treated water for drinking and cooking each day assuming that a person consumes 6 litre of water per day.

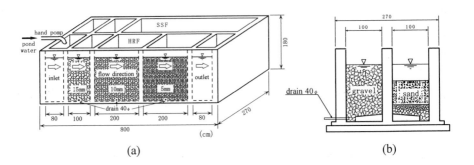

(a)   (b)

Figure 5. Structure of PSF

4.2 Model Test of HRF

The performance of decreasing suspended solid articles by HRF was examined using a model of reduced scale 1:2 in length, height and width at the laboratory of Miyazaki University. The raw water was artificially made by mixing kaolin with water, and the kaolin concentration of the water in the outlet was measured after the water flowed through the filter bed for about five hours. The kaolin concentration was reduced to 1/10 as shown in Fig.6. Since the turbidity of the pond water for PSF in Samta is less than 50mg/l, HRF was considered to work well as 5mg/l would be a suitable load for the SSF.

4.3 PSF Tests Results in Samta

The PSF was constructed in Samta during October 1998 and January 1999. The pond 4, which contained a little arsenic in the water, was donated for PSF by the owner. The drain from a shallow tubewell nearby was changed and therefore no wasted tubewell water flows into the pond 4. The re-excavation of the pond is scheduled for May 1999. We measured the arsenic concentration of the pond just after the completion of PSF construction, at the end of January, namely three months later after the inflow of wasted tubewell water was stopped. There was no arsenic in the pond water. It indicates that the same phenomenon occurred during the three months as the pond 12 mentioned previously.

The quality of the treated water of PSF was as follows. Microorganism multiplied on the surface zone in the sand layer of the SSF during the month after the completion of PSF, and coliform bacteria and general bacteria have decreased as shown in Fig.7. The number of coliform bacteria groups in outlet of HRF, measured once (12 March, 1999) till now, was only 4 to show the excellent performance of HRF.   After 40 days from the completion of PSF, the beginning of March, 1999, the treated water is now being supplied. Currently, water of 220 litre per hour is flowing from a single tap. Two taps are fixed on the PSF at the moment, and another two or three taps are needed to satisfy the design condition of 1000 litre per hour as mentioned above.

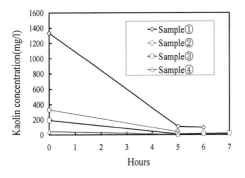
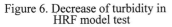

Figure 6. Decrease of turbidity in
HRF model test

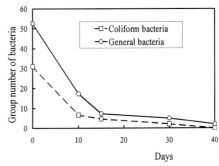

Figure 7. Decrease of bacteria of PSF in
Samta

5. CONCLUSIONS

The following results were obtained through this research:

1) Both reduction action and oxidation action in aquifer are considered as the causes of arsenic contamination of groundwater

2) The major cause of arsenic pollution of ponds is considered as drainage from arsenic-contaminated shallow tubewells.

3) The anaerobic condition at the bottom of ponds is also considered to play a major role in arsenic contamination of pond water. This will be investigated when the pond is re-excavated in May 1999.

4) It was recognized that HRF, an alternative pre-treatment system for SSF, had the ability to decrease the turbidity and the number of bacteria of raw water very well.

5) The PSF installed in Samta is working very well. After 40 days from the completion of PSF construction, the treated water began to be supplied. Water of 440 litre per hour is flowing from the two taps and another two or three taps are needed to satisfy the design condition of 1000 litre per hour.

6) The mechanism of arsenic contamination of groundwater is yet to be clarified after the analyses of drilled core samples.

REFERENCES

1) Sk.Akhtar Ahmad, et al., "Health effects due to arsenic toxicity in Bangladesh", The 3rd Forum on Arsenic Contamination of Groundwater in Asia", Miyazaki, Japan, November 1998.

2) H.Yokota, et al., "The arsenic pollution of groundwater in Samta, Jessor, Bangladesh", Bilateral Consultation between Bangladesh and India on Arsenic in Drinking Water, WHO/SEARO, New Delhi, April 1997.

3) K.Tanabe, et al., "Arsenic concentration of groundwater in Samta village and the applicability of a field kit by Hironaka to quantify arsenic", International Conference on Arsenic Pollution of Groundwater in Bangladesh, Dhaka, Bangladesh, February 1998.

4) K. J. Nath, et al., "Horizontal Roughing Filter – An Appropriate Pretreatment Method for Upgradation of Traditional Surface Water Sources", *Bulletin on Ground Water*, pp81-86.